Diagnostic Electron Microscopy – A Practical Guide to Interpretation and Technique

Current and future titles in the Royal Microscopical Society – John Wiley Series

Published

Principles and Practice of Variable Pressure/Environmental Scanning Electron Microscopy (VP-ESEM)
Debbie Stokes

Aberration-Corrected Analytical Electron Microscopy
Edited by Rik Brydson

Diagnostic Electron Microscopy – A Practical Guide to Interpretation and Technique
Edited by John W. Stirling, Alan Curry & Brian Eyden

Forthcoming

Low Voltage Electron Microscopy: Principles and Applications
Edited by David C. Bell & Natasha Erdman

Atlas of Images and Spectra for Electron Microscopists
Edited by Ursel Bangert

Understanding Practical Light Microscopy
Jeremy Sanderson

Focused Ion Beam Instrumentation: Techniques and Applications
Dudley Finch & Alexander Buxbaum

Electron Beam-Specimen Interactions and Applications in Microscopy
Budhika Mendis

Diagnostic Electron Microscopy – A Practical Guide to Interpretation and Technique

Edited by

John W. Stirling
The Centre for Ultrastructural Pathology, Adelaide, Australia

Alan Curry
Manchester Royal Infirmary, Manchester, UK

and

Brian Eyden
Christie NHS Foundation Trust, Manchester, UK

Published in association with the Royal Microscopical Society

Series Editor: Susan Brooks

A John Wiley & Sons, Ltd., Publication

This edition first published 2013

Registered office
John Wiley & Sons Ltd, The Atrium, Southern Gate, Chichester, West Sussex, PO19 8SQ, United Kingdom

For details of our global editorial offices, for customer services and for information about how to apply for permission to reuse the copyright material in this book please see our website at www.wiley.com.

Library of Congress Cataloging-in-Publication Data

Diagnostic electron microscopy : a practical guide to interpretation and technique / edited by John W. Stirling, Alan Curry, and Brian Eyden.
p. ; cm.
Includes bibliographical references and index.
ISBN 978-1-119-97399-7 (cloth)
I. Stirling, John W. II. Curry, Alan. III. Eyden, Brian.
[DNLM: 1. Diagnostic Imaging--methods. 2. Microscopy, Electron, Transmission. WN 180]
616.07′54 – dc23

2012027835

A catalogue record for this book is available from the British Library.

ISBN: 978-1-119-97399-7
Set in 10.5/13pt Sabon by Laserwords Private Limited, Chennai, India

Printed in the UK

Acknowledgements and Dedication

All three editors wish to thank the many individuals who have helped to make this volume possible. Firstly, they would like to express their appreciation to all the authors for their hard work and generosity in sharing their professional experience, as well as all the 'behind-the-scenes' staff and colleagues without whom this book could not have been produced.

John Stirling thanks the staff of the Centre for Ultrastructural Pathology, SA Pathology, Adelaide, for their support and photographic contributions – especially Alvis Jaunzems and Jeffrey Swift – and Dr Sophia Otto of the Department of Surgical Pathology, SA Pathology, for her advice and for proofreading.

Alan Curry acknowledges the contributions to his work of the pathologists, particularly Dr Helen Denley and Dr Lorna McWilliam, and technical staff of the Manchester Royal Infirmary, as well as two inspirational organisations – the Public Health Laboratory Service Electron Microscopy network and the Manchester Electron Microscope Society.

Brian Eyden wishes to thank all of the Pathology Department staff at the Christie NHS Foundation Trust (Manchester), without whose technical and light microscopic input the interpretation of tumour ultrastructure would be compromised, if not, in some instances, impossible.

Secondly, the editors wish to recognise the support and encouragement of their families in this endeavour. John Stirling thanks his partner, Jill, and expresses a special appreciation of his teachers and mentors, particularly Alec Macfarlane who helped him achieve his dream of a career in biology and Andrew Dorey who introduced him to electron microscopy and the wonders of cell ultrastructure. Alan Curry thanks his wife, Collette (particularly for her exceptional computer skills), and Brian Eyden thanks his wife, Freda, for understanding the needs of a writing scientist.

Finally, the editors dedicate this book to diagnostic electron microscopists – wherever they may be – who continue to make uncertain diagnoses more precise as a result of their labours, which, in turn, help clinicians to treat their patients better, the ultimate purpose of our work.

Contents

List of Contributors xvii
Preface – Introduction xxi

1 Renal Disease 1
John W. Stirling and Alan Curry
1.1 The Role of Transmission Electron Microscopy (TEM) in Renal Diagnostics 1
1.2 Ultrastructural Evaluation and Interpretation 2
1.3 The Normal Glomerulus 3
1.3.1 The Glomerular Basement Membrane 4
1.4 Ultrastructural Diagnostic Features 5
1.4.1 Deposits: General Features 5
1.4.2 Granular and Amorphous Deposits 6
1.4.3 Organised Deposits: Fibrils and Tubules 7
1.4.4 Nonspecific Fibrils 11
1.4.5 General and Nonspecific Inclusions and Deposits 11
1.4.6 Fibrin 12
1.4.7 Tubuloreticular Bodies (Tubuloreticular Inclusions) 12
1.4.8 The Glomerular Basement Membrane 13
1.4.9 The Mesangial Matrix 14
1.4.10 Cellular Components of the Glomerulus 14
1.4.11 Parietal Epithelium 16
1.5 The Ultrastructural Pathology of the Major Glomerular Diseases 16
1.5.1 Diseases without, or with Only Minor, Structural GBM Changes 16
1.5.2 Diseases with Structural GBM Changes 19

1.5.3 Diseases with Granular Deposits 25
1.5.4 Diseases with Organised Deposits 40
1.5.5 Hereditary Metabolic Storage Disorders 46
References 47

2 Transplant Renal Biopsies 55
John Brealey
2.1 Introduction 55
2.2 The Transplant Renal Biopsy 55
2.3 Indications for Electron Microscopy of Transplant Kidney 56
2.3.1 Transplant Glomerulopathy 56
2.3.2 Recurrent Primary Disease 64
2.3.3 De Novo Glomerular Disease 72
2.3.4 Donor-Related Disease 74
2.3.5 Infection 74
2.3.6 Inconclusive Diagnosis by LM and/or IM 79
2.3.7 Miscellaneous Topics 81
References 84

3 Electron Microscopy in Skeletal Muscle Pathology 89
Elizabeth Curtis and Caroline Sewry
3.1 Introduction 89
3.1.1 The Biopsy Procedure 90
3.1.2 Sampling 90
3.1.3 Tissue Processing 90
3.1.4 Artefacts 91
3.2 Normal Muscle 91
3.3 Pathological Changes 96
3.3.1 Sarcolemma 96
3.3.2 Myofibrils 99
3.3.3 Glycogen 102
3.3.4 Cores 104
3.3.5 Target Fibres 105
3.3.6 Myonuclei 105
3.3.7 Mitochondria 106
3.3.8 Reticular System 108
3.3.9 Vacuoles 109
3.3.10 Capillaries 110

3.3.11 Other Structural Defects 111
References 113

4 **The Diagnostic Electron Microscopy of Nerve** **117**
Rosalind King
4.1 Introduction 117
4.2 Tissue Processing 118
4.2.1 Preparation of Nerve Biopsy Specimens 118
4.3 Normal Nerve Ultrastructure 120
4.3.1 Axons 120
4.3.2 Schwann Cells 120
4.3.3 The Myelin Sheath 120
4.3.4 Node of Ranvier 122
4.3.5 Paranode 123
4.3.6 Juxtaparanode 123
4.3.7 Internode 123
4.3.8 Schmidt–Lanterman Incisures 124
4.3.9 Remak Fibres 124
4.3.10 Fibroblasts 124
4.3.11 Renaut Bodies 125
4.4 Pathological Ultrastructural Features 125
4.4.1 Axonal Degeneration 125
4.4.2 Axonal Regeneration 126
4.4.3 Remak Fibre Abnormalities 128
4.4.4 Polyglucosan Bodies 128
4.4.5 Nonspecific Axonal Inclusions 128
4.4.6 Demyelination and Remyelination 130
4.4.7 Specific Schwann Cell Inclusions 135
4.4.8 Nonspecific Schwann Cell Inclusions 136
4.4.9 Fibroblasts 142
4.4.10 Perineurial Abnormalities 142
4.4.11 Cellular Infiltration 143
4.4.12 Endoneurial Oedema 143
4.4.13 Connective Tissue Abnormalities 143
4.4.14 Endoneurial Blood Vessels 145
4.4.15 Mast Cells 145
4.5 Artefact 145
4.6 Conclusions 147
References 148

5 The Diagnostic Electron Microscopy of Tumours **153**
Brian Eyden
5.1 Introduction 153
5.2 Principles and Procedures for Diagnosing Tumours by Electron Microscopy 154
5.2.1 The Objective of Tumour Diagnosis 154
5.2.2 The Intellectual Requirements for Tumour Diagnosis by Electron Microscopy 155
5.2.3 Technical Considerations 156
5.2.4 Identifying Good Preservation 158
5.2.5 Distinguishing Reactive from Neoplastic Cells 162
5.3 Organelles and Groups of Cell Structures Defining Cellular Differentiation 162
5.3.1 Rough Endoplasmic Reticulum 162
5.3.2 Melanosomes 165
5.3.3 Desmosomes 167
5.3.4 Tonofibrils 167
5.3.5 Basal Lamina 169
5.3.6 Glandular Epithelial Differentiation and Cell Processes 171
5.3.7 Neuroendocrine Granules 171
5.3.8 Smooth-Muscle Myofilaments 173
5.3.9 Sarcomeric Myofilaments (Thick-and-Thin Filaments with Z-Disks) 176
References 178

6 Microbial Ultrastructure **181**
Alan Curry
6.1 Introduction 181
6.2 Practical Guidance 182
6.3 Viruses 183
6.4 Current Use of EM in Virology 185
6.5 Viruses in Thin Sections of Cells or Tissues 186
6.6 Bacteria 191
6.7 Fungal Organisms 194
6.8 Microsporidia 196
6.9 Parasitic Protozoa 206
6.9.1 Cryptosporidium 207
6.9.2 Isospora belli 211
6.10 Examples of Non-enteric Protozoa 212

6.11 Parasitic Amoebae 213
6.12 Conclusions 214
Acknowledgements 214
References and Additional Reading 214

7 The Contemporary Use of Electron Microscopy in the Diagnosis of Ciliary Disorders and Sperm Centriolar Abnormalities 221

P. Yiallouros, M. Nearchou, A. Hadjisavvas and K. Kyriacou

7.1 Introduction 221
7.2 Ultrastructure of Motile Cilia 224
7.3 Genetics of PCD 226
7.4 Current Diagnostic Modalities 228
7.5 Clinical Features 229
7.6 Procurement and Assessment of Ciliated Specimens 230
7.7 Centriolar Sperm Abnormalities 231
7.8 Discussion 232
Acknowledgements 234
References 234

8 Electron Microscopy as a Useful Tool in the Diagnosis of Lysosomal Storage Diseases 237

Joseph Alroy, Rolf Pfannl and Angelo A. Ucci

8.1 Introduction 237
8.2 Morphological Findings 247
8.3 Conclusion 261
References 262

9 Cerebral Autosomal Dominant Arteriopathy with Subcortical Infarcts and Leukoencephalopathy (CADASIL) 269

John W. Stirling

9.1 Introduction 269
9.2 Diagnostic Strategies – Comparative Specificity and Sensitivity 271
9.3 Diagnosis by TEM 271
References 274

10 Diagnosis of Platelet Disorders by Electron Microscopy 277

Hilary Christensen and Walter H.A. Kahr

10.1 Introduction 277

10.2 TEM Preparation of Platelets 278
10.3 Whole-Mount EM Preparation of Platelets 280
10.4 EM Preparation of Bone Marrow 281
10.5 Pre-embed Immunogold Labelling of Von Willibrand Factor in Platelets 282
10.6 Ultrastructural Features of Platelets 282
10.7 Normal Platelets 283
10.8 Grey Platelet Syndrome 285
10.9 Arthrogryposis, Renal Dysfunction and Cholestasis Syndrome 285
10.10 Jacobsen Syndrome 285
10.11 Hermansky–Pudlak Syndrome, Chediak–Higashi Syndrome and Other Dense-Granule Deficiencies 287
10.12 Type 2B von Willebrand Disease and Platelet-Type von Willebrand Disease 288
References 290

11 Diagnosis of Congenital Dyserythropoietic Anaemia Types I and II by Transmission Electron Microscopy 293

Yong-xin Ru

11.1 Introduction 293
11.2 Preparation of Bone Marrow and General Observation Protocol 294
11.3 CDA Type I 294
11.3.1 Proerythroblasts and Basophilic Erythroblasts 294
11.3.2 Polychromatic and Orthochromatic Erythroblasts 295
11.3.3 Reticulocytes and Erythrocytes 299
11.4 CDA Type II 299
11.4.1 Erythroblasts 301
11.4.2 Erythrocytes 306
11.5 Summary 306
Acknowledgements 307
References 307

12 Ehlers–Danlos Syndrome 309

Trinh Hermanns-Lê, Marie-Annick Reginster, Claudine Piérard-Franchimont and Gérald E. Piérard

12.1 Introduction 309
12.2 Collagen Fibrils 310

12.3 Elastic Fibers 310
12.4 Nonfibrous Stroma and Granulo-Filamentous Deposits 311
12.5 Connective Tissue Disorders 311
12.5.1 Ehlers–Danlos Syndrome 311
12.5.2 Spontaneous Cervical Artery Dissection 317
12.5.3 Recurrent Preterm Premature Rupture of Fetal Membrane Syndrome 319
References 319

13 Electron Microscopy in Occupational and Environmental Lung Disease 323

Victor L. Roggli

13.1 Introduction 323
13.2 Asbestos 324
13.2.1 Preparatory Techniques 324
13.2.2 Analytical Methodology 326
13.2.3 Asbestos-Related Diseases 326
13.2.4 Exposure Categories 330
13.3 Hypersensitivity Pneumonitis and Sarcoidosis 330
13.3.1 Preparatory Techniques and Analytical Methodology 331
13.4 Silicosis 331
13.4.1 Preparatory Techniques and Analytical Methodology 333
13.5 Silicate Pneumoconiosis 333
13.5.1 Talc Pneumoconiosis 333
13.5.2 Kaolin Worker's Pneumoconiosis 334
13.5.3 Mica and Feldspar Pneumoconiosis 334
13.5.4 Mixed Dust Pneumoconiosis 335
13.5.5 Preparatory Techniques and Analytical Methodology 335
13.6 Metal-Induced Diseases 335
13.6.1 Siderosis 336
13.6.2 Aluminosis 336
13.6.3 Hard Metal Lung Disease 336
13.6.4 Berylliosis 337
13.6.5 Preparatory Techniques and Analytical Methodology 337
13.7 Rare-Earth Pneumoconiosis 338
13.8 Miscellaneous Disorders 338
References 339

14 General Tissue Preparation Methods **341**
John W. Stirling
14.1 Introduction 341
14.1.1 Specimens Suitable for Diagnostic TEM 341
14.2 Tissue Collection and Dissection 342
14.2.1 Tissue Cut-Up 343
14.3 Tissue Processing 345
14.3.1 Fixatives and Fixation 345
14.3.2 Primary Fixation: Glutaraldehyde 347
14.3.3 Secondary Fixation (Post-fixation): Osmium Tetroxide 347
14.3.4 Fixative Vehicles and Wash Buffers 347
14.3.5 En Bloc Staining with Uranyl Acetate 348
14.3.6 Dehydrant and Transition Fluids 348
14.3.7 Resin Infiltration and Embedding Media 349
14.3.8 Tissue Embedding 352
14.4 Tissue Sectioning 352
14.4.1 Ultramicrotomy 352
14.4.2 Sectioning Technique and Ultramicrotome Setup 355
14.4.3 Common Sectioning Problems and Artefacts 356
14.4.4 Section Staining 362
14.4.5 Section Contamination and Staining Artefacts 363
Protocol 364
Processing Schedules 364
References 379

15 Ultrastructural Pathology Today – Paradigm Change and the Impact of Microwave Technology and Telemicroscopy **383**
Josef A. Schroeder
15.1 Diagnostic Electron Microscopy and Paradigm Shift in Pathology 383
15.2 Standardised and Automated Conventional Tissue Processing 385
15.3 Microwave-Assisted Sample Preparation 390
15.4 Cyberspace for Telepathology via the Internet 397
15.5 Conclusions and Future Prospects 400
Acknowledgements 404
References 404

16 Electron Microscopy Methods in Virology **409**
Alan Curry
16.1 Biological Safety Precautions 409
16.2 Collection of Specimens 410
16.3 Preparation of Faeces, Vomitus or Urine Samples 410
16.4 Viruses in Skin Lesions 410
16.5 Reagents and Methods 411
16.5.1 Negative Stains 411
16.6 Coated Grids 412
16.7 Important Elements in the Negative Staining Procedure 412
16.8 TEM Examination 413
16.9 Immunoelectron Microscopy 413
16.9.1 Immune Clumping 413
16.9.2 Solid-Phase Immunoelectron Microscopy 413
16.9.3 Immunogold Labelling 414
16.9.4 Particle Measurement 414
16.10 Thin Sectioning of Virus-Infected Cells or Tissues 414
16.11 Virology Quality Assurance (QA) Procedures 415
16.11.1 External QA 415
16.11.2 Internal QA 415
Acknowledgements 415
References 416

17 Digital Imaging for Diagnostic Transmission Electron Microscopy **419**
Gary Paul Edwards
17.1 Introduction 419
17.2 Camera History 419
17.3 The Pixel Dilemma 420
17.4 Camera Positioning 421
17.5 Resolution 422
17.6 Fibre Coupled or Lens Coupled? 423
17.7 Sensitivity, Noise and Dynamic Range 424
17.8 CCD Chip Type (Full Frame or Interline) 426
17.9 Binning and Frame Rate 426
17.10 Software 427
17.11 Choosing the Right Camera 428
References 429

18 Uncertainty of Measurement **431**
Pierre Filion
18.1 Introduction 431
18.2 Purpose 432
18.2.1 Diagnostic Value 432
18.2.2 Internal Quality Control 432
18.2.3 External Quality Control and Accreditation 432
18.3 Factors That Influence Quantitative Measurements 433
18.3.1 Sources of Variation 433
18.3.2 Alteration of the Intrinsic Dimension of the Structure 434
18.3.3 Variation Due to the Analytical Equipment and Method 436
18.3.4 Variation Due to Selection Bias 438
18.3.5 Measurement Using a Digital Camera 439
18.4 How to Calculate the UM 440
18.4.1 Steps Required to Analyse and Calculate the UM 440
18.4.2 Type of Error and Distribution of Measurements 440
18.4.3 Calculating the UM 442
18.4.4 Precision of Measurement and Biological Significance 443
18.4.5 The Electronic Spread Sheet as an Aid to Calculating UM 443
18.4.6 Reporting the UM 444
18.5 Worked Examples 444
18.5.1 Diameter of Fibrils in a Glomerular Deposit 444
18.5.2 Thickness of the Glomerular Basement Membrane 445
18.6 Conclusion 446
References 447

Index **449**

List of Contributors

Joseph Alroy, Department of Pathology, Tufts University Cumming's School of Veterinary Medicine, Grafton, Massachusetts, United States and Department of Pathology and Laboratory Medicine, Tufts Medical Center and Tufts University School of Medicine, Boston, Massachusetts, United States

John Brealey, Centre for Ultrastructural Pathology, Surgical Pathology – SA Pathology (RAH), Adelaide, Australia

Hilary Christensen, Program in Cell Biology, The Hospital for Sick Children, Toronto, Ontario, Canada

Alan Curry, Health Protection Agency, Clinical Services Building, Manchester Royal Infirmary, Manchester, United Kingdom

Elizabeth Curtis, Muscle Biopsy Service/Electron Microscope Unit, Department of Cellular Pathology, Queen Elizabeth Hospital Birmingham, Birmingham, United Kingdom

Gary Paul Edwards, Chelford Barn, Stowmarket, Suffolk, United Kingdom

Brian Eyden, Department of Histopathology, Christie NHS Foundation Trust, Manchester, United Kingdom

Pierre Filion, Electron Microscopy Section, Division of Anatomical Pathology, PathWest Laboratory Medicine, QE II Medical Centre, Nedlands, Australia

A. Hadjisavvas, Department of Electron Microscopy/Molecular Pathology, The Cyprus Institute of Neurology and Genetics, Nicosia, Cyprus

Trinh Hermanns-Lê, Department of Dermatopathology, University Hospital of Liège, Liège, Belgium

Walter H.A. Kahr, Division of Haematology/Oncology, Program in Cell Biology, The Hospital for Sick Children, Toronto, Ontario, Canada and Departments of Paediatrics and Biochemistry, University of Toronto, Toronto, Ontario, Canada

Rosalind King, Institute of Neurology, University College London, London, United Kingdom

K. Kyriacou, Department of Electron Microscopy/Molecular Pathology, The Cyprus Institute of Neurology and Genetics, Nicosia, Cyprus

M. Nearchou, Department of Electron Microscopy/Molecular Pathology, The Cyprus Institute of Neurology and Genetics, Nicosia, Cyprus

Rolf Pfannl, Department of Pathology and Laboratory Medicine, Tufts Medical Center and Tufts University School of Medicine, Boston, Massachusetts, United States

Gérald E. Piérard, Department of Dermatopathology, University Hospital of Liège, Liège, Belgium

Claudine Piérard-Franchimont, Department of Dermapathology, University Hospital of Liège, Liège, Belgium

Marie-Annick Reginster, Department of Dermatopathology, University Hospital of Liège, Liège, Belgium

Victor L. Roggli, Department of Pathology, Duke University Medical Center, Durham, North Carolina, United States

Yong-xin Ru, Institute of Haematology & Blood Diseases Hospital, Chinese Academy of Medical Sciences and Peking Union Medical College, Tianjin, China

Josef A. Schroeder, Zentrales EM-Labor, Institut für Pathologie, Klinikum der Universität Regensburg, Regensburg, Germany

Caroline Sewry, Wolfson Centre for Inherited Neuromuscular Diseases, RJAH Orthopaedic Hospital, Oswestry, United Kingdom and Dubowitz Neuromuscular Centre, Institute of Child Health and Great Ormond Street Hospital, London, United Kingdom

John W. Stirling, Centre for Ultrastructural Pathology, IMVS – SA Pathology, Adelaide, Australia

Angelo A. Ucci, Department of Pathology and Laboratory Medicine, Tufts Medical Center and Tufts University School of Medicine, Boston, Massachusetts, United States

P. Yiallouros, Cyprus International Institute, Cyprus University of Technology, Limassol, Cyprus

Preface – Introduction

John W. Stirling, Alan Curry and Brian Eyden

DIAGNOSTIC ELECTRON MICROSCOPY

Science progresses as a result of a variety of factors. Critical to progress, however, is the invention and availability of appropriate tools and techniques that can completely transform our ability to investigate and understand the world around us – without such tools our ability to investigate even basic phenomena would be severely restricted. One such 'transformational' technology is the electron microscope. Although transmission electron microscopy (TEM) is now taken for granted, its application to the biological and medical sciences in the late 1950s and early 1960s ranks as one of the single most important factors that has impacted on our knowledge in biology and medicine. The resolving power of the transmission electron microscope (~0.2 nm as compared with the light microscope with a resolution of ~200 nm) made two important things possible for the first time, these being the visualisation of: (1) cell organelles and cytoplasmic structures at the macromolecular level (both useful indicators of cell differentiation) and; (2) viruses and microorganisms in general. Thus, TEM gave us new fundamental insights into cell structure and function, histogenesis and differentiation, and, following from this, our understanding of disease and disease processes.

TEM was quickly taken up as a diagnostic tool. In the clinical setting, electron microscopy has been used to improve diagnostic precision and confidence in many fields, including renal disease, neuromuscular disease, microbiology (particularly virology), tumour pathology, skin diseases, industrial diseases, haematology, metabolic storage diseases and conditions involving abnormalities of cilia and sperm. A number of encyclopaedic atlases of normal and pathological tissues quickly followed the introduction of electron microscopy and the medical literature contains many articles describing diagnostic applications of TEM in a wide range of conditions and specialist areas. Diagnostic TEM reached a zenith during the 1980s; however, since then, the introduction of new methodologies (particularly molecular techniques and affinity labelling systems) has reduced the need for TEM, particularly in tumour diagnosis. Despite this, TEM continues to play a significant and important role in pathology, and techniques continue to develop and improve. For example, the introduction of microwave processing and digital cameras has transformed tissue processing and screening so that 'same-day' reporting is easily achieved.

THE PURPOSE AND USE OF TEM

The purpose of TEM is to diagnose disease based on the ultrastructural features of the tissue. These features include:

1. The presence (or sometimes the absence) of specific or characteristic cellular structures or organelles that indicate cell differentiation
2. The general ultrastructural architecture, including the identity, location and morphology of specific structural features that may be associated with pathology, or indicate disease.

In general, the use of TEM will be predetermined either as a stand-alone protocol (e.g., CADASIL) or as part of a broad integrated diagnostic strategy (e.g., renal biopsies). However, TEM can also be applied on an *ad hoc* basis whenever there is a chance it will give an improved diagnosis (and therefore better patient care). The general criteria indicating the use of TEM may be summarised simply as follows:

1. When it provides useful (complementary) structural, functional or compositional information in respect to diagnosis, differential diagnoses or disease staging
2. When only atypical features or minor abnormalities are visible by light microscopy despite clear clinical evidence of disease (e.g. some renal diseases)
3. When affinity labelling results are equivocal (e.g. renal disease and tumours)

4. When there is no realistic alternative diagnostic technique or a 'simple' test is not available or feasible (e.g. genetic diseases with multiple mutations such as CADASIL and primary ciliary dyskinesia)
5. The investigation and diagnosis of new diseases and microorganisms
6. When it is time and/or cost effective in respect to alternative techniques.

THE AIM AND PURPOSE OF THIS BOOK

The prime aim and purpose of this book is to summarise the current interpretational applications of TEM in diagnostic pathology. In this respect, we have not attempted to reproduce previous encyclopaedic texts but to provide what we regard as a working guide to the main, or most useful, applications of the technique given the limited space available in a text of this size. In addition, we have also included practical topics of concern to laboratory scientists, including brief guides to traditional tissue and microbiological preparation techniques, microwave processing, digital imaging and measurement uncertainty.

1

Renal Disease

John W. Stirling[1] **and Alan Curry**[2]

[1] *Centre for Ultrastructural Pathology, IMVS – SA Pathology, Adelaide, Australia*

[2] *Health Protection Agency, Clinical Sciences Building, Manchester Royal Infirmary, Manchester, United Kingdom*

1.1 THE ROLE OF TRANSMISSION ELECTRON MICROSCOPY (TEM) IN RENAL DIAGNOSTICS

The ultrastructural examination of renal biopsies has made a significant contribution to our understanding of renal disease and is fundamental to accurate diagnosis. For overall tissue evaluation, light microscopy (LM), immunolabelling and transmission electron microscopy (TEM) are generally combined as an integrated protocol. LM is used to make an assessment of overall tissue morphology and to identify the major pathological processes present. Immunolabelling (preferably using immunofluorescence or by the immunoperoxidase technique) is used to determine the composition and location of glomerular immune deposits. Local practices vary, but an antibody panel can contain antibodies directed against IgG, IgA, IgM, complement (C3, C1q and sometimes C4), κ and λ light chains and albumin. TEM can play a major role when LM and immunolabelling findings are normal, only mildly atypical or equivocal and difficult to interpret, particularly in respect to conditions where there may be similar LM or immunolabelling findings. Thus, the technique is particularly useful in the setting of familial disease where the

Diagnostic Electron Microscopy: A Practical Guide to Interpretation and Technique, First Edition. Edited by John W. Stirling, Alan Curry and Brian Eyden.

structural abnormalities in the glomerular basement membrane (GBM) cannot be resolved by LM (e.g. Alport's syndrome). TEM can also provide critical information not revealed by the other methodologies to identify underlying primary disease and unexpected concomitant disease. Similarly with immunolabelling, the full classification and staging of deposits require ultrastructural analysis. Some transplant biopsies can also benefit from ultrastructural evaluation (see Chapter 2); however, TEM rarely contributes to the diagnosis of tubular, vascular or interstitial disease. Overall, ultrastructural screening is essential; it can change the diagnosis in ~25% of cases and provides 'useful' information in ~66% of cases (Pearson *et al.*, 1994; Elhefnawy, 2011).

1.2 ULTRASTRUCTURAL EVALUATION AND INTERPRETATION

Examination of glomeruli (and other areas, if necessary) should be thorough and systematic with all components being evaluated for possibly significant features or changes. During screening, a range of representative images should be taken. These should include low-power images to show overall glomerular morphology, plus a representative selection of higher power images to show the specific and critical diagnostic features. In some instances, it may also be important to show that certain features are, in fact, absent (e.g. deposits) or normal (e.g. foot processes). The principal elements that should be examined are (i) the location, size and morphology of immune-related deposits and other inclusions; (ii) the thickness, overall morphology and texture of the GBM; (iii) the size and morphology of the mesangial matrix and (iv) the number and morphology of the cellular components of the glomerulus (Stirling *et al.*, 2000). Sclerotic glomeruli should be avoided, and only well-preserved functional (or significantly functional) glomeruli should be examined. It is also important to ensure that the glomeruli screened are representative of the LM findings: this means that, ideally, the choice of glomeruli to be screened (from semithin sections) should be done in collaboration with the reporting pathologist. Finally, it should be stressed that screening should be unbiased, although some knowledge of the pathology and immunolabelling results may be useful if the features expected are minor or uncommon. The vascular pole should be avoided during ultrastructural evaluation as it may contain misleading nonpathologic deposits, and likewise Bowman's capsule which has no real diagnostic value, although the presence of crescents can be confirmed.

Following evaluation, representative images and findings should be communicated to the reporting pathologist, the latter verbally or in a concise written report. If the initial evaluation does not correspond with the LM evaluation (e.g. the electron microscopy (EM) samples only a tiny fraction of the available tissue), then the specimen should be re-examined or additional glomeruli observed to increase diagnostic confidence.

A critical question is 'How many glomeruli should be examined, and for how long?' Unfortunately, there is no definitive answer to this dilemma except to say that enough tissue should be examined to answer the diagnostic question posed and to ensure that no additional or unexpected pathology is present. A single glomerulus (or even part of one) may be adequate in respect to diffuse disease and/or when the glomerulus screened is typical of the disease process identified by LM. In contrast, several glomeruli, or possibly glomeruli from different blocks, may be required to capture the full range of pathological changes in focal disease. Perhaps the final word on this issue is to say that the tissue must be screened thoroughly; it is bad practice to stop screening once the features that were expected have been located because additional findings that affect the accuracy of the diagnosis may be missed.

1.3 THE NORMAL GLOMERULUS

The glomerulus (Figure 1.1) is composed of a tuft of branching capillaries that originate from the afferent arteriole at the vascular pole to form a series of lobules (segments) that ultimately rejoin at the vascular pole and exit the glomerulus via the efferent arteriole. At the core of each lobule is the mesangium which supports the capillary loops; capillary loops are lined by endothelial cells (Figure 1.1). The mesangial matrix principally consists of collagen IV and is populated by mesangial cells (usually 1–3 in normal mesangium) plus a small number of immune-competent cells and rare transient cells of the monocyte–macrophage lineage (Sterzel *et al.*, 1982). The entire capillary tuft is enclosed within Bowman's capsule, the inner aspect of which is lined by a thin layer of epithelial cells (the parietal epithelial cells); a second inner population of epithelial cells (the visceral epithelial cells or podocytes) is closely associated with the capillary tufts, and extensions of these cells form the foot processes (pedicels) that cover the outer aspect of the capillary walls (Figure 1.1). The podocytes are the sole source of the collagen IV $\alpha 3$, $\alpha 4$ and $\alpha 5$ subtypes that form the bulk of the GBM (Abrahamson *et al.*, 2009), and the foot processes play a major role in ultrafiltration and the

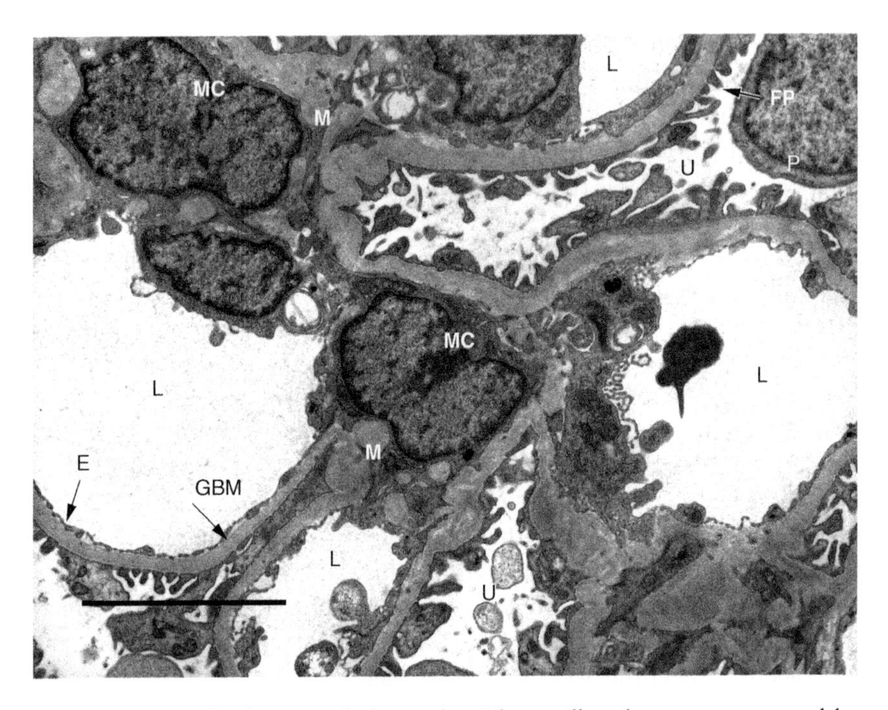

Figure 1.1 Detail of a normal glomerulus. The capillary loops are supported by the mesangium (M). Mesangial cells with nuclei (MC); capillary lumens (L); urinary space (U); podocyte (P) (epithelial cell) and foot processes (FP). Here, the overall width of Overall, the glomerular basement membrane (GBM) averages ~380 nm in width. Loops are lined with fenestrated endothelial cells (E). Bar = 5 μm.

maintenance of the filtration barrier. As a result, podocyte dysfunction plays a major role in a wide range of glomerular diseases (Wiggins, 2007; Haraldsson, Nystrom and Deen, 2008). Opposite the vascular pole, Bowman's capsule is continuous with the proximal tubule which drains filtrate from the glomerulus (the urinary pole). Overall, filtration is said to be a function of size, shape and charge selection, although the nature and contribution of charge selection are debated (Harvey *et al.*, 2007; Haraldsson, Nystrom and Deen, 2008; Goldberg *et al.*, 2009). The capillary wall as a whole is responsible for the filtration process, and it appears that the capillary endothelium, the GBM and the podocyte foot processes must all be intact for normal filtration to occur (Patrakka and Tryggvason, 2010).

1.3.1 The Glomerular Basement Membrane

The GBM (Figure 1.1) is made of three layers: (i) the lamina rara interna, the electron-lucent layer immediately adjacent to the endothelium;

(ii) the lamina densa, the central layer and (iii) the lamina rara externa, the outer electron-lucent area immediately adjacent to the foot processes. The lamina densa makes up the bulk of the GBM and is its main structural element; it has a felt-like fibrillar construction, and knowledge of its molecular makeup is helpful in understanding and interpreting familial and autoimmune disease. The principal component is collagen IV, which consists of six subtypes ($\alpha1$–$\alpha6$) (Patrakka and Tryggvason, 2010). In the developing kidney, the GBM is initially formed of the $\alpha1$ and $\alpha2$ subtypes with the $\alpha3, \alpha4$ and $\alpha5$ subtypes forming later (the additional subtype, $\alpha6$, is restricted to Bowman's capsule and some tubular basement membranes) (Harvey *et al.*, 1998; Miner, 1998). In the mature kidney, the $\alpha1$ and $\alpha2$ subtypes are restricted to a narrow band immediately adjacent to the capillary endothelium; the $\alpha3$, $\alpha4$ and $\alpha5$ subtypes form the remaining bulk of the GBM (extending out to the foot processes). The core of the mesangial matrix is composed of the $\alpha1$ and $\alpha2$ subtypes (continuous with the inner aspect of the GBM), while the outer peripheral layer is made up of the $\alpha3$, $\alpha4$ and $\alpha5$ subtypes (continuous with the outer layer of the GBM) (Butkowski *et al.*, 1989; Harvey *et al.*, 1998; Miner, 1998). The $\alpha3$, $\alpha4$ and $\alpha5$ subtypes are essential for the maintenance of normal glomerular function, and mutations in the genes for these subtypes are responsible for the various forms of membrane-related hereditary nephritis. The structural abnormalities of the GBM in hereditary disease are caused by the absence of the $\alpha3$ and $\alpha5$ subtypes, because without either of these, the membrane fails to form correctly (Kalluri *et al.*, 1997; LeBleu *et al.*, 2010; Miller *et al.*, 2010). The $\alpha3$ subtype has been identified as the Goodpasture epitope (Saus *et al.*, 1988). However, it appears that both the $\alpha3$ and $\alpha5$ subtypes are targeted in anti-GBM disease, while in Alport's post-transplantation nephritis, only the $\alpha5$ subtype is involved (Pedchenko *et al.*, 2010).

1.4 ULTRASTRUCTURAL DIAGNOSTIC FEATURES

1.4.1 Deposits: General Features

Immune-related material accumulates as discrete or linear deposits of finely granular electron-dense material within or adjacent to the GBM and/or mesangium in several diseases. Deposits may also be 'organised' as tubules and fibrils of various diameters, as crystals and as whorls with a fingerprint-like appearance (Herrera and Turbat-Herrera, 2010). The identity and content of specific deposits vary and must be confirmed by immunolabelling.

Note that scattered deposits may sometimes be an incidental finding with no obvious pathological or diagnostic relevance. Approximately 4–16% of normal individuals have mesangial IgA deposits (without IgG or C3) (Coppo, Feehally and Glassock, 2010), and small numbers of discrete deposits are occasionally seen in individuals with naturally high levels of antigenic challenge.

1.4.2 Granular and Amorphous Deposits

1.4.2.1 Subepithelial Deposits

Subepithelial deposits are finely granular, medium-density deposits located on the outer surface of the GBM and the mesangium. Foot processes that lie over the surface of the deposit are generally effaced.

- Large oval or dome-like deposits (humps). True humps are not usually associated with a GBM reaction (spikes). Seen typically in post-infectious glomerulonephritis (PIGN) (Figure 1.14).
- Flat or nodular deposits. There may be an associated GBM reaction with 'spikes' of new membrane forming adjacent to the deposits, a process that may ultimately lead to the deposits becoming incorporated into the GBM (chain-link appearance by LM). Seen typically in stage I membranous glomerulonephritis (Figure 1.10).

1.4.2.2 Intramembranous Discrete Deposits

These are finely granular, medium-density deposits that lie completely within the lamina densa. The material may be uniform in appearance, or patchy and irregular. Resorption of deposits results in irregular electron-lucent areas surrounded by thickened GBM; badly damaged membrane may become laminated and similar in appearance to the 'basket weave' pattern seen in Alport's syndrome. Seen typically in stage III membranous glomerulonephritis (Figure 1.12).

1.4.2.3 Intramembranous Linear Deposits

- Linear transformation: a uniform dense amorphous transformation of the GBM that results in a dark, 'ribbon-like' appearance. The material is essentially sited within the subendothelial layer of GBM matrix

and may penetrate the GBM for some distance. Seen typically in mesangiocapillary glomerulonephritis (MCGN) type II (dense deposit disease (DDD)) (Figure 1.17). A similar effect may be seen around the periphery of loops when subepithelial deposits merge to form a continuous or semicontinuous band.
- Medium-density linear deposits of finely granular or powdery material within the GBM. Seen typically in κ light-chain disease (Figure 1.21).

1.4.2.4 *Subendothelial Deposits*

These are linear or plaque-like, finely granular, medium-density deposits located between the inner aspect of the GBM and the capillary endothelium. Large deposits may be visible by LM as nodular hyaline 'thrombi' or as 'wire-loop' capillary wall thickening. Seen typically in MCGN type I (Figure 1.16).

1.4.2.5 *Mesangial Deposits*

- Mesangial: finely granular, medium-density deposits within the central mesangial matrix. Seen typically in IgA disease (Figure 1.15).
- Paramesangial: finely granular, medium-density deposits around the outer periphery of the mesangial matrix, especially at the junction of the capillary loop and the mesangium. Seen typically in IgA disease (Figure 1.15).

1.4.3 Organised Deposits: Fibrils and Tubules

Many normal and pathological fibrils are found in the glomerulus, and the deposition of proteins, such as monoclonal immunoglobulins or their light-chain or heavy-chain subunits, can produce several glomerular diseases (see reviews by Furness, 2004; Basnayake *et al.*, 2011). The accurate identification of fibrils can be problematic, and distinguishing between normal and pathological types may require the correlation of ultrastructural, immunolabelling and LM-staining characteristics (Table 1.1). A number of algorithms have been published to aid in the diagnosis of renal diseases containing organised deposits (Figure 1.2) (see Ivanyi and Degrell, 2004; Herrera and Turbat-Herrera, 2010).

Table 1.1 Characteristics of immune and non-immune glomerular fibrils.

Fibril type	Characteristics
Amyloid (Figure 1.22)	**Transmission electron microscopy (TEM) characteristics** Extracellular nonbranching fibrils Random orientation, occasionally in parallel arrays Fibrils may appear to penetrate, or flow (cascade) from, adjacent cells Distribution diffuse or focal; includes GBM Diameter: ~7–10 nm (Ghadially, 1988) or ~8–12 nm (Rosenstock *et al.*, 2003) **Light microscopy (LM) characteristics** Periodic acid–Schiff (PAS) negative; non-argyrophilic Congo red positive (also Sirius red and Thioflavin T and S positive in fresh tissue) Positive Congo red reaction requires a reasonable amount of amyloid to be present. Sections should be 8–10 µm thick. Apple green birefringence under polarised light (Vowles, 2008). **Immunolabelling** Negative for immunoglobulins, complement components, fibrinogen and albumin. Composition variable. Immunolabelling may be used to identify precursor proteins in order to identify amyloid type (Vowles, 2008).
'Immune' fibrils (fibrillary glomerulonephritis) (Figure 1.23)	**TEM characteristics** Extracellular nonbranching fibrils Random, sometimes parallel orientation Distribution diffuse; includes GBM Diameter: 13–29 nm (mean 20.1 nm ±0.4) (Rosenstock *et al.*, 2003) or 15–25 nm (Herrera and Turbat-Herrera, 2010) **LM characteristics** Glassy and weakly eosinophilic; weakly PAS positive; grey-purple on trichrome stain; non-argyrophilic (Rosenstock *et al.*, 2003) PAS–methenamine–silver: weak (reticular) staining Congo red negative **Immunolabelling** Smudgy, ribbon-like to granular staining for IgG, C3 and κ and λ light chains (sometimes with light-chain restriction) in mesangium and peripheral capillary walls. IgG4 is dominant in most cases, but rarely IgG1 (Herrera and Turbat-Herrera, 2010).

Table 1.1 (*continued*)

Fibril type	Characteristics
Immunotactoid glomerulopathy ('immune' tubules) (Figure 1.24)	**TEM characteristics** Extracellular nonbranching tubules Orientation random, or in parallel arrays; sometimes in a background matrix (Herrera and Turbat-Herrera, 2010 Distribution diffuse; includes mesangium and GBM Diameter: 20–55 nm (mean 38.2 nm ± 5.7) (Rosenstock *et al.*, 2003) or 10–90 nm (commonly more than 30 nm) (Herrera and Turbat-Herrera, 2010) **LM characteristics** Silver stains negative **Immunolabelling** Variable, IgG and C3 in mesangium and peripheral capillary walls with a granular or pseudolinear pattern; IgA, IgM and C1q variable or negative; light-chain restriction (κ rather than λ) in some cases (Herrera and Turbat-Herrera, 2010)
Microfibrils	**TEM characteristics** Extracellular nonbranching fibrils Parallel, 'bundled' or sometimes random orientation Normal microfibrillar glomerular components include (i) collagen fibrils ~30 nm in diameter and greater; (ii) 'large' fibrils ~18–20 nm in diameter; (iii) 'small' fibrils ~10 nm in diameter and (iv) 'thin filaments' ~3–5 nm in diameter (Coleman and Seymour, 1992). Nonspecific fibrils ~12 nm in diameter may be found in several conditions, especially sclerosing glomerular diseases (Hsu and Churg, 1979; Kronz, Nue and Nadasdy, 1998). 12 nm diameter non-immune fibrils are a major component of Kimmelstiel–Wilson nodules (Figure 1.8) (Yasuda *et al.*, 1992). **LM characteristics** Generally PAS and PAS–methenamine–silver strongly positive Congo red negative **Immunolabelling** Negative or nonspecific (Kronz, Nue and Nadasdy, 1998)

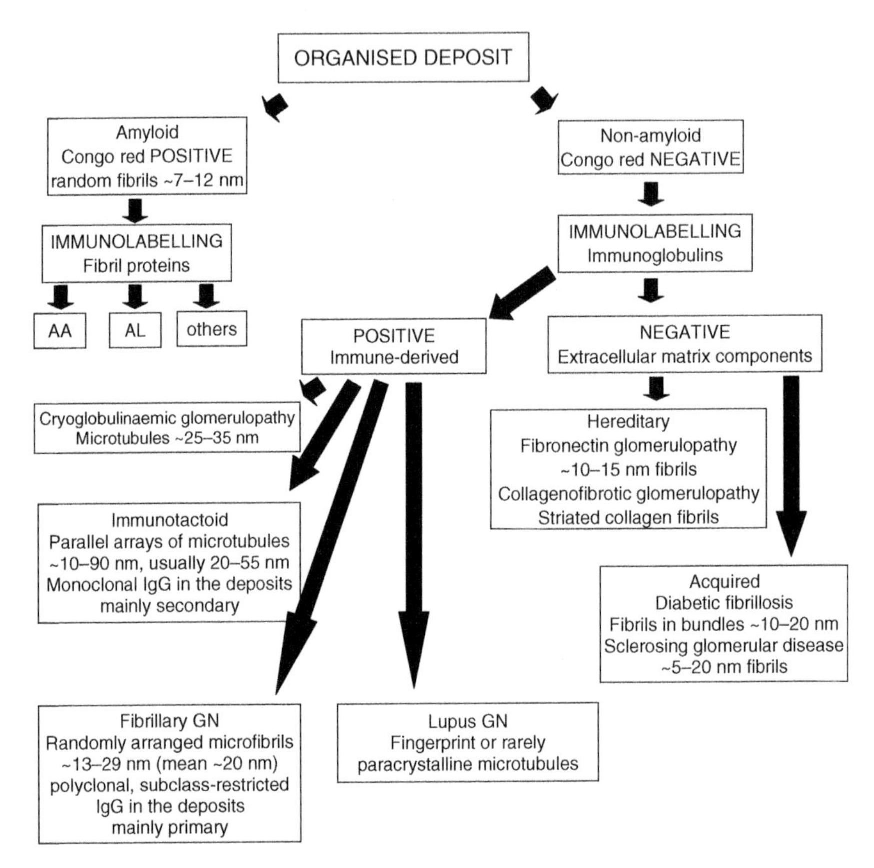

Figure 1.2 Algorithm for diagnosis of organised deposits. An algorithm to aid in the diagnosis of diseases with organised deposits (based on Ivanyi and Degrell, 2004; Herrera and Turbat-Herrera, 2010). Measurements given are fibril and tubule diameters (see Table 1.1 for source references). This strategy may be combined with silver methenamine staining to further define matrix-derived components (Herrera and Turbat-Herrera, 2010).

1.4.3.1 *Amyloid*

Amyloid is composed of insoluble fine fibrils made up of low-molecular-weight proteins of various types in a β-pleated sheet conformation (Dember, 2006; Vowles, 2008). Amyloid deposits may form anywhere in the glomerulus (and other tissues) as extracellular nonbranching fibrils of indeterminate length and without cross-striations (Table 1.1; Figure 1.22). Immunofluorescence can sometimes indicate the possibility of amyloid deposition (dull green colouration), but this can be difficult to differentiate from diabetic changes (increase in matrix).

1.4.3.2 Non-amyloid Fibrils and Tubules

Both fibrils and tubules of immune-related material may be found in glomeruli; they may be randomly orientated or organised in parallel arrays. Reported diameters for both fibrils and tubules vary greatly, presumably because of their individual physiochemical makeup, the range of cases sampled and variability in laboratory processing regimes (Table 1.1). Fingerprint deposits (parallel or linear arrays of fibrils – usually curved in a fingerprint-like pattern; Figure 1.20) may also be found, particularly in systemic lupus erythematosus (SLE) and cryoglobulinaemia. There is some dispute as to whether diseases featuring immune fibrils (fibrillary glomerulonephritis; Figure 1.23) and those featuring tubules (immunotactoid glomerulopathy; Figure 1.24) should be treated together or separately (Schwartz, Korbet and Lewis, 2002; Ivanyi and Degrell, 2004; Herrera and Turbat-Herrera, 2010). Here, they are treated as separate entities as there are important clinical differences between them, that is, immunotactoid is normally associated with a lymphoproliferative disorder, but typically, fibrillary glomerulonephritis is not.

1.4.4 Nonspecific Fibrils

A variety of nonspecific fibrils that may be confused with immune deposits have been found in glomeruli (see Herrera and Turbat-Herrera, 2010; and review by Coleman and Seymour, 1992). Some normal matrix components may be more prominent in pathological conditions. Mesangial microfibrils may be particularly well developed in MCGN and diabetic glomerulosclerosis (Table 1.1). Fibrillar collagen is found within the GBM in nail–patella syndrome (Coleman and Seymour, 1992).

1.4.5 General and Nonspecific Inclusions and Deposits

Several nonspecific inclusions such as microparticles (small electron-dense granules), fibrils and membrane-like material may be found within the GBM and mesangial matrix – mostly with doubtful or unknown diagnostic significance. Most common are accumulations of spherical particles and vesicles (often referred to as inclusions or virus-like particles): these may be found in the mesangium but are often seen in a subepithelial location, especially in the angle (or 'notch') between two adjoining capillary loops. Microparticles are sometimes seen in areas of laminated GBM in Alport's syndrome (Figure 1.5); similar granules are also found in thickened loops in diabetic sclerosis. Local accumulations of moderately electron-dense material similar to immune-related deposits may be observed, most commonly in a subendothelial location. In the absence of positive immunolabelling, these may relate to the insudation of plasma proteins and/or the development of 'fibrin' caps.

1.4.6 Fibrin

Fibrin may be present in numerous diseases in almost any location in the glomerulus. It may be found as irregular, angular or needle-like accumulations of amorphous material of medium electron density, or it may show cross-striations with a characteristic periodicity. The periodicity of normal fibrin is reported as 22.5 nm (Standeven, Ariëns and Grant, 2005); however, periodicities of ~19–35 nm are reported in pathological tissues in general (Ghadially, 1988). Fibrin is easily distinguished from fibrillar collagen which has an axial banding periodicity of 64–68 nm.

1.4.7 Tubuloreticular Bodies (Tubuloreticular Inclusions)

Tubuloreticular bodies (TRBs) are small clusters of fine anastomosing tubules that arise within the cisternae of the rough endoplasmic reticulum; in the glomerulus, they are most commonly found in endothelial cells (Figure 1.18b). The formation of TRBs has been linked to α- and ß-interferon activity (Hammar *et al.*, 1992). In renal disease, TRBs are most commonly associated with SLE (Figure 1.18b) (and collagen vascular diseases in general), with viral infections (especially HIV/AIDS and hepatitis B) and as a result of α-interferon treatment (Hammar *et al.*, 1992; Haas *et al.*, 2000; Haas, 2007; Yang *et al.*, 2009b). However, TRBs are not specific to these conditions as they have also been found in renal transplants and have been linked to diabetes, lymphoma and *Helicobacter pylori* infection (Yang *et al.*, 2009b). The presence of TRBs in idiopathic conditions can act as an indicator of underlying systemic disease, and their presence should prompt additional investigations (Yang *et al.*, 2009b).

1.4.8 The Glomerular Basement Membrane

1.4.8.1 GBM Width

Reported measurements for adult GBM vary greatly, but mean widths generally fall in the range of 300–400 nm (Figure 1.1) (see Coleman *et al.*, 1986; review by Dische, 1992; Bonsib, 2007). The GBM of children is reported as being slightly thinner up to age 9 (Morita *et al.*, 1988). A review by Marquez *et al.* (1999) found that in addition to the variability in values reported for GBM width in adults, there are conflicting data regarding age- and sex-related differences: some studies indicate that the GBM in males and females is similar, but in others, males are said to have slightly thicker GBM than females. Similarly, some studies suggest that the GBM continues to increase in width with age, whereas others do not. Overall, this variation is presumed to reflect the type of tissue used to obtain 'normal' data, the differing morphometric techniques employed and the variation inherent in tissue processing and choice of reagents (Marquez *et al.*, 1999; Edwards *et al.*, 2009). With these variations in mind, it is clear that statements about GBM width must be treated with caution. For measurements to be meaningful, a set of normal values must be established based on local processing methodology.

1.4.8.2 Thickness, Texture and General Morphology

- **Thickness**: the GBM may be:
  - **Thin**: the membrane may be evenly thinned to ~20–30% of normal width (±100 nm) (Figure 1.6).
  - **Thickened**: the membrane may be evenly thickened up to four times normal width or more (±1000 nm or more) (Figure 1.7).
  - **Irregular**: the membrane may be highly irregular in thickness, although the overall mean may be close to normal (Figure 1.5).
- **Texture**: the GBM may appear to be split or laminated. Lamination may be irregular with a 'basket weave pattern' (Figure 1.5). Small areas of irregular lamination combined with poorly defined electron-lucent areas may be caused by the resorption of deposits.

Faint circumferential lamination is sometimes seen in thickened loops in diabetic glomerulosclerosis.

- **Folding:** loops may become folded or show concertina-like wrinkling as a result of ischaemic collapse. Folds may consolidate to form a thickened area of GBM in which the original folds remain visible. Similar folding may be seen around the periphery of the mesangium (Figure 1.4).
- **Double contouring (interposition):** the GBM becomes duplicated. The duplication is caused by the interposition of mesangial and inflammatory cells, matrix and (in some cases) deposits between the original GBM and the endothelium with the eventual formation of a new inner (luminal) layer of basement membrane material ('tram-tracking' by LM) (Figure 1.16).
- **Subendothelial widening:** the endothelium and the GBM may become separated with the formation of a subendothelial space filled with flocculent material (Figure 1.9) or, more rarely, microfibrils and cellular elements from the blood.
- **Gaps:** small breaks are sometimes seen in the GBM. Despite speculation that these are linked to haematuria, breaks are rare, even in cases of macroscopic haematuria.

1.4.9 The Mesangial Matrix

The principal pathological change seen in the mesangium is enlargement of the extracellular matrix, a process that may lead to the obliteration of part or all of the glomerulus (sclerosis). Matrix expansion may be accompanied by an increase in cellularity, immune-related deposits and several nonspecific inclusions such as microfibrils and granular and/or vesicular material. Rarely, the mesangium may be dissolved or attenuated (mesangiolysis) so that the capillary loops are able to fuse and expand (Morita and Churg, 1983).

1.4.10 Cellular Components of the Glomerulus

1.4.10.1 Capillary Endothelium

Endothelial swelling and hypercellularity may occur in many conditions. Significant numbers of TRBs may be found in several conditions and may contribute to the identification and understanding of underlying pathology (Yang *et al.*, 2009b).

1.4.10.2 *Epithelium*

1.4.10.2.1 *Visceral Epithelium (Podocytes)*

In proteinuric diseases, the normal structure and arrangement of the podocyte foot processes are often lost: this occurs when the actin cytoskeleton is reorganised, and the processes merge to form a continuous or semicontinuous cytoplasmic layer (foot process effacement, obliteration or fusion) (D'Agati, 2008; Haraldsson, Nystrom and Deen, 2008). Numerous cytoplasmic inclusions are found in podocytes, most notably lysosomes, lipid vesicles and protein droplets. Abnormal lysosomes may be present because of an inherited storage disorder such as Fabry's disease or as a result of drug treatment (e.g. amiodarone and gold therapy). Podocytes may develop numerous long microvilli in a process known as microvillous transformation (Figure 1.3).

1.4.10.2.2 *Parietal Epithelium*

The parietal epithelium may proliferate to form cellular 'crescents' that range from small groups of cells to large masses that completely surround

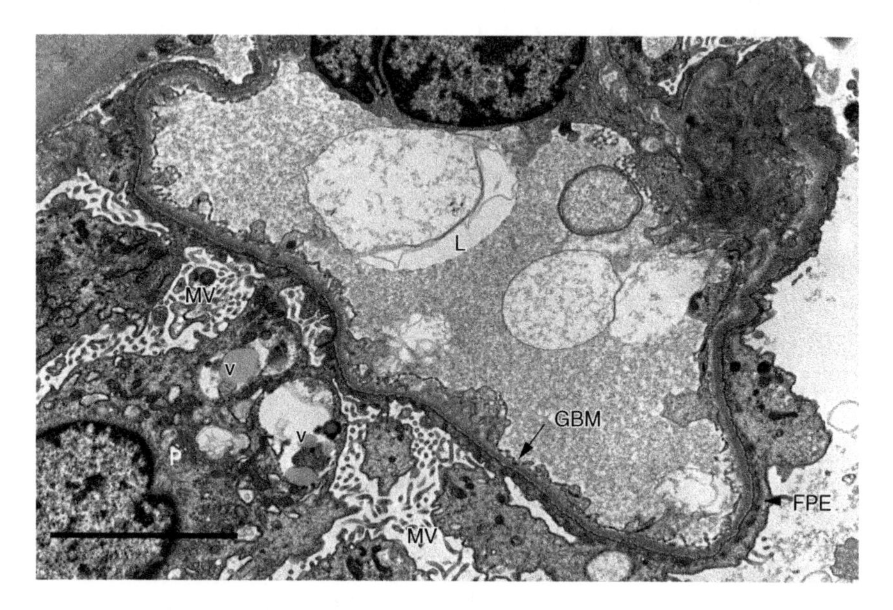

Figure 1.3 Minimal change disease. Foot processes are almost completely effaced (foot process effacement), (FPE). The podocyte (P) contains several vesicles, some of which contain lipid (V); there is also significant microvillous transformation (MV). The glomerular basement membrane (GBM) is slightly thin (190 nm in the thinnest areas). L = capillary lumen. Bar = 5 μm.

the capillary tuft. Fibrin and inflammatory cells such as polymorphs and macrophages may be intermingled with the proliferating cells; in older crescents, fibroblasts and myofibroblasts may also be present. Crescents are nonspecific and a marker of significant glomerular damage; some degree of scarring always remains. At the urinary pole, the junction of the columnar epithelium of the proximal tubule and the parietal epithelial cells is normally sharply marked. Rarely, the tubular epithelium encroaches into the glomerulus, replacing the parietal epithelium in the immediate vicinity of the tubule (termed tubularisation).

1.4.10.3 Mesangial Cells

Mesangial cells proliferate in a number of conditions; this may be accompanied by matrix expansion.

1.4.11 The Capillary Lumen

The capillary lumen may be partially or wholly occluded (stenosed) because of the presence (or effect) of immune-related deposits, insudate, mesangial interposition, endothelial swelling, infiltrating inflammatory cells, capillary wall collapse and/or GBM thickening or folding. Fibrin tactoids and platelets may also be present. On rare occasions, tubular epithelial cells and cell fragments may be found within capillary loops (and the urinary space) – presumably a biopsy artefact.

1.5 THE ULTRASTRUCTURAL PATHOLOGY OF THE MAJOR GLOMERULAR DISEASES

1.5.1 Diseases without, or with Only Minor, Structural GBM Changes

1.5.1.1 Minimal Change Disease (Minimal Change Nephrotic Syndrome or Lipoid Nephrosis)

Minimal change disease (MCD) is a diagnosis based on morphological features, clinical presentation (nephrotic syndrome) and response to therapy (steroid responsive). Primary disease appears to be linked to T-cell dysfunction and type 2 cytokines, while secondary disease results from a wide range of neoplasms, infections, allergens and drugs (Glassock, 2003; Grimbert *et al.*, 2003; Mathieson, 2003).

By LM, the glomeruli appear normal. The most significant ultrastructural feature is foot process effacement, which generally affects greater than 75% of the capillary loop surface (Figure 1.3). Foot process effacement is reversible so that during treatment, only partial effacement or 'smudging' may be observed. Podocytes may contain cytoplasmic vesicles and show significant microvillous transformation (Figure 1.3). Occasionally, mesangial cellularity may be slightly increased. There are no other significant pathological features except that the GBM may be slightly thin (Figure 1.3) (Coleman and Stirling, 1991). Immunolabelling is negative except for weak localised mesangial positivity for IgM and C3 that may indicate segmental sclerosis.

1.5.1.2 Focal Segmental Glomerulosclerosis

Focal segmental glomerulosclerosis (FSGS) is a term that covers five histopathological variants as defined by the Columbia classification (collapsing, glomerular tip lesion, cellular, perihilar and not otherwise specified) which can be applied to both primary and secondary disease (D'Agati *et al.*, 2004). In addition to primary (idiopathic) disease, FSGS has a wide variety of causes including mutations in podocyte proteins, viral infections, toxins and drugs; it can also result from adaptive structural-functional responses in association with numerous insults (D'Agati *et al.*, 2004). The disease recurs in ~20–30% of grafts, and there is a poor graft survival rate in patients with recurrent disease (Ponticelli, 2010). Historically, FSGS has been grouped with MCD, and it has been argued that primary FSGS is an autoimmune T-cell-driven disease – a concept that is increasingly out of favour. To date, no T-cell association or circulating permeability factor has been found, and the true nature of the underlying pathophysiology of primary nephrotic FSGS remains elusive (Meyrier, 2009).

FSGS is characterised by segmental sclerosis caused by expanding mesangial matrix with capillary loop folding and collapse that lead to progressive consolidation. Foot processes are extensively effaced in sclerosing areas with >50% effacement of capillary loop foot processes elsewhere (in secondary disease, there may be <50% effacement) (Figure 1.4). In some cases, the foot processes may be completely lost so that the outer surface of the GBM is exposed; in such areas,

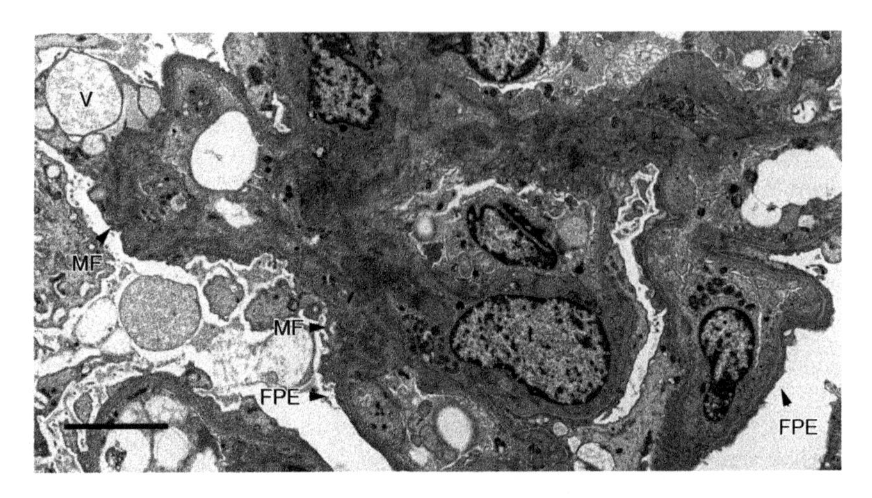

Figure 1.4 Focal segmental glomerulosclerosis. Foot processes are extensively effaced (FPE) with no intact processes remaining, and podocytes contain large vesicles with a variety of appearances (V). Microvillous change and mesangial consolidation are not prominent in this example, but areas of glomerular basement membrane and peripheral mesangium are folded and consolidated, presumably because of ischaemic collapse (MF). Bar = 5 μm.

the underlying matrix may be patchy or laminated in appearance. The podocytes often show significant microvillous transformation and may contain large numbers of cytoplasmic vesicles of varying density (Figure 1.4). In advanced disease, there may be collections of spherical particles and vesicles plus electron-dense granular deposits within the affected segments, as well as large areas of dense material corresponding to the hyaline seen by LM. There may be foam cells present in the glomerular tip lesion and cellular variants, and in the collapsing variant, the capillary loops may be folded, wrinkled and collapsed. In HIV-associated nephropathy, which mainly presents as the collapsing variant, there are generally numerous TRBs in endothelial cells. Immunolabelling generally shows reactions for IgM and C3 in segmental lesions and irregular masses corresponding to hyaline deposits. Diffuse mesangial positivity for IgM and C3 is seen in HIV-associated nephropathy. Nonsclerotic segments may also show some IgM positivity.

1.5.1.3 Pauci-Immune Glomerulonephritis (ANCA-Associated Glomerulonephritis)

Pauci-immune crescentic glomerulonephritis is the most common form of crescentic disease (50–60% of cases). Approximately 80–90% of patients have circulating anti-neutrophil cytoplasmic antibodies

(ANCA). By light and electron microscopy, pauci-immune and anti-GBM glomerulonephritis are similar (see Section 1.5.1.4). Some ANCA-positive cases may show positive immunolabelling and/or electron-dense deposits: these represent concurrent anti-GBM disease (linear labelling) or immune-complex mediated disease (granular labelling and electron-dense deposits), respectively (Rogers, Rakheja and Zhou, 2009). As vascular injury progresses, loops may show endothelial swelling with the development of a subendothelial lucent zone and fibrin deposition; there may also be breaks in the GBM and loss of mesangial matrix, also with fibrin deposition.

1.5.1.4 Anti-GBM Glomerulonephritis (Goodpasture Syndrome)

Anti-GBM disease is characterised by rapidly progressive glomerulonephritis with two peaks of onset, the first at 20–30 years with a male preponderance and a second at 50–70 years with a female preponderance. The disease is linked to cigarette smoking and exposure to hydrocarbons. The target antigens of the circulating autoantibodies are principally the noncollagenous NC1 domains of collagen IV $\alpha3$ and, to a lesser extent, collagen IV $\alpha5$. Antibodies to $\alpha3$ generally appear first, followed by a gradual increase in antibodies to $\alpha5$ (Pedchenko *et al.*, 2010). Ultimately, antibodies may develop that react with all of the $\alpha1$–$\alpha5$ subtypes (Zhao *et al.*, 2009; Pedchenko *et al.*, 2010). High levels of circulating autoantibodies and the presence of antibodies against a broad range of collagen IV subtypes are associated with more severe disease (Yang *et al.*, 2009a; Zhao *et al.*, 2009).

LM shows widespread crescents (often more than 80%) and focal segmental necrotising glomerulonephritis with infiltrating neutrophils. There are no distinct diagnostic ultrastructural features of anti-GBM disease beyond those of necrotising glomerulonephritis: there may be breaks in the GBM, segmental necrosis and inflammatory cells in capillary loops. Fibrin within necrotic areas may mimic electron-dense deposits. Immunolabelling shows diffuse, global, linear staining for IgG; some IgM, IgA and C3 may also be present.

1.5.2 Diseases with Structural GBM Changes

1.5.2.1 Alport's Syndrome

Alport's syndrome is a progressive, genetically and phenotypically diverse disease that affects ~1 : 50 000 live births (Levy and Feingold,

2000). The predominant form is X linked; however, there are also autosomal recessive and autosomal dominant forms (Jais *et al.*, 2000, 2003; Hudson *et al.*, 2003). The mutations involved result in defective basement membrane assembly that affects renal function and causes deafness, ocular changes and leiomyomatosis. The X-linked form (~80% of cases) is associated with a large number of mutations in the collagen IV $\alpha5$ gene (Barker *et al.*, 1990; Martin *et al.*, 1998). Affected individuals show microscopic haematuria, proteinuria (with or without nephrotic syndrome) and hypertension; progressive renal failure develops in males. A small number of males (~2.5%) develop anti-GBM glomerulonephritis on transplantation (Jais *et al.*, 2000; Adler and Salant, 2003). Female carriers generally have haematuria, but fewer progress to end-stage renal failure and this occurs at a later age (Jais *et al.*, 2003). The autosomal recessive form (~15% of cases) involves mutations to the collagen IV $\alpha3$ and $\alpha4$ genes and is clinically similar to the X-linked type except that both males and females are equally affected (Lemmink *et al.*, 1994; Mochizuki *et al.*, 1994; Adler and Salant, 2003). Autosomal dominant disease (~5% of cases) arises from heterozygous mutations in the collagen IV $\alpha3$ or $\alpha4$ genes (Longo *et al.*, 2002; Pescucci *et al.*, 2004). Again, the clinical and pathological features are similar to those of X-linked disease except that the symptoms are less severe and renal function deteriorates slower.

Disease expression is highly variable so that at the ultrastructural level, the GBM may show several changes in contour, width and texture (Figure 1.5) (Jais *et al.*, 2000, 2003). In individual cases, the majority of loops may be (i) thick and multilayered with laminations and splitting (basket weave appearance), (ii) irregular in contour with alternating thin and thick (laminated) areas or (iii) diffusely thin. GBM thickness and texture may also vary between loops so that thin, highly irregular and laminated loops may be found within the same glomerulus. In children, membrane thinning may be the most significant change, and a small number of X-linked female carriers have normal GBM (Jais *et al.*, 2003; Gubler, Heidet and Antignac, 2007). Microparticles are often seen in laminated areas (Figure 1.5).

Immunolabelling has limited usefulness: occasional positivity for IgG, IgM or C3 may be seen, but the meaning of this finding is uncertain. Labelling for the collagen IV $\alpha1$, $\alpha3$ and $\alpha5$ subtypes can determine the molecular makeup of the GBM and help in disease analysis. Similarly, the same panel of antibodies can be used to label epidermal basement membrane to investigate X-linked disease. For a full discussion of labelling patterns, see Hennigar and Tumlin (2009).

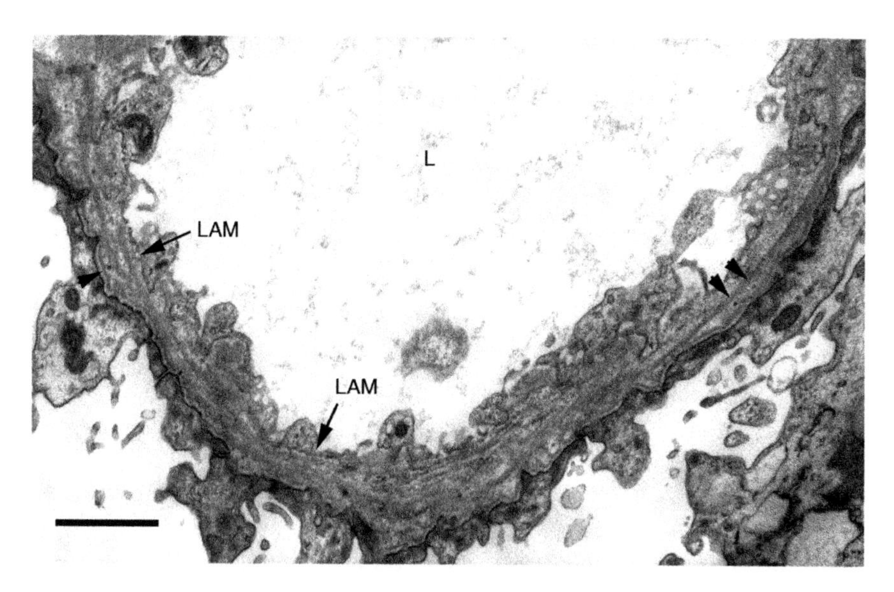

Figure 1.5 Alport's syndrome. The glomerular basement membrane shows internal laminations (LAM) and is irregular in width and contour. Microparticles (arrow heads) are seen here in laminated areas. L = capillary lumen. Bar = 1.5 μm.

1.5.2.2 Thin Basement Membrane Disease

Thin basement membrane disease (TBMD), otherwise known as benign essential haematuria or benign familial haematuria, was historically regarded as a nonprogressive renal disease with recurrent haematuria. With a greater understanding of the genetics underlying Alport's syndrome, TBMD is increasingly seen as a continuum with Alport's syndrome, and ~40% of families with TBMD have haematuria that segregates with the collagen IV $\alpha 3$–collagen IV $\alpha 4$ locus (Savige *et al.*, 2003). Indeed, misdiagnosis of Alport's syndrome as TBMD may be common, and some individuals diagnosed with TBMD are likely to be carriers of Alport's syndrome (Buzza, Wilson and Savige, 2001; Gregory, 2004; Haas, 2006).

By LM, TBMD is characterised by near-normal glomeruli with only slight mesangial expansion and proliferation; in some cases, there may be premature glomerular sclerosis. Small deposits of IgM, C3 and C1q may also be present, but these are nonspecific (Savige *et al.*, 2003). At the ultrastructural level, there is diffuse GBM thinning without splitting or lamination (Figure 1.6). Overall, the degree of thinning in TBMD cases varies. Although in some individuals there may be areas of extreme thinning (~25–30% of normal width, i.e. ~100–150 nm), overall, the

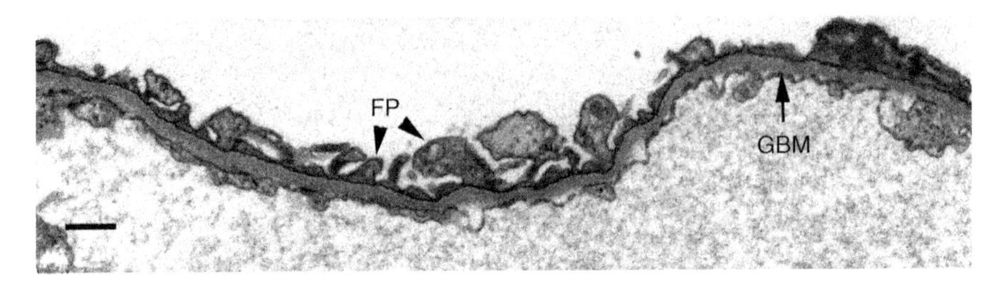

Figure 1.6 Thin basement membrane disease. The glomerular basement membrane is evenly thinned without any lamination (GBM). Here, the thinnest areas are ~90 nm in width and ~170–200 nm overall. Foot processes (FP) are essentially intact with no significant effacement. Bar = 750 nm.

GBM should be less than the normal range as established by the reporting laboratory (published values are in the range of <200–250 nm) (Savige *et al.*, 2003). Haas (2006) suggests that average GBM thickness should be less than the lower limit of the normal range (i.e. less than two standard deviations below the mean of 50 control biopsies from individuals of the same sex) and that at least 50% of individual GBM measurements should be below the lower limit of the normal range. For a diagnosis of TBMD to be made, IgA disease, minimal change, mesangiocapillary and some forms of lupus glomerulonephritis must be excluded as these may also show a degree of membrane thinning (Savige *et al.*, 2003). Note that TBMD cannot be reliably diagnosed using tissue retrieved from wax because of artefactual thinning of the GBM (Nasr *et al.*, 2007).

1.5.2.3 Diabetic Nephropathy

Approximately 40% of diabetics develop diabetic nephropathy, which is the most common cause of end-stage renal failure worldwide (Estacio and Schrier, 2001; Alsaad and Herzenberg, 2007). The principal changes that occur in both type I and type II diabetes are GBM thickening, mesangial expansion and sclerosis (Figure 1.7). Similar changes have also been reported in individuals with metabolic syndrome and glucose intolerance but no clinical manifestation of diabetes (Sanai *et al.*, 2007). Note that diabetic changes may be superimposed on those of other (concomitant) renal diseases, a situation that is increasingly common.

Based on histological appearances, diabetic nephropathy is typed as nodular, diffuse or exudative diabetic glomerulosclerosis. The nodular lesion is typified by nodules of expanded mesangial matrix containing nonspecific fibrils, collagen (including types I, III and VI), small

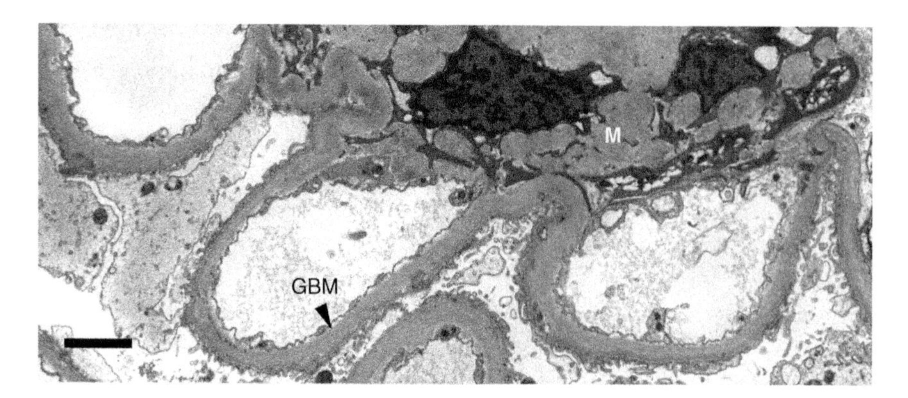

Figure 1.7 Diabetic nephropathy. Here, the glomerular basement membrane (GBM) is extensively thickened and is between ~775 and 800 nm in width. The mesangial matrix (M) shows significant expansion and consolidation. Bar = 2 um.

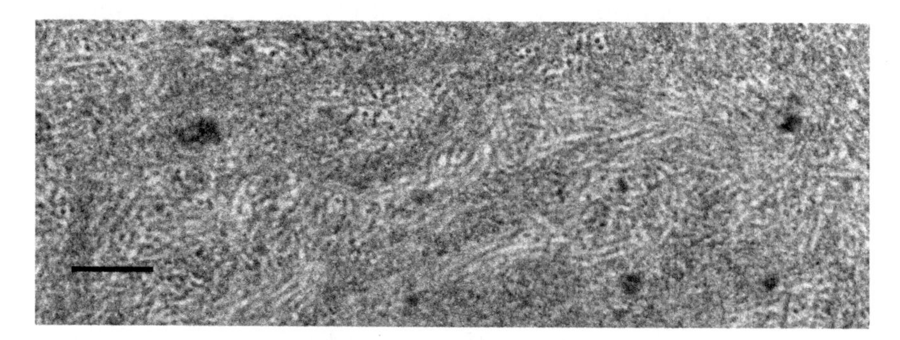

Figure 1.8 Diabetic nephropathy: Kimmelstiel–Wilson nodule. Nodules may contain ~12 nm diameter fibrils (as seen here) that may be confused with immune deposits or amyloid. Bar = 200 nm.

lipid vesicles and miscellaneous cell debris (Nishi *et al.*, 2000). Kimmelstiel–Wilson nodules may contain nonspecific ~12 nm diameter fibrils (Figure 1.8) (diabetic fibrillosis) that sometimes extend into the subendothelial space and may be confused with fibrillar immune deposits (Figure 1.23) (Yasuda *et al.*, 1992; Alsaad and Herzenberg, 2007). Loops are commonly displaced to the periphery of the glomerulus, and capillary lumina may contain homogeneous electron-dense material. The diffuse lesion exhibits global mesangial matrix expansion (containing some lipid and collagen fibrils) with or without mesangial hypercellularity. In contrast to the nodular type, the structure of the glomerular tufts is generally better preserved. The exudative lesion is commonly found in advanced disease. In this type, heterogeneous electron-dense material and small lipid vesicles accumulate in the subendothelial zone to form a thickened area that encroaches on the capillary lumen (fibrin cap by LM) (Nishi *et al.*, 2000). In general, GBM thickening occurs early and, as the disease progresses, may become extreme (up to four times the normal width) (Figure 1.7). Thickened GBM may also be finely laminated with occasional microparticle inclusions. Rarely, the outer GBM may show small focal areas of irregular lamination and disruption. Foot process effacement is common, and intramembranous, paramesangial and mesangial electron-dense deposits are occasionally observed. Immunolabelling may show mild to moderate linear capillary loop staining for IgG (κ and λ) and albumin (C3 negative); there may also be some nonspecific labelling for IgM and C3 in mesangium and nodules (Alsaad and Herzenberg, 2007).

1.5.2.4 Thrombotic Microangiopathy: Haemolytic Uraemic Syndrome and Thrombotic Thrombocytopaenic Purpura

Thrombotic microangiopathy (TMA) is applied to a heterogeneous group of diseases in which there is microvasculature injury with endothelial damage. Haemolytic uraemic syndrome (HUS) and thrombotic thrombocytopaenic purpura (TTP) are the two most significant forms (often with overlapping clinical features); HUS occurs as two forms, typical (classical or diarrhoea-positive HUS) and atypical (diarrhoea-negative HUS). The typical form of HUS occurs mainly in young children and accounts for 90–95% of paediatric cases. Most cases are associated with Shiga or Vero toxin-secreting *Escherichia coli* (principally strain 0157:H7), *Shigella dysenteriae* or *Streptococcus pneumoniae*. Recurrence in transplants is rare (Sánchez-Corral and Melgosa, 2010). The atypical form occurs in both adults and children, and 40–60% of patients have a mutation affecting the complement system. Thrombomodulin gene mutations have also been reported, and a few cases are related to anti–factor H autoantibodies. Recurrence in transplants is common (Sánchez-Corral and Melgosa, 2010). By contrast, TTP occurs mainly in adults and is relatively rare. Most cases are acquired; some result from acquired or congenital ADAMTS13 deficiency (George, 2009).

LM usually shows thickening of the capillary walls and a 'double contour' appearance. Fragmented red blood cells, platelets and fibrin deposition can also be found in capillary loops; there may also be

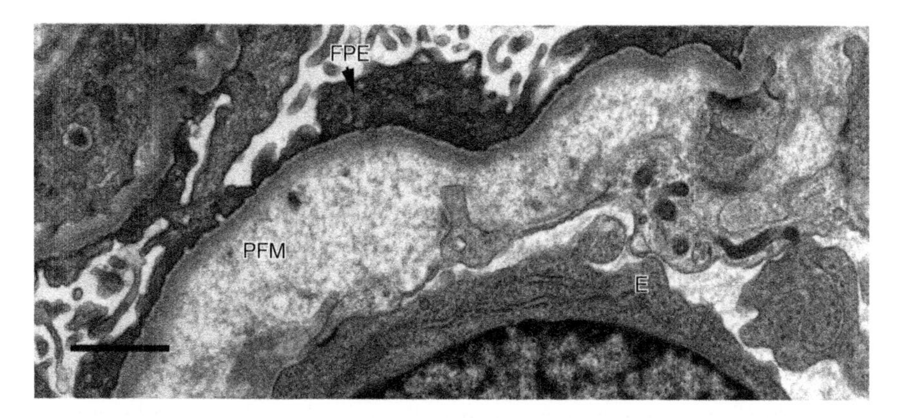

Figure 1.9 Thrombotic microangiopathy: haemolytic uraemic syndrome. The subendothelial aspect of the membrane is expanded by an accumulation of pale 'fluffy' material (PFM), and the endothelial cells (E) are swollen. Foot processes are effaced (FPE) in the area illustrated. Bar = 1 μm.

microvascular injury in arterioles (myxoid change and fibrinoid necrosis) which results in glomerular ischaemia. TEM shows swelling of the glomerular endothelium and accumulation of lucent granular, or 'fluffy', material along the subendothelial aspect of the GBM (Figure 1.9). Lucent areas can be extensive and wide and can lead to GBM reduplication. Foot processes may be partially or wholly effaced (Figure 1.9). Immunolabelling is not specific for any particular type of TMA and is commonly positive for fibrinogen (mesangium, loops and intracapillary thrombi), IgG, IgM, C3 and, rarely, IgA (loops).

1.5.3 Diseases with Granular Deposits

1.5.3.1 Membranous Glomerulonephritis

Membranous glomerulonephritis (membranous GN) is the most common cause of idiopathic nephrotic syndrome in white adults. The primary disease is autoimmune in nature; targets antigens implicated in the disease process include neutral endopeptidase, M-type phospholipase A2 receptor (PLA2R) (Beck and Salant, 2010), aldose reductase and manganese superoxide dismutase (Prunotto *et al.*, 2010). Over 75% of patients have circulating IgG4 autoantibodies to PLA2R, a podocyte transmembrane protein. Disease is initiated when antibodies bind to the PLA2R receptor on the foot processes, thereby creating subepithelial deposits, activating the complement system and injuring the podocytes. The anti-PLA2R antibody is not found in normal individuals or those with secondary membranous GN (or other glomerular diseases), and it may therefore have a role in diagnosis and disease monitoring (Beck and Salant, 2010). Secondary disease is caused by a wide range of factors. Infectious diseases (hepatitis B, malaria, syphilis and schistosomiasis) are the most important; however, neoplasms, autoimmune diseases and a variety of drugs also play a significant role (Beck and Salant, 2010).

Membranous GN is characterised by a pattern of subepithelial and intramembranous granular electron-dense deposits with associated GBM reactions. Disease progression is staged I–IV (Ehrenreich and Churg, 1968); however, the pattern is usually mixed with multiple stages occurring together (e.g. stages II–III or III–IV). Ehrenreich and Churg (1968) define the stages as follows:

- **Stage I (early stage):** small subepithelial deposits with no or minimal membrane reaction. Deposits may be scattered or in small clumps; some loops may have no deposits. Foot processes over deposits are effaced. The GBM may be normal or slightly uneven and irregular; it may also be thickened (depending on the amount of deposit present) (Figure 1.10).
- **Stage II (fully developed):** numerous and extensive (sometimes contiguous) deposits that may cover entire loops. Significant membrane reaction with 'spike' formation. The GBM is irregular but only slightly and focally thickened (Figure 1.11).

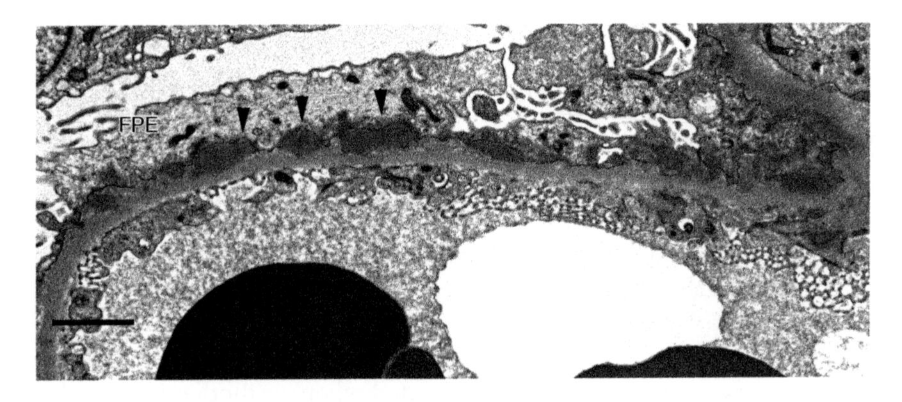

Figure 1.10 Membranous glomerulonephritis stage I (well-developed). Small to medium-sized subepithelial deposits (arrow heads) are present with minimal membrane reaction. There is foot process effacement (FPE) over deposits. Bar = 2 μm.

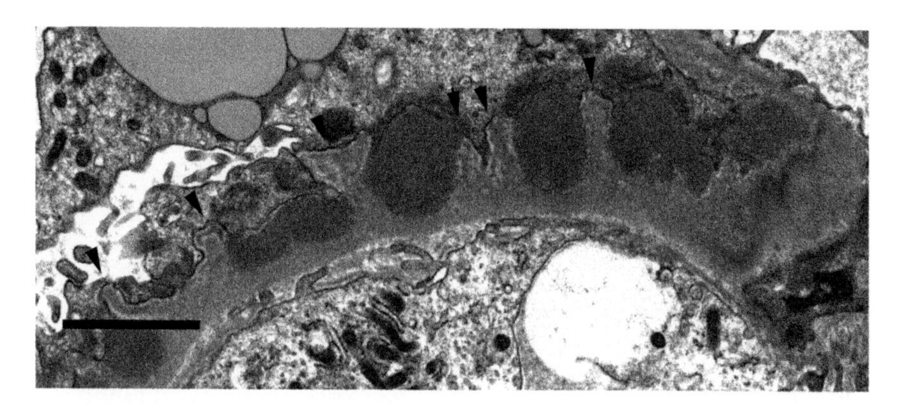

Figure 1.11 Membranous glomerulonephritis stage II. Deposits are larger, there is a significant membrane reaction to form 'spikes' (arrow heads) and there is also some membrane thickening and extensive foot process effacement. Bar = 2 μm.

- **Stage III (advanced stage)**: numerous intramembranous deposits present. Some deposits are electron dense and well demarcated, while others are pale and irregular in contour and texture (suggesting resorption). The GBM is often irregular and focally thickened (Figure 1.12).
- **Stage IV (late stage)**: intramembranous deposits lie within a significantly irregular and disorganised GBM which may be up to 10 times the normal thickness. The surface of the GBM is smooth and

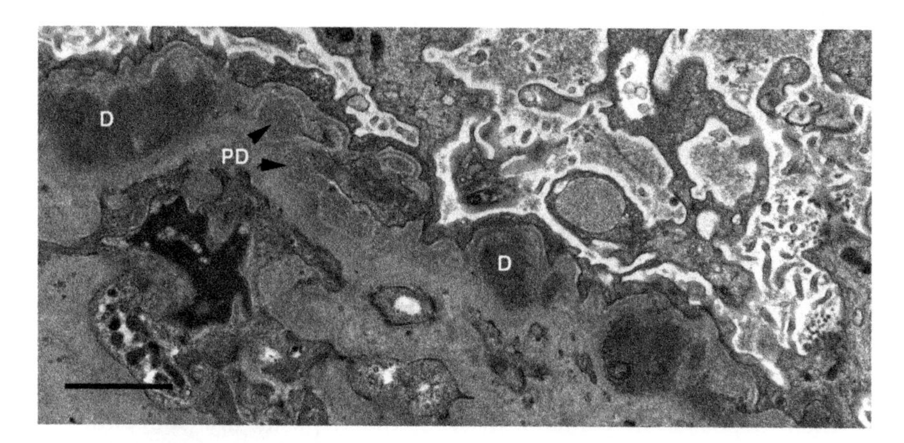

Figure 1.12 Membranous glomerulonephritis stage III. Numerous intramembranous deposits are present – seen here peripheral to mesangium. Deposits (D) vary in density; some are pale (PD) or show peripheral lucency, suggesting resorption. Bar = 2 μm.

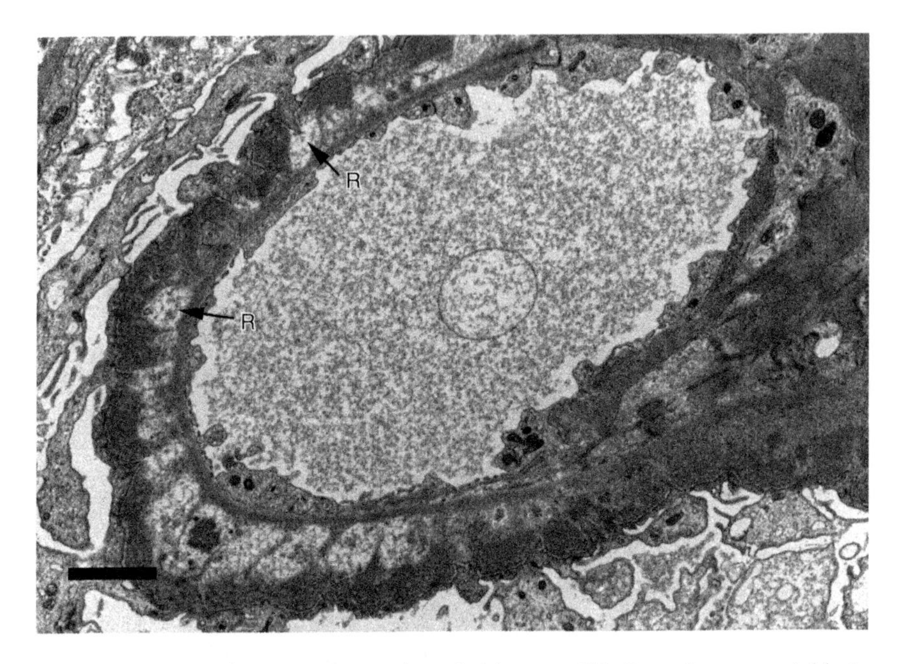

Figure 1.13 Membranous glomerulonephritis stage IV. Deposits are variable in density and texture with significant areas of resorption (R); the membrane is also highly disrupted and irregular in width and texture. Bar = 2 μm.

wavy; spikes may still be present. Deposits are highly variable in size, density and texture, suggesting varying degrees of resorption (Figure 1.13).

Ehrenreich and Churg (1968) suggest that an additional fifth, late or end stage (i.e. stage V) may be added to this classification: this final stage is characterised by capillary collapse, focal mesangial hypercellularity and sclerosis. In contrast, Coleman and Seymour (1992) include disease remission as stage V. In this system, remission is indicated by the presence of deposits that are significantly resorbed, although the GBM remains highly irregular in contour and thickness with poorly defined zones of lucency and lamination.

Finally, it is important to note that primary and secondary disease may differ in respect to both the pattern and immunological makeup of their deposits. In primary disease, the deposits are subepithelial and intramembranous, while in secondary disease there may also be mesangial and subendothelial deposition. Furthermore, the presence of TRB suggests viral infection or lupus-related disease (Beck and Salant,

2010). The finding of deposits with a fibrillar or tubular pattern suggests the presence of cryoglobulins. In respect to the immunological makeup of deposits, in primary disease there are characteristic diffuse global granular peripheral reactions to IgG (principally IgG4), often with C3 which is less intense; staining for C1q, IgM and IgA is uncommon. In secondary disease (lupus and malignancy associated), IgG2 and IgG3 are common (Beck and Salant, 2010).

1.5.3.2 Acute Post-infectious Glomerulonephritis

Acute post-infectious glomerulonephritis (PIGN) is associated with transient infection caused by numerous organisms, particularly bacteria (Rodríguez-Iturbe and Batsford, 2007). The most significant are the nephritogenic Group A β-haemolytic streptococci which cause acute post-streptococcal glomerulonephritis.

PIGN is characterised by the presence of subepithelial dome-shaped deposits (humps) around the capillary loops and mesangium (Figure 1.14). The number of deposits corresponds approximately to the level of inflammation and decreases with time. Note that humps are not specific to PIGN, and similar rounded subepithelial deposits may also occur in membranous GN (in the latter, the deposits are usually associated with a membrane reaction and are smaller and flatter than those in PIGN). Mesangial, subendothelial and intramembranous deposits may also occur, and the GBM may be disrupted because of the presence of deposits and resorbed deposits, as well as slightly irregular in width and contour. Typically, there is an endocapillary inflammatory infiltrate composed mainly of macrophages and polymorphs that may completely occlude capillary lumina; fibrin may also be present. Foot processes are commonly effaced (up to 100%), particularly over deposits (Figure 1.14). In a few cases, there is epithelial proliferation with crescent formation. Immunolabelling shows coarsely granular positivity along capillary loops for IgG and C3 and occasionally for IgM, IgA, light chains, C1q and C4. Large numbers of subepithelial deposits may create a linear or band-like effect.

Persistent infections that cause an acute diffuse proliferative glomerulonephritis with a pattern similar to that of PIGN include infectious endocarditis, bacterial infections of cerebrospinal fluid shunts (shunt nephritis), osteomyelitis and deep-seated abscess. In particular, hump-like deposits may be seen in infectious endocarditis and shunt nephritis (Rogers, Rakheja and Zhou, 2009).

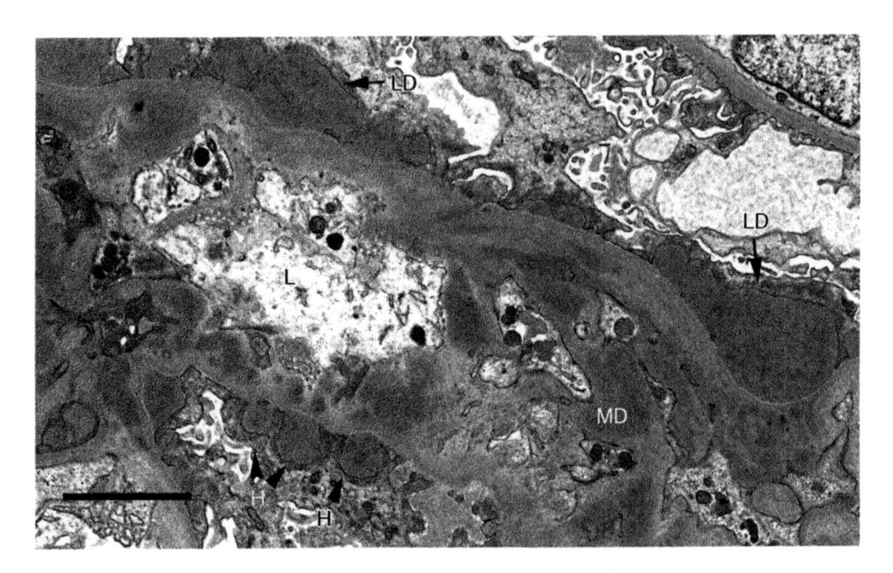

Figure 1.14 Acute post-infectious glomerulonephritis. In this case, large irregular subepithelial deposits (LD) as well as dome-shaped 'humps' (H) and mesangial deposits (MD) are present. Foot processes are effaced over the deposits. L = capillary lumen. Bar = 2.5 μm.

1.5.3.3 IgA Nephropathy (Berger's Disease) and Henoch-Schonlein Purpura Nephritis

1.5.3.3.1 IgA Nephropathy (Berger's Disease)

IgA nephropathy (IgAN) is the most common form of glomerulonephritis worldwide, and reported rates may be underestimated because of subclinical disease (McGrogan, Franssen and de Vries, 2011). The disease is heterogeneous in respect to aetiology and clinicopathological features, and its incidence varies geographically, as well as with age and ethnicity – it is also likely that the diagnosis encompasses several disease subsets (Kiryluk *et al.*, 2010). Primary disease appears to have a genetic basis. Males are affected more frequently than females (in the ratio of 6 : 1), and a significant number of IgAN patients have circulating galactose-deficient IgA1 that appears to act as an autoantigen (Kiryluk *et al.*, 2010). The mechanisms underlying the formation of IgA1 immune complexes and subsequent deposition are probably multifaceted; however, the predominant view is that the IgA1 interacts with an autoantibody (oligoclonal IgG1 and possibly IgA) to create circulating immune complexes that subsequently lodge within the mesangium

(Coppo *et al.*, 2010). Secondary disease has been linked to a wide range of aetiologic agents, and the mechanism underlying the formation of deposits is unknown. There have been several systems introduced for the classification of IgAN. The most recent of these is the Oxford system, the goal of which is to identify pathological features that predict risk of progression (Cattran *et al.*, 2009; Coppo *et al.*, 2010). Since its introduction, the utility of the Oxford system has been validated by Herzenberg *et al.* (2011).

IgAN is characterised by mesangial IgA deposition accompanied by a variable increase in mesangial hypercellularity that is usually segmental rather than global. The mesangial matrix may also be variably increased. The GBM is sometimes thinned: this may be focal or more widespread and similar to that seen in thin basement membrane nephropathy. Deposits are often large, nodular and situated around the periphery of the mesangium; with advancing disease and mesangial enlargement, deposits may also be found within the mesangial core (Figure 1.15). Subendothelial deposits are common, but subepithelial and intramembranous deposits are rare. The presence of numerous subepithelial deposits suggests an unrelated concomitant immune-related process. IgA deposits are, in fact, found in association with a wide range of disease states (which may be classed as secondary IgAN) as well as in 4–16% of normal

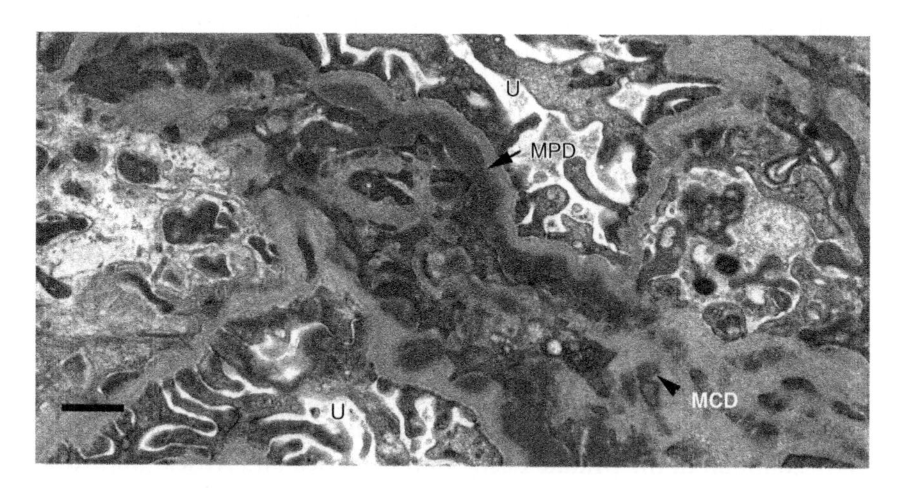

Figure 1.15 IgA nephropathy. Deposits are present within the paramesangial zone (mesangial peripheral deposits) (MPD) with some small deposits within the mesangial core proper (MCD). U = urinary space. Bar = 1 μm.

individuals (Coppo, Feehally and Glassock, 2010). Important differential diagnoses are infectious and post-infectious glomerulonephritis where the presence of IgA may mimic IgAN, and MCD where a finding of IgA may be coincidental or represent mixed disease. Additionally, significant mesangial deposits are seen in C1q nephropathy. In this disease, which may have the morphological features of MCD or FSGS, capillary wall deposits are scarce and TRBs are absent. In secondary IgAN, deposits are usually restricted to the mesangium and the mesangioproliferative process is not pronounced. Immunolabelling shows dominant or co-dominant granular diffuse mesangial IgA deposition, commonly with C3 and IgG, and sometimes with IgM. The presence of C1q suggests secondary IgAN, lupus or C1q nephropathy.

1.5.3.3.2 Henoch–Schonlein Purpura Nephritis

Henoch–Schonlein purpura nephritis (HSPN) and IgAN are related diseases sharing similar pathological and biological abnormalities. HSPN principally occurs in children (IgAN has a peak age range of 15–30 years) and is the most common vasculitis of children and young adults. In addition to the extra-renal clinical signs restricted to HSPN, the disease also has larger circulating IgA-containing complexes and a greater incidence of increased plasma IgE levels (Davin, Ten Berge and Weening, 2001).

At the ultrastructural level, HSPN is similar to IgAN except that subepithelial, intramembranous and subendothelial deposits are more common and large subepithelial deposits are frequently associated with crescents and synechiae. Mesangial deposits are principally of IgA; these commonly extend to the capillary walls. Mesangial IgG, IgM and C3 may also be present. Fibrin deposits are more frequent in HSPN than in IgAN.

1.5.3.4 Mesangiocapillary Glomerulone-phritis (Membranoproliferative Glomerulonephritis)

Primary idiopathic mesangiocapillary glomerulonephritis (MCGN) accounts for <5% of primary glomerulonephritis, and it mainly affects children and young adults. The disease is declining worldwide but is relatively common in the Middle East, South America and Africa (30–40% of cases) (Rogers, Rakheja and Zhou, 2009). MCGN is not a single entity: the term denotes a common underlying pathological mechanism and pattern of glomerular injury typified by mesangial hypercellularity with matrix expansion, structural alterations to the capillary walls (with subendothelial extension of the mesangium) and hypocomplementemia (but with different mechanisms of complement activation). The primary forms are grouped into three types (I, II and III) based on pathological features. Secondary forms with an MCGN pattern are linked to systemic and infectious disorders and mainly occur in adults. Chronic hepatitis C infection is now known to cause a significant number of cases previously thought to be idiopathic (Alchi and Jayne, 2010). Some authorities consider type III MCGN to be a variant of type I, and some consider type II (DDD) to be distinct from MCGN based on epidemiological, morphological and aetiopathogenic criteria (Rogers, Rakheja and Zhou, 2009).

1.5.3.4.1 Type I MCGN (Classical MCGN)

Type I MCGN is defined by the deposition of immune complexes associated with the activation of the classical complement pathway (low or normal C3, low C4 and CH_{50}) – the precise identity of the putative antigen is unknown in most cases (Alchi and Jayne, 2010). Typically, there is a lobular appearance with global mesangial hypercellularity and diffuse endocapillary proliferation with infiltrating monocytes and neutrophils accompanied by an increase in mesangial matrix. Completely sclerotic mesangium may mimic diabetic-type Kimmelstiel–Wilson nodules. Deposits, which may be numerous and large, are principally subendothelial, although scattered subepithelial and mesangial deposits may also be present (Figure 1.16). Overall, the GBM becomes thickened and duplicated because of mesangial interposition ('tram-tracking' by LM) (Figure 1.16). Foot process effacement is widespread, and there may be focal microvillous transformation. Immunolabelling is strongly positive for C3 in a fine to coarse, or broad, granular pattern along capillary loops and occasionally the mesangium. There may also be labelling for IgG, IgM and, more rarely, IgA in a similar pattern; positivity for C1q, C4 and properdin may also occur.

1.5.3.4.2 Type II MCGN (Dense Deposit Disease)

Dense deposit disease (DDD) principally affects children aged 5–15 years and is associated with a serum immunoglobulin, C3 nephritic factor (C3NeF), that stabilises C3 convertase (C3bBb), thereby activating the alternative complement pathway (low C3, normal C4 and low CH_{50}) (Appel *et al.*, 2005; Alchi and Jayne, 2010). The morphological features of DDD are variable, and the glomerular hypercellularity and lobulation typical of MCGN type I are seen in only ~25% of cases. The ultrastructural appearance is distinctive and provides a definitive diagnosis.

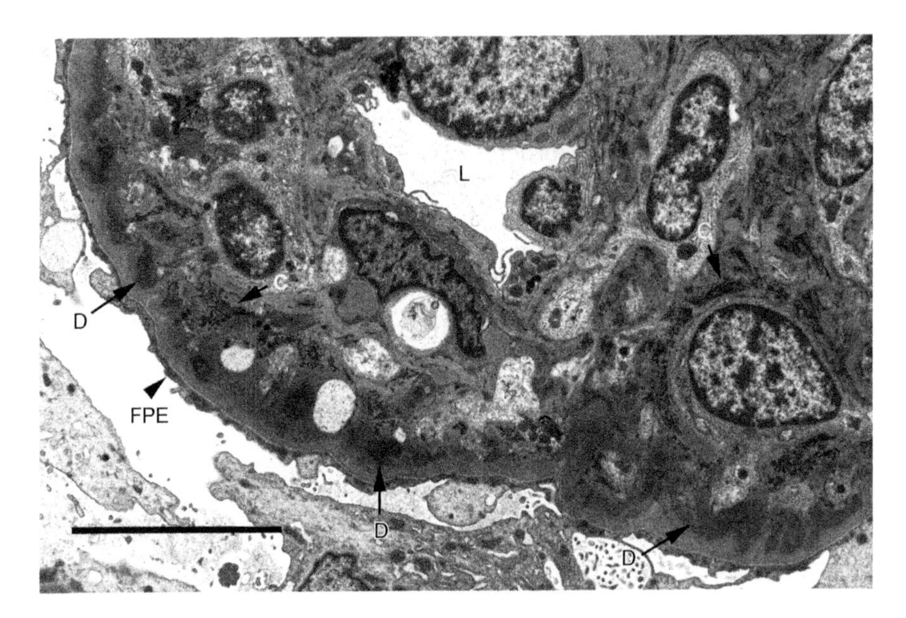

Figure 1.16 Mesangiocapillary glomerulonephritis type I. Large heterogeneous deposits (D) are seen around the inner aspect of the capillary loop and the paramesangial zone. These deposits were originally subendothelial in location but have been incorporated into the capillary wall, which is significantly thickened because of mesangial cell interposition and matrix formation. Fibrillar collagen (C) is also seen within the thickened wall and mesangium. There are no subepithelial deposits present in this example, but foot processes are totally effaced (FPE). L = capillary lumen. Bar = 7.5 μm.

Capillary loops have continuous electron-dense homogeneous deposits along the subendothelial aspect of the GBM; these may be within the inner portion of the membrane or across its entire width (Figure 1.17). The effect is a band or ribbon-like appearance that is often referred to as a linear 'transformation', although occasionally the deposits may be interrupted, in which case they have the appearance of a string of sausages (sometimes called bacon rash deposits) (Alchi and Jayne, 2010). Deposits may also be found in the paramesangium, mesangium, Bowman's capsule, tubular basement membranes and blood vessel walls. Subepithelial and subendothelial deposits may be found in cases with the LM features of acute proliferative glomerulonephritis and membranous glomerulopathy. Mesangial interposition and hypercellularity may also be present. Immunolabelling shows intense linear capillary loop and

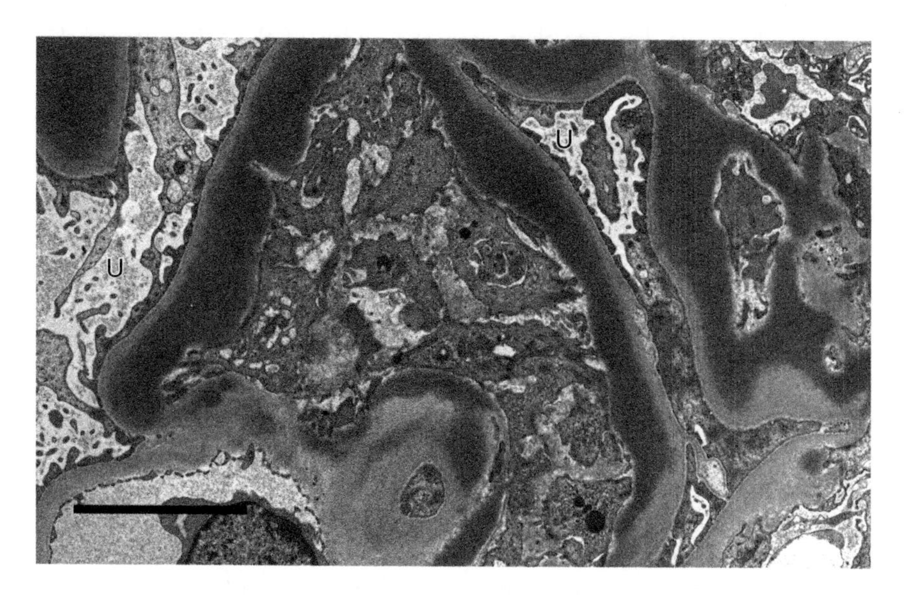

Figure 1.17 Mesangiocapillary glomerulonephritis type II (dense deposit disease). A typical continuous linear electron-dense homogeneous deposit is seen around the inner aspect of the capillary basement membrane (the feature in the centre of the image may be an occluded loop or mesangium). U = urinary space. Bar = 5 μm.

mesangial positivity for C3, generally without C1q and C4. Deposits are not typical immune complexes, so immunoglobulins are therefore generally absent, although IgM, IgG and IgA have been reported (Alchi and Jayne, 2010).

1.5.3.4.3 Type III MCGN

In type III MCGN, both the alternative and terminal complement pathways are activated (low C3, normal C4 and low C5–C9) (Alchi and Jayne, 2010).

The gross features of MCGN types I and III are similar but, although mesangial hypercellularity may be less, there is often a more lobular appearance and GBM thickening is more pronounced with significant tram tracking. Two types of type III MCGN are recognised, that of Burkholder, Marchand and Krueger (1970) and that of Strife *et al.* (1977) and Anders *et al.* (1977). The characteristic features of these subtypes are large subepithelial deposits with a membrane reaction similar to 'spikes' (type III of Burkholder), and subendothelial and intramembranous deposits that span the width of the GBM (from the subendothelial to the subepithelial zone) with associated membrane disruption and lamination (type III of Strife and Anders). Immunolabelling reveals coarse granular staining for C3 along capillary loops and in mesangium in ~50% of cases. In the other 50%, the C3 labelling is combined with staining for IgG with lesser and more variable reactions for IgM and/or IgA.

1.5.3.5 Systemic Lupus Erythematosus

Systemic lupus erythematosus (SLE) is a complex autoimmune disease with a wide range of clinical presentations. SLE causes inflammation in multiple organ systems, particularly the kidney, skin, serosal membranes, central nervous system and joints. Approximately 15–55% of individuals with SLE develop lupus nephritis (Siso *et al.*, 2010). SLE occurs at all ages and in both sexes (with a female-to-male ratio of ~10:1), but it particularly affects premenopausal women with a predilection for those of Asian and African American descent. Lupus nephritis is generally more severe in children, males and patients of Asian, Hispanic and African descent (Venuturupalli and Wallace, 2007; Stokes, Nasr and D'Agati, 2009). Some cases of SLE are idiopathic, but the disease may be precipitated or aggravated by numerous agents including drugs, viral infection (particularly HIV), ultraviolet light and hormones – there may also be genetic factors involved (Stokes, Nasr and D'Agati, 2009). A wide variety of circulating autoantibodies have been found in SLE; the profile of these antibodies persists despite the level of disease activity and state of remission (Fattal *et al.*, 2010). Levels of immunoglobulins (IgG) directed against single- and double-stranded DNA (antinuclear antibodies), Epstein–Barr virus and hyaluronic acid all increase; however, some specific IgM levels decrease, and this profile is highly specific (>88%) for SLE (Fattal *et al.*, 2010). Antibodies against double-stranded DNA predominate in glomerular immune deposits. The site of immune complex deposition within the glomerulus may be determined by the type, specificity and physicochemical characteristics of the immunoglobulins involved and the circulating immune complex load (Stokes, Nasr and D'Agati, 2009).

To improve diagnosis and the interpretation of the renal pathology associated with SLE, several classification systems have been developed. The current system is the International Society of Nephrology/Renal Pathology Society (ISN/RPS) 2003 classification of lupus nephritis, which is based on previous World Health Organisation classification systems (Weening *et al.*, 2004; D'Agati, 2007). The ISN/RPS classification is based on an integrated analysis of the LM and immunolabelling findings and, while TEM is not required for classification, it is highly recommended.

The range of TEM findings in SLE is summarised in Table 1.2. Most cases have finely granular, medium-density mesangial deposits, usually in combination with subepithelial and/or subendothelial deposits; organised deposits also occur (Figures 1.18 and 1.19). Organised material may consist of fingerprint deposits (Figure 1.20) or lattice-like, fibrillar or tubular deposits, the latter sometimes organised in parallel arrays. Organised deposits may be associated with circulating type III mixed cryoglobulin (Stokes, Nasr and D'Agati, 2009). Some cases of SLE also have large numbers of TRBs in capillary endothelial cells (Figure 1.18), although numbers vary and only a few may be present, particularly during treatment. Characteristically, immunolabelling is positive for IgG (generally the strongest), IgA, IgM, C3 and C1q ('full-house' labelling). Nuclei may show staining for IgG because of the presence of circulating antinuclear antibodies.

1.5.3.6 Monoclonal Immunoglobulin Deposition Disease

Monoclonal immunoglobulin deposition disease (MIDD) is a rare systemic disease that involves the deposition of monoclonal immunoglobulins in several organs, with most patients having kidney involvement. Three types of MIDD have been described: light-chain deposition disease (LCDD), heavy-chain deposition disease (HCDD) and mixed light- and heavy-chain deposition disease (LHCDD) (Lin *et al.*, 2001; Basnayake *et al.*, 2011). LCDD and HCDD appear to be similar clinically; LCDD is the most common type, and only a few LHCDD and HCDD cases have been reported.

Morphologically, LCDD and HCDD are similar. Early cases may show little change but, as the disease progresses, mesangial proliferation is followed by a membranoproliferative pattern. Finally, the characteristic appearance is that of nodular sclerosis. Immunolabelling shows a characteristic linear pattern of positivity along the basement membranes of glomerular capillaries and tubules. In LCDD, monoclonal κ or λ light chains may be present, while in HCDD, α heavy chains predominate with a small number of cases involving γ or u heavy chains. (Gokden, Barlogie and Liapis, 2008; Basnayake *et al.*, 2011). By TEM, deposits are finely granular or powdery in appearance and can usually be found in the

Table 1.2 Abbreviated ISN/RPS 2003 classification of SLE: immunolabelling and TEM features (based on Weening *et al.*, 2004; Stokes, Nasr and D'Agati, 2009).

Class I: minimal mesangial lupus nephropathy

No change by LM. Mesangial immune deposits by immunolabelling and TEM. Minimal foot process effacement.

Class II: mesangial proliferative lupus nephropathy

Mesangial hypercellularity. Mesangial deposits with rare isolated and small subepithelial or subendothelial deposits involving the capillary walls in some cases (but not visible by LM). The presence of subendothelial deposits visible by LM warrants designation as class III/IV depending on the percentage of glomeruli affected.

Class III: focal lupus nephropathy (involving less than 50% of glomeruli)

(A) active lesions; (A/C) active and chronic lesions; (C) chronic lesions

Focal disease: in assessing the extent of the lesions, both active and sclerotic lesions must be included. Focal or diffuse mesangial alterations, including proliferation and deposits. Subendothelial deposits are also present; usually segmental. Small subepithelial deposits may be present; if these affect >50% of glomerular surface area in a minimum of 50% of glomeruli (by LM or IF), an additional (combined) diagnosis of membranous lupus nephritis (class V) may be given.

Class IV: diffuse lupus nephropathy (involving 50% or more of glomeruli)

Diffuse segmental (IV-S) or global (IV-G)

(A) active lesions; (A/C) active and chronic lesions; (C) chronic lesions

Typically shows diffuse subendothelial deposits, with or without mesangial alterations. Scattered subepithelial deposits are often found in class IV disease: a combined diagnosis of IV/V is if these deposits involve at least 50% of the surface of capillary loops in a minimum of 50% of glomeruli by LM or immunolabelling.

IV-S: segmental endocapillary proliferation encroaching on capillary lumina ± necrosis

IV-G: diffuse and global endocapillary, extracapillary or mesangiocapillary proliferation or widespread wire loops. Includes cases with extensive (diffuse and global) subendothelial deposits with little or no proliferation.

Class V: membranous lupus nephropathy

Membranous disease with global or segmental continuous granular subepithelial immune deposits (with membrane thickening and spikes in established disease), often combined with mesangial deposits ± mesangial hypercellularity. Scattered subendothelial deposits by TEM or immunolabelling. However, if deposits are present by LM, then a combined diagnosis of class III/V or class IV/V is warranted, depending on their distribution.

Class VI: advanced sclerosing lupus nephropathy (≥90% **sclerosed glomeruli without residual activity)**

Biopsies showing ≥90% global glomerulosclerosis with clinical or pathological evidence that the sclerosis is caused by lupus nephritis, but with no evidence of ongoing active glomerular disease. Residual deposits in sclerosing glomeruli. May represent the advanced stage of chronic class III, IV or V disease.

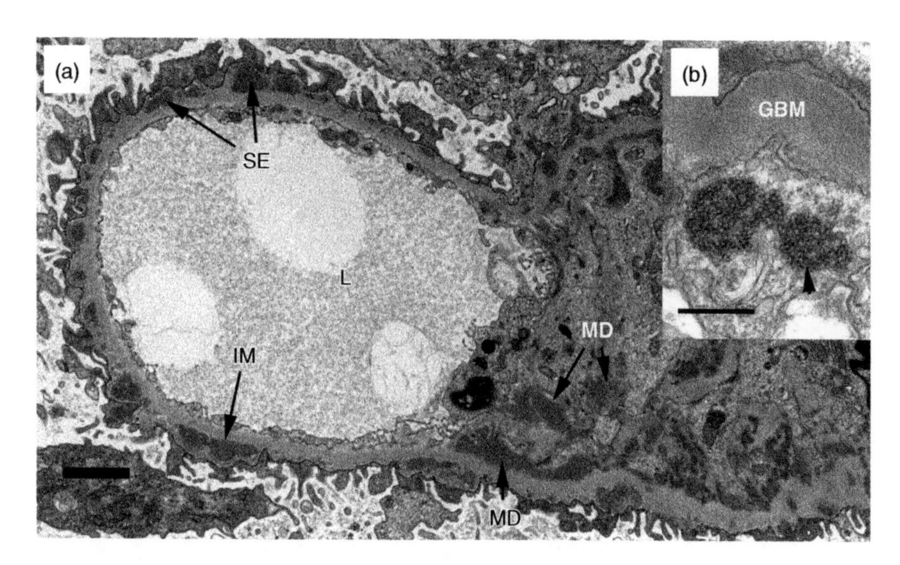

Figure 1.18 Systemic lupus erythematosus. (a) In addition to mesangial and paramesangial deposits (MD), subepithelial (SE) and intramembranous deposits (IM) are present around the capillary loop. There is only minor effacement of capillary loops. L = capillary lumen. Bar = 2 μm. (b, inset) A tubuloreticular body (arrow head) in an endothelial cell: these are sometimes found in large numbers in SLE. GBM = glomerular basement membrane. Bar = 500 nm.

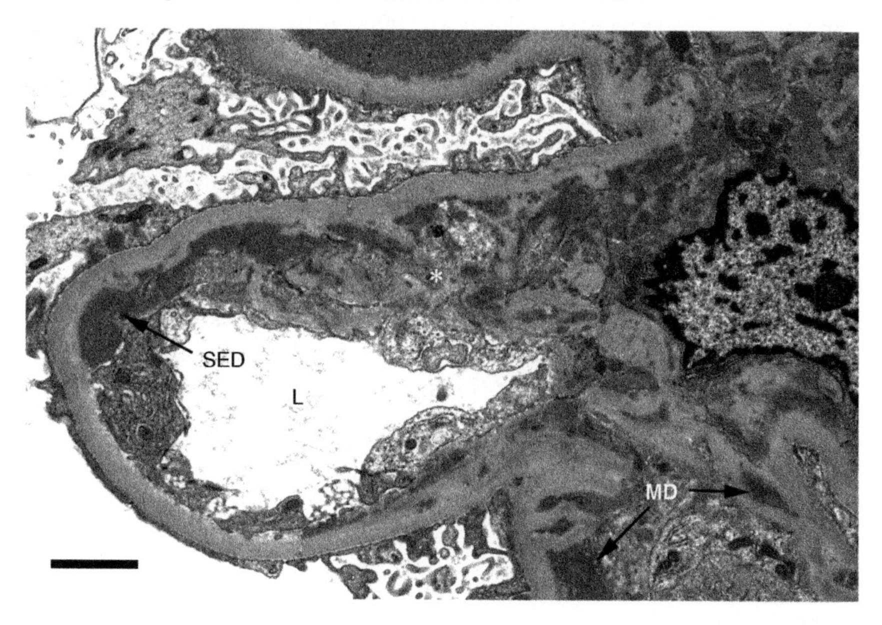

Figure 1.19 Systemic lupus erythematosus. In addition to mesangial and paramesangial deposits (MD), there are extensive subendothelial deposits (SED) with endothelial cell swelling and possible mesangial interposition (*). In addition, the GBM appears to be slightly thickened and there is some foot process effacement. L = capillary lumen. Bar = 2 μm.

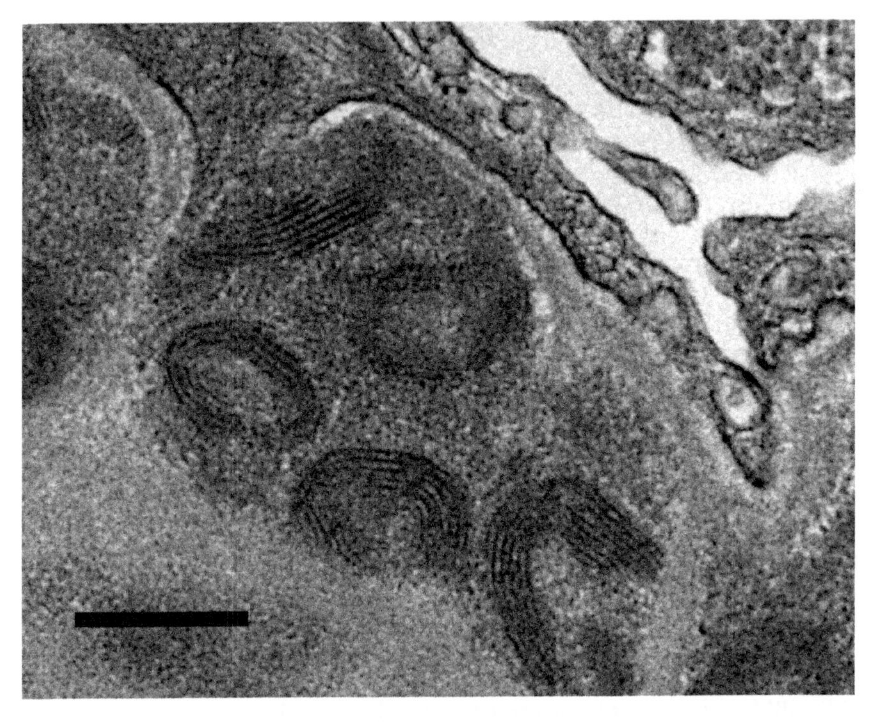

Figure 1.20 Systemic lupus erythematosus – organised deposit. Organised deposits with a whorl or 'fingerprint' arrangement of this type are sometimes seen in cases of SLE.

mesangium and along the GBM, tubular basement membranes and Bowman's capsule. In the glomerulus, linear deposition usually occurs along the lamina rara interna of the GBM and sometimes within the lamina densa (Figure 1.21), a pattern that may mimic Type II MCGN (DDD) (Figure 1.17). In tubules, deposition is between the basement membrane and the interstitium. Sometimes, deposition is not widespread and careful searching is required (Gokden, Barlogie and Liapis, 2008). LCDD commonly recurs in transplants (Leung *et al.*, 2004).

1.5.4 Diseases with Organised Deposits

1.5.4.1 Amyloidosis

Deposition of amyloid can occur in many organs, and at least 25 types of amyloid have been identified, all with different precursor proteins. All amyloid deposits have been found to contain up to 15% of amyloid P, a nonfibrillary glycoprotein identical to serum amyloid P (Vowles,

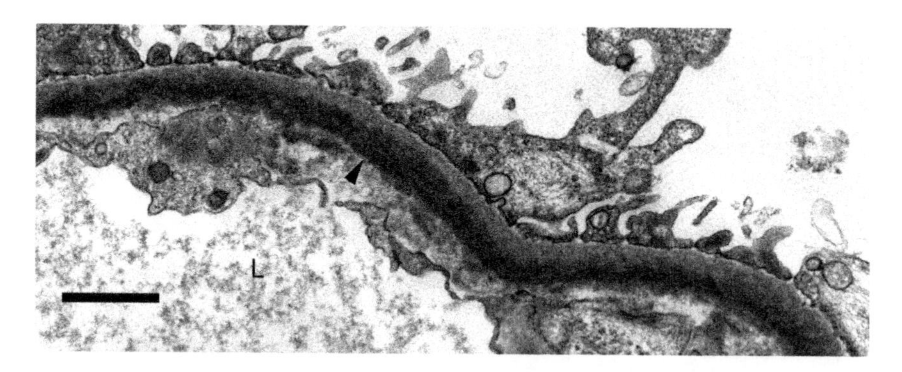

Figure 1.21 Light-chain deposition disease (κ light-chain disease). κ light chains are seen as a powdery linear deposit (arrow head) that may be regarded as either intramembranous (within the lamina densa) or within the lamina rara interna of the GBM. There is also minor foot process effacement. L = capillary lumen. Bar = 1 μm.

2008). Many amyloid types may be found in the kidney, most commonly AL amyloid (derived from immunoglobulin light chains; primary and myeloma-associated amyloidosis), and more rarely AA (derived from serum amyloid A; reactive secondary amyloidosis) and AH (derived from immunoglobulin heavy chains; primary and myeloma-associated amyloidosis).

Amyloid may infiltrate all areas of the glomerulus, including the mesangium, the GBM, Bowman's capsule and the urinary space. By LM, amyloid deposition appears as amorphous hyaline material which is Congo red positive (Table 1.1). However, note that small deposits may be missed by LM, and TEM examination is critical. At the ultrastructural level, amyloid deposits consist of randomly arranged, long, straight, nonbranching fibrils ~7–12 nm in width (Table 1.1, Figure 1.22). Fibrils may sometimes appear to penetrate cell membranes (Figure 1.22). More organised fibrils may be seen as expanding, almost parallel arrays (cascades) (Figure 1.22); this spiculated type of amyloid deposition can mimic the appearance of GBM 'spikes' with silver staining. TEM does not differentiate the various amyloid types. Immunolabelling may show monotypic light-chain staining in some cases of AL-associated amyloidosis; labelling for immunoglobulins, complement components, fibrinogen and albumin is negative.

1.5.4.2 Fibrillary Glomerulonephritis

Fibrillary glomerulonephritis (fibrillary GN) is rare, predominantly affects Caucasians and has a peak incidence in the fifth to sixth decades of

Figure 1.22 Amyloid. Fibrils with a random organisation (R) and in parallel arrays (P) are seen in this image, the latter appearing to penetrate (or cascade or flow from) the overlying epithelial cell (E). Bar = 750 nm.

life; it is associated with a wide range of conditions that result in chronic immune stimulation, including systemic disorders and viral infections such as HIV (Haas *et al.*, 2000). Association with a lymphoproliferative disorder is uncommon (<5% of cases). The disease can recur in renal transplants (Herrera and Turbat-Herrera, 2010).

LM can show many changes including mesangial expansion and proliferation, capillary wall thickening (segmental or diffuse), segmental collapse and crescents. TEM is definitive with the finding of fibrillar deposits in the mesangium and capillary walls (subepithelial, subendothelial and transmembrane) (Figure 1.23). Fibrils are ~13-29 nm in diameter and must be distinguished from amyloid and nonspecific fibrillar deposits using a combination of TEM, LM and immunolabelling criteria (Table 1.1). Characteristically, mesangial and/or peripheral capillary walls show linear or pseudo-linear positivity for IgG (mostly IgG4), C3 and κ and λ light chains (Herrera and Turbat-Herrera, 2010). Amyloid-P component may co-localise with the fibrils.

1.5.4.3 Immunotactoid Glomerulonephritis

Immunotactoid glomerulonephritis (immunotactoid GN) is less common than fibrillary GN (~100 times rarer), predominantly affects Caucasians and has a peak incidence after 60 years of age. Importantly, immunotactoid GN is normally associated with lymphoproliferative disorders and,

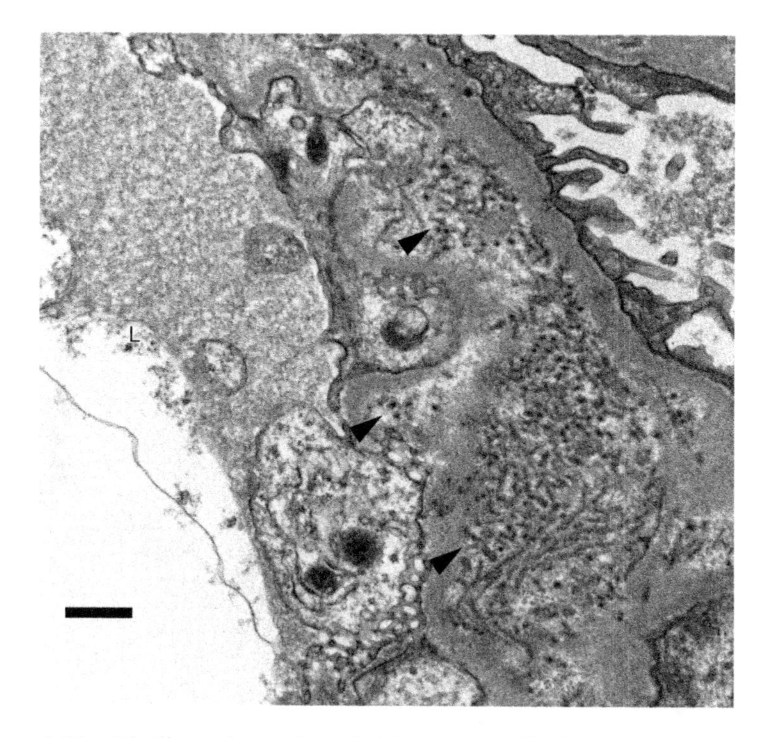

Figure 1.23 Fibrillary glomerulonephritis. Immune fibrils (arrow heads) ~12–18 nm in diameter are seen in the capillary basement membrane. L = capillary lumen. Bar = 500 nm.

more rarely, viral infection (Markowitz *et al.*, 1998; Haas *et al.*, 2000; Herrera and Turbat-Herrera, 2010). Like fibrillary GN, immunotactoid GN recurs in transplants (Herrera and Turbat-Herrera, 2010).

The LM appearance is variable and nonspecific, so TEM is required for diagnosis. Ultrastructurally the deposits are seen as extracellular, long, nonbranching tubules (Figure 1.24), sometimes with a lattice-like pattern; tubules may be similar to those seen in cryoglobulinaemia (Figure 1.25) (Ivanyi and Degrell, 2004). Tubules are usually densely packed in parallel arrays, but are sometimes random. Overall tubule diameter ranges from 10 to 90 nm but is usually >30 nm (Table 1.1, Figure 1.24). The tubules may be embedded in an amorphous or granular material or within the background matrix. All cases generally show mesangial deposits, but the capillary loops can also be involved. Immunolabelling commonly shows patchy granular or pseudolinear

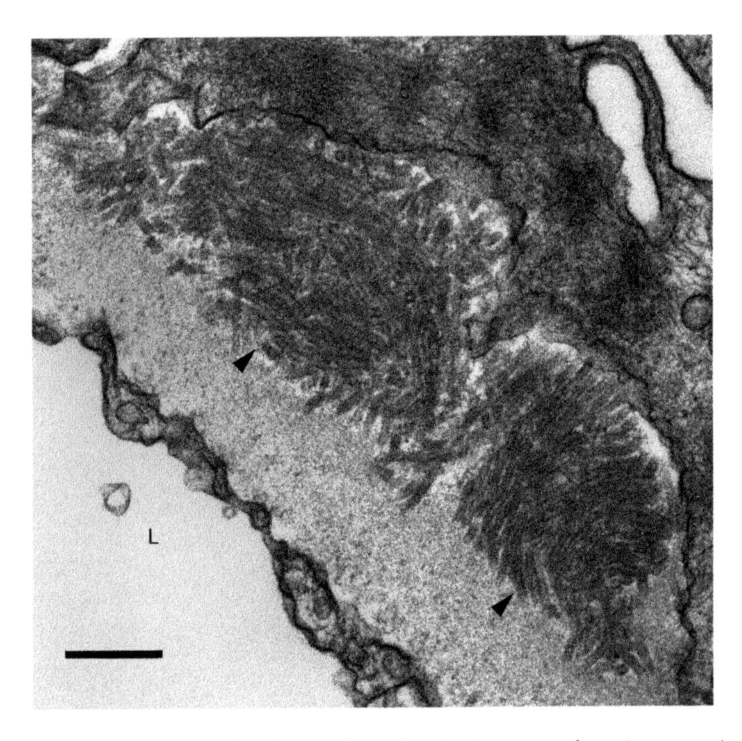

Figure 1.24 Immunotactoid glomerulonephritis. Immune deposits organised as densely packed and randomly organised hollow tubules (arrow heads) ~35 nm in diameter are seen here in a subepithelial location. L = capillary lumen. Bar = 350 nm.

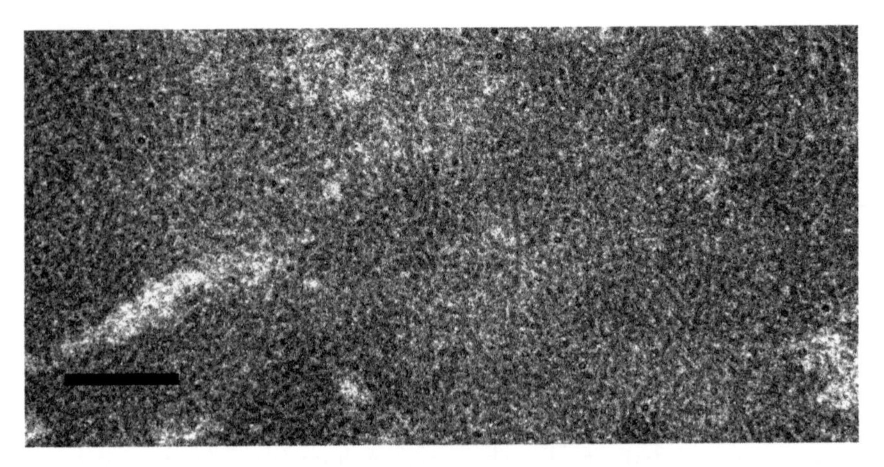

Figure 1.25 Cryoglobulinaemic glomerulonephritis. The appearance of cryoglobulin deposits varies considerably. In this case, the deposits are seen as randomly organised straight or slightly curved tubules ~30 nm in diameter. The tubules are set in a finely granular amorphous background and are not quite as distinctly formed as the tubules seen in immunotactoid. Bar = 500 nm.

positivity for IgG and C3 (peripheral and mesangial), and there may be light-chain restriction (κ more than λ). Labelling for IgM, IgA and C1q is uncommon (Herrera and Turbat-Herrera, 2010).

1.5.4.4 Cryoglobulinaemic Glomerulonephritis

Cryoglobulins are mixtures of rheumatoid factor, immunoglobulins and hepatitis C virus RNA that undergo reversible precipitation below 37 °C. Three types have been identified: type I, which contains monoclonal immunoglobulins only (usually IgM); type II, a mixture of monoclonal immunoglobulins (usually IgM) and polyclonal IgG and type III, a mixture of polyclonal immunoglobulins (usually IgM) and polyclonal IgG. Types II and III contain rheumatoid factor; type I does not. 'Mixed' cryoglobulinaemia represents a mixture of types II and III, with Hepatitis C being responsible for >90% of such cases (mainly type II, with a small number of type III) (Charles and Dustin, 2009). Additional cases are principally caused by infectious agents (mainly viral), autoimmune diseases and neoplasms; this leaves ~5% of cases in which no cause is identified, and these are termed essential (Charles and Dustin, 2009; Matignon *et al.*, 2009). Approximately 20% of individuals with cryoglobulinaemia develop renal disease.

In cryoglobulinaemic glomerulonephritis, the pattern of glomerular changes is diverse and ranges from mild mesangial hypercellularity through to proliferative disease with an MCGN type I pattern. Immunolabelling depends on the cryoglobulin involved. Deposits may be found in any part of the glomerulus, but particularly in the subendothelial region, mesangium and thrombi. Thrombi often show the most characteristic ultrastructural features. Overall, the appearance varies considerably from coarsely granular to fibrillar or tubular (with a random orientation) (Figure 1.25), depending on the cryoglobulins involved and the orientation of the material within the section. Classical granular immune deposits may also be present. Typically, deposits of mixed cryoglobulins are described as consisting of paired, curved microtubular or annular structures (with spokes) ~25–40 nm in diameter. Monoclonal cryoglobulins are described as fibrils (seen singly or in bundles, and sometimes with cross-hatching) or tubules with a fingerprint organisation.

1.5.5 Hereditary Metabolic Storage Disorders

A number of rare hereditary storage diseases cause renal complications. Here we include only Fabry's disease and lecithin cholesterol acyltransferase (LCAT) deficiency as illustrative examples.

1.5.5.1 Fabry's Disease

Fabry's disease is an X-linked lysosomal storage disorder that affects both sexes. The incidence is ~1:40 000–1 : 117 000 (Torra, 2008). Affected individuals have no, or significantly reduced, α-galactosidase A activity that results in the lysosomal accumulation of glycosphingolipids, mainly globotriaosylceramide (GL-3). Nephropathy develops by age ~27 years followed by end-stage renal disease and death, generally by the age of 60. Enzyme replacement therapy can stabilise glomerular filtration in early cases (Öqvist *et al.*, 2009).

By LM, most podocytes have a vacuolated (or honeycomb) appearance, or may contain myelin-like inclusions. Unlike LCAT deficiency, the mesangial and endothelial cells may not be affected. At the ultrastructural level, podocytes are seen to contain large numbers of abnormal secondary lysosomes with a lamellar or myelin-like appearance (zebra bodies) (Figure 1.26) (Torra, 2008). The lamellae may be loosely arranged. Lysosomal deposits similar to those seen in Fabry's disease have also been reported in silicosis, and as the result of various drug

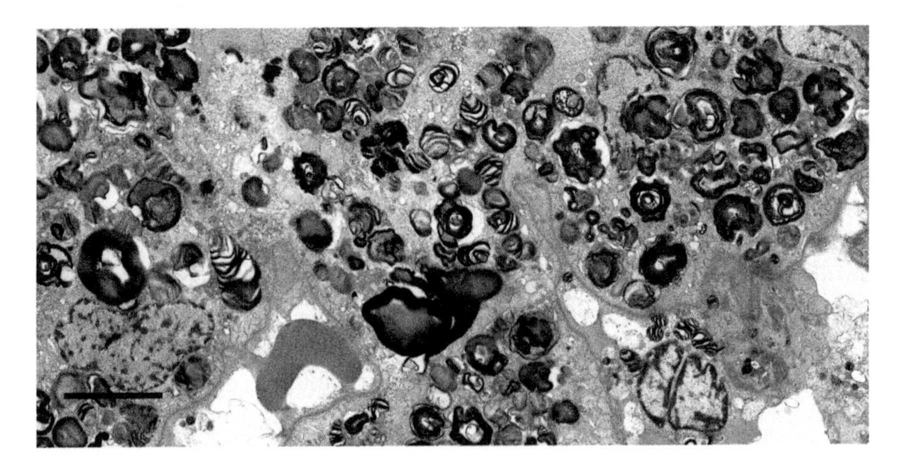

Figure 1.26 Fabry's disease. Podocytes contain large abnormal secondary lysosomes that are either solid or lamellar (myelin-like) in appearance. Bar = 5 μm.

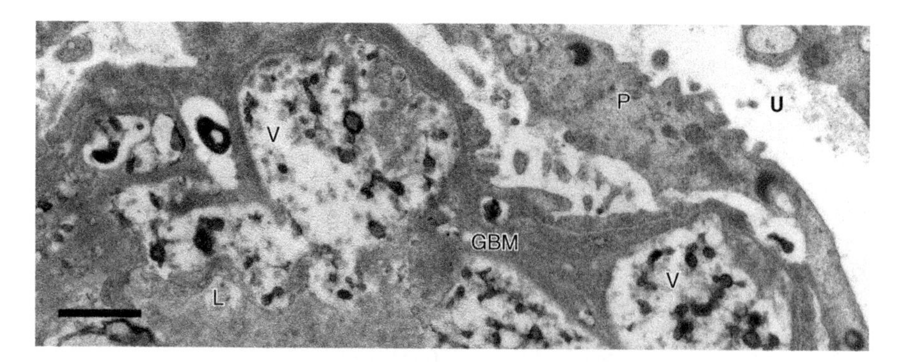

Figure 1.27 Lecithin cholesterol acyltransferase deficiency. Electron-lucent vesicles (V) with electron-dense cores of various types are seen within the glomerular basement membrane (GBM). L = capillary lumen; U = urinary space and P = podocyte. Bar = 1.5 μm.

treatments (amiodarone, chloroquine and hydroxychloroquine) (Öqvist *et al.*, 2009). Immunolabelling is negative.

1.5.5.2 Lecithin Cholesterol Acyltransferase Deficiency

LCAT deficiency is a very rare hereditary autosomal recessive disorder currently described in ~50 families world-wide. Patients in their third and fourth decades usually develop renal failure, with glomerulosclerosis being a major cause of morbidity and mortality (Miarka *et al.*, 2011).

By LM, glomeruli have a honeycomb appearance in all compartments because of the extraction of lipid deposits during processing. At the ultrastructural level, heterogeneous lipid deposits are seen in the GBM (subepithelial, intramembranous and subendothelial locations); these have the appearance of electron-lucent vacuoles with a dense core consisting of curvilinear fibrils or lamellae (Figure 1.27). Mesangial deposits appear as large lucent areas with irregular dense cores. Immunolabelling is usually negative or nonspecific, but silver stains can show spikes in the GBM, mimicking membranous GN.

REFERENCES

Abrahamson, D.R., Hudson, B.G., Stroganova, L. *et al.* (2009) Cellular origins of type IV collagen networks in developing glomeruli. *Journal of the American Society of Nephrology*, 20, 1471–1479.

Adler, S.G. and Salant, D.J. (2003) An outline of essential topics in glomerular pathophysiology, diagnosis, and treatment for nephrology trainees. *American Journal of Kidney Diseases*, **42** (2), 395–418.

Alchi, B. and Jayne, D. (2010) Membranoproliferative glomerulonephritis. *Pediatric Nephrology*, **25**, 1409–1418.

Alsaad, K.O. and Herzenberg, A.M. (2007) Distinguishing diabetic nephropathy from other causes of glomerulosclerosis: an update. *Journal of Clinical Pathology*, **60**, 18–26.

Anders, D., Agricola, B., Sippel, M. and Thoenes, W. (1977) Basement membrane changes in membranoproliferative glomerulonephritis. II. Characterisation of a third type by silver impregnation of ultra thin sections. *Virchow Archives A: Pathological Anatomy and Histology*, **376** (1), 1–19.

Appel, G.B., Cook, H.T., Hageman, G. *et al.* (2005) Membranoproliferative glomerulonephritis type II (dense deposit disease): an update. *Journal of the American Society of Nephrology*, **16**, 1392–1404.

Barker, D.F., Hostikka, S.L., Zhou, J. *et al.* (1990) Identification of mutations in the COL4A5 collagen gene in Alport syndrome. *Science*, **248** (4960), 1224–1227.

Basnayake, K., Stringer, S.J., Hutchison, C.A. and Cockwell, P. (2011) The biology of immunoglobulin free light chains and kidney injury. *Kidney International*, **79**, 1289–1301.

Beck, J.H. and Salant, D.J. Jr (2010) Membranous nephropathy: recent travels and new roads ahead. *Kidney International*, **77**, 765–770.

Bonsib, S.M. (2007) Renal anatomy and histology, in *Heptinstall's Pathology of the Kidney*, vol. 1, 6th edn (eds J.C. Jennette, J.L. Olson, M.M. Schwartz and F.G. Silva), Lippincott Williams and Wilkins, Philadelphia, PA, pp. 1–70.

Burkholder, P.M., Marchand, A. and Krueger, R.P. (1970) Mixed membranous and proliferative glomerulonephritis: a correlative light, immunofluorescence, and electron microscopic study. *Laboratory Investigation*, **23** (5), 459–479.

Butkowski, R.J., Wieslander, J., Kleppel, M. *et al.* (1989) Basement membrane collagen in the kidney: regional localization of novel chains related to collagen IV. *Kidney International*, **35**, 1195–1202.

Buzza, M., Wilson, D. and Savige, J. (2001) Segregation of hematuria in thin basement membrane disease with haplotypes at the loci for Alport syndrome. *Kidney International*, **59**, 1670–1676.

Cattran, D.C., Coppo, R., Cook, H.T. *et al.* (2009) The Oxford classification of IgA nephropathy: rationale, clinicopathological correlations, and classification. *Kidney International*, **76**, 534–545.

Charles, E.D. and Dustin, L.B. (2009) Hepatitis C virus-induced cryoglobulinemia. *Kidney International*, **76**, 818–824.

Coleman, M., Haynes, W.D.G., Dimopoulos, P. *et al.* (1986) Glomerular basement membrane abnormalities associated with apparently idiopathic hematuria: ultrastructural morphometric analysis. *Human Pathology*, **17**, 1022–1030.

Coleman, M. and Seymour, A.E. (1992) The kidney, in *Diagnostic Ultrastructure of Non-neoplastic Diseases* (eds J.M. Papadamitiou, D.W. Henderson and D.V. Spagnolo), Churchill Livingstone, Edinburgh, pp. 374–416.

Coleman, M. and Stirling, J.W. (1991) Glomerular basement membrane thinning is acquired in minimal change disease. *American Journal of Nephrology*, **11**, 437–438.

Coppo, R., Feehally, J. and Glassock, R.J. (2010) IgA nephropathy at two score and one. *Kidney International*, 77, 181–186.

Coppo, R., Troyanov, S., Camilla, R. *et al.* (2010) The Oxford IgA nephropathy clinicopathological classification is valid for children as well as adults. *Kidney International*, 77, 921–927.

D'Agati, V.D. (2007) Renal disease in systemic lupus erythematosus, mixed connective tissue disease, Sjögren's syndrome, and rheumatoid arthritis, in *Heptinstall's Pathology of the Kidney*, vol. 1, 6th edn (eds J.C. Jennette, J.L. Olson, M.M. Schwartz and F.G. Silva), Lippincott Williams and Wilkins, Philadelphia, PA, pp. 518–612.

D'Agati, V.D. (2008) Podocyte injury in focal segmental glomerulosclerosis: lesson from animal models (a play in five acts). *Kidney International*, **73**, 399–406.

D'Agati, V.D., Fogo, A.B., Bruijn, J.A. and Jennette, J.C. (2004) Pathologic classification of focal segmental glomerulosclerosis: a working proposal. *American Journal of Kidney Disease*, **43** (2), 368–382.

Davin, J-C., Ten Berge, I.J. and Weening, J.J. (2001) What is the difference between IgA nephropathy and Henoch–Schönlein purpura nephritis? *Kidney International*, **59**, 823–834.

Dember, L.M. (2006) Amyloidosis-associated kidney disease. *Journal of the American Society of Nephrology*, **17**, 3458–3471.

Dische, F.E. (1992) Measurement of glomerular basement membrane thickness and its application to the diagnosis of thin-membrane nephropathy. *Archives of Pathology and Laboratory Medicine*, **116** (1), 43–49.

Edwards, K., Griffiths, D., Morgan, J. *et al.* (2009) Can the choice of intermediate solvent or resin affect glomerular basement membrane thickness? *Nephrology Dialysis Transplantation*, **24**, 400–403.

Ehrenreich, T. and Churg, J. (1968) Pathology of membranous nephropathy. *Pathology Annual*, **3**, 145–186.

Elhefnawy, N.G. (2011) Contribution of electron microscopy to the final diagnosis of renal biopsies in Egyptian patients. *Pathology and Oncology Research*, **17**, 121–125.

Estacio, R.O. and Schrier, R.W. (2001) Diabetic nephropathy: pathogenesis, diagnosis, and prevention of progression. *Advances in Internal Medicine*, **46**, 359–408.

Fattal, I., Shental, N., Mevorach, D. *et al.* (2010) An antibody profile of systemic lupus erythematosus detected by antigen microarray. *Immunology*, **130**, 337–343.

Furness, P. (2004) Paraproteinaemia and renal disease. *Current Diagnostic Pathology*, **10**, 52–60.

George, J.N. (2009) The thrombotic thrombocytopenic purpura and hemolytic uremic syndromes: overview of pathogenesis (Experience of The Oklahoma TTP-HUS Registry, 1989–2007). *Kidney International*, **75** (Suppl. 112), S8–S10.

Ghadially, F.N. (1988) *Ultrastructural Pathology of the Cell and Matrix*, 3rd edn, Butterworths, London.

Glassock, R.J. (2003) Secondary minimal change disease. *Nephrology Dialysis Transplantation*, **18** (Suppl. 6), vi52–vi58.

Gokden, N., Barlogie, B. and Liapis, H. (2008) Morphologic heterogeneity of renal light-chain deposition disease. *Ultrastructural Pathology*, **32**, 17–24.

Goldberg, S., Harvey, S.J., Cunningham, J. *et al.* (2009) Glomerular filtration is normal in the absence of both agrin and perlecan-heparan sulphate from the glomerular basement membrane. *Nephrology Dialysis Transplantation*, **24**, 2044–2052.

Gregory, M.C. (2004) Alport syndrome and thin basement membrane nephropathy: unraveling the tangled strands of type IV collagen. *Kidney International*, **65**, 1109–1110.

Grimbert, P., Audard, V., Remy, P. *et al.* (2003) Recent approaches to the pathogenesis of minimal-change nephrotic syndrome. *Nephrology Dialysis Transplantation*, **18**, 245–248.

Gubler, M-C., Heidet, L. and Antignac, C. (2007) Alport's syndrome, thin basement membrane nephropathy, nail–patella syndrome, and type III collagen glomerulopathy, in *Heptinstall's Pathology of the Kidney*, 6th edn (eds J.C. Jennette, J.L. Olson, M.M. Schwartz and F.G. Silva), Lippincott Williams and Wilkins, Philadelphia, PA, pp. 487–515.

Haas, M. (2006) Thin glomerular basement membrane nephropathy: incidence in 3471 consecutive renal biopsies examined by electron microscopy. *Archives of Pathology and Laboratory Medicine*, **130**, 699–706.

Haas, M. (2007) Electron microscopy in renal biopsy interpretation – when and why we still need it. *US Nephrology*, **1** (1), 19–22.

Haas, M., Rajaraman, S., Ahuja, T. *et al.* (2000) Fibrillary/immunotactoid glomerulonephritis in HIV-positive patients: a report of three cases. *Nephrology Dialysis Transplantation*, **15** (10), 1679–1683.

Hammar, S.P., Luu, J., Bockus, D.E. *et al.* (1992) Induction of tubuloreticular structures in cultured human endothelial cells by recombinant interferon alpha and beta. *Ultrastructural Pathology*, **16**, 211–218.

Haraldsson, B., Nystrom, J. and Deen, W.M. (2008) Properties of the glomerular barrier and mechanisms of proteinuria. *Physiological Reviews*, **88**, 451–487.

Harvey, S.J., Jarad, G., Cunningham, J., Rops, A.L. *et al.* (2007) Disruption of glomerular basement membrane charge through podocyte-specific mutation of agrin does not alter glomerular permselectivity. *American Journal of Pathology*, **171**, 139–152.

Harvey, S.J., Zheng, K., Sado, Y. *et al.* (1998) Role of distinct type IV collagen networks in glomerular development and function. *Kidney International*, **54**, 1857–1866.

Hennigar, R.A. and Tumlin, J.A. (2009) Glomerular diseases associated primarily with asymptomatic or gross haematuria, in *Silva's Diagnostic Renal Pathology* (eds X.J. Zhou, Z. Laszik, T. Nadasdy *et al.*), Cambridge University Press, Cambridge, pp. 127–177.

Herrera, G.A. and Turbat-Herrera, E.A. (2010) Renal diseases with organized deposits: an algorithmic approach to classification and clinicopathologic diagnosis. *Archives of Pathology and Laboratory Medicine*, **134**, 512–531.

Herzenberg, A.M., Fogo, A.B., Reich, H.N. *et al.* (2011) Validation of the Oxford classification system of IgA nephropathy. *Kidney International*, **80**, 310–317.

Hsu, H-C. and Churg, J. (1979) Glomerular microfibrils in renal disease: a comparative electron microscopic study. *Kidney International*, **16**, 497–504.

Hudson, B.G., Tryggvason, K., Sundaramoorthy, M. and Neilson, E.G. (2003) Alport's syndrome, Goodpasture's syndrome and type IV collagen. *New England Journal of Medicine*, **348**, 2543–2556.

Ivanyi, B. and Degrell, P. (2004) Fibrillary glomerulonephritis and immunotactoid glomerulopathy. *Nephrology Dialysis Transplantation*, **19** (9), 2166–2170.

Jais, J.P., Knebelmann, B., Giatras, I. *et al.* (2000) X-linked Alport syndrome: natural history in 195 families and genotype–phenotype correlations in males. *Journal of the American Society of Nephrology*, **11** (4), 649–657.

Jais, J.P., Knebelmann, B., Giatras, I. *et al.* (2003) X-linked Alport syndrome: natural history and genotype–phenotype correlations in girls and women belonging to 195 families: a 'European Community Alport Syndrome Concerted Action' study. *Journal of the American Society of Nephrology*, **14**, 2603–2610.

Kalluri, R., Shield, C.F., Todd, P. *et al.* (1997) Isoform switching of type IV collagen is developmentally arrested in X-linked Alport syndrome leading to increased susceptibility of renal basement membranes to endoproteolysis. *Journal of Clinical Investigation*, **99** (10), 2470–2478.

Kiryluk, K., Gharavi, A.G., Izzi, C. and Scolari, F. (2010) IgA nephropathy – the case for a genetic basis becomes stronger. *Nephrology Dialysis Transplantation*, **25**, 336–338.

Kronz, J.D., Nue, A.M. and Nadasdy, T. (1998) When noncongophilic glomerular fibrils do not represent fibrillary glomerulonephritis: nonspecific mesangial fibrils in sclerosing glomeruli. *Clinical Nephrology*, **50** (4), 218–223.

LeBleu, V., Sund, M., Sugimoto, H. *et al.* (2010) Identification of the NC1 domain of $\alpha 3$ chain as critical for $\alpha 3 \alpha 4 \alpha 5$ type IV collagen network assembly. *The Journal of Biological Chemistry*, **285** (53), 41874–41885.

Lemmink, H.H., Mochizuki, T., van den Heuvel, L.P. *et al.* (1994) Mutations in the type IV collagen alpha 3 (COL4A3) gene in autosomal recessive Alport syndrome. *Human Molecular Genetics*, **3** (8), 1269–1273.

Leung, N., Lager, D.J., Gertz, M.A. *et al.* (2004) Long-term outcome of renal transplantation in light-chain deposition disease. *American Journal of Kidney Diseases*, **43** (1), 147–153.

Levy, M. and Feingold, J. (2000) Estimating prevalence in single-gene kidney diseases progressing to renal failure. *Kidney International*, **58** (3), 925–943.

Lin, J., Markowitz, G.S., Valeri, A.M. *et al.* (2001) Renal monoclonal immunoglobulin deposition disease: the disease spectrum. *Journal of the American Society of Nephrology*, **12**, 1482–1492.

Longo, I., Porcedda, P., Mari, F. *et al.* (2002) COL4A3/COL4A4 mutations: from familial hematuria to autosomal-dominant or recessive Alport syndrome. *Kidney International*, **61** (6), 1947–1956.

Markowitz, G.S., Cheng, J-T., Colvin, R.B. *et al.* (1998) Hepatitis C viral infection is associated with fibrillary glomerulonepritis and immunotactoid glomerulopathy. *Journal of the American Society of Nephrology*, **9**, 2244–2252.

Marquez, B., Stavrou, F., Zouvani, I. *et al.* (1999) Thin glomerular basement membranes in patients with hematuria and minimal change disease. *Ultrastructural Pathology*, **23**, 149–156.

Martin, P., Heiskari, N., Zhou, J. *et al.* (1998) High mutation detection rate in the COL4A5 collagen gene in suspected Alport syndrome using PCR and direct DNA sequencing. *Journal of the American Society of Nephrology*, **9** (12), 2291–2301.

Mathieson, P.W. (2003) Immune dysregulation in minimal change nephropathy. *Nephrology Dialysis Transplantation*, **18** (Suppl. 6), vi26–vi29.

Matignon, M., Cacoub, P., Colombat, M. *et al.* (2009) Clinical and morphological spectrum of renal involvement in patients with mixed cryoglobulinemia without evidence of hepatitis C infection. *Medicine*, **88**, 341–348.

McGrogan, A., Franssen, C.F.M. and de Vries, C.S. (2011) The incidence of primary glomerulonephritis worldwide: a systematic review of the literature. *Nephrology Dialysis Transplantation*, **26**, 414–430.

Meyrier, A.Y. (2009) Treatment of focal segmental glomerulosclerosis with immunophilin modulation: when did we stop thinking about pathogenesis? *Kidney International*, **76**, 487–491.

Miarka, P., Idzior-Waluś, B., Kuźniewski, M. *et al.* (2011) Corticosteroid treatment of kidney disease in a patient with familial lecithin–cholesterol acyltransferase deficiency. *Clinical and Experimental Nephrology*, **15**, 424–429.

Miller, C.A., Gattone, V.H. II, McLaughlin, H. *et al.* (2010) Identification of the NC1 domain of $\alpha 3$ chain as critical for $\alpha 3\alpha 4\alpha 5$ Type IV collagen network assembly. *Journal of Biological Chemistry*, 285 (53), 41874–41885.

Miner, J.H. (1998) Developmental biology of glomerular basement membrane components. *Current Opinion in Nephrology and Hypertension*, 7, 13–19.

Mochizuki, T., Lemmink, H.H., Mariyama, M. *et al.* (1994) Identification of mutations in the alpha 3(IV) and alpha 4(IV) collagen genes in autosomal recessive Alport syndrome. *Nature Genetics*, 8 (1), 77–81.

Morita, T. and Churg, J. (1983) Mesangiolysis. *Kidney International*, **24**, 1–9.

Morita, M., White, R.H.R., Raafat, F. *et al.* (1988) Glomerular basement membrane thickness in children: a morphometric study. *Pediatric Nephrology*, **2**, 190–195.

Nasr, S.H., Markowitz, G.S., Valeri, A.M. *et al.* (2007) Thin basement membrane nephropathy cannot be diagnosed reliably in deparaffinized, formalin-fixed tissue. *Nephrology Dialysis Transplantation*, **22**, 1228–1232.

Nishi, S., Ueno, M., Hisaki, S. *et al.* (2000) Ultrastructural characteristics of diabetic nephropathy. *Medical Electron Microscopy*, **33**, 65–73.

Öqvist, B., Brenner, B.M., Oliveira, J.P. *et al.* (2009) Nephropathy in Fabry disease: the importance of early diagnosis and testing in high-risk populations. *Nephrology Dialysis Transplantation*, **24**, 1736–1743.

Patrakka, J. and Tryggvason, K. (2010) Molecular make-up of the glomerular filtration barrier. *Biochemical and Biophysical Research and Communications*, **396**, 164–169.

Pearson, J.M., McWilliam, L.J., Coyne, J.D. and Curry, A. (1994) Value of electron microscopy in the diagnosis of renal disease. *Journal of Clinical Pathology*, **47**, 126–128.

Pedchenko, V., Bondar, O., Fogo, A.B. *et al.* (2010) Molecular architecture of the Goodpasture autoantigen in anti-GBM nephritis. *New England Journal of Medicine*, **363**, 343–354.

Pescucci, C., Mari, F., Longo, I. *et al.* (2004) Autosomal-dominant Alport syndrome: natural history of a disease due to COL4A3 or COL4A4 gene. *Kidney International*, **65** (5), 1598–1603.

Ponticelli, C. (2010) Recurrence of focal segmental glomerular sclerosis (FSGS) after renal transplantation. *Nephrology Dialysis Transplantation*, **25**, 25–31.

Prunotto, M., Carnevale, M.L., Candiano, G. *et al.* (2010) Autoimmunity in membranous nephropathy targets aldose reductase and SOD2. *Journal of the American Society of Nephrology*, **21** (3), 507–519.

Rodríguez-Iturbe, B. and Batsford, S. (2007) Pathogenesis of poststreptococcal glomerulonephritis a century after Clemens von Pirquet. *Kidney International*, **71**, 1094–1104.

Rogers, T.E., Rakheja, D. and Zhou, X.J. (2009) Glomerular diseases associated with nephritic syndrome and/or rapidly progressive glomerulonephritis, in *Silva's Diagnostic Renal Pathology* (eds X.J. Zhou, Z. Laszik, T. Nadasdy *et al.*), Cambridge University Press, Cambridge, pp. 178–228.

Rosenstock, J.L., Markowitz, G.S., Valeri, A.M. *et al.* (2003) Fibrillary and immunotactoid glomerulonephritis: distinct entities with different clinical and pathologic features. *Kidney International*, **63**, 1450–1461.

Sanai, T., Okuda, S., Yoshimitsu, T. *et al.* (2007) Nodular glomerulosclerosis inpatients without any manifestation of diabetes mellitus. *Nephrology*, **12**, 69–73.

Sánchez-Corral, P. and Melgosa, M. (2010) Advances in understanding the aetiology of atypical haemolytic uraemic syndrome. *British Journal of Haematology*, **150**, 529–542.

Saus, J., Wieslander, J., Langeveld, J.P.M. *et al.* (1988) Identification of the Goodpasture antigen as the $\alpha 3$ (IV) chain of collagen IV. *Journal of Biological Chemistry*, **263** (26), 13374–13380.

Savige, J., Rana, K., Tonna, S. *et al.* (2003) Thin basement membrane nephropathy. *Kidney International*, **64**, 1169–1178.

Schwartz, M.M., Korbet, S.M. and Lewis, E.J. (2002) Immunotactoid glomerulopathy. *Journal of the American Society of Nephrology*, **13**, 1390–1397.

Siso, A., Ramos-Casals, M., Bové, A. *et al.* (2010) Outcomes in biopsy-proven lupus nephritis: evaluation of 190 white patients from a single centre. *Medicine*, **89** (5), 300–307.

Standeven, K.F., Ariëns, R.A.S. and Grant, P.J. (2005) The molecular physiology and pathology of fibrin structure/function. *Blood Reviews*, **19**, 275–288.

Sterzel, R.B., Lovett, H.D., Stein, H.D. and Kashgarian, M. (1982) The mesangium and glomerulonephritis. *Wiener Klinische Wochenschrift*, **60** (18), 1077–1094.

Stirling, J.W., Coleman, M., Thomas, A. and Woods, A.E. (2000) Role of transmission electron microscopy in tissue diagnosis: diseases of the kidney, skeletal muscle and myocardium. *Journal of Cellular Pathology*, **4** (4), 223–243.

Stokes, M.B., Nasr, S.H. and D'Agati, V.D. (2009) Systemic lupus erythematosus and other autoimmune diseases (mixed connective tissue disease, rheumatoid arthritis and Sjogren's syndrome), in *Silva's Diagnostic Renal Pathology* (eds X.J. Zhou, Z. Laszik, T. Nadasdy *et al.*), Cambridge University Press, Cambridge, pp. 229–272.

Strife, C.F., McEnery, P.T., McAdams, A.J. and West, C.D. (1977) Membranoproliferative glomerulonephritis with disruption of the glomerular basement membrane. *Clinical Nephrology*, 7 (2), 65–72.

Torra, R. (2008) Renal manifestations in Fabry disease and therapeutic options. *Kidney International*, **74** (Suppl. 111), S29–S32.

Venuturupalli, R.S. and Wallace, D.J. (2007) Lupus nephritis – an update. *Touch Briefings*, 63–66, http://www.touchbriefings.com/pdf/2968/wallace.pdf (accessed July 2012).

Vowles, G.H. (2008) Amyloid, in *Theory and Practice of Histological techniques*, 6th edn (eds J.D. Bancroft and M. Gamble), Churchill Livingstone Elsevier, Edinburgh, pp. 261–281.

Weening, J.J., D'Agati, V.D., Schwartz, M.M. *et al.* (2004) The classification of glomerulonephritis in systemic lupus erythematosus revisited. *Kidney International*, **65**, 521–530.

Wiggins, R.C. (2007) The spectrum of podocytopathies: a unifying view of glomerular diseases. *Kidney International*, **71**, 1205–1214.

Yang, Y., Hellmark, T., Zhao, J. *et al.* (2009a) Levels of epitope-specific autoantibodies correlate with renal damage in anti-GBM disease. *Nephrology Dialysis Transplantation*, **24**, 1838–1844.

Yang, A-H., Lin, B-S., Kuo, K-L. *et al.* (2009b) The clinicopathological implications of endothelial tubuloreticular inclusions found in glomeruli having histopathology of idiopathic membranous nephropathy. *Nephrology Dialysis Transplantation*, **24**, 3419–3425.

Yasuda, T., Imai, H., Nakamoto, Y. *et al.* (1992) Significance of fibrils in the formation of the Kimmelstiel–Wilson nodule. *Virchows Archives A: Pathological Anatomy and Histology*, **421**, 297–303.

Zhao, J., Cui, Z., Yang, R. *et al.* (2009) Anti-glomerular basement membrane autoantibodies against different target antigens are associated with disease severity. *Kidney International*, **76**, 1108–1115.

2

Transplant Renal Biopsies

John Brealey

Centre for Ultrastructural Pathology, Surgical Pathology – SA Pathology (RAH), Adelaide, Australia

2.1 INTRODUCTION

The use of transmission electron microscopy (TEM) in the diagnosis of native renal disease is well established; however, its use in the diagnosis of transplant renal disease is variable among pathology institutions. Performing TEM on all transplant renal biopsies is unnecessary, but omitting TEM altogether risks misdiagnosis in a small percentage of cases. TEM can be critical in providing information unsuspected at the time of biopsy; therefore, having a small sample of tissue reserved for this modality provides flexibility in the diagnostic approach, while optimising laboratory efficiency and patient outcomes. Alternatively, tissue can be retrieved from the paraffin block, although the preservation of the deparaffinised specimen will be sub-optimal for ultrastructural assessment.

2.2 THE TRANSPLANT RENAL BIOPSY

The renal biopsy is considered the best investigative tool in determining the cause of graft dysfunction episodes, which occur in 30–60% of recipients following transplantation (Colvin and Nickeleit, 2007). The (needle-core) biopsy is the most common histological specimen received

Diagnostic Electron Microscopy: A Practical Guide to Interpretation and Technique, First Edition. Edited by John W. Stirling, Alan Curry and Brian Eyden.

from the renal transplant, and is usually performed to investigate graft dysfunction or check the status of a graft at a given time post insertion (often referred to as a protocol or surveillance biopsy). For the purposes of light microscopy (LM), the 'Banff 97 Working Classification of Renal Allograft Pathology' defines an adequate biopsy as containing two cores of cortical tissue in which a total of 10 glomeruli and two arterial profiles are present (Racusen *et al.*, 1999). Additionally, three or four glomeruli for immunolabelling (IM) techniques (usually immunofluorescence microscopy) and one or two glomeruli for TEM are usually adequate for these modalities. A less common but equally important specimen is the insertion biopsy, performed at implantation (0-hour) or reperfusion (1-hour), which provides for a baseline assessment of the donor kidney at the time of insertion. Other transplant-related renal specimens include the transplant nephrectomy (TEM is seldom used for the examination of such specimens), the pre-transplant cadaveric biopsy for potential suitability (TEM is of little benefit as the time required for a result is too great) and the native kidney biopsy of a prospective living donor (refer to Chapter 1).

2.3 INDICATIONS FOR ELECTRON MICROSCOPY OF TRANSPLANT KIDNEY

The commonly encountered indications for TEM in transplant renal biopsies are transplant glomerulopathy (TG), recurrent primary disease, *de novo* glomerular disease, donor-related disease, infection and inconclusive diagnosis by LM and/or IM. These investigations are discussed throughout the remainder of this section, and the chapter concludes with four miscellaneous topics: post-transplant nephrotic syndrome (PTNS), LM versus TEM, neoplasia associated with renal transplantation and the future of TEM. Conditions such as hyper-acute and acute rejection, acute tubular necrosis, tubulointerstitial nephritis, interstitial fibrosis, hypertension, vascular sclerosis, thrombosis, mechanical obstruction, reflux nephropathy, reperfusion injury and pyelonephritis are commonly investigated by LM without the need for TEM.

2.3.1 Transplant Glomerulopathy

The pathogenesis of TG is not fully understood. Histologically, TG is considered to represent a distinct pattern of chronic graft injury involving the entire renal capillary network that manifests from the effects of one or more aetiologies including C4d-associated antibody-mediated rejection, donor-specific antibodies, thrombotic microangiopathy (TMA) and hepatitis C virus infection (Cosio *et al.*, 2008; Baid-Agrawal *et al.*, 2011; Haas, 2011). TG affects approximately 20% of grafts by 5 years post insertion (Gloor *et al.*, 2007) and is the primary cause of late graft loss following renal transplantation. For recipients with proteinuria, TEM is an important adjunct to LM and IM in distinguishing TG from recurrent and *de novo* glomerulonephritis (Herrera, Isaac and Turbat-Herrera, 1997). Additionally, TEM can assist in differentiating TG from entities having a similar appearance by LM; for example, TG, mesangiocapillary (membranoproliferative) glomerulonephritis, TMA and ischaemia can variably display thickened glomerular basement membranes (GBMs), 'double contours' of the GBM and mesangial expansion by LM (Cosio *et al.*, 2008). It has been proposed that TEM can demonstrate changes indicative of TG before they are evident by LM (ultrastructural evidence of TG can be identified within 1 to 3 months post insertion) (Wavamunno, O' Connell and Vitalone, 2007), although some investigators contend the early identification of such changes does not imply overt TG is inevitable (Haas and Mirocha, 2011). More than one glomerulus may need examination by TEM as the degree of ultrastructural changes in TG can vary greatly between glomeruli.

The ultrastructural features of TG are:

- Reduplicated GBM with or without mesangial interposition. Reduplication and mesangial interposition may be circumferential in severe cases (Figure 2.1).
- Expanded zones of subendothelial electron lucency containing flocculent or lace-like material (Figure 2.2).
- Reduplication and/or splitting of the peritubular capillary basal lamina (transplant capillaropathy) (Figure 2.3).
- Foot process effacement, the degree of which is generally in synchrony with the degree of proteinuria. Foot process effacement can be severe in advanced TG, and distinction from focal segmental glomerulosclerosis (FSGS) by TEM alone may not be possible (Herrera, Isaac and Turbat-Herrera, 1997).
- Lack of glomerular immune-type deposit in nonsclerotic zones.
- Endothelial hypertrophy and decreased endothelial cell fenestrations of glomerular and peritubular capillaries. The degree of endothelial

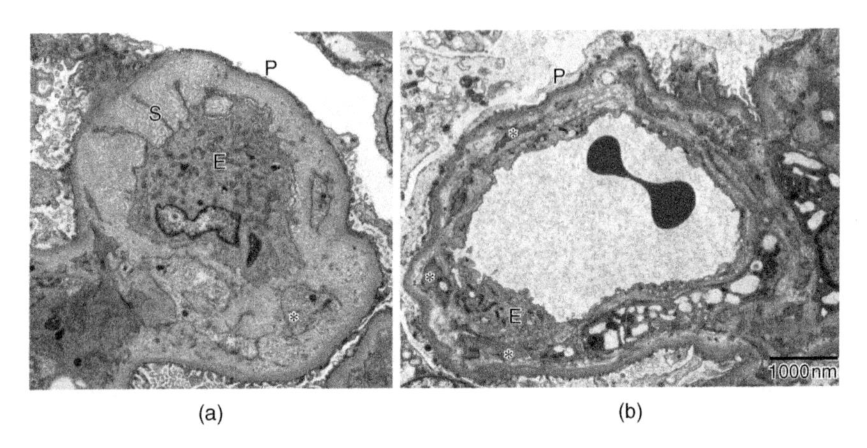

Figure 2.1 Transplant glomerulopathy – glomerular basement membrane (GBM). (a) The GBM is grossly thickened as a result of reduplication of basement membrane material. Note the endothelial serrations (S). (b) The GBM is grossly thickened through circumferential mesangial interposition. Both capillary loops exhibit mesangial interposition (*), endothelial hypertrophy (E) with decreased endothelial fenestrations and flattening of podocytes (P) with widespread foot process effacement.

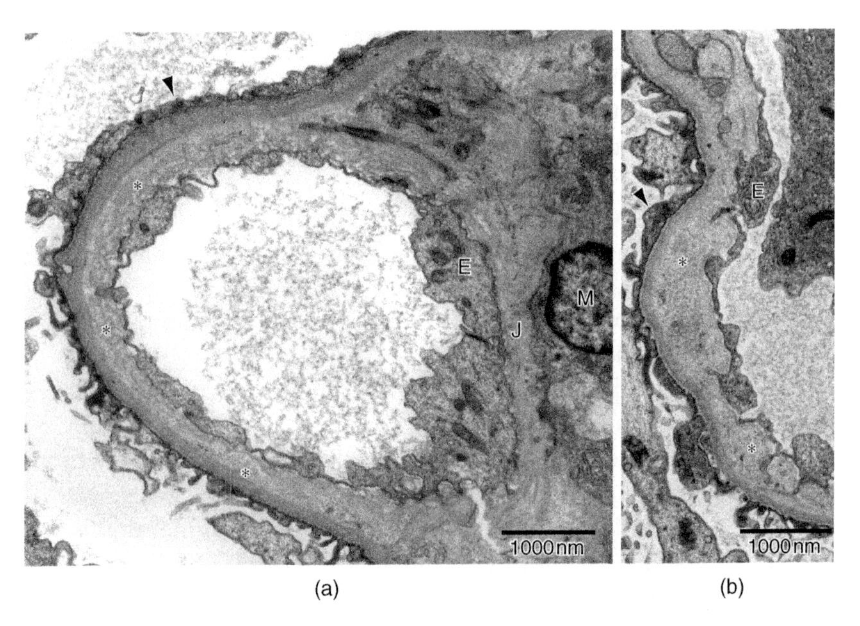

Figure 2.2 Transplant glomerulopathy – subendothelial lucency. (a) The subendothelial zone is widened and contains reduplicated membrane material and electron-lucent flocculent material (asterisks). There is endothelial hypertrophy (E) with decreased endothelial fenestrations. The juxtamesangial zone (J) is widened by reduplicated membrane material. There is focal foot process effacement (arrowheads). Mesangial cell (M). (b) Similar features to A are depicted in another case of transplant glomerulopathy.

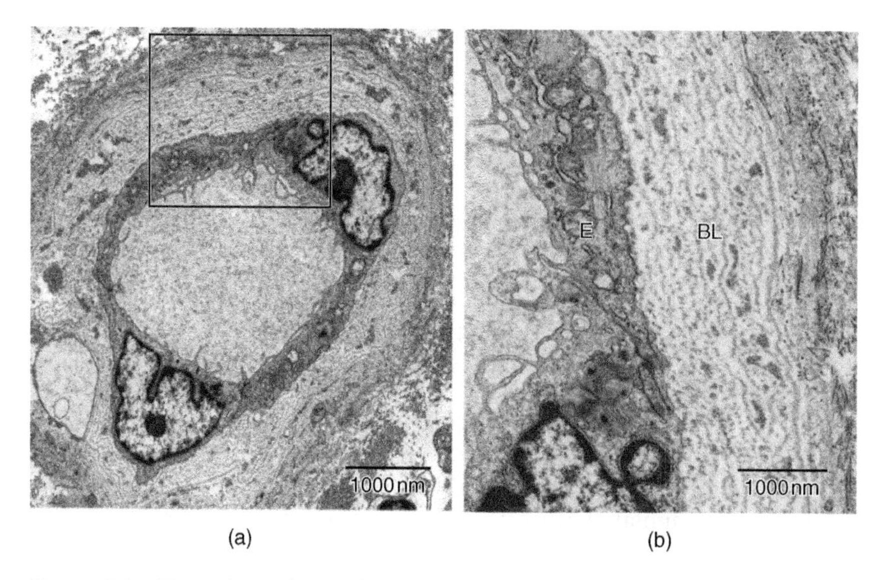

Figure 2.3 Transplant glomerulopathy – peritubular capillary. (a) A peritubular capillary in advanced transplant glomerulopathy. (b) Boxed area represented in (a). There is severe reduplication (greater than six layers) of the basal lamina (BL) and endothelial hypertrophy (E) with decreased endothelial fenestrations.

change is generally in synchrony with the severity of TG. Interdigitations (serrations) of the endothelial cytoplasm may be evident (Wavamunno, O' Connell and Vitalone, 2007) (Figure 2.1a).

- Mesangia may display hypercellularity and expanded matrix.
- Glomerulitis – numerous mononucelocytes may be present in capillary loops.
- Reduplicated GBM material in the juxtamesangial zone (Chicano *et al.*, 2006). In contrast, the basal lamina of the endothelium in the juxtamesangial zone of the normal glomerulus (and in TMA) is unilayered and often imperceptible.
- Translocation of endothelial cell nuclei from the juxtamesangial zone to the periphery of capillary loops (Nickeleit, 2009).

Reduplication and/or splitting of the peritubular capillary basal lamina in the appropriate clinical context of proteinuria and rising creatinine is considered a highly predictive marker for TG, and a scoring system devised for the classification of this ultrastructural change demonstrated that severe reduplication and/or splitting is specific for

Table 2.1 Classification of basal lamina change in the peritubular capillary (Drachenberg *et al.*, 1997)

Class	Basal lamina
I	Normal, unilayered (50–70 nm)
II	Thick, unilayered (>200 nm)
III	Mild splitting and reduplication (2–3 layers) and/or lace-like (up to 500 nm)
IV	Moderate splitting and reduplication (4–6 layers) and/or lace-like (500–1000 nm)
V	Severe splitting and reduplication (>6 layers) and/or lace-like (>1000 nm)

TG (Drachenberg *et al.*, 1997) (Table 2.1). Reduplication is commonly present in the angulated areas of any vascular contour; therefore, for the purposes of the study, reduplication was deemed significant when it was present in more than 60% of a vessel's circumference and such changes were present in the majority of vessels.

Significant reduplication of the peritubular capillary basal lamina was demonstrated in 91% of TG cases (Sis *et al.*, 2007), and the demonstration of reduplication within glomerular and peritubular capillary basement membranes doubled the frequency of the diagnosis of chronic rejection (Ivanyi *et al.*, 2001). Reduplicated and/or lace-like tubular basement membrane can mimic peritubular capillary basal lamina reduplication when the two structures are closely apposed. Close examination of the basal lamina of the peritubular capillary and subjacent interstitial collagen can differentiate a normal capillary from tubular basement membrane changes (Figure 2.4).

TEM commonly demonstrates small amounts of glomerular electron-dense granular material in transplant renal biopsies, and distinguishing significant deposits from those that are insignificant or nonspecific can be difficult. Immune-type deposits in nonsclerotic zones are significant as they are indicative of a disease process other than TG and manifest as uniformly electron-dense granular material, usually in the absence of cytoplasmic debris. In contrast, nonspecific deposits associated with sclerosis manifest as variably electron-dense granular material (insudate), and often contain collagen fibrils and cytoplasmic debris such as vesicular structures and membranous ribbons (Figure 2.5). Advanced sclerotic segments may contain large nodules of insudate (hyaline by LM). By IM, insudate is commonly found to contain IgM and complement components. A study of 118 consecutive transplant renal

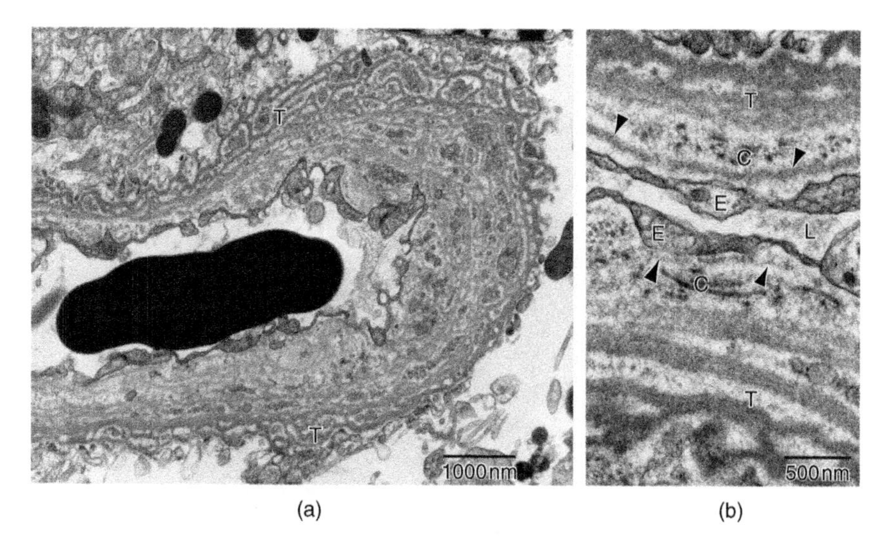

Figure 2.4 False reduplication of a peritubular capillary in transplant kidney. (a) The reduplicated basement membrane of a necrotic proximal convoluted tubule (T) surrounds a peritubular capillary. (b) Higher magnification of another peritubular capillary from the same specimen depicts two zones, either side of the capillary lumen (L), in which interstitial collagen fibrils (C) separate the single (normal) basal lamina of the capillary (arrowheads) from the reduplicated tubular basement membrane (T).

biopsies using LM, IM and TEM demonstrated immune-type deposits in approximately 20% of biopsies, of which 39% were IgM type (Gough *et al.*, 2005). Another study investigating mesangial C1q-dominant or co-dominant deposits in transplant kidneys found that the presence of C1q was not clinically significant in the majority of recipients (Said *et al.*, 2010).

2.3.1.1 Differential for Transplant Glomerulopathy

2.3.1.1.1 Mesangiocapillary Glomerulonephritis

TEM distinguishes mesangiocapillary glomerulonephritis (MCGN) types I, II and III from TG by the characteristic distribution of electron-dense deposits in each type of MCGN (refer to Chapter 1) and the lack of immune-type deposits (in nonsclerotic zones) in TG. Additionally, there is minimal widening of the subendothelial zone by flocculent material in MCGN. Immune-type deposits may be diminished or absent in MCGN associated with the hepatitis C virus, possibly because of immunosuppression, making the distinction from TG difficult in such

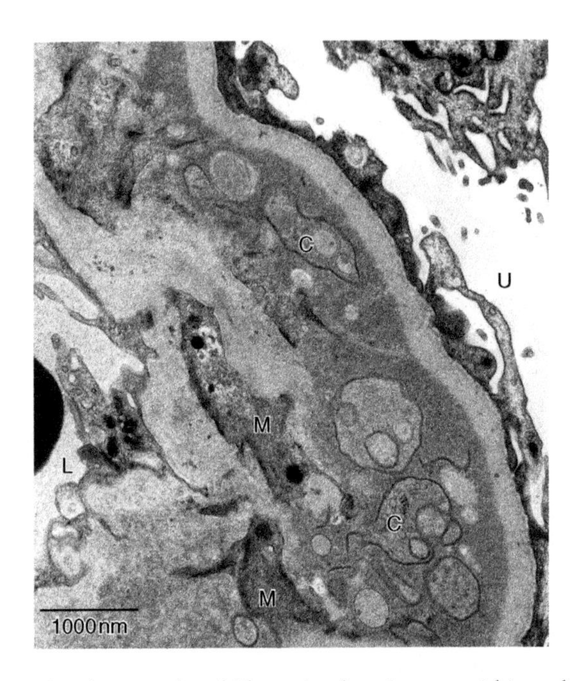

Figure 2.5 Insudate in transplant kidney. A sclerotic zone within a glomerulus contains variably electron-dense granular material, cytoplasmic debris and membranous ribbons (C). IM revealed segmental staining for IgM and complement components. Mesangial cell (M). Capillary lumen (L). Urinary space (U).

cases (Baid-Agrawal *et al.*, 2011). IM staining for IgM is more intense than C3 in TG, whereas IgM is less intense than C3 in recurrent MCGN (type I) (Andresdottir *et al.*, 1998).

2.3.1.1.2 Thrombotic Microangiopathy

TEM of TMA demonstrates widely expanded subendothelial and mesangial regions that may contain flocculent material, fibrin and blood cells (Figure 2.6). The extent of subendothelial expansion is generally greater than that of TG. Reduplication may be present in the peripheral segments of the GBM; however, circumferential reduplication is rare (in contrast to TG) (Chicano *et al.*, 2006). Capillary loops may be greatly stenosed as a result of endothelial detachment, swelling and hyperplasia. Luminal platelets and fibrin may be present (capillary thrombi are rare in TG). At low magnification, fibrin can mimic immune-type deposit; however, at high magnification, it is usually identifiable by its filamentous substructure and faint periodicity of 20–25 nm (Figure 2.7). Occasionally,

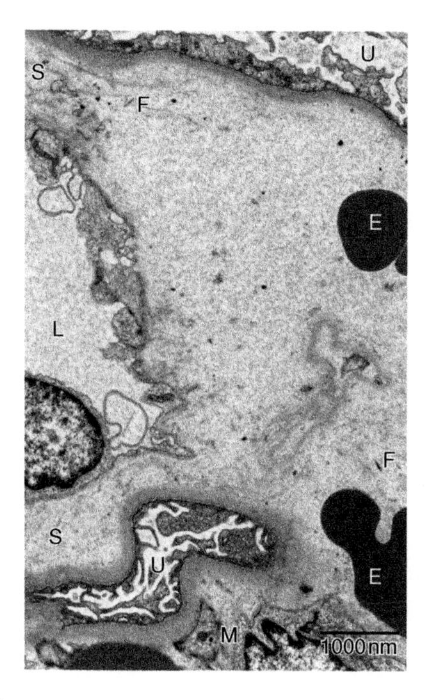

Figure 2.6 Thrombotic microangiopathy in transplant kidney. Widely expanded subendothelial (S) and mesangial (M) zones contain flocculent material, erythrocytes (E) and wisps of fibrin (F). Capillary lumen (L). Urinary space (U).

it is extremely difficult to distinguish TMA from TG based solely on ultrastructural assessment of the subendothelial zone; in such instances, the examination of more glomeruli may facilitate this distinction as the degree of changes present in both diseases can vary greatly between glomeruli. TMA may indicate calcineurin inhibitor toxicity.

2.3.1.1.3 Ischaemia

TEM of ischaemic glomeruli demonstrates thickening and wrinkling of mesangial matrix, especially in the overlying regions (perimesangial segments of the GBM) and paramesangia (mesangial angles) (Figure 2.8). In more severe cases, thickening and wrinkling extend to the GBM of the capillary loops. Electron lucency is present within the mesangial matrix and expanded subendothelial zone of the GBM. The extent of subendothelial expansion and subendothelial flocculence in ischaemia is generally less than in TG. Ischaemic changes may indicate a renovascular cause of graft dysfunction.

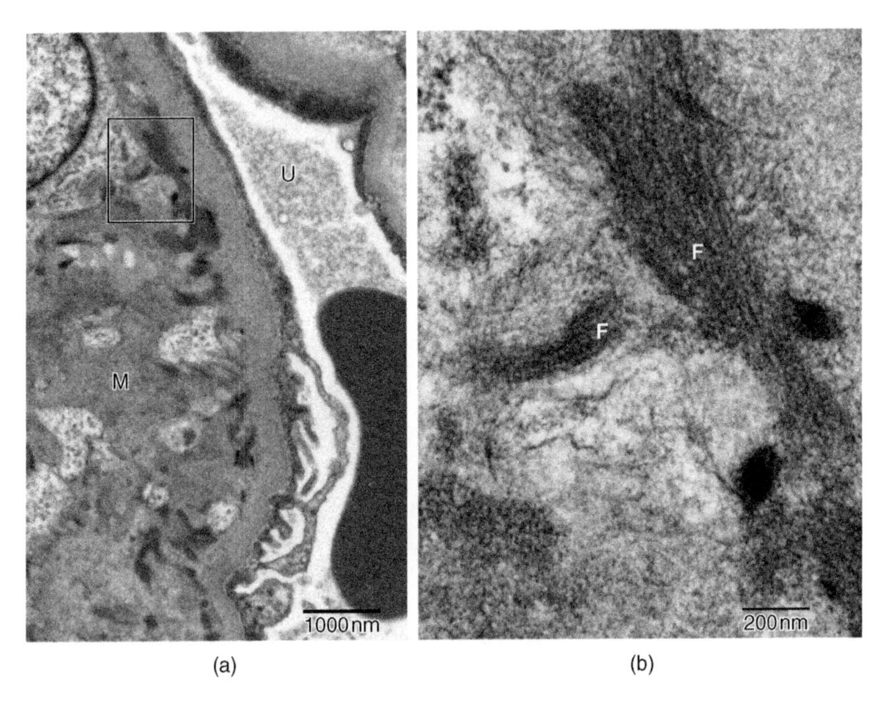

Figure 2.7 Thrombotic microangiopathy, fibrin. (a) The electron-dense material within the mesangium (M) is difficult to characterise at low magnification. Urinary space (U). (b) Boxed area represented in (a). The material is identified as fibrin (F) by its filamentous substructure.

2.3.2 Recurrent Primary Disease

A disease is designated as 'recurrent' when the same form of disease that affected the native kidney develops in the recipient's transplant kidney. Recurrent disease is the third most common cause of graft failure, after chronic rejection and death with a functioning graft (Fairhead and Knoll, 2010). Many renal diseases can recur in the transplant kidney, and each disease has its own clinical recurrence rate (Table 2.2). Histological signs of recurrent disease may be evident in the absence of clinical signs. Graft failure due to recurrent glomerular disease was found to occur in approximately 8% of recipients by 10 years post insertion (Briganti *et al.*, 2002), and recurrent glomerulonephritis was found to be the cause of 13% of graft dysfunction episodes, with 7% leading to graft failure (Gourishankar *et al.*, 2010). TEM is particularly useful in identifying early recurrence as the ultrastructural changes often precede

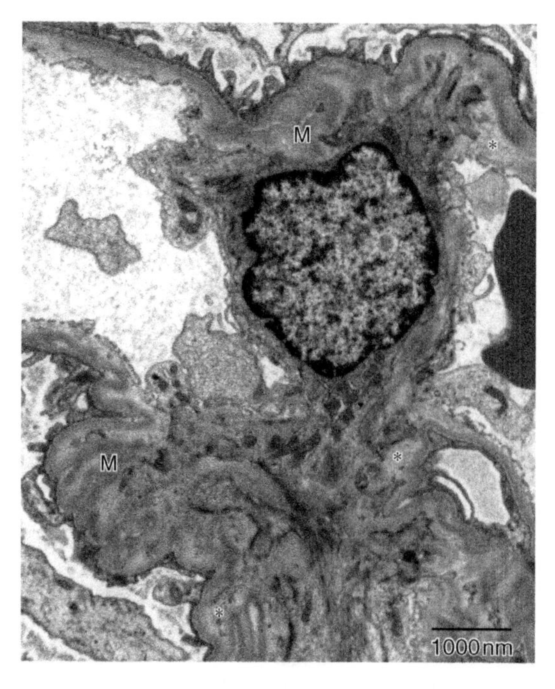

Figure 2.8 Mild ischaemia in transplant kidney. The mesangium (M) is folded upon itself in the overlying regions (perimesangial segment of the GBM) and paramesangia (mesangial angles). Electron lucency is evident within the mesangial matrix (asterisks).

light microscopical changes. In cases of suspected recurrent disease, it is useful to examine the corresponding insertion biopsy by TEM to eliminate any possibility of pre-existing donor-related disease; for example, living-related donors of patients with a hereditary nephropathy may donate a similarly affected kidney.

Certain diseases cannot recur as they are intrinsic to the native kidney itself – most notably, Alport nephritis and autosomal dominant polycystic kidney disease (Colvin and Nickeleit, 2007).

It is not always possible to designate transplant disease as either recurrent or *de novo* as the original disease may not be known or may never have been proven by native biopsy. The ultrastructure of recurrent and *de novo* glomerular diseases is generally similar to that found in the corresponding native diseases but may be superimposed upon changes attributable to rejection or drug toxicity.

The more frequent TEM investigations for recurrent primary disease in the transplant kidney are described in this section.

Table 2.2 Recurrent diseases in renal transplants

Disease	Clinical recurrence rate (% of recipients)	Onset of recurrence	Additional comments
Focal segmental glomerulosclerosis	~30% of first grafts, ~80% of subsequent grafts	Can recur immediately. Adult: 7.5 months average. Child: 2 weeks average.	Secondary FSGS, arising as a late complication of other native renal diseases, does not recur.
Membranous nephropathy	10–30%	28% of recurrent cases develop within 4 months. Can recur within 1–2 weeks.	May signify underlying hepatitis C infection. Recurrence of idiopathic membranous nephropathy in up to 40% of recipients (Fairhead and Knoll, 2010).
IgA nephropathy	15–50% (Fairhead and Knoll, 2010)	Range: months–10 years.	Recurrence rate is higher in subsequent grafts (Choy, Chan and Lai, 2006).
Diabetic nephropathy	38% by 7 years	Average: 8 years.	High recurrence rate at 20 years (Mathew, 1988). Histological recurrence time averages 97 months (range: 41–154 months) (Hariharan *et al.*, 1996).
MCGN (MPGN), Type I	30–50% (Lorenz *et al.*, 2010)	Usually within first 4 years. As early as 2 weeks.	May signify underlying hepatitis C infection. Recurrence rate is higher in subsequent grafts. Graft loss to MCGN (Type I) at 10 years is approximately 15%.
MCGN, Type II* (DDD)	Over 80% (Lorenz *et al.*, 2010)	As early as 12 days.	Dense deposits in the GBM and paramesangium. Additional deposits may be seen in tubular basement membranes and Bowman's capsule. Graft loss to DDD at 5 years is approximately 30%.

Table 2.2 (*continued*)

Disease	Clinical recurrence rate (% of recipients)	Onset of recurrence	Additional comments
MCGN, Type III*	100%	16 months.	Data from a single case.
Systemic lupus erythematosus	1–9%	Average: 1–6 years. Range: 6 days–15 years.	Studies estimate the histological recurrence rate to be 30% (Goral *et al.*, 2003) and 19% (at an average of 5 years post insertion) (Yu *et al.*, 2012).
Amyloidosis	10–30%	Average: 1–10 years.	AL-type recurs frequently but rarely causes graft failure (Ponticelli, Moroni and Glassock, 2011). AA-type recurs infrequently; one case recurred at 17 years post insertion (Sethi *et al.*, 2011).
HUS/TMA	1–15% (Satoskar *et al.*, 2010)	Usually within 6 months. As early as 1 day.	Highest rates of recurrence in familial forms and scleroderma. HUS associated with *Escherichia coli* infection does not recur. It is difficult to distinguish recurrent HUS/TMA from other causes of HUS/TMA in transplants (e.g. induced by calcineurin inhibitors, sirolimus or OKT3).
MIDD	High	Range: 2–45 months (median: 33 months).	On rare occasions, immunoglobulin deposits may recur without typical histological signs of disease. Disease will recur if the production of monoclonal immunoglobulins cannot be stopped.

(*continued overleaf*)

Table 2.2 (*continued*)

Disease	Clinical recurrence rate (% of recipients)	Onset of recurrence	Additional comments
ANCA-mediated diseases	9% (Gera *et al.*, 2007)	Range: 5 days–89 months.	Transplantation may be withheld until the disease becomes inactive. Few recur within 5 years with current drug regimens.
Henoch–Schonlein purpura	20–35%	Usually within 9 months.	Mesangial IgA deposits were demonstrated in 53% of follow-up biopsies; however, the percentage showing histological evidence of recurrent HSP nephritis was not stated (Meulders, Pirson and Cosyns, 1994).
Anti-GBM disease	Up to 5%	Can be years after transplantation.	Transplantation is withheld until treatment removes the antibody and the disease is no longer active.
Fibrillary/immunotactoid GN	50%	Range: 2–9 years.	Limited data.
LCAT deficiency	High	Possibly as early as 2 days with definite recurrence at 6 weeks (Strom *et al.*, 2011).	Mesangial matrix and GBM have a 'moth-eaten' appearance due to deposits of electron-lucent and electron-dense material. Capillary loops may contain foam cells. Graft function tends to be maintained despite the presence of deposits.
Fabry disease	Rare (especially with ERT)	–	Laminated osmiophilic cytoplasmic inclusions may be evident in podocytes, mesangium and endothelium. Enzyme replacement therapy (ERT) is now able to clear deposits from the tissue.

Table 2.2 (*continued*)

Disease	Clinical recurrence rate (% of recipients)	Onset of recurrence	Additional comments
Other	–	–	Cryoglobulinaemia, fibronectin glomerulopathy, sickle cell nephropathy, oxalosis, cystinosis, adenosine phosphoribosyl transferase deficiency, thrombotic thrombocytopaenic purpura, complement factor H-related protein 5 nephropathy, C3 glomerulopathy.

Modified from Colvin and Nickeleit (2007) and Ivanyi (2008).
*Possibly included in the C3-related glomerulopathies.
FSGS: focal segmental glomerulosclerosis; MCGN: mesangiocapillary glomerulonephritis; MPGN: membranoproliferative glomerulonephritis; DDD: dense deposit disease; HUS: haemolytic uraemic syndrome; TMA: thrombotic microangiopathy; MIDD: monoclonal immunoglobulin deposition disease; ANCA: anti-neutrophil cytoplasmic antibody; HSP: Henoch–Schonlein purpura; GBM: glomerular basement membrane; GN: glomerulonephritis; and LCAT: lecithin–cholesterol acyltransferase.

2.3.2.1 Recurrent Focal Segmental Glomerulosclerosis

Primary FSGS can recur rapidly, and diagnosis requires identification of extensive foot process effacement (Figure 2.9). Additional podocyte changes variably include hyperplasia, hypertrophy, detachment from the GBM, villous transformation, vacuolation and aggregation of actin filaments adjacent to the GBM. Foot process effacement and villous transformation generally precede podocyte detachment, segmental tuft sclerosis and the accumulation of foam cells. The ultrastructural changes may be evident within days of transplantation, whereas LM changes can take 4–6 weeks (Ivanyi, 2008). Glomeruli from insertion wedges and biopsies performed shortly after insertion often exhibit foot process damage due to peri-operative stresses placed on the kidney during the transplantation process; therefore, it is important to consider these nonspecific changes when estimating foot process effacement in cases of possible recurrent FSGS. The difficulty in estimating foot process

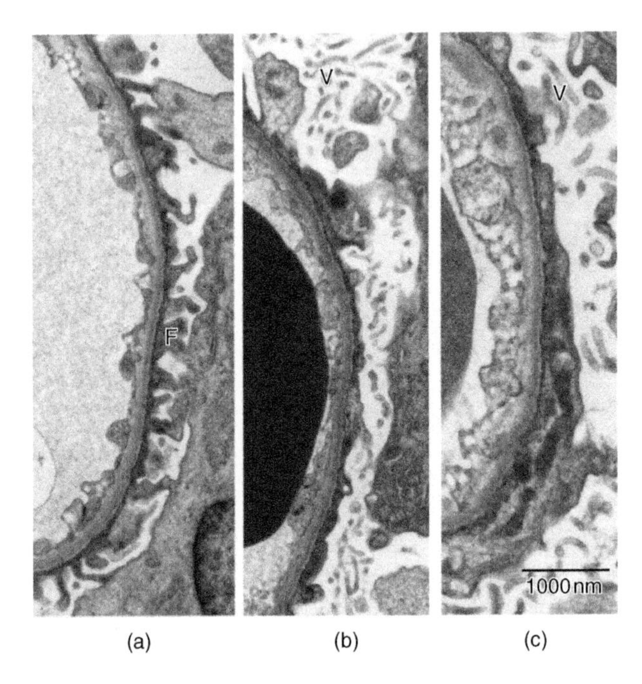

Figure 2.9 Recurrent focal segmental glomerulosclerosis (FSGS). Development series of recurrent FSGS from the same transplant kidney (identical magnification). Biopsies were performed at insertion (a), 5 days (b) and 6 weeks (c). (a) Some peri-operative podocyte damage is present; however, foot processes (F) are generally intact. (b) There is extensive foot process effacement and prominent villous transformation (V). (c) The extensive foot process effacement and prominent villous transformation remain, and there is very mild subendothelial lucency of the GBM.

effacement is compounded when examining glomeruli retrieved from paraffin wax, as the tissue processing for LM creates additional artefactual changes to the podocytes and their foot processes.

2.3.2.2 Recurrent Membranous Nephropathy

As with native membranous nephropathy, epimembranous immune-type deposits of the recurrent form are evident ultrastructurally before 'spike' formation is evident by LM (Figure 2.10). The deposits may be sparse or present in only a few peripheral capillary loops. Recurrent membranous nephropathy usually develops within a few months of transplantation and is ultrastructurally indistinguishable from the *de novo* form. Repeat

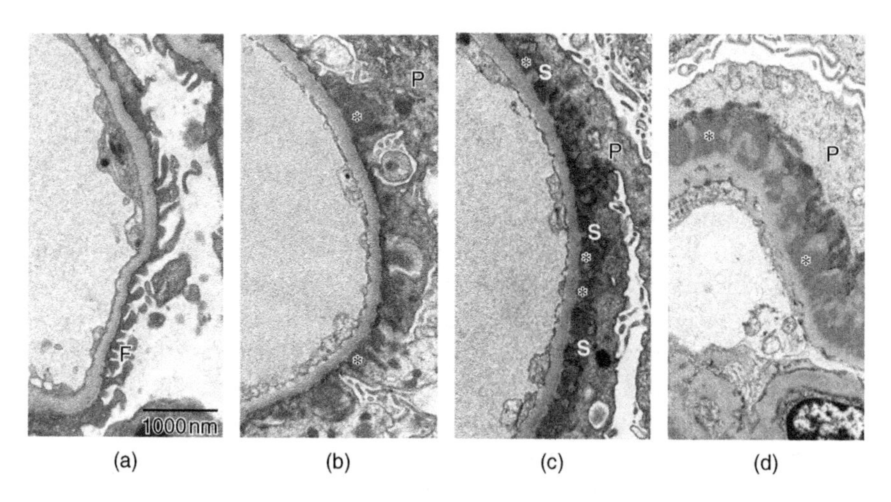

Figure 2.10 Recurrent membranous nephropathy. Development series of recurrent membranous nephropathy from the same transplant kidney (identical magnification). Biopsies were performed at insertion (a), 2 months (b), 4 months (c) and 13 months (d). (a) Some peri-operative podocyte damage is present; however, foot processes (F) are generally intact. The GBM is within normal limits. (b) Epimembranous and small hump-like immune-type deposits (*) in the absence of spike formation are present. (c) Deposits (*) are bordered laterally by spike formation (S). (d) The GBM is markedly thickened, and deposits (*) are incorporated within the membrane. Note the progressive loss of foot processes and swelling of podocytes (P) over time.

biopsies indicate that deposits tend to persist and progress rather than resolve (Monga *et al.*, 1993).

2.3.2.3 Recurrent IgA Nephropathy

The main ultrastructural feature of recurrent IgA nephropathy is mesangial immune-type deposit. Mesangial hypercellularity is generally mild or absent (Colvin and Nickeleit, 2007). The recurrence rates for IgA nephropathy vary greatly between investigative series, and the possible reasons for this include differing lengths of follow-up, differing intervals between protocol biopsies and diagnostic criteria used in assessing recurrence (clinical versus histological) (Ivanyi, 2008). Identification rates of recurrent IgA nephropathy are higher in recipients who receive protocol biopsies (with IM and/or TEM) over those who receive biopsies on purely clinical grounds, indicating IM and TEM can demonstrate IgA deposition before clinical symptoms appear. The recurrence rate for

IgA staining (intensity of 1+ or greater by IM) at 2 years post insertion was found to be 32% (Ortiz *et al.*, 2011).

2.3.2.4 *Recurrent Diabetic Glomerulosclerosis*

Diabetic glomerulosclerosis was found to recur in 38% of grafts by 7 years post insertion (Kim and Cheigh, 2001). Mesangial enlargement precedes thickening of the GBM (thickening becomes evident at 2–3 years post insertion) (Wilczek *et al.*, 1995); however, Kimmelstiel–Wilson nodules are not evident until 5–15 years post insertion (Colvin and Nickeleit, 2007). Arteriolar hyalinosis is a diagnostic feature common to diabetic glomerulosclerosis, calcineurin inhibitor toxicity and hypertension; TEM may assist in differentiating these entities by the demonstration of thickened GBM in cases of hyalinosis due to diabetic glomerulosclerosis.

2.3.3 De Novo Glomerular Disease

Transplant kidney disease is designated as *de novo* (new onset) when it differs from that which affected the native kidney. TEM is useful in the investigation of *de novo* glomerular diseases, and the most commonly encountered entities (Ivanyi, 2008) are described in this section.

2.3.3.1 *De Novo Membranous Nephropathy*

Membranous nephropathy is the most common *de novo* glomerulonephritis following transplantation and usually develops after the first year post insertion (and may occur many years later), but has been reported as early as 3 months post insertion (Gough *et al.*, 2005). *De novo* membranous nephropathy was detected by biopsy in 2% of 848 renal transplant recipients at an average of 63 months post insertion (range: 6–157 months) (Schwarz *et al.*, 1994). By 8 years post insertion, the incidence was found to be 5.3%. The pathogenesis of *de novo* membranous nephropathy is not fully understood. Studies suggest the disease is due to antibody production against podocyte proteins, although the target antigens are yet to be identified (Debiec *et al.*, 2011). None of nine cases of *de novo* membranous nephropathy demonstrated antibodies against the phospholipase A_2 receptor, which is the target antigen in most cases of idiopathic membranous nephropathy of the native kidney. The antibodies involved in *de novo* membranous nephropathy are thought to represent a subset of the humoral response involved in the pathogenesis of TG – hence, the association of these two diseases in some transplant renal biopsies. The epimembranous immune-type deposits may be sparse or present in only a few peripheral capillary loops.

2.3.3.2 De Novo Focal Segmental Glomerulosclerosis

De novo FSGS is generally a secondary phenomenon resulting from other pathology within the graft and usually develops after the first year post insertion. The TEM is ultrastructurally indistinguishable from that of the recurrent form. The disease has been described in several settings (modified from Colvin and Nickeleit, 2007):

- As the result of glomerular hyperfiltration injury, which occurs in longstanding transplants and paediatric kidneys transplanted into adult recipients
- In association with TG
- As a manifestation of calcineurin inhibitor toxicity (Morozumi *et al.*, 2004)
- As a manifestation of sirolimus immunosuppression (Letavernier *et al.*, 2007)
- In grafts with severe vascular disease resulting in reduced glomerular perfusion and collapsing FSGS
- In recipients infected with human immunodeficiency virus (Ivanyi, 2008)
- As *de novo* primary FSGS (rare).

2.3.3.3 Alport Anti-GBM Nephritis

Two to three percent of Alport patients who receive a transplant kidney develop anti-GBM nephritis as antibodies are raised against collagen antigens in the graft GBM, to which the recipient is not immunologically tolerant. Approximately 75% of cases occur within the first year post insertion, and the disease appears earlier in subsequent grafts. The reason why only a small percentage of recipients develop anti-GBM nephritis is not fully understood. It is thought that partial expression of collagen antigens occurs in some forms of Alport syndrome, and these recipients are, therefore, immunologically tolerant to the particular collagen antigens (Byrne *et al.*, 2002). For example, women with the X-linked form of Alport syndrome partially express the type IV alpha-5 collagen chain. As a corollary, the prevalence of Alport anti-GBM nephritis is greater in males than females, and the difference is partly attributable to the predominance among Alport syndrome cases of the X-linked form (85%) over the autosomal forms (15%). As with anti-GBM nephritis of the native kidney, TEM is regarded as having minimal contribution to the diagnosis of Alport anti-GBM nephritis, although a thin centrimembranous linear zone of electron-dense, granular deposit-like material within the GBM has been observed in one case (author's personal observation).

2.3.3.4 Drug-Induced Thrombotic Microangiopathy

Drug-induced TMA usually occurs in the early post-transplantation period and develops in approximately 3.5% (weighted mean) of recipients on calcineurin inhibitor–based immunosuppression (Schwimmer *et al.*, 2003). The ultrastructure of TMA is described in Section 2.3.1.1.

2.3.4 Donor-Related Disease

When pathology is demonstrated in a transplant renal biopsy, it is important to establish whether or not the disease process was pre-existing in the donor. TEM can play a confirmatory role in demonstrating the presence or absence of donor-related disease. A lack of ultrastructural pathology in the insertion biopsy would indicate the development of either recurrent or *de novo* disease in the graft. Two common investigations for donor-related disease are diabetes (TEM demonstrates thickened GBM and mesangial expansion) and IgA nephropathy (TEM demonstrates mesangial immune-type deposit). IgA deposition was demonstrated in 11% of living donors (Rosenberg *et al.*, 1990). Donor-related electron-dense deposits tend to clear from the transplant over time.

2.3.5 Infection

The immunocompromised transplant kidney recipient is susceptible to new infection or reactivation of latent infection. BK virus and microsporidia are the microorganisms most commonly found by TEM in transplant renal biopsies, while bacteria, protozoa and fungi (other than the related microsporidia) are rarely observed.

2.3.5.1 Viruses

The identification of viruses by TEM in transplant renal biopsies is an infrequent occurrence due to low infection rates, sampling problems relating to the small area of examination and the focal nature of some viral diseases. A low-magnification scan of the tissue section for cytopathic changes such as electron-dense cells, undulating nuclear contours and unusual chromatin patterns may expedite the search for viruses. Tubular epithelial cells and sloughed cells within tubule lumina are the structures most likely to contain viruses. Occasionally, viruses are identified in glomeruli. The presence of tubuloreticular bodies in endothelial cells may be an indicator of viral infection in the graft and should prompt close scrutiny of tubule cells for viruses. Aside from the kidney biopsy, TEM can demonstrate viruses in urine using the negative staining technique (refer to Chapter 16). Respiratory viruses, hepatitis B and C viruses, polyomaviruses and herpesviruses are the most common causes of viral infection of the transplant kidney recipient. Of these, polyomaviruses (usually BK virus), some herpesviruses (usually cytomegalovirus) and adenovirus are most likely to be identified by TEM in transplant renal biopsies.

Polyomavirus is a genus of non-enveloped icosahedral DNA viruses, the members of which are morphologically identical by TEM, having a round to slightly polygonal shape and measuring approximately 40 nm in diameter. The BK and JC species are the most important members of the genus in human pathology, and species identification requires immunological and/or genetic techniques. Polyomavirus-associated nephropathy (PVAN) is a cause of graft failure in 1–10% of renal transplant recipients (Hirsch *et al.*, 2005), and the infectious agent is usually the BK species. Careful examination is required to distinguish polyomaviruses from other electron-dense structures of similar size such as ribosomes, glycogen and chromatin granules. Intranuclear polyomaviruses may be distributed singly or aggregated in a lattice arrangement (Figure 2.11). Cytoplasmic polyomaviruses may be distributed singly, within linear and tubuloreticular arrays or within vacuoles (Figure 2.12). Glomerular polyomaviruses are usually found within parietal epithelial cells, although viruses within Bowman's capsule, podocytes, subepithelial immune-type deposits and the subepithelial aspect of the GBM have been identified (Brealey, 2007) (Figures 2.13 and 2.14).

Herpesviridae is a family of enveloped icosahedral DNA viruses, the members of which are morphologically identical by TEM. Intranuclear capsids are round to polygonal and contain a core composed of nucleic acid and proteins that is round to polygonal or rod-shaped. The capsid becomes a mature enveloped virus through a process of budding with nuclear and/or cytoplasmic membranes. Capsids are approximately

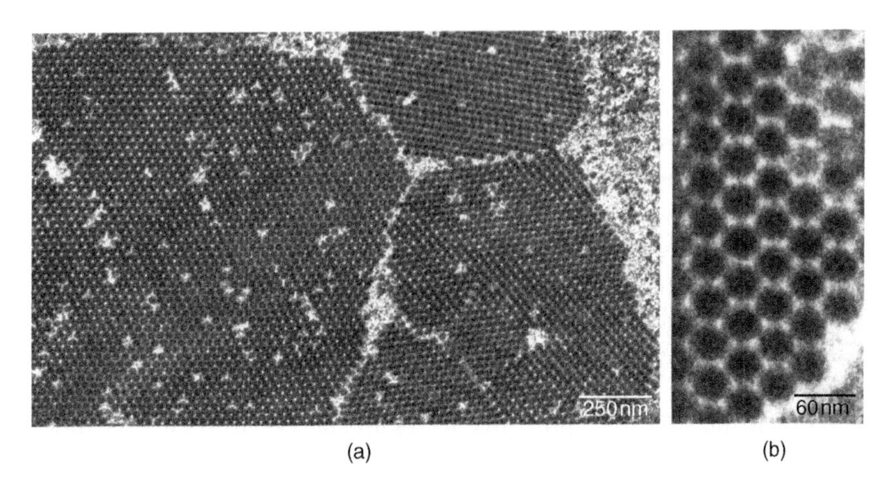

Figure 2.11 BK virus in transplant kidney (intranuclear lattice). (a) The regular arrangement of virions is demonstrated in the intranuclear lattice of BK virus capsids. (b) At higher magnification, the round to hexagonal shape of the capsids is evident.

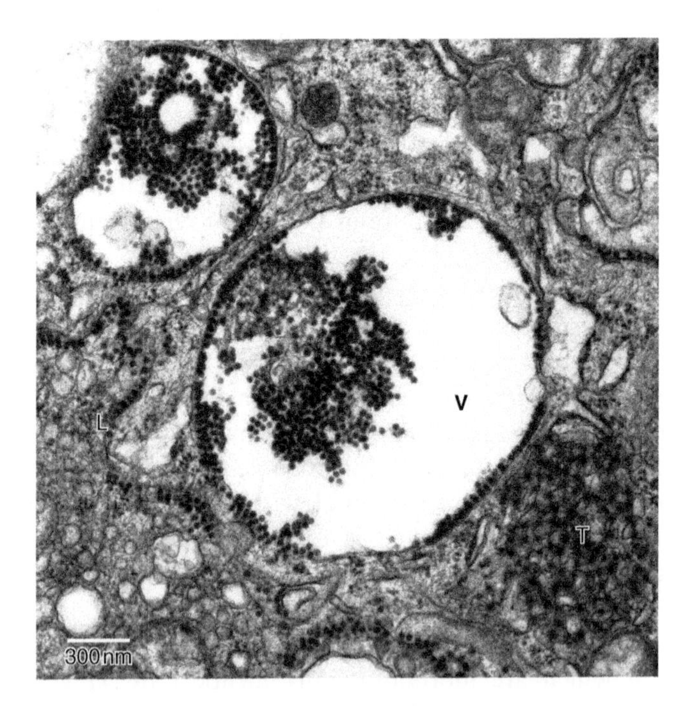

Figure 2.12 BK virus in transplant kidney (tubular epithelium). BK virus in a proximal convoluted tubule cell. Viruses are present within vacuoles (V), tubuloreticular arrays (T) and linear arrays (L).

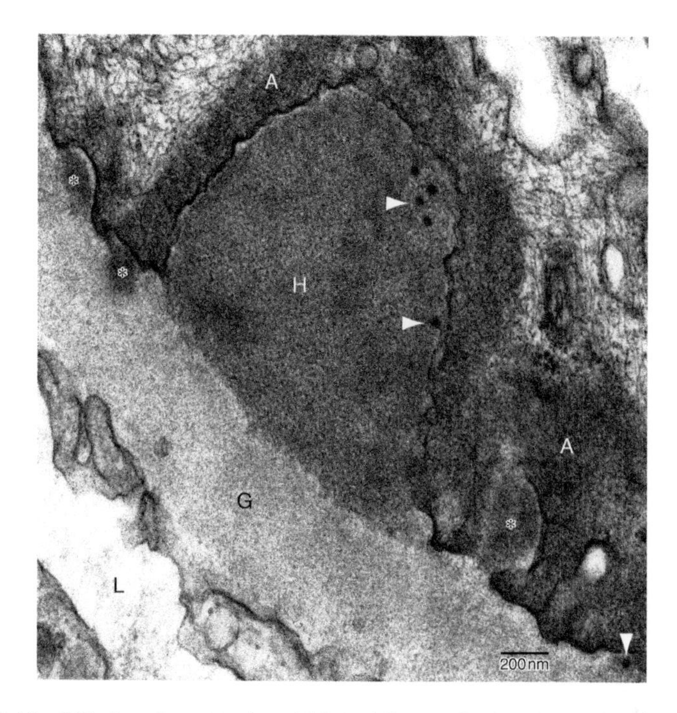

Figure 2.13 BK virus in transplant kidney (glomerular immune-type deposit). BK virus (arrows) within a subepithelial hump of immune-type deposit (H) and within the GBM (G). Smaller epimembranous deposits (*) are also present. The single virus within the GBM is associated with a small amount of deposit. Note the foot process effacement and aggregation of actin filaments (A) within the podocyte. Capillary lumen (L).

100 nm in diameter, and the mature enveloped viruses are approximately 120–150 nm (Figures 2.15 and 2.16).

Adenoviridae is a family of non-enveloped icosahedral DNA viruses. By TEM, adenoviruses are hexagonal and measure approximately 70–75 nm. Lattices of adenovirus may be observed in tubular cell nuclei and cytoplasm (Lim, Parsons and Ierino, 2005).

2.3.5.2 Microsporidia

Microsporidia are unicellular obligate parasites related to fungi (Didier and Weiss, 2011). At least 18 species are currently implicated in human disease, and immunocompromised individuals are especially vulnerable to infection (refer to Chapter 6). The organisms have been identified in numerous tissues and cell types including those of the eye, ear, brain, muscle, skin and urinary, gastrointestinal and respiratory tracts

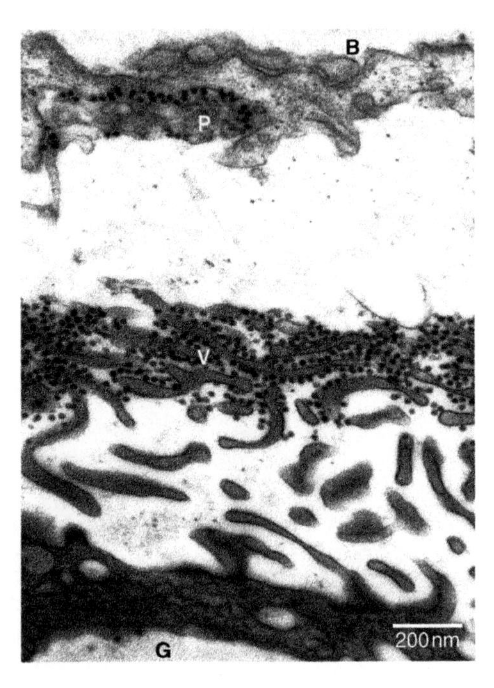

Figure 2.14 BK virus in transplant kidney (glomerular urinary space). BK viruses within the urinary space are in close association with a zone of podocyte villous transformation (V). Note the viruses in the GBM (G), parietal epithelial cell (P) and Bowman's capsule (B).

(Orenstein *et al.*, 2005). Symptoms of infected transplant recipients include chronic diarrhoea, vomiting, dyspepsia and weight loss (Gumbo *et al.*, 1999). Infection of the transplant kidney is usually caused by *Encephalitozoon* species. During the infectious stage, microsporidia produce spores 1–4 μm across (for species identified in human disease) that, by TEM are round to ovoid, depending on the plane of section. Microsporidia are unique in their mode of infection by the injection of sporoplasm into the host cell via an extrusion apparatus, the tube of which is visible by TEM. The number of turns of the coiled polar tube within the mature spore varies among the different species, and the cross-sectioned tube is observable on both sides of the organism. The *Encephalitozoon* species *intestinalis, hellem* and *cuniculi* display a polar tube that coils four to eight times (usually in single-row alignment on cross-section), whereas *Enterocytozoon bieneusi* displays a polar tube that coils approximately six times (usually in double-row alignment) (Didier *et al.*, 2004). This distinction is clinically important because the drug albendazole is effective against *Encephalitozoon* species but

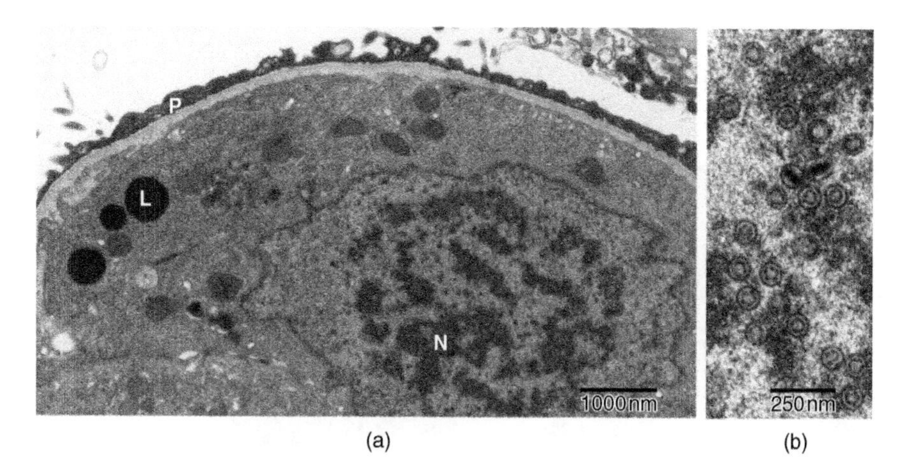

Figure 2.15 Cytomegalovirus in transplant kidney. (a) The glomerular endothelial cell displays hypertrophy and prominent lysosomes (L). The nucleus (N) displays an undulating contour of the nuclear membrane and a dispersed chromatin pattern. Immature cytomegalovirus particles are just discernible within the nucleus. The podocyte (P) displays the electron-dense cytopathic effect and extensive foot process effacement. (b) At higher magnification, nucleocapsids are dispersed amongst the chromatin.

of limited efficacy against *E. bieneusi*, which is susceptible to the drug fumagillin (Orenstein, 2003). *Encephalitozoon* species develop within an intracellular parasitophorous vacuole (Figure 2.17), with that of *E. intestinalis* being divided into thin-walled (septate) chambers (Didier *et al.*, 2004). The microsporidian spore has a chitinous wall which can be difficult to infiltrate completely by resin during tissue processing; hence, separation of the spore from the supporting resin can occur during sectioning when standard processing methods are used.

2.3.6 Inconclusive Diagnosis by LM and/or IM

TEM can be useful in situations where findings from LM and/or IM are atypical, equivocal, negative, noncontributory or unavailable. For example, positive IM in a case of suspected TG may indicate mixed glomerular pathology, and the identification of ultrastructural features for both TG and immune-complex-mediated disease can provide valuable diagnostic information. TEM can play a confirmatory role when IM is weakly positive, for example in some cases of viral infection and low-level (possibly early) immune-complex-mediated disease. Similarly, TEM can be useful when the IM is atypical, which can occur in cases

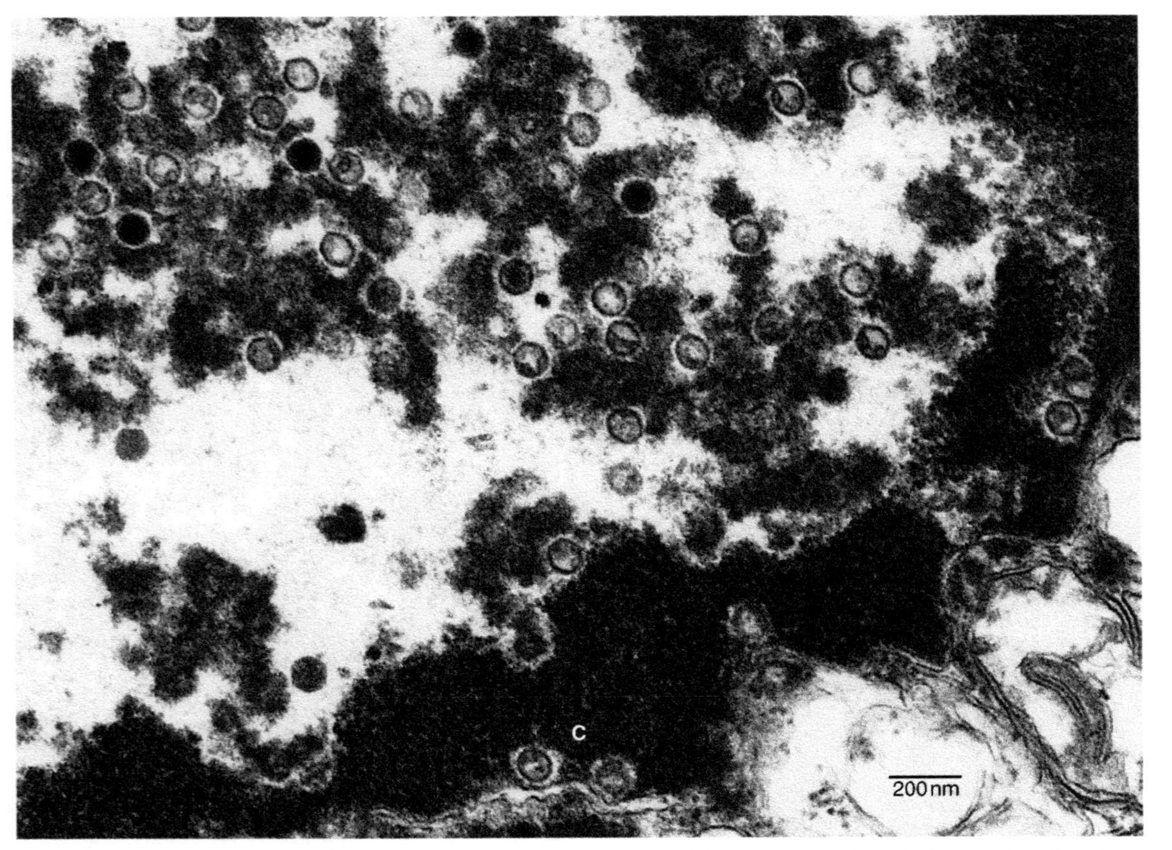

Figure 2.16 Varicella zoster virus in transplant kidney: the nucleus of a cell from a transplant kidney infected with varicella zoster virus. Capsids are in the process of maturation by budding with the inner nuclear membrane (C). Autopsy specimen from a child who developed chickenpox around the time of transplantation (retrieved from the paraffin block).

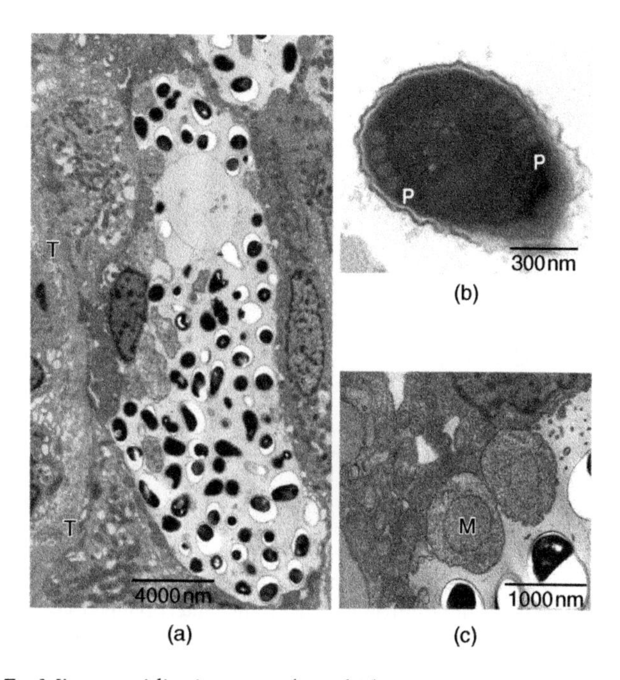

Figure 2.17 Microsporidia in transplant kidney. (a) Parasitophorous vacuoles within the tubular epithelium of a renal transplant infected with an *Encephalitozoon* species. Less electron-dense immature forms (meronts) are generally located at the periphery of the vacuoles. Tubular basement membrane (T). (b) Microsporidian spore. Note that the single row of the coiled polar tube (P) is evident on both sides of the organism. (c) Meronts (M) are less electron dense than mature spores (lower right). Note the separation of the mature spores from the supporting resin.

of recurrent and *de novo* glomerular disease. Occasionally, insufficient glomeruli are present in the LM and/or IM specimens; TEM alone may provide valuable diagnostic information, and the semithin survey sections may provide sufficient information to make a diagnosis by LM. Rarely, the only diagnostic material present is in the tissue sent for TEM – for example, a single blood vessel in survey sections may demonstrate vascular rejection (Figure 2.18).

2.3.7 Miscellaneous Topics

2.3.7.1 Post-Transplant Nephrotic Syndrome

Approximately 13% of transplant kidney recipients develop PTNS, a condition characterised clinically by persistent nephrotic-range proteinuria (Yakupoglu *et al.*, 2004). The syndrome is correlated with decreased

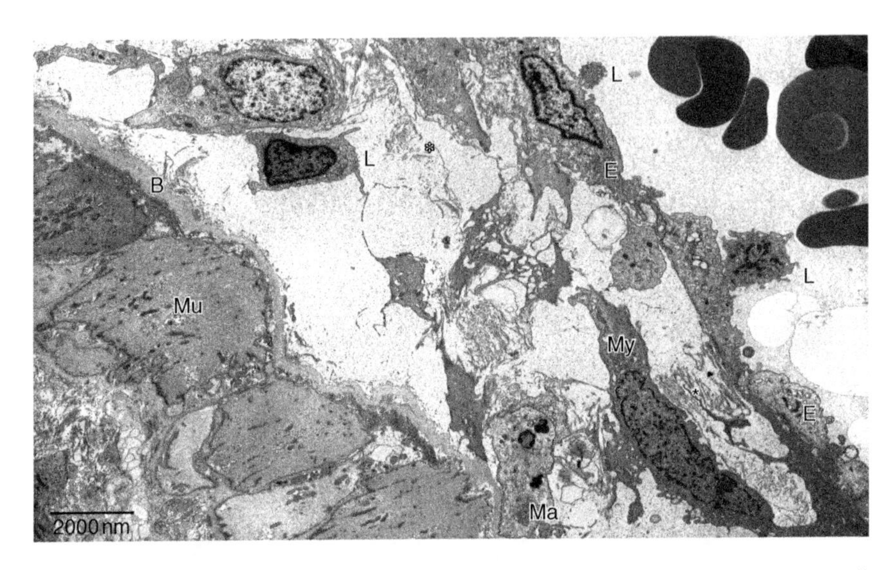

Figure 2.18 Vascular rejection in transplant kidney. Endarteritis within a small artery. A lymphocyte (L), myofibroblast (My) and macrophage (Ma) are identified in the subintimal zone between the endothelium (E) and basal lamina (B), and remodelling of a new basal lamina is evident (*). Luminal lymphocytes are adherent with the endothelium. Smooth muscle cells of the arterial wall (Mu). Luminal erythrocytes are at the upper right.

graft survival rate. PTNS can develop in association with chronic allograft nephropathy (with or without features of TG or secondary FSGS), recurrent glomerular disease or *de novo* glomerular disease, and shows no particular predisposition with regard to primary renal disease (Stokes and De Palma, 2006). The most common causes of PTNS among the recurrent and *de novo* glomerular diseases are FSGS and membranous nephropathy, respectively.

Transplant recipients with congenital nephrotic syndrome of the Finnish type as the original disease can develop PTNS as antibodies are raised against nephrin, a transmembrane protein of the foot process slit diaphragm, to which the recipient is not immunologically tolerant. TEM demonstrates endothelial swelling, foot process effacement and decreased slit diaphragms in podocyte pores (Patrakka *et al.*, 2002).

De novo minimal change disease is a rare cause of PTNS that can develop shortly after transplantation, and TEM demonstrates effacement of foot processes, the only histological change within the glomeruli (Ivanyi, 2008).

Some paediatric grafts transplanted into adult recipients develop diffuse irregular lamellation along the subepithelial aspect of the GBM

(Nasdasdy *et al.*, 1999). The GBM lamellation was evident in 9% of cases where the donor was younger than 10 years of age and developed as early as 10 weeks post insertion. Recipients had PTNS, and mesangia were variably expanded.

2.3.7.2 *LM versus TEM*

Many of the morphologies identifiable by LM are also demonstrable by TEM; however, it is impractical to use TEM routinely in most of these situations owing to the critical need for examination of multiple levels of tissue. For example, tubulitis (intra-epithelial lymphocytes in non-atrophic tubules) as an indicator of rejection is readily demonstrable by TEM. Similarly, hyaline arteriolar beading in calcineurin inhibitor toxicity is identifiable by TEM (Figure 2.19). The beads are composed of variably electron-dense granular material and form at the adventitial aspect of the arteriolar wall; their presence may be indicative of chronic toxicity. In contrast, hyaline material in the subendothelial zone or basal lamina of arterioles is considered nonspecific as it can also occur in diabetic glomerulosclerosis and hypertension (Nadasdy *et al.*, 2009). Acute (early) calcineurin inhibitor toxicity is generally diagnosed by LM through the identification of isometric vacuolation in proximal tubular epithelium, and TEM may assist by the demonstration of giant mitochondria in the proximal tubular epithelium, endothelial swelling and TMA.

2.3.7.3 *Neoplasia Associated with Renal Transplantation*

There is a prevalence of lymphoproliferative disorders, Kaposi's sarcoma and certain carcinomas among renal transplant recipients. Additionally, there is an increased risk of renal cell carcinoma arising in end-stage native kidneys. TEM may be helpful in identifying these neoplasms, although diagnosis is usually made by LM and IM. Post-transplant lymphoproliferative disorder is commonly associated with Epstein–Barr virus.

2.3.7.4 *The Future of TEM*

The need for TEM in the diagnosis of TG, recurrent disease and *de novo* glomerular disease is likely to increase as graft survival time increases. Renal transplant medicine is a dynamic field of study, and, as innovative transplantation techniques and immunosuppressive therapies

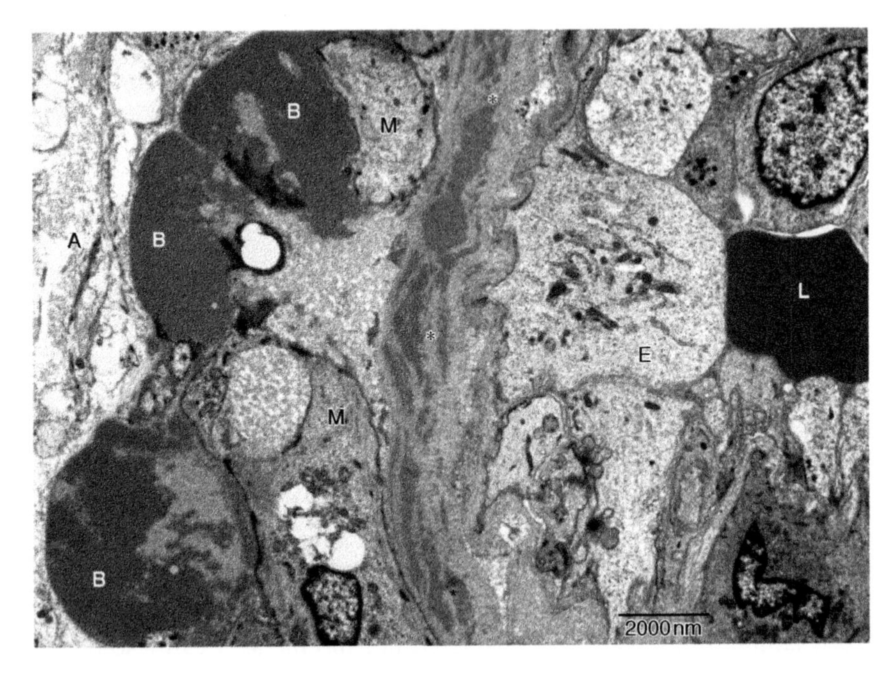

Figure 2.19 Calcineurin inhibitor toxicity in transplant kidney. Hyaline arteriolar beading (B) on the adventitial aspect of the arteriolar wall is evident ultrastructurally as variably electron-dense granular material. Nondiagnostic electron-dense granular material is present within the basal lamina (*). Adventitia (A). Smooth muscle cells (M). Endothelium (E). Luminal erythrocyte (L).

are developed, it is likely that new transplant renal pathology will be identified. LM, IM and TEM will each have their role in characterising and defining these new pathological entities.

REFERENCES

Andresdottir, M.B., Assmann, K.J., Koene, R.A. and Wetzels, J.F. (1998) Immunohistological and ultrastructural differences between recurrent Type I membranoproliferative glomerulonephritis and chronic transplant glomerulopathy. *American Journal of Kidney Diseases*, 32 (4), 582–588.

Baid-Agrawal, S., Farris, A.B. III, Pascual, M. *et al.* (2011) Overlapping pathways to transplant glomerulopathy: chronic humoral rejection, hepatitis C infection, and thrombotic microangiopathy. *Kidney International*, 80 (8), 879–885.

Brealey, J.K. (2007) Ultrastructural observations in a case of BK virus nephropathy with viruses in glomerular subepithelial humps. *Ultrastructural Pathology*, 31 (1–3), 1–7.

Briganti, E.M., Russ, G.R., McNeil, J.J. *et al.* (2002) Risk of renal allograft loss from recurrent glomerulonephritis. *New England Journal of Medicine*, **347** (2), 103–109.

Byrne, M.C., Budisavljevic, M.N., Fan, Z. *et al.* (2002) Renal transplant patients with Alport's syndrome. *American Journal of Kidney Diseases*, **39** (4), 769–775.

Chicano, S.L., Cornell, L.D., Selig, M.K. *et al.* (2006) Distinctive ultrastructural features of chronic allograft glomerulopathy: new formation of circumferential glomerular basement membrane. *Laboratory Investigation*, **86** (Suppl. 1), 260A.

Choy, B.Y., Chan, T.M. and Lai, K.M. (2006) Recurrent glomerulonephritis after kidney transplantation. *American Journal of Transplantation*, **6**, 2535–2542.

Colvin, R.B. and Nickeleit, V. (2007) Renal transplant pathology, in *Heptinstall's Pathology of the Kidney*, 6th edn (eds J.C. Jennette and R.H. Heptinstall), Lippincott Williams & Wilkins, Philadelphia, PA, pp. 1409–1540.

Cosio, F.G., Gloor, J.M., Sethi, S. and Stegall, M.D. (2008) Transplant glomerulopathy. *American Journal of Transplantation*, **8** (3), 492–496.

Debiec, H., Martin, L., Jouanneau, C. *et al.* (2011) Autoantibodies specific for the phospholipase A_2 receptor in recurrent and de novo nephropathy. *American Journal of Transplantation*, **11** (10), 2144–2152.

Didier, E.S., Stovall, M.E., Green, L.C. *et al.* (2004) Epidemiology of microsporidiosis: sources and modes of transmission. *Veterinary Parasitology*, **126** (1–2), 145–166.

Didier, E.S. and Weiss, L.M. (2011) Microsporidiosis: not just in AIDS patients. *Current Opinion in Infectious Diseases*, **24** (5), 490–495.

Drachenberg, C.B., Steinberger, E., Hoehn-Saric, E. *et al.* (1997) Specificity of intertubular capillary changes: comparative ultrastructural studies in renal allografts and native kidneys. *Ultrastructural Pathology*, **21** (3), 227–233.

Fairhead, T. and Knoll, G. (2010) Recurrent glomerular disease after kidney transplantation. *Current Opinion in Nephrology and Hypertension*, **19**, 578–585.

Gera, M., Griffin, M.D., Specks, U. *et al.* (2007) Recurrence of ANCA-associated vasculitis following transplantation in the modern era of immunosuppression. *Kidney International*, **71** (12), 1296–1301.

Gloor, J.M., Sethi, S., Stegall, M.D. *et al.* (2007) Transplant glomerulopathy: subclinical incidence and association with alloantibody. *American Journal of Transplantation*, 7 (9), 2124–2132.

Goral, S., Ynares, C., Shappell, S.B. *et al.* (2003) Recurrent lupus nephritis in renal transplant recipients revisited: it is not rare. *Transplantation*, **75** (5), 651–656.

Gough, J., Yilmaz, A., Yilmaz, S. and Benediktsson, H. (2005) Recurrent and *de novo* glomerular immune-complex deposits in renal transplant biopsies. *Archives of Pathology and Laboratory Medicine*, **129**, 231–233.

Gourishankar, S., Leduc, R., Connett, J. *et al.* (2010) Pathological and clinical characterization of the 'troubled transplant': data from the DeKAF study. *American Journal of Transplantation*, **10** (2), 324–330.

Gumbo, T., Hobbs, R.E., Carlyn, C. *et al.* (1999) Microsporidia infection in transplant patients. *Transplantation*, **67** (3), 482–484.

Haas, M. (2011) Transplant glomerulopathy: it's not always about chronic rejection. *Kidney International*, 80 (8), 801–803.

Haas, M. and Mirocha, J. (2011) Early ultrastructural changes in renal allografts: correlation with antibody-mediated rejection and transplant glomerulopathy. *American Journal of Transplantation*, **11** (10), 2123–2131.

Hariharan, S., Smith, R.D., Viero, R. and First, M.R. (1996) Diabetic nephropathy after renal transplantation. Clinical and pathological features. *Transplantation*, **62** (5), 632–635.

Herrera, G.A., Isaac, J. and Turbat-Herrera, E.A. (1997) Role of electron microscopy in transplant renal pathology. *Ultrastructural Pathology*, **21** (6), 481–498.

Hirsch, H.H., Brennan, D.C., Drachenberg, C.B. *et al.* (2005) Polyomavirus-associated nephropathy in renal transplantation: interdisciplinary analyses and recommendations. *Transplantation*, **79** (10), 1277–1286.

Ivanyi, B. (2008) A primer on recurrent and *de novo* glomerulonephritis in renal allografts. *Nature Clinical Practice Nephrology*, **4** (8), 446–457.

Ivanyi, B., Kemeny, E., Szederkenyi, E. *et al.* (2001) The value of electron microscopy in the diagnosis of chronic renal allograft rejection. *Modern Pathology*, **14** (12), 1200–1208.

Kim, H. and Cheigh, J.S. (2001) Kidney transplantation in patients with type 1 diabetes mellitus: long-term prognosis for patients and grafts. *Korean Journal of Internal Medicine*, **16** (2), 98–104.

Letavernier, E., Bruneval, P., Mandet, C. *et al.* (2007) High sirolimus levels may induce focal segmental glomerulosclerosis de novo. *Clinical Journal of the American Society of Nephrology*, **2** (2), 326–333.

Lim, A.K., Parsons, S. and Ierino, F. (2005) Adenovirus tubulointerstitial nephritis presenting as a renal allograft space-occupying lesion. *American Journal of Transplantation*, **5** (8), 2062–2066.

Lorenz, E.C., Sethi, S., Leung, N. *et al.* (2010) Recurrent membranoproliferative glomerulonephritis after kidney transplantation. *Kidney International*, **77** (8), 721–728.

Mathew, T.H. (1988) Recurrence of disease following renal transplantation. *American Journal of Kidney Diseases*, **12** (2), 85–96.

Meulders, Q., Pirson, Y. and Cosyns, J.P. (1994) Course of Henoch-Schonlein nephritis after renal transplantation. Report on ten patients and review of the literature. *Transplantation*, **58** (11), 1179–1186.

Monga, G., Mazzucco, G., Basolo, B. *et al.* (1993) Membranous glomerulonephritis (MGN) in transplanted kidneys: morphologic investigation on 256 renal allografts. *Modern Pathology*, **6** (3), 249–258.

Morozumi, K., Takeda, A., Uchida, K. and Mihatsch, M.J. (2004) Cyclosporine nephrotoxicity: how does it affect renal allograft function and transplant morphology? *Transplantation Proceeding*, **36** (Suppl. 2), 251S–256S.

Nasdasdy, T., Abdi, R., Pitha, J. *et al.* (1999) Diffuse glomerular basement membrane lamellation in renal allografts from pediatric donors to adult recipients. *American Journal of Surgical Pathology*, **23** (4), 437–442.

Nadasdy, T., Satoskar, A., Nadasdy, G. *et al.* (2009) Pathology of renal transplantation, in *Silva's Diagnostic Renal Pathology* (eds X.J. Zhou, Z. Lasnik and T. Nadasdy), Cambridge University Press, New York, p. 545.

Nickeleit, V. (2009) The pathology of kidney transplantation, in *Transplantation Pathology* (ed. P. Ruiz), Cambridge University Press, New York, p. 69.

Orenstein, J.M. (2003) Diagnostic pathology of microsporidiosis. *Ultrastructural Pathology*, **27** (3), 141–149.

Orenstein, J.M., Russo, P., Didier, E.S. *et al.* (2005) Fatal pulmonary microsporidiosis due to *Encephalitozoon cuniculi* following allogenic bone marrow transplantation for acute myelogenous leukemia. *Ultrastructural Pathology*, **29** (3–4), 269–276.

Ortiz, F., Gelpi, R., Koskinen, P. *et al.* (2011) IgA nephropathy recurs early in the graft when assessed by protocol biopsy. *Nephrology Dialysis Transplantation*, **27** (6), 2553–2558.

Patrakka, J., Ruotsalainen, V., Reponen, P. *et al.* (2002) Recurrence of nephrotic syndrome in kidney grafts of patients with congenital nephrotic syndrome of the Finnish type: role of nephrin. *Transplantation*, **73** (3), 394–403.

Ponticelli, C., Moroni, G. and Glassock, R.J. (2011) Recurrence of secondary glomerular disease after renal transplantation. *Clinical Journal of the American Society of Nephrology*, **6** (5), 1214–1221.

Racusen, L.C., Solez, K., Colvin, R.B. *et al.* (1999) The Banff 97 working classification of renal allograft pathology. *Kidney International*, **55** (2), 713–723.

Rosenberg, H.G., Martinez, P.S., Vaccarezza, A.S. *et al.* (1990) Morphological findings in 70 kidneys of living donors for renal transplant. *Pathology Research and Practice*, **186** (5), 619–624.

Said, S.M., Cornell, L.D., Valeri, A.M. *et al.* (2010) C1q deposition in the renal allograft: a report of 24 cases. *Modern Pathology*, **23** (8), 1080–1088.

Satoskar, A.A., Pelletier, R., Adams, P. *et al.* (2010) *De novo* thrombotic microangiopathy in renal allograft biopsies – role of antibody-mediated rejection. *American Journal of Transplantation*, **10** (8), 1804–1811.

Schwarz, A., Krause, P., Offermann, G. and Keller, F. (1994) Impact of de novo glomerulonephritis on the clinical course after transplantation. *Transplantation*, **58** (6), 650–654.

Schwimmer, J., Nadasdy, T.A., Spitalnik, P.F. *et al.* (2003) De novo thrombotic microangiopathy in renal transplant recipients: a comparison of hemolytic uremic syndrome with localized renal thrombotic microangiopathy. *American Journal of Kidney Diseases*, **41** (2), 471–479.

Sethi, S., El Ters, M., Vootukuru, M. and Qian, Q. (2011) Recurrent AA amyloidosis in a kidney transplant. *American Journal of Kidney Diseases*, **57** (6), 941–944.

Sis, B., Campbell, P.M., Mueller, T. *et al.* (2007) Transplant glomerulopathy, late antibody-mediated rejection and the ABCD tetrad in kidney allograft biopsies for cause. *American Journal of Transplantation*, **7** (7), 1743–1752.

Stokes, M.B. and De Palma, J. (2006) Post-transplantation nephrotic syndrome. *Kidney International*, **69** (6), 1088–1091.

Strom, E.H., Sund, S., Reier-Nilsen, M. *et al.* (2011) Lecithin:cholesterol acyltransferase (LCAT) deficiency: renal lesions with early graft recurrence. *Ultrastructural Pathology*, **35** (3), 139–145.

Wavamunno, M.D., O'Connell, P.J. and Vitalone, M. (2007) Transplant glomerulopathy: ultrastructural abnormalities occur early in longitudinal analysis of protocol biopsies. *American Journal of Transplantation*, **7** (12), 2757–2768.

Wilczek, H.E., Jeremko, G., Tyden, G. and Groth, C.G. (1995). Evolution of diabetic nephropathy in kidney grafts. Evidence that a simultaneously transplanted pancreas exerts a protective effect. *Transplantation*, **59** (1), 51–57.

Yakupoglu, U., Baranowska-Daca, E., Rosen, D. *et al.* (2004) Post-transplant nephritic syndrome: a comprehensive clinicopathologic study. *Kidney International*, **65** (6), 2360–2370.

Yu, T.M., Wen, M.C., Li, C.Y. *et al.* (2012) Impact of recurrent lupus nephritis on lupus kidney transplantation: a 20-year single center experience. *Clinical Rheumatology*, **31**(4), 705–710.

3

Electron Microscopy in Skeletal Muscle Pathology

Elizabeth Curtis[1] **and Caroline Sewry**[2,3]

[1]*Muscle Biopsy Service/Electron Microscope Unit, Department of Cellular Pathology, Queen Elizabeth Hospital Birmingham, Birmingham, United Kingdom*
[2]*Wolfson Centre for Inherited Neuromuscular Diseases, RJAH Orthopaedic Hospital, Oswestry, United Kingdom*
[3]*Dubowitz Neuromuscular Centre, Institute of Child Health and Great Ormond Street Hospital, London, United Kingdom*

3.1 INTRODUCTION

Ultrastructural abnormalities in diseased human muscle can be observed in any organelle of a fibre, and also in other components of the tissue such as the extracellular matrix, blood vessels or nerves. Transmission electron microscopy (TEM) provides valuable information by helping to determine if a sample is normal or abnormal, clarifying the identity of structures observed at the light microscope (LM) level and visualising structures not apparent with LM. Although changes are not disease specific, ultrastructural examination can help to focus genetic screening.

Diagnostic Electron Microscopy: A Practical Guide to Interpretation and Technique, First Edition. Edited by John W. Stirling, Alan Curry and Brian Eyden.

3.1.1 The Biopsy Procedure

Muscle biopsies are of two types, open and needle. Needle biopsies may provide a smaller sample, but with careful handling, useful information can be obtained and a range of tests applied. Open biopsy provides a larger sample, orientation is more easily achieved and fewer fibres suffer mechanical trauma. However, there are times when a needle biopsy will be the favoured or only option.

On rare occasions, a sample may be received which has been removed by a surgeon using diathermy. This technique must be strongly discouraged, as the tissue is damaged beyond usefulness. Similarly, contact with too much saline results in artefactual swelling of organelles and affects interpretation.

3.1.2 Sampling

A biopsy should be dealt with as soon as possible after it is taken as a delay of longer than about 20 minutes will compromise results for both ultrastructural analysis and biochemical assays.

TEM can be a time consuming technique, and as only a small amount of tissue is examined, sampling has to be considered. Interpretation must always be related to LM results and clinical phenotype.

The TEM sample should be a thin strip with a maximum diameter of about 2 mm, taken from an area least likely to be affected by handling trauma. Areas adjacent to fascia and myotendinous junctions should be avoided as myofibres in these areas contain a high number of internal nuclei and the sarcolemma is invaginated and associated with abundant collagen. A layer of electron-dense material can be seen immediately below the sarcolemma which forms from bundles of fine filaments arising from myofibrils, and rod-like structures may also be present. The strip of muscle should be pinned to a matchstick or clamped before being removed to prevent contraction of the myofilaments, but this is not possible with needle biopsies or open biopsies taken through a small incision. Some laboratories no longer fix at resting length, but the contraction that occurs does not unduly influence interpretation, although it may produce less aesthetically pleasing images.

3.1.3 Tissue Processing

Several protocols are suitable for muscle, using glutaraldehyde alone or in combination with paraformaldehyde, and the choice is a matter of personal preference. However, methods may have to be adjusted to optimise results for immunolabelling. After fixation under tension for approximately 1.5–2.0 hours, the sample is cut into smaller blocks of about 1 mm^3 or slightly larger.

Tissue should be orientated in 'coffin' moulds, micromoulds or Beem capsules so that longitudinal sections can be cut. More information, particularly on myofibrillar structure, is obtained from longitudinal sections. A minimum of five pieces of tissue should be blocked out. Examination of semithin survey sections stained with toluidine blue determines the number of blocks needed for ultrastructural investigation.

3.1.4 Artefacts

A summary of various artefacts that can arise are shown in Figure 3.1. These include contraction artefact caused by rough handling, loss of glycogen during processing or through staining *en bloc* with uranyl acetate, swelling of mitochondria and/or sarcoplasmic reticulum (SR) caused by contact with too much saline and, lastly, alterations in mitochondrial matrix caused by delay in fixation. Artefacts can arise for several reasons, but the most common are rough handling and delay in fixation. A subjective assessment of glycogen content may be difficult as several factors govern levels in fibres (see Section 3.2). Assessment can also be hampered by artefactual redistribution of particles and may give a false impression of greater amounts. Large subsarcolemmal blebs of glycogen particles may be seen in partially contracted fibres, for example those at the periphery of a block or after rough handling.

3.2 NORMAL MUSCLE

Interpreting pathological samples relies on an understanding of the variability of normal muscle and correlation with all other techniques, especially LM and clinical phenotype. Fibre-type proportions vary between different muscles, nutritional status may affect lipid and glycogen content and the amount of physical exertion that has been placed on the muscle may affect the number and appearance of the various organelles (discussed further in this chapter).

A fibre is encased by the sarcolemma which consists of a plasmalemma and an overlying external lamina. This in turn comprises an outer amorphous lamina densa (open arrow) and an internal lucent lamina rara (Figure 3.4). Myofibrils, compact bundles of filaments that are

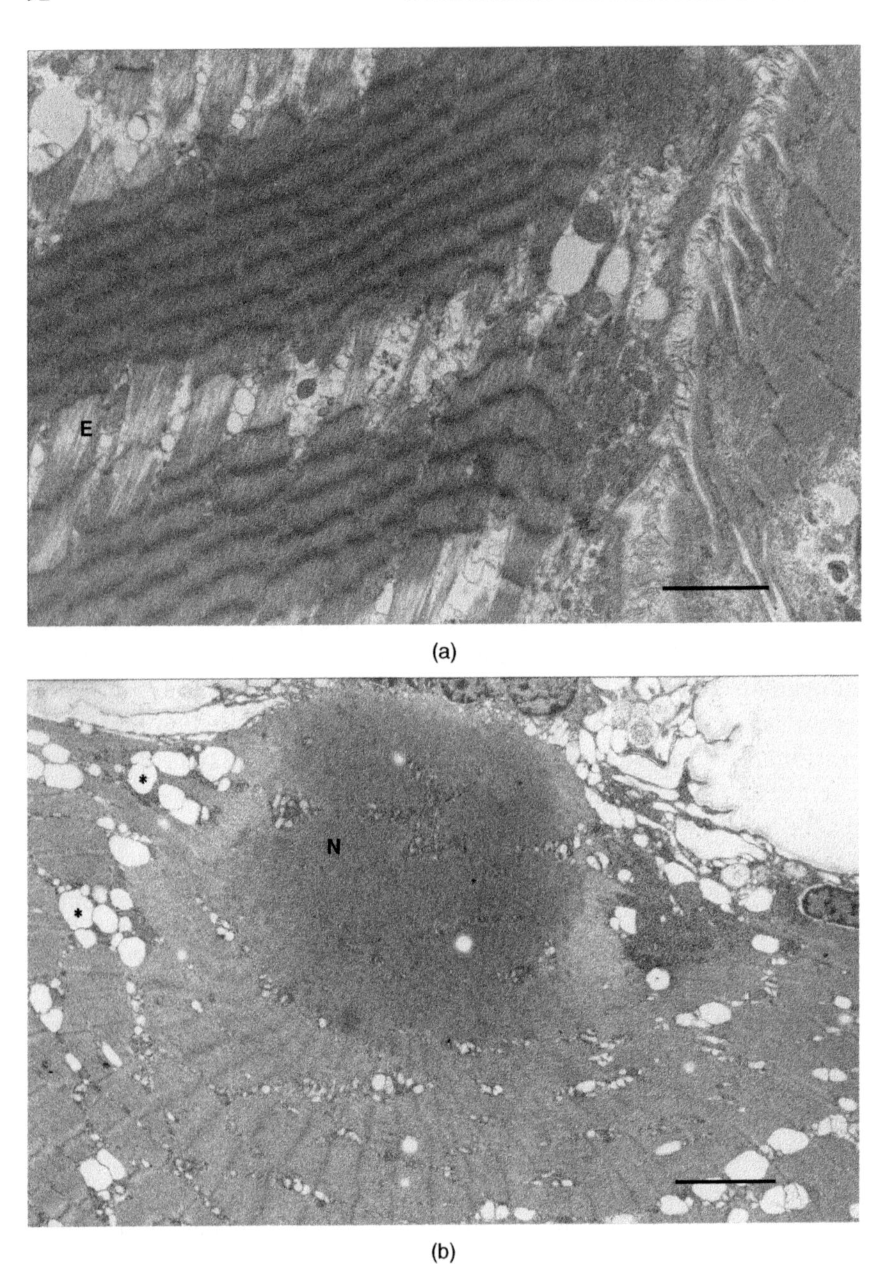

Figure 3.1 (a) Rough handling can lead to the formation of contraction bands where dense over-contracted fibre segments alternate with electron-lucent areas of over-extended fibrils (E). (b) A focal contraction node (N) and multiple swollen T-tubule cisternae (*).

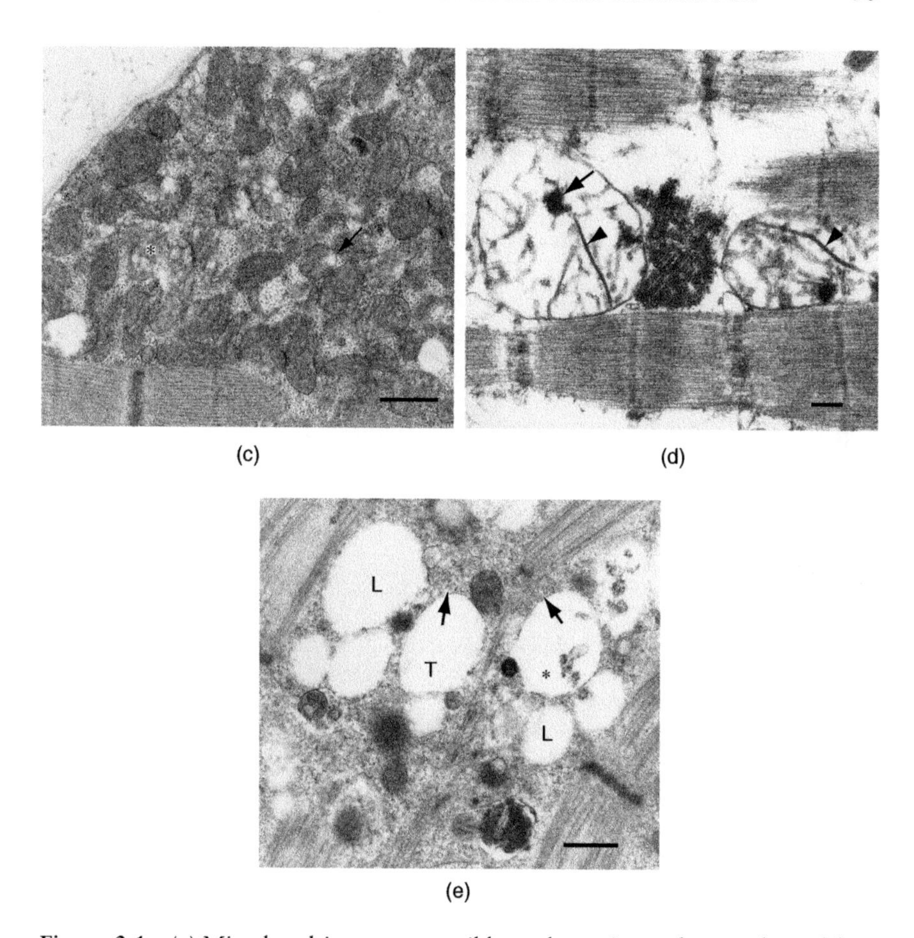

Figure 3.1 (c) Mitochondria are susceptible to hypoxia, and even short delays in fixation can lead to morphological change. Mild mitochondrial swelling is seen as discrete areas of matrical pallor (arrow), but severe swelling may lead to ballooning or even rupture of mitochondrial membranes (*). (d) Severe delays in tissue fixation result in appearances akin to those seen in post-mortem muscle. The sarcoplasm contains only sparse glycogen, often in the form of 'crystalline' arrays which may be mistaken by the unwary for virus particles. Mitochondria show features of anoxia including woolly densities (arrow) and of 'zipper'-like intracristal inclusions (arrowheads). (e) Swollen T-tubules (T) arising from a fixation delay may be misinterpreted as lipid at lower magnifications. Vacuoles may be of a similar size to lipid droplets (L), but the T-tubules are membrane bound (arrows). Particles may be washed into dilated T-tubule cisternae giving the impression of lysosomes (*). Magnification bars: (a) 2 μm; (b) 2 μm; (c) 500 nm; (d) 200 nm and (e) 500 nm.

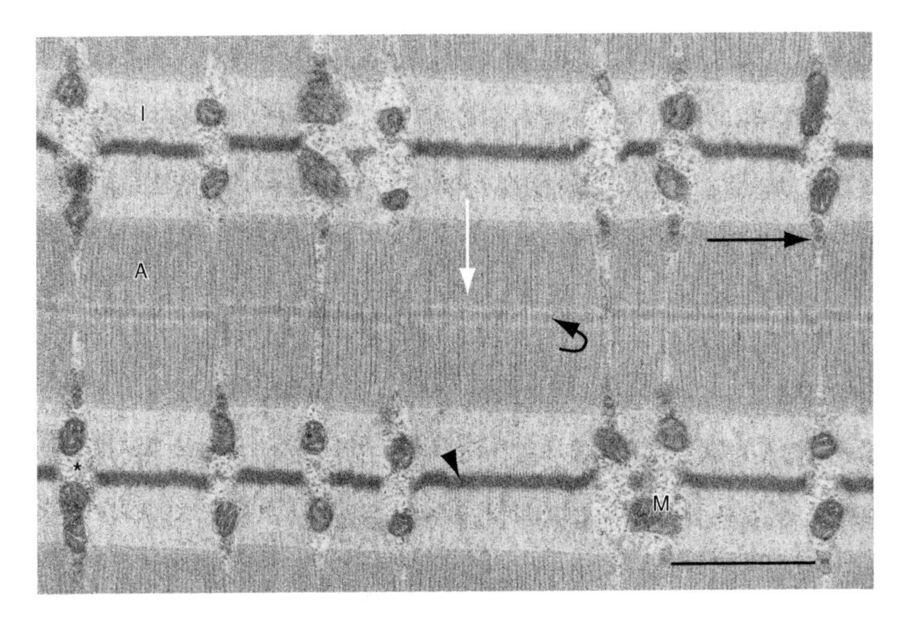

Figure 3.2 In normal muscle, the darkly staining Z line (arrowhead), composed largely of actin and α-actinin, bisects the pale I (*isotropic*) band (I), which consists mainly of fine actin filaments. The darker A (*anisotropic*) band (A), adjacent to the I band, consists of thick myosin filaments. The A band is divided by a pale region, or H zone (open arrow), where there are no myosin heads, and this is transected by the M line (curved arrow) composed of three or five lines, depending on the fibre type. Many large proteins of the myofibrils are not visible with conventional TEM. Mitochondrion (M); triad (arrow); and glycogen (*). Magnification bar: 1 µm.

aligned in parallel but interspersed with varying amounts of cytoplasm (sarcoplasm), make up the majority of a muscle fibre (Figure 3.2). It is the regular organisation of the filaments which provides the distinctive striated architecture of the sarcomere.

Specialised areas of the sarcolemma may be encountered. At the neuromuscular junction, between a terminal axon of an intramuscular nerve and a muscle fibre, the sarcolemma is highly convoluted, forming junctional folds and secondary synaptic clefts. Junctional folds contain numerous organelles including specialised nuclei, small vesicles, tubules, filaments and ribosomes. At a myotendinous junction, the interface between a muscle cell and tendon, the sarcolemma is also invaginated (see Section 3.1.2).

The length of a sarcomere is defined as the distance between two Z lines and is dependent on the state of contraction of the muscle fibre. The length of the A (*anisotropic*) band is constant at 1.5–1.6 µm, but the I (*isotropic*) band length is determined by the degree of the actin filaments of the I band sliding between the myosin of the A band. The thickness of the Z line varies with fibre type (type 2A fibres are thickest, type 2B are narrowest and type I have intermediate thickness), but the differences are slight and not a recommended way of differentiating between types at the ultrastructural level, particularly in human muscle. The number of M lines (3 or 5) transecting the centre of the A band also varies with fibre type. The intermediate filament desmin with plectin link myofibrils to each other and to the sarcolemma through interaction with the Z line and areas of specialisation on the sarcolemma known as costameres (Capetanaki *et al.*, 2007). Other major proteins include titin and nebulin. Titin, the largest protein known, extends from the M line to the Z line.

The sarcoplasm contains a number of organelles including lipid droplets, mitochondria, T-tubules and SR plus various intermediate filaments and micro tubules.

In a normal muscle fibre, the nuclei of striated muscle cells (myonuclei) occur beneath the sarcolemma, although they may occasionally occur internally. They are aligned parallel to the main axis of the myofibrils (Figure 3.5b). The nuclear envelope is often smooth, but may become convoluted in contracted fibres. Nuclear chromatin has a granular appearance, and the relative proportions of the darkly stained, often peripheral, heterochromatin and the paler euchromatin vary with nuclear activity. Euchromatin is metabolically active, and large pale nuclei can be seen in regenerating fibres. There are usually one or two nucleoli.

Satellite cells lie beneath the external lamina of the muscle fibre and are separated from the fibre by their own plasma membrane and a slender gap of 50 nm or less. They are dormant cells that become activated during regeneration and give rise to the formation of new fibres. They are usually small cells with an oval, heterochromatic nucleus, sparse cytoplasm with few mitochondria, a few endoplasmic reticulum (ER) profiles, ribosomes, occasional glycogen particles and a Golgi apparatus. When activated, the proportions of organelles change to reflect increased metabolism.

The amount of glycogen also varies with fibre type, being greater in type 2 fibres than in type 1, but differences cannot be used to determine fibre type at the EM level. Glycogen content varies with diet, is higher in individuals who have been tube fed and is also dependent on the amount of exertion that has been placed on the muscle (Schrauwen-Hinderling *et al.*, 2006a).

Lipid content also varies with fibre type, and is greater in type I fibres, correlating with the higher number of mitochondria. Droplets are on average 0.5 μm in diameter and are not membrane bound (Figure 3.8b). They may look slightly grey or glassy or appear as vacant areas where the lipid has been lost during processing. They are usually sparse but may be more numerous in obese or tube-fed patients, or in physically trained individuals (Schrauwen-Hinderling *et al.*, 2006a, b).

Mitochondria occur in intermyofibrillar spaces (Figure 3.2) and also in small groups beneath the sarcolemma, often in the hof region close to myonuclei. In our experience, mitochondria are frequently larger in gastrocnemius muscle than in the deltoids or quadriceps. Mitochondria may also be larger or more numerous in individuals who are physically trained or endurance athletes (Hoppeler and Fluck, 2003).

The SR, the specialised ER of muscle and the transverse tubular system (T-system) function in the excitation–contraction coupling process and are usually more abundant in type 2 fibres. Two lateral sacs or terminal cisternae of the SR are visible at the junction of the A and I bands, where they come into close contact with a tubule of the T-system and form a triad (Figures 3.2 and 3.3b). At high magnification, dense junctional foot processes, the ryanodine receptors, can be seen bridging the gap between the lateral sacs and T-system.

3.3 PATHOLOGICAL CHANGES

As many of the ultrastructural changes seen in muscle are nonspecific and not related to particular disorders, we discuss here the changes related to each organelle and refer to conditions where such changes may be encountered.

3.3.1 Sarcolemma

The sarcolemma comprises the plasmalemma (the myocyte's limiting membrane) and an overlying external lamina. The external lamina of an atrophic fibre may be thrown into redundant folds which can extend into the endomysium, the fine layer of connective tissue surrounding each muscle fibre. This can be a useful way to distinguish between atrophic and hypotrophic fibres which have never reached their correct diameter and have a close fitting external lamina (Gazzerro *et al.*, 2010). Hypotrophic fibres can be seen in congenital myopathies such as centronuclear myopathies, nemaline rod myopathy and multicore disease.

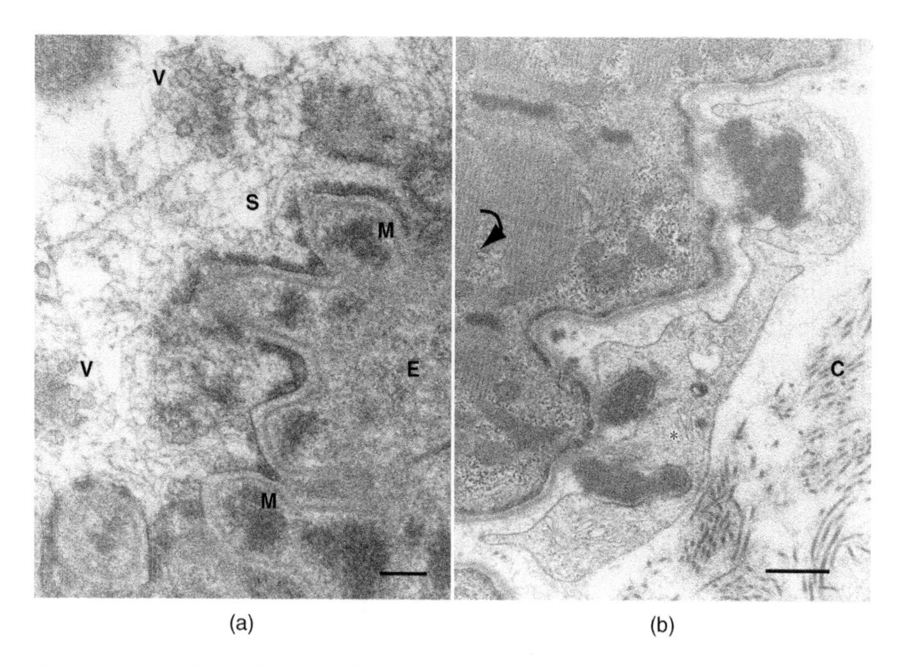

Figure 3.3 (a) In dysferlinopathies, clumps of moderately electron-dense material (M) near the endomysial surface of some myocytes may be extruded from fibres as a result of the failure of the membrane repair process. Clusters of vesicles (V) occur near defective membrane. Sarcoplasm (S); and endomysium (E). (b) A phagocyte (*) can be observed engulfing this debris (M). A triad is visible at the junction of the A and I bands (curved arrow). Collagen fibrils (C). Magnification bars: (a) 200 nm and (b) 500 nm.

Multilayering of the myofibre external lamina and collections of small vesicles beneath regions of the plasmalemma occur in dysferlinopathies where there are defects in the membrane repair system (Selcen, Stilling and Engel, 2001) (Figure 3.3).

Light-chain deposits may very occasionally be seen as a continuous, moderately electron-dense layer of variable thickness, finely flecked with more darkly staining material, over the basal lamina around vessels and occasionally along the surface of myocytes in some immune-mediated conditions (in contrast to the larger, multiple discrete clumps in dysferlinopathies).

Markedly thickened capillary basal lamina and, to a lesser degree, myofibre external lamina can be seen in biopsies from diabetic patients and patients with rheumatoid arthritis. Thickening can increase with age, and multilayering may occur where there is increased turnover of endothelial cells.

Focal loss or breaks in the plasmalemma can be seen in conditions characterised by necrosis such as muscular dystrophies, particularly Duchenne and Becker and dysferlinopathies.

Caveolae (Figure 3.4) have an important role in maintaining normal function of the myocyte through regulation of signalling pathways. Caveolin-3 is a major constituent of caveolae, and reduced protein expression is associated with several phenotypes affecting skeletal muscle, including limb girdle muscular dystrophy (LGMD) 1C, rippling muscle disease, idiopathic hyperCKaemia and autoimmunity to caveolin-3 (Gazzerro *et al.*, 2010). In addition, a reduction of caveolin-3 and/or the presence of caveolae can sometimes be seen as secondary consequence of defects in other genes such as dysferlin and cavin 1 (Matsuda *et al.*, 2001; Hayashi *et al.*, 2009).

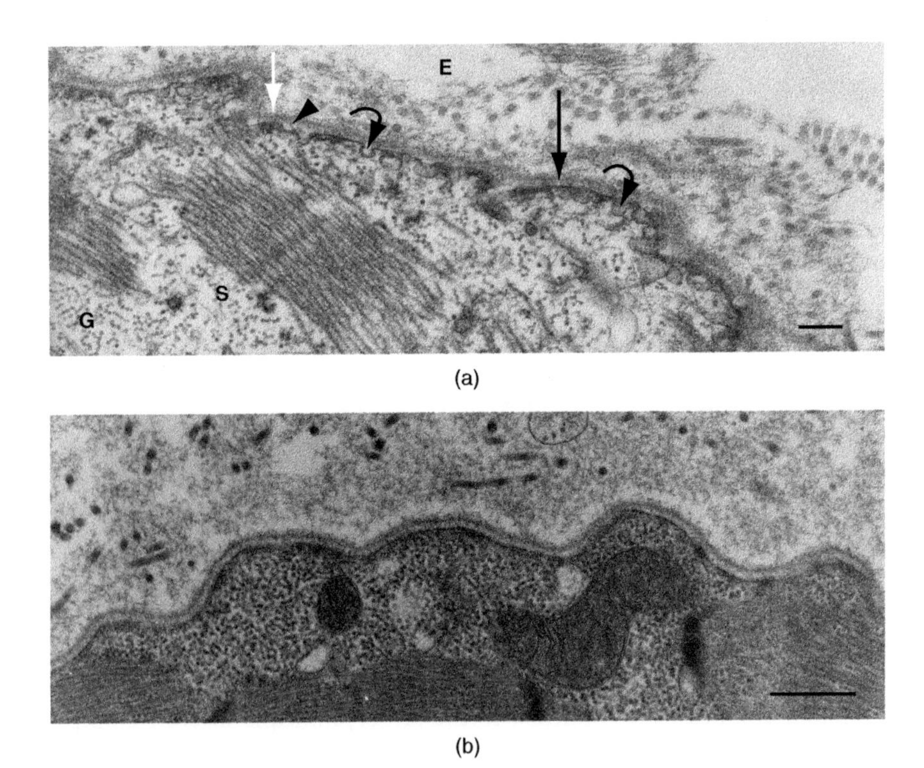

Figure 3.4 (a) The plasmalemma (arrow) has numerous pits and small vesicles termed caveolae (curved arrows). Lamina densa (open arrow); lamina rara (arrowhead); endomysium (E); sarcoplasm (S); and glycogen (G). (b) An absence or reduction in caveolae in a case with a mutation in the cavin–PTRF gene. Magnification bars: (a) 200 nm and (b) 500 nm.

3.3.2 Myofibrils

The appearance of myofibrils can be altered either pathologically or artefactually (see Section 3.1.4). Irregular Z lines and Z line streaming (Figure 3.8a) are common and can occur in normal muscle fibres, particularly adjacent to capillaries (see also Sections 3.3.4 and 3.3.5). Myofibrillar disarray is a frequent, nonspecific finding but may be particularly prominent in dermatomyositis. Thickening of the Z line may be widespread, particularly where myofibrillar structure is disorganised in degenerating or regenerating fibres; when grossly enlarged, they may form rod-like structures (nemaline rods) that may be in register within the striations, show continuity with the Z line or form clusters (Figure 3.5).

Small clusters of nemaline rods, similar to collections of thickened Z lines, are a nonspecific finding, but in nemaline myopathies they are usually more numerous. Rods originate from Z lines, have a similar lattice structure to Z lines and contain similar proteins such as α-actinin and

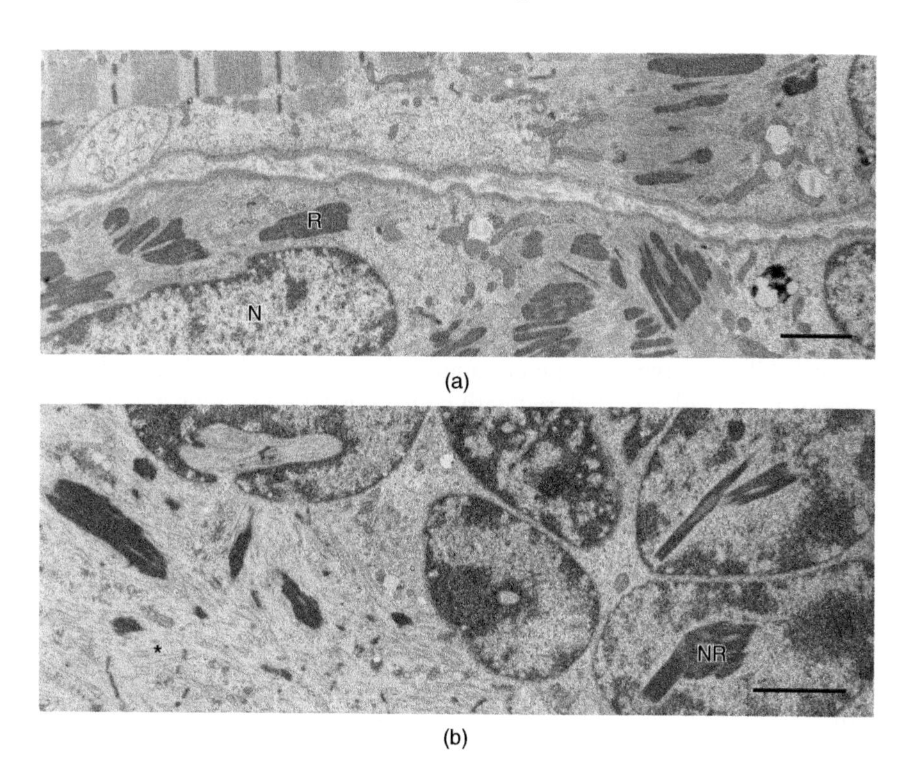

Figure 3.5 (a) Nemaline rods (R) in the sarcoplasm adjacent to a myonucleus (N). (b) Nuclear nemaline rods (NRs); area of myofibrillar atrophy and disorganisation (*). Magnification bars: (a) 2 μm and (b) 2 μm.

actin. They may be several micrometres in length and appear as dark red staining rods or speckles with the Gomori trichrome stain by LM. The heterogeneous forms of nemaline myopathy differ in clinical phenotype, age of onset, severity and inheritance. At least seven genes are implicated in the childhood forms: skeletal actin (*ACTA1*), nebulin (*NEB*), slow troponin T1 (*TNNT1*), α-tropomyosin (*TPM3*), β-tropomyosin (*TPM2*), cofilin (*CFL2*) and kelch repeat and BTB (POZ) domain containing 13 (*KBTBD13*). Mutations in *ACTA1* and *NEB* are the most common. Most adult cases with rods are unresolved at the molecular level, although some are of autoimmune origin with a phenotype different from that of childhood nemaline myopathy (Wallgren-Pettersson *et al.*, 2011).

Rods are usually cytoplasmic, but rare cases show cytoplasmic and/or intranuclear nemaline rods (Figure 3.5b). Nuclear rods and actin accumulation (discussed further in this section) are usually associated with *ACTA1* mutations, although there are some unresolved cases. Nuclear rods have also been observed in a few cases in association with other gene defects.

Thickened Z lines associated with very focal disruption of subsarcolemmal myofibrils, involving considerable loss of myosin filaments, are a characteristic feature in paediatric muscle of cap myopathy or 'cap disease'. These peripheral areas or 'caps' can be seen in a high proportion of fibres and are associated with mutations in the α-actin (*ACTA1*) and the β- and α-tropomyosin genes (*TPM2* and *TPM3*) (Kee and Hardeman, 2008; Clarke *et al.*, 2009; Waddell *et al.*, 2010).

A selective loss of myosin filaments may be seen in a small number of fibres in severe cases of dermatomyositis. However, it can be striking in critical illness myopathy (acute quadriplegic myopathy) (Figure 3.6a). Patients in critical care, sedated with neuromuscular blocking agents and given intravenous corticosteroids, may show a flaccid paresis that usually affects all limbs and facial muscles plus muscles of the diaphragm; these patients may become ventilator dependent.

Myosin accumulation in the form of hyaline bodies can occur in some cases with a mutation of the slow myosin heavy-chain gene *MYH7* (Goebel and Laing, 2009). Material is granular in appearance and can distort large areas of a fibre. By LM, these areas label with antibodies to slow myosin but are not stained for NADH–tetrazolium reductase (NADH-TR), in contrast to caps.

Large swathes of fine actin filaments are termed actin masses and are a feature of one of the actinopathies (or protein aggregate myopathies) resulting from a mutation in the *ACTA1* gene (Goebel *et al.*, 1997). Most cases are neonates who are hypotonic from birth, but they have also

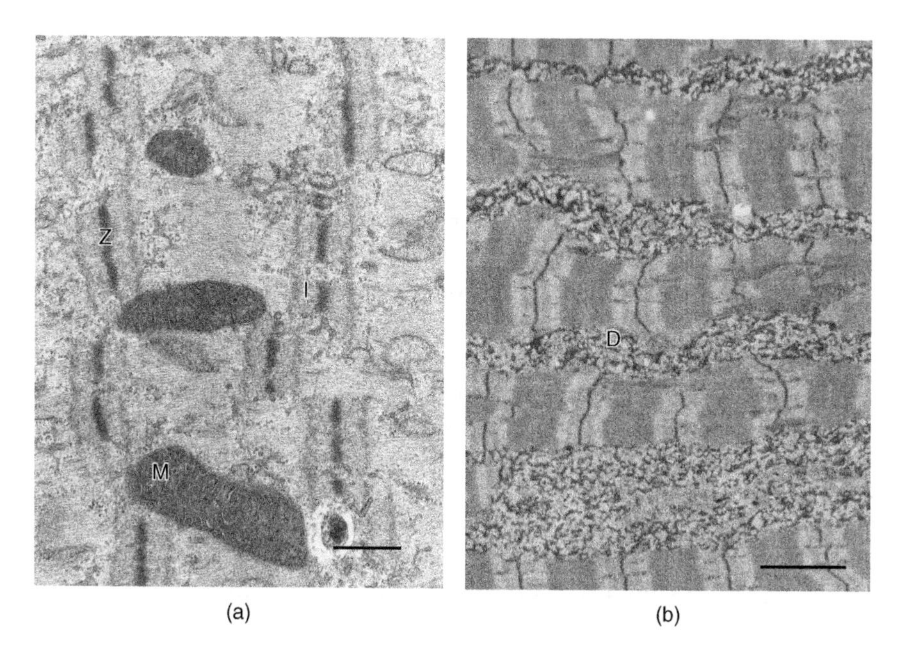

Figure 3.6 (a) Nearly all myofibrils show selective loss of myosin, and the remaining Z lines (Z) may be thickened and I bands (I) can be disorganised. Mitochondrion (M). (b) Large areas of a fibre containing abundant abnormal granulomatous desmin (D) which can be highlighted by immunolabelling. Magnification bars: (a) 500 nm and (b) 2 μm.

been seen in the case of zebra body myopathy, which is also caused by a mutation in *ACTA1* (Sewry *et al.*, 2009) and in a few molecularly unresolved cases. These cases also show structures resembling nemaline rods and can be considered as part of the *ACTA1* nemaline myopathy clinical and pathological spectrum, rather than as separate disease entities.

Damage to myofibrils through hypercontraction can be an artefact (see Section 3.3.2), but it may also be a pathological change seen in dystrophic muscle.

Myofibrillar structure is almost completely absent in necrotic fibres, depending on the stage of the process. An amorphous felt-like or finely granular material punctuated by lysosomes and degenerating mitochondria may be all that remains of a fibre, and if the degeneration is advanced, debris-laden phagocytes may be numerous (see also the discussion of vacuolar changes in Section 3.3.9).

A ring fibre is best seen in transverse section. It consists of a peripheral band of myofibrils which encircles the bulk of the fibre at 90° to its main axis. Ring fibres can occur in a number of conditions including many dystrophies and can be numerous in myotonic dystrophies, but their origin is not clear.

Small amounts of desmin may often be encountered where there is significant myofibrillar disruption, but it must be distinguished from fragmented Z lines. In myofibrillar myopathies (MFM), it can be one of the most prominently expressed ectopic proteins, identifiable as dense granulomatous material at the TEM level (Figure 3.6b).

The MFM are a group of disorders characterised by prominent morphological changes including vacuole formation, dissolution of myofibrillar structure and an accumulation of excess proteins, including desmin, αB-crystallin and myotilin (Selcen, 2011). There is pathological overlap between them and other disorders such as inclusion body myositis (IBM). It is rarely possible to determine the defective myofibrillar gene from TEM. but subtle differences have been reported (Claeys *et al.*, 2009). Several genes are involved in this group of disorders, including those encoding desmin, αB-crystallin, Z-band alternatively spliced PDZ-motif protein (ZASP), filamin C, BAG3 and myotilin; FHL1 is also often included (Schroder and Schoser, 2009).

3.3.3 Glycogen

Defects in enzymes of the glycolytic pathway give rise to a number of glycogen storage disorders (GSDs) which have childhood- and adult-onset forms. Several are associated with characteristic ultrastructural changes which include a lysosomal form (Pompe disease or GSD type II – see Section 3.3.9) (Figure 3.7a,b) and several nonlysosomal forms, for example McArdle disease (GSD type V). In McArdle disease (myophosphorylase deficiency), excessive amounts of normal glycogen particles accumulate in the sarcoplasm and subsarcolemmal areas; glycogen 'lakes' may be observed and be visible by LM. Whilst ultrastructural changes can be striking, diagnosis must be confirmed histochemically and at the molecular level. Affected individuals suffer cramps on exercise and sometimes rhabdomyolysis. More than 65 mutations have been described in the myophosphorylase gene *PYGM*, but R50X is the most common in the Caucasian population (Quinlivan *et al.*, 2010).

Brancher enzyme deficiency (GSD type IV) affects the liver, and severe neuromuscular forms have been described (Nolte *et al.*, 2008). Abnormal glycogen may accumulate as periodic acid–Schiff (PAS) staining positive, diastase-resistant polyglucosan material that has a filamentous appearance (Figure 3.7c). Paracrystalline structures may be embedded within the accumulated material (Figure 3.7d).

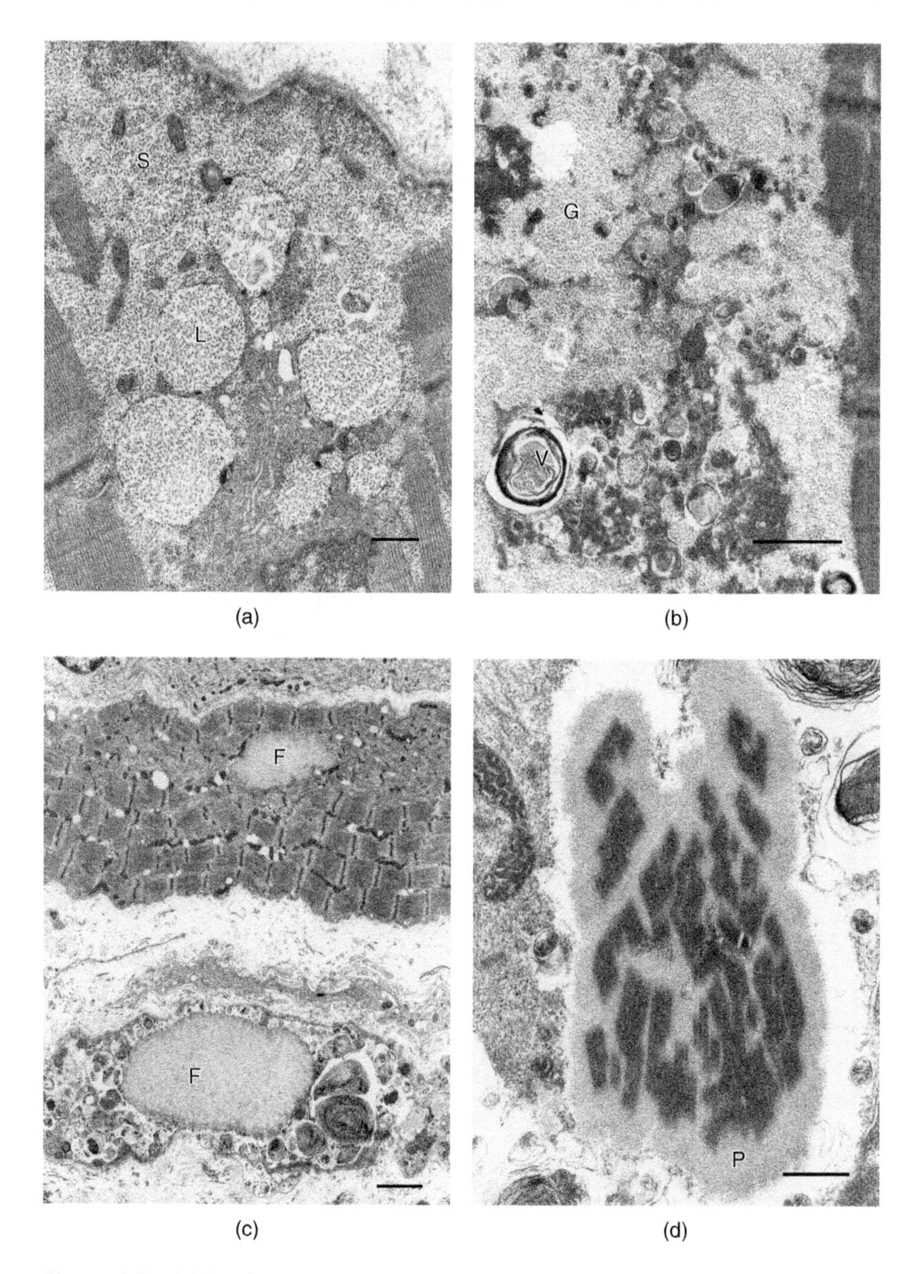

Figure 3.7 (a) In glycogen storage diseases, glycogen particles occur free in the sarcoplasm (S) and also in lysosomes (L) in cases of acid maltase deficiency. (b) Lysosomes aggregate in autophagic vacuoles (V), and abundant glycogen is present in the sarcoplasm (G). (c) Abnormally structured glycogen appears as fine filamentous material (F) in atrophic muscle fibres in the polyglucosan disorder branching enzyme deficiency (GSD type IV). (d) Darkly staining paracrystalline structures can occur within polyglucosan deposits (P). Magnification bars: (a) 500 nm; (b) 1 μm; (c) 2 μm and (d) 500 nm.

3.3.4 Cores

Cores are areas of variable size that lack mitochondria, glycogen and sarcotubular cisternae, and show variable degrees of myofibrillar disruption (Figure 3.8). TEM is particularly useful in distinguishing small core areas, which may be only a few micrometres in diameter, from myofibrillar loss. In some cores, the area with loss of mitochondria may be greater than that showing myofibrillar disruption, or there may be only misalignment of myofibrils without sarcomeric disruption. Structured cores are defined as areas in which the striation pattern is maintained, whereas in unstructured cores the myofibrillar organisation is severely disrupted and ATPase staining is lost at the LM level. Cores that extend an appreciable length down a fibre are characteristic of central core disease caused by mutations in the *RYR1* gene. Multiple cores within a fibre can occur in association with several gene defects and phenotypes, including *ACTA1,MYH7,CFL,Col 6,RYR1* and *SEPN1*, although the latter has given its name to 'multi minicore myopathy'. There is considerable pathological overlap in disorders characterised by cores, which has led to the term core myopathies rather than the historic description

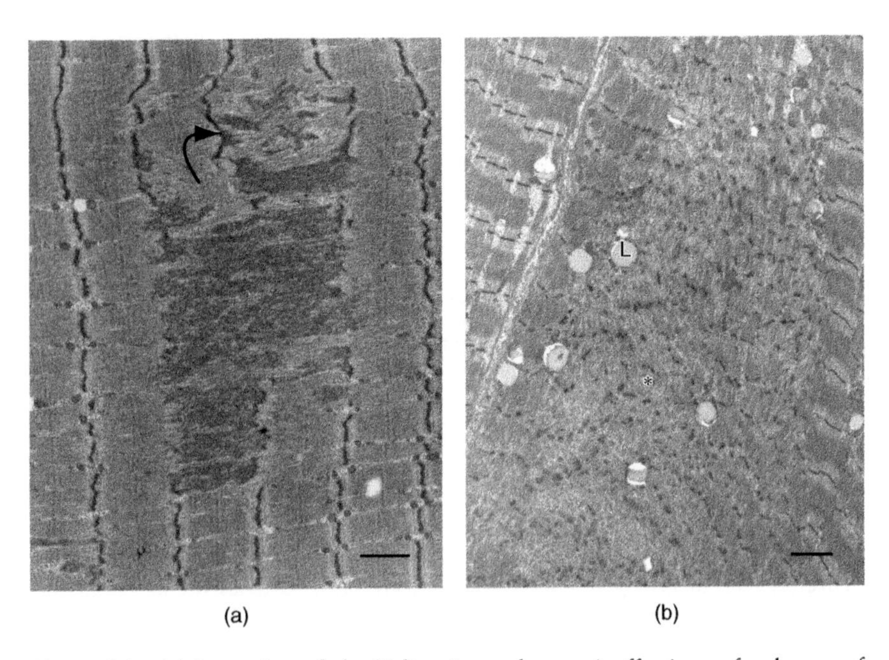

Figure 3.8 (a) Streaming of the Z line (curved arrow) affecting a focal area of sarcomeres (minicore). (b) Marked disorganisation of myofibrils in a large core (*). Lipid droplets (L). Magnification bars: (a) 1 μm and (b) 2 μm.

of central or minicore disease (Sewry, 2008; Jungbluth, Sewry and Muntoni, 2011). Histopathological diagnosis is complicated in these disorders, as pathology can change over time and a biopsy taken early in a condition may differ from one taken later in a disease process.

3.3.5 Target Fibres

Target fibres have similar elements of myofibrillar disruption to cores and minicores, but they get their name from the arrangement of the disrupted zones which can be seen with oxidative enzyme stains by LM. There are usually three zones, but if only two can be identified, the structure is termed targetoid; however, this difference probably has no clinical significance. Targets usually extend only a short distance along the length of a fibre and are thought to result from denervation followed by reinnervation.

3.3.6 Myonuclei

In Marinesco–Sjogren syndrome (MSS), myonuclei in some cases demonstrate an electron-dense rim around the nuclear envelope (Sewry, Voit and Dubowitz, 1988). This feature is unique to MSS and can be observed only with TEM. Patients with the autosomal recessive multisystem disorder demonstrate cerebellar ataxia, early bilateral cataracts and variable mental retardation. Recent studies have located the defect to the *SIL1* gene, but not all affected individuals have this defect (Senderek *et al.*, 2005; Takahata *et al.*, 2010).

Apoptosis or 'programmed cell death' can be identified in certain myonuclei and is seen as dense, clumped peripheral heterochromatin that is not attached to the inner nuclear membrane.

Various intranuclear filamentous inclusions may occur in cases of inclusion body myopathy (see Section 3.3.9) (Figure 3.9a).

There are references to fine intranuclear filaments being observed in dermatomyositis (Figure 3.9b), but although they have been regarded as nonspecific they may relate to autoimmune myositic conditions. In our experience, they are most often seen where there is a strong clinical suspicion of inflammatory myopathy. These intranuclear filaments should prompt careful examination of vascular endothelial cells for tubuloreticular inclusions (see Section 3.3.10).

Collections of fine intranuclear filaments (~8 nm in diameter) forming characteristic tangles and palisades confined to myonuclei are

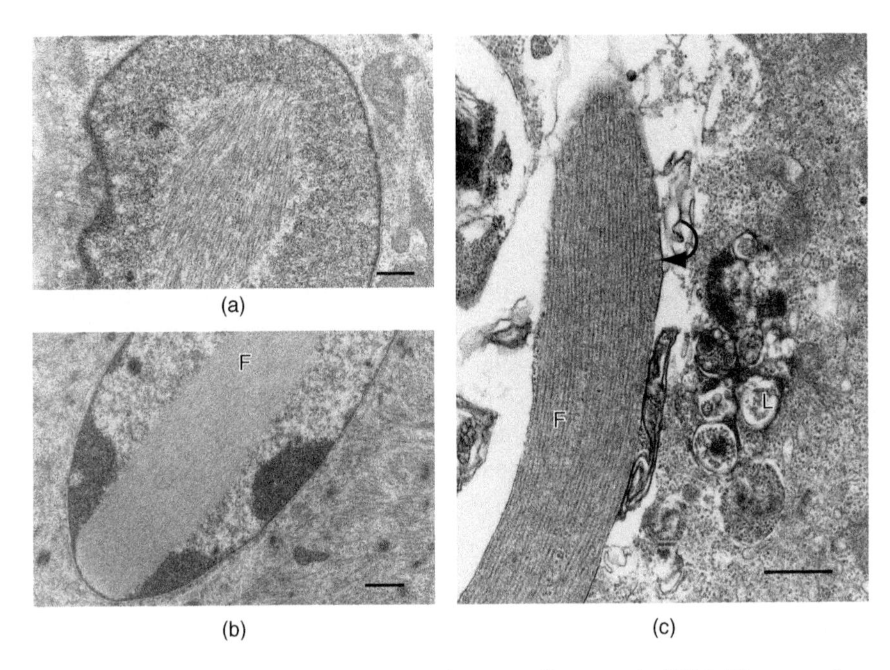

Figure 3.9 (a) Intranuclear 18–22 nm diameter filaments in IBM. These are less common than those found in the sarcoplasm. (b) Compact bundles of fine pale staining intranuclear 8–10 nm filaments (F) in a case of myositis. (c) A bundle of sarcoplasmic filaments (F) (18–22 nm) adjacent to membranous whorls and lysosomes (L) is surrounded by a clear space and partially by a possible limiting membrane (curved arrows) (IBM). Magnification bars: (a) 500 nm; (b) 500 nm and (c) 500 nm.

pathognomonic for oculopharyngeal muscular dystrophy (OPMD) (Minami *et al.*, 2001). OPMD is often diagnosed genetically, but finding these filaments is one of the rare instances where an ultrastructural finding in skeletal muscle is diagnostic. These collections of filaments can be seen on sections stained with toluidine blue as nuclei with unusually pale centres. The much rarer condition oculopharyngodistal myopathy, characterised clinically by cranial and distal limb muscle weakness, also demonstrates intranuclear 'tubulofilaments' as well as rimmed vacuoles. The reported diameter of nuclear filaments is more variable, and the morphology is unlike that seen in OPMD (Thevathasan *et al.*, 2011).

3.3.7 Mitochondria

Biochemical abnormalities in the mitochondrial respiratory chain enzymes may sometimes be reflected in abnormal mitochondrial morphology (Figure 3.10). Extensive subsarcolemmal and intermyofibrillar

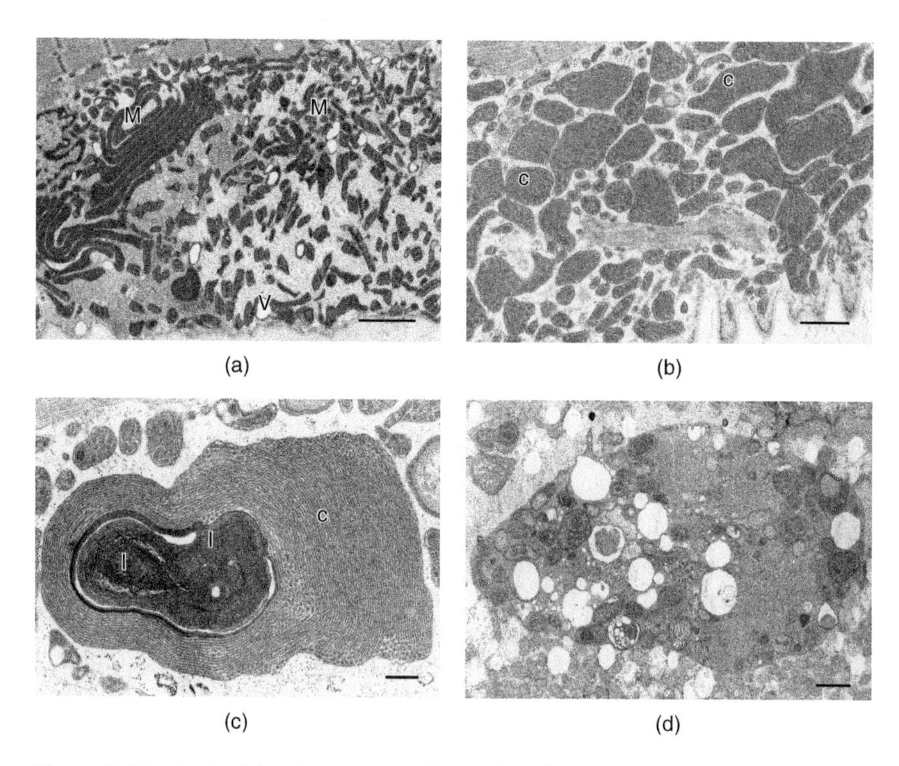

Figure 3.10 (a) Proliferating abnormal mitochondria (M) accumulate beneath the sarcolemma with many paracrystalline inclusions and several with vacuoles (V). (b) Subsarcolemmal mitochondria showing abundant compact cristae (C), some with bizarre orientation. (c) A single mitochondrion with numerous concentrically arranged cristae (C) and several inclusions (possibly lipid) (I) towards the centre. (d) Bizarre mitochondrion several micrometres in diameter, possibly formed from the fusion of multiple mitochondria. Magnification bars: (a) 2 μm; (b) 1 μm; (c) 500 nm and (d) 1 μm.

accumulations of mitochondria correspond to the 'red' in ragged red fibres seen on Gomori trichrome–stained sections at the LM level. However, in adult muscle the correlation between biochemical and morphological abnormality is generally too weak to permit reliance on the presence of morphological changes alone to indicate a mitochondrial myopathy.

Mitochondrial DNA deletions increase with age, particularly after the fifth decade of life (Sleigh, Ball and Hilton, 2011), and can lead to the presence of fibres devoid of cytochrome activity (cytochrome-c oxidase (COX)) and structurally abnormal mitochondria. COX negative fibres can occur in the inflammatory myopathies (Blume *et al.*, 1997), especially IBM (Oldfors *et al.*, 2006).

Collections of morphologically normal mitochondria occurring beneath the sarcolemma and extending into the body of the fibre, forming 'wedges' between myofibrils, are the characteristic pattern seen in NADH-TR-stained sections in lobulated fibres. Lobulated fibres can be seen in a variety of situations but can be a feature of LGMD 2A. They are rare in paediatric muscle, but fibres resembling lobulated fibres may occur in Ullrich congenital muscular dystrophy.

3.3.8 Reticular System

3.3.8.1 T-Tubules

Proliferation of the T-system can occur in regenerating fibres, and collections of T-tubules may form honeycomb structures or T-system networks. It is thought that these structures may contribute to membrane formation, for example around autophagic vacuoles (Engel and Franzini-Armstrong, 2004).

Membrane-bound vacuoles thought to be derived from the T-system may be seen in the absence of inflammation in some cases of periodic paralyses and may have a finely granular content. Tubular aggregates may also be present (see Section 3.3.8.2).

3.3.8.2 Tubular Aggregates

Cisternae of the SR usually appear empty or demonstrate an amorphous or finely granular content; swelling is a common nonspecific abnormality and has to be distinguished from artefacts (see Section 3.1.4).

Tubular aggregates are closely packed collections of 50–70 nm diameter parallel tubules which have been shown to be continuous with the SR and express proteins of the SR such as sarcoendoplasmic reticulum calcium ATPase (SERCA). The duplication and remodelling of SR may be extensive, with the resulting aggregates replacing a large proportion of the fibre profile. Where aggregates are visible with the Gomori trichrome, they stain a bright red or pink, differing slightly from the red of mitochondrial proliferation. The appearance of the aggregates in transverse section can vary, and several types have been identified, although they are not currently assigned any clinical significance (Dubowitz and Sewry, 2007).

Small tubular aggregates can be an incidental finding and often occur in subsarcolemmal areas. More extensive and numerous aggregates have been described in muscle biopsies from patients suffering frequent cramps, myalgia and, in some periodic paralyses. They are more common in type 2 fibres and in males, and occur in a number of animal models. Tubular aggregates are common in some familial myasthenic patients, particularly those with a mutation in the *GFPT1* gene (Senderek *et al.*, 2011). Numerous aggregates have been reported in rare unresolved tubular aggregate myopathies (Goebel and Bonnemann, 2011).

3.3.9 Vacuoles

Gross swelling of the T-tubule system distorts normal myofibrillar architecture in otherwise intact regions of fibres undergoing segmental necrosis. These spaces are usually empty except for small amounts of membranous debris and are evident in inflammatory conditions. (Where vacuoles are smaller, they may be confused with lipid droplets when examined at low magnification; see Section 3.1.4.)

An autophagic vacuole is a collection of partially degraded cytoplasmic constituents bound by a single membrane. These can be seen in any biopsy but are common in fibre segments adjacent to an area of necrosis and in some cases of myositis. A vacuole near the sarcolemma may fuse its limiting membrane with that of the plasmalemma and exocytose the debris into the endomysial space. Debris can sometimes be seen trapped between loops of external lamina.

Characteristic filaments approximately 18–22 nm in diameter occur in the sarcoplasm amongst dense membranous whorls and lysosomes in 'rimmed vacuoles' (Askanas and Engel, 2011; Dalakas, 2011) (see also Section 3.3.6). Randomly orientated filaments may form large masses, or they may occur in partially membrane bound bundles often encircled by halos of clear space (Figure 3.9c). The collections of small dense bodies, vesicles, debris and membranous whorls are not membrane bound as is the case for the autophagic vacuole.

Rimmed vacuoles with their thick filaments are characteristic of IBM, but they can also occur in a wide variety of disorders including acid maltase deficiency, distal myopathies, MFM, several muscular dystrophies and occasionally neurogenic disorders (Kojima *et al.*, 2009; Udd, 2011).

Autophagic vacuoles with sarcolemmal features (AVSFs) can be located within the fibre or beneath the sarcolemma and the basal lamina. The limiting membranes may display immunostaining for most sarcolemmal proteins, including laminins, dystrophin, caveolin and sarcoglycans, and vacuoles may contain myeloid bodies, cytoplasmic debris and calcium. Multiple layers of basal lamina are also present. Acid phosphatase staining is variable.

The autophagic vacuolar myopathies are a phenotypically diverse group of conditions characterised by AVSF (Nishino, 2006). They include Danon disease, X-linked myopathy with excessive autophagia (XMEA), infantile and childhood autophagic vacuolar myopathies and adult autophagic vacuolar myopathy with multi-organ involvement. Many other myopathies, especially MFMs and various forms of LGMD, also show them. Danon disease (primary lysosomal-associated membrane protein 2 (LAMP-2) deficiency), originally described as a lysosomal glycogen storage disease with normal acid maltase, is an X-linked condition characterised by varying degrees of mental retardation, hypertrophic cardiomyopathy and skeletal myopathy (Malicdan *et al.*, 2008). An abnormal accumulation of lysosomes in autophagic vacuoles is also a feature of XMEA which is restricted to skeletal muscle and caused by mutations in the VMA21 gene (Ramachandran *et al.*, 2009).

In the lysosomal storage disorder Pompe disease (lysosomal α-1,4 glucosidase or acid maltase deficiency), the distribution of excessive glycogen particles in skeletal muscle is characteristic for this condition, and a vacuolar myopathy may be observed at the LM level (Malicdan *et al.*, 2008). These are lysosomes which may, depending on the severity of the disease in the biopsied muscle, be the only clue as to the presence of the condition (Figure 3.7a). Where the disease is more severe, the lysosomes merge to form larger, membrane-bound autophagic vacuoles containing collections of morphologically diverse degenerate material, some of which may be exocytosed (Figure 3.7b).

3.3.10 Capillaries

Occasional capillary endothelial cells may show evidence of clear swelling where the cytoplasm has a 'washed-out' appearance and few organelles remain. This change can occur in many situations and is common, but it is not diagnostic of inflammatory conditions such as dermatomyositis. Endothelial cell shrinkage, which is identified by a marked increase in electron density of the cytoplasm, making organelle identification difficult, may also occur nonspecifically, and these cells can alternate with normal or swollen endothelial cells in the same vessel. In some inflammatory conditions, endothelial cell hypertrophy may be marked. Unlike the swollen cells, hypertrophied cells have a normal granular cytoplasm and complement of organelles.

Empty loops of basal lamina material may occasionally be seen in the endomysium. The thickness of the material may identify it as a remnant of a degenerate capillary rather than a redundant loop of myofibre external lamina. Small numbers of these loops may be insignificant, but they have been described in long-term, partially denervated muscle (Engel and Franzini-Armstrong, 2004). Tubuloreticular inclusions in capillary endothelial cells (see Figure 1.18b) are an early characteristic feature of dermatomyositis that may be found in the absence of other pathology (De Visser, Emslie-Smith and Engel, 1989). There is currently interest in the role of interferon and the relationship with tubuloreticular inclusions in the pathophysiology of dermatomyositis (Greenberg, 2010; Kao, Chung and Fiorentino, 2011). However, where there is little clinical suspicion, tubuloreticular inclusions alone are not specific and may occur in capillaries in other disorders including Sjogren's disease, scleroderma and the myositis of systemic lupus erythematosus (SLE).

3.3.11 Other Structural Defects

A number of additional ultrastructural defects have been reported, and some are illustrated in Figure 3.11a–f. These include cylindrical spirals,

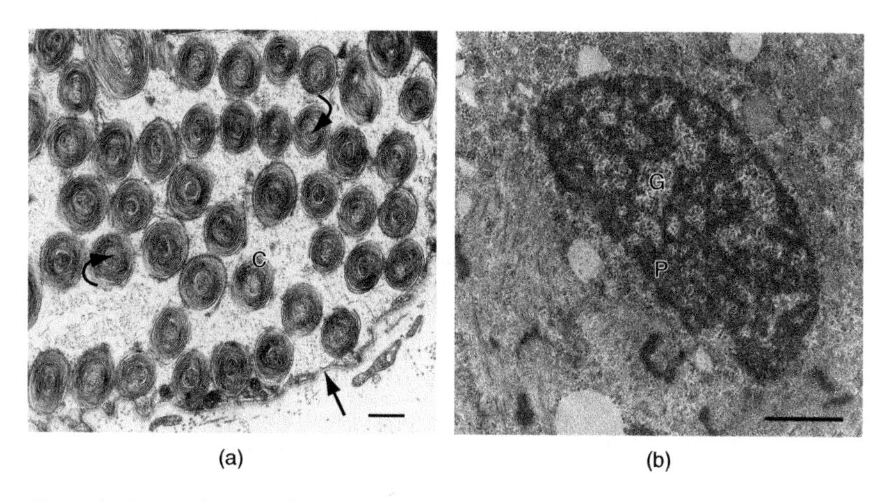

Figure 3.11 (a–f) Examples of other structural defects that may be observed; (a–c) rarely and (d–f) more often. (a) A collection of cylindrical spirals (C) beneath the sarcolemma (arrow) with glycogen in the centre of each spiral (curved arrows). (b) Reducing bodies which stain with menadione–nitroblue tetrazolium and appear as electron-dense cytoplasmic aggregates of material are caused by mutations in the *FHL1* gene; glycogen (G).

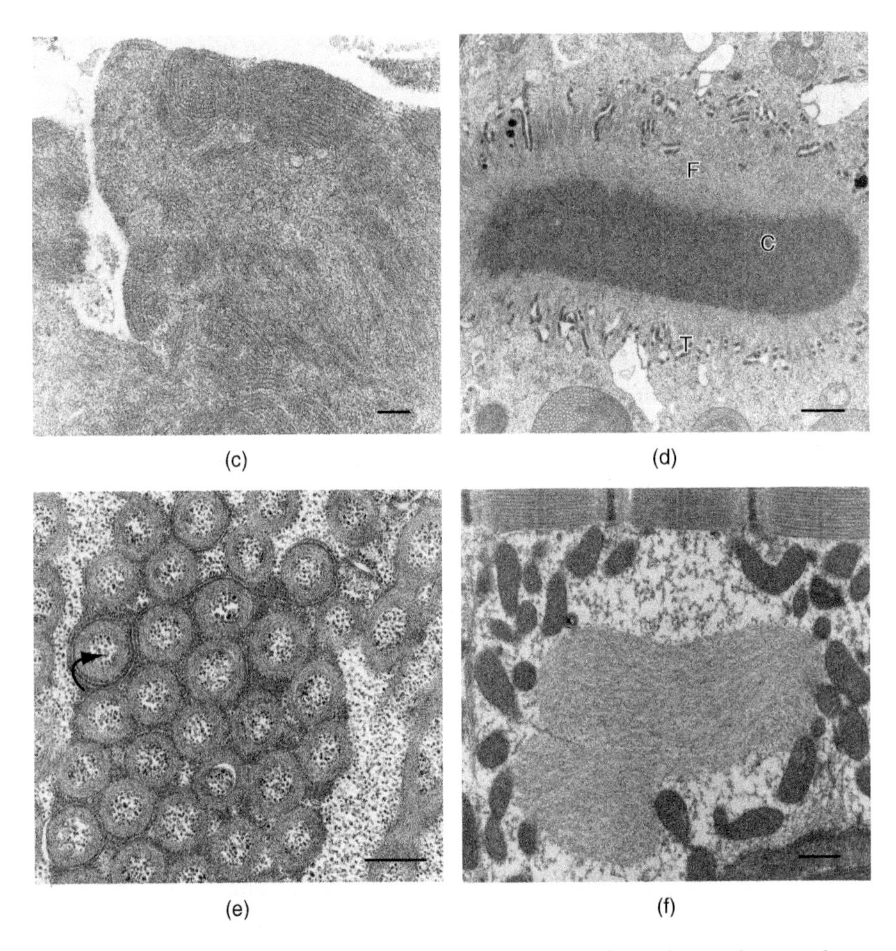

Figure 3.11 (c) Fingerprint body. (d) Cytoplasmic body with an electron-dense centre (C) and desmin filaments (F) radiating from it. Note also the ring of triads (T) around this example. (e) Concentric laminated bodies; note also the presence of glycogen particles at the centre of each body (curved arrow). (f) Filamentous body with swirls of actin-like filaments, often in subsarcolemmal regions. Magnification bars: (a) 500 nm, (b) 1 μm, (c) 200 nm, (d) 1 μm, (e) 500 nm and (f) 500 nm.

reducing bodies, fingerprint bodies, cytoplasmic bodies, concentric laminated bodies and filamentous bodies. The molecular causes for several are known (e.g. reducing bodies), and some are common in various disorders (e.g. cytoplasmic bodies which occur in IBM, MFMs and reducing body myopathy) and concentric laminated bodies, but it is not clear if some relate to a disease entity or if they are of genetic origin (Goebel and Bonnemann, 2011).

REFERENCES

Askanas, V. and Engel, W.K. (2011) Sporadic inclusion-body myositis: conformational multifactorial ageing-related degenerative muscle disease associated with proteasomal and lysosomal inhibition, endoplasmic reticulum stress, and accumulation of amyloid-beta42 oligomers and phosphorylated tau. *Presse Medicale*, **40**, 219–235.

Blume, G., Pestronk, A., Frank, B. and Johns, D.R. (1997) Polymyositis with cytochrome oxidase negative muscle fibres: early quadriceps weakness and poor response to immunosuppressive therapy. *Brain*, **120** (Pt 1), 39–45.

Capetanaki, Y., Bloch, R.J., Kouloumenta, A. *et al.* (2007) Muscle intermediate filaments and their links to membranes and membranous organelles. *Experimental Cell Research*, **313**, 2063–2076.

Claeys, K.G., Van Der Ven, P.F., Behin, A. *et al.* (2009) Differential involvement of sarcomeric proteins in myofibrillar myopathies: a morphological and immunohistochemical study. *Acta Neuropathologica*, **117**, 293–307.

Clarke, N.F., Domazetovska, A., Waddell, L. *et al.* (2009) Cap disease due to mutation of the beta-tropomyosin gene (TPM2). *Neuromuscular Disorders*, **19**, 348–351.

Dalakas, M.C. (2011) Review: an update on inflammatory and autoimmune myopathies. *Neuropathology and Applied Neurobiology*, **37**, 226–242.

De Visser, M., Emslie-Smith, A.M. and Engel, A.G. (1989) Early ultrastructural alterations in adult dermatomyositis: capillary abnormalities precede other structural changes in muscle. *Journal of the Neurological Sciences*, **94**, 181–192.

Dubowitz, V. and Sewry, C.A. (2007) *Muscle Biopsy: A Practical Approach*, 3rd edn, Elsevier, Edinburgh.

Engel, A.G. and Franzini-Armstrong, C. (eds) (2004) *Myology*, McGraw-Hill, New York.

Gazzerro, E., Sotgia, F., Bruno, C. *et al.* (2010) Caveolinopathies: from the biology of caveolin-3 to human diseases. *European Journal of Human Genetics*, **18**, 137–145.

Goebel, H.H., Anderson, J.R., Hubner, C. *et al.* (1997) Congenital myopathy with excess of thin myofilaments. *Neuromuscular Disorders*, **7**, 160–168.

Goebel, H.H. and Bonnemann, C.G. (2011) 169th ENMC International Workshop Rare Structural Congenital Myopathies, 6–8 November 2009, Naarden, the Netherlands. *Neuromuscular Disorders*, **21**, 363–374.

Goebel, H.H. and Laing, N.G. (2009) Actinopathies and myosinopathies. *Brain Pathology*, **19**, 516–522.

Greenberg, S.A. (2010) Dermatomyositis and type 1 interferons. *Current Rheumatology Reports*, **12**, 198–203.

Hayashi, Y.K., Matsuda, C., Ogawa, M. *et al.* (2009) Human PTRF mutations cause secondary deficiency of caveolins resulting in muscular dystrophy with generalized lipodystrophy. *Journal of Clinical Investigation*, **119**, 2623–2633.

Hoppeler, H. and Fluck, M. (2003) Plasticity of skeletal muscle mitochondria: structure and function. *Medicine and Science in Sports and Exercise*, **35**, 95–104.

Jungbluth, H., Sewry, C.A. and Muntoni, F. (2011) Core myopathies. *Seminars in Paediatric Neurology*, **18**, 239–249.

Kao, L., Chung, L. and Fiorentino, D.F. (2011) Pathogenesis of dermatomyositis: role of cytokines and interferon. *Current Rheumatology Reports*, **13**, 225–232.

Kee, A.J. and Hardeman, E.C. (2008) Tropomyosins in skeletal muscle diseases. *Advances in Experimental Medicine and Biology*, **644**, 143–157.

Kojima, Y., Sakai, K., Ishida, C. *et al.* (2009) Hereditary rimmed vacuole myopathy showing interstitial amyloid deposition in muscle tissue. *Muscle & Nerve*, **40**, 472–475.

Malicdan, M.C., Noguchi, S., Nonaka, I. *et al.* (2008) Lysosomal myopathies: an excessive build-up in autophagosomes is too much to handle. *Neuromuscular Disorders*, **18**, 521–529.

Matsuda, C., Hayashi, Y.K., Ogawa, M. *et al.* (2001) The sarcolemmal proteins dysferlin and caveolin-3 interact in skeletal muscle. *Human Molecular Genetics*, **10**, 1761–1766.

Minami, N., Ikezoe, K., Kuroda, H. *et al.* (2001) Oculopharyngodistal myopathy is genetically heterogeneous and most cases are distinct from oculopharyngeal muscular dystrophy. *Neuromuscular Disorders*, **11**, 699–702.

Nishino, I. (2006) Autophagic vacuolar myopathy. *Seminars in Pediatric Neurology*, **13**, 90–95.

Nolte, K.W., Janecke, A.R., Vorgerd, M. *et al.* (2008) Congenital type IV glycogenosis: the spectrum of pleomorphic polyglucosan bodies in muscle, nerve, and spinal cord with two novel mutations in the GBE1 gene. *Acta Neuropathologica*, **116**, 491–506.

Oldfors, A., Moslemi, A.R., Jonasson, L. *et al.* (2006) Mitochondrial abnormalities in inclusion-body myositis. *Neurology*, **66**, S49–S55.

Quinlivan, R., Buckley, J., James, M. *et al.* (2010) McArdle disease: a clinical review. *Journal of Neurology, Neurosurgery, and Psychiatry*, **81**, 1182–1188.

Ramachandran, N., Munteanu, I., Wang, P. *et al.* (2009) VMA21 deficiency causes an autophagic myopathy by compromising V-ATPase activity and lysosomal acidification. *Cell*, **137**, 235–246.

Schrauwen-Hinderling, V.B., Hesselink, M.K., Moonen-Kornips, E. *et al.* (2006a) Short-term training is accompanied by a down regulation of ACC2 mRNA in skeletal muscle. *International Journal of Sports Medicine*, **27**, 786–791.

Schrauwen-Hinderling, V.B., Hesselink, M.K., Schrauwen, P. and Kooi, M.E. (2006b) Intramyocellular lipid content in human skeletal muscle. *Obesity (Silver Spring)*, **14**, 357–367.

Schroder, R. and Schoser, B. (2009) Myofibrillar myopathies: a clinical and myopathological guide. *Brain Pathology*, **19**, 483–492.

Selcen, D. (2011) Myofibrillar myopathies. *Neuromuscular Disorders*, **21**, 161–171.

Selcen, D., Stilling, G. and Engel, A.G. (2001) The earliest pathologic alterations in dysferlinopathy. *Neurology*, **56**, 1472–1481.

Senderek, J., Krieger, M., Stendel, C. *et al.* (2005) Mutations in SIL1 cause Marinesco-Sjogren syndrome, a cerebellar ataxia with cataract and myopathy. *Nature Genetics*, **37**, 1312–1314.

Senderek, J., Muller, J.S., Dusl, M. *et al.* (2011) Hexosamine biosynthetic pathway mutations cause neuromuscular transmission defect. *American Journal of Human Genetics*, **88**, 162–172.

Sewry, C.A. (2008) Pathological defects in congenital myopathies. *Journal of Muscle Research and Cell Motility*, **29**, 231–238.
Sewry, C., Holton, J., Dick, D.J. *et al.* (2009) Zebra body myopathy resolved. *Neuromuscular Disorders*, **19**, 637–628.
Sewry, C.A., Voit, T. and Dubowitz, V. (1988) Myopathy with unique ultrastructural feature in Marinesco–Sjogren syndrome. *Annals of Neurology*, **24**, 576–580.
Sleigh, K., Ball, S. and Hilton, D.A. (2011) Quantification of changes in muscle from individuals with and without mitochondrial disease. *Muscle & Nerve*, **43**, 795–800.
Takahata, T., Yamada, K., Yamada, Y. *et al.* (2010) Novel mutations in the SIL1 gene in a Japanese pedigree with the Marinesco–Sjogren syndrome. *Journal of Human Genetics*, **55**, 142–146.
Thevathasan, W., Squier, W., Maciver, D.H. *et al.* (2011) Oculopharyngodistal myopathy – a possible association with cardiomyopathy. *Neuromuscular Disorders*, **21**, 121–125.
Udd, B. (2011) Distal muscular dystrophies. *Handbook of Clinical Neurology*, **101**, 239–262.
Waddell, L.B., Kreissl, M., Kornberg, A. *et al.* (2010) Evidence for a dominant negative disease mechanism in cap myopathy due to TPM3. *Neuromuscular Disorders*, **20**, 464–466.
Wallgren-Pettersson, C., Sewry, C.A., Nowak, K. and Laing, N.G. (2011) Nemaline myopathies. *Seminars in Pediatric Neurology*, **18**, 230–238.

4

The Diagnostic Electron Microscopy of Nerve

Rosalind King
Institute of Neurology, University College London, London, United Kingdom

4.1 INTRODUCTION

Peripheral nerves consist of several fascicles containing axons, Schwann cells (SCs) and collagen. Each fascicle is defined by a continuous layer of flattened cells that form the perineurium (Shanthaveerappa and Bourne, 1962) creating a barrier separating the endoneurium from the surrounding tissue, the epineurium, which is composed mainly of fibrous collagen and fat. Blood vessels and small lymph vessels run longitudinally in the epineurium.

Axons are electrically active extensions of neurons whose cell bodies lie in the central nervous system (CNS) (motor neurons) or dorsal root ganglia (sensory neurons). They transmit impulses from the cell body to peripheral end organs by means of sodium and potassium channels in the axolemma. SCs are specialised axonal support cells that produce a myelin sheath around larger axons. The myelin sheath is interrupted by nodes of Ranvier, where adjacent SCs meet and the nervous impulse jumps from one node to the next. Smaller axons do not have a one-to-one relationship with their supporting SC and do not possess a myelin sheath, and conduction is much slower.

Diagnostic Electron Microscopy: A Practical Guide to Interpretation and Technique, First Edition. Edited by John W. Stirling, Alan Curry and Brian Eyden.

4.2 TISSUE PROCESSING

4.2.1 Preparation of Nerve Biopsy Specimens

Fresh specimens of nerve need extremely careful handling after removal from the patient and should be fixed or frozen as soon as possible. Their fragility mainly stems from the myelin component, as this is almost liquid and readily flows when pinched or squeezed, thereby rendering myelin and axons very susceptible to handling and fixation artefact; protocols used for most other tissues are not suitable for nerves. Chopping a nerve into 1 mm cubes causes severe damage.

On removal from the patient, the specimen should be laid out straight on a piece of saline-soaked gauze or dental wax and carefully cut with a sharp scalpel blade using a gentle sawing motion with as little downward pressure as possible. Pressing downward on a hard surface produces enlarged, distorted myelinated fibres in the side of the fascicle adjacent to the dish. The specimen is then placed carefully on a small piece of card and put into a pot of 2.5% buffered, best quality glutaraldehyde. 0.1 M Piperazine-*N*-*N*′-bis (2-ethane sulfonic acid) (PIPES) buffer (Sigma) (Baur and Stacey, 1977) produces the best results when fixing nerve specimens. Phosphate buffers are not so good, and inadequate washing prior to osmication may produce a fine electron-dense precipitate; sodium cacodylate is toxic and also unsuitable for prolonged fixation.

After at least 2 hours, the nerve can be cut carefully into smaller pieces; if teased fibre preparations are planned, a suitable length should be left for this. It is preferable to fix overnight. Nerves can be stored in a 2.5% glutaraldehyde–PIPES buffer for several days without deleterious effects.

After fixation, the nerve is washed before being post-fixed in buffered 1% osmium tetroxide. Two percent sucrose should be added to the wash buffer to match its osmolality to that of the fixative. It is best to use the same buffer for all the solutions.

The addition of 1.5% potassium ferricyanide to the osmium tetroxide improves myelin staining, and the addition of 3% sodium iodate improves penetration. Overnight fixation is preferable. Dehydration is by using increasing concentrations of ethanol. It is important to give sufficient time in dried absolute ethanol to ensure extraction of all water from the myelin sheath. A suitable schedule is shown in Table 4.1.

Table 4.1 Suggested processing schedule for resin embedding

Step	Solution	Temperature	Time
1° fix	2.5% glutaraldehyde/0.1 M PIPES buffer	4 °C	3 h – O/N
Wash	2% sucrose/0.1 M PIPES	4 °C	30 min
Second fix	1% OsO_4 + 1.5%$K_3Fe(CN)_6$ + 3%$NaIO_3$/0.1 M PIPES	4 °C	O/N
Dehydration	15% ethanol	RT	2 × 5 min
	30% ethanol	RT	2 × 10 min
	50% ethanol	RT	2 × 15 min
	70% ethanol	RT	2 × 30 min
	100% ethanol	RT	3 × 20 min
	100% ethanol	RT	2 × 1 h
Embedding	1,2 epoxy propane	RT	2 × 15 min
	1 : 1 epoxy propane: epoxy resin	RT	1 h
	1 : 3 epoxy propane: epoxy resin	RT	O/N
	Epoxy resin	RT	24 h
	Embed	60 °C	24 h

Note: For 100% ethanol, dry over molecular sieve 3A (with indicator) (e.g. Sigma Aldritch, Fluka 69828).
For teasing, use toluene instead of 1,2 epoxy propane.
For solids % = weight/volume.
RT: room temperature; and O/N: overnight.

Embedding can be in any of the commercially available epoxy resins via a suitable intermediary; note that 1,2 epoxy propane renders nerves too brittle for teased fibre studies, and toluene is recommended. Teased specimens can be processed into epoxy resin without accelerator and stored at −20 °C. To prepare teased nerve preparations, the nerve is placed in a drop of resin on a slide, the perineurium is cut and individual nerve fibres are carefully extracted using fine forceps. They are then laid out straight on a clean slide and cover-slipped using complete resin mixture.

Silver to grey sections for transmission electron microscopy (TEM) are stained with lead citrate and uranyl acetate. Methanolic uranyl acetate is quick and easy but does not stain collagen. This, however, may be advantageous in some circumstances, as it allows better definition of basal laminae and extracellular material such as amyloid.

N.B.: All percentages given in this section are weights/volume (W/V).

4.3 NORMAL NERVE ULTRASTRUCTURE

4.3.1 Axons

These are long, thin extensions of the neuronal cell bodies in the CNS and in dorsal root ganglia. The axoplasmic cytoskeleton consists of 85% neurofilaments plus microtubules and a microtrabecular matrix (Hirokawa, 1982). Neurofilaments are specialised intermediate filaments that provide mechanical strength to the axon cytoskeleton and support the less robust microtubules. See Berthold and Rydmark (1995) for details of axonal structure. Microtubules readily depolymerise in suboptimal fixation conditions, so inability to visualise them must be interpreted with care. The optimal requirements for best preservation have been extensively studied (Ohnishi, O'Brien and Dyck, 1976a,b).

4.3.2 Schwann Cells

SCs have a basal lamina on their outer surfaces composed of collagens IV and V, laminin and glycosaminoglycans. TEM shows that this can persist in the endoneurium if the SC moves or dies (Figure 4.1a). The other cells in the endoneurium such as fibroblasts, mast cells (Figure 4.1b), macrophages and lymphocytes do not have a basal lamina.

Typical lysosomes are found only in SCs associated with unmyelinated axons. Reich granules are conspicuous inclusions with a composite lamellar and amorphous structure that occur in myelinating SCs (Figure 4.1c). Their acid phosphatase activity suggests they are equivalent to lysosomes. It is unusual and probably abnormal to find small, dense lysosomes in SCs of myelinated fibres (Figure 4.1d).

4.3.3 The Myelin Sheath

The presence of a myelin sheath increases conduction speeds in a myelinated fibre from 2 to 60 m/s or more in large fibres. Myelin is formed from SC membranes during development by spiralling of SC processes around a centrally located axon (Kleitman and Bunge, 1995). During compaction of the consecutive layers, cytoplasm is excluded and additional proteins incorporated (myelin basic protein (MBP), myelin protein zero (MPZ) and peripheral myelin protein 22 (PMP22)). The latter has an importance out of proportion to its abundance, being associated with the commonest inherited neuropathy, Charcot–Marie–Tooth disease 1A (CMT 1A).

The inner surface of the SC plasma membrane forms the major dense line and the outer surface, the interperiod or less dense line. The repeat distance in fixed myelin is about 14 nm. In addition, myelin-associated glycoprotein (MAG) is localised in the peripheral nervous system (PNS) to the paranodes, outer mesaxons and Schmidt–Lanterman incisures (Quarles, 2007) and is thought to determine the spacing of the major dense line.

Mesaxons are extensions of the SC plasma membrane that connect the myelin lamellae to the inner and outer surfaces of the SC. These are stabilised by localised tight junctions (Balice-Gordon, Bone and Scherer, 1998).

In humans, nerve maturation is not complete until about 12 years of age (Jacobs and Love, 1985). In the adult, the thickness of the sheath and the internodal length are directly related to the diameter of the axon. Myelin thickness is often expressed as a ratio of axon diameter over total fibre diameter (the g ratio). The mean value of this is ~0.7 in the human sural nerve. A larger g ratio implies a relatively thin myelin sheath, and a small value implies a thicker one.

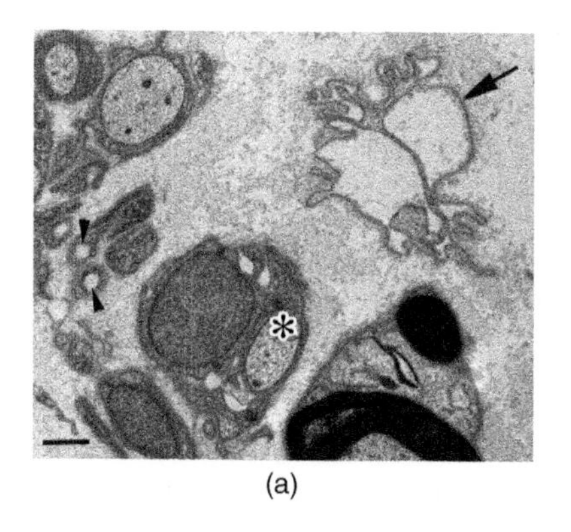
(a)

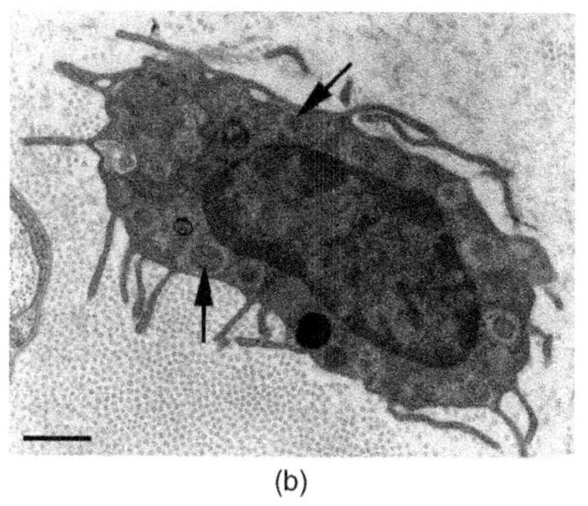
(b)

Figure 4.1 (a) This tube of SC basal lamina (arrow) contains only a very small SC process. The apparently separate myelin figure (asterisk) is a sectioning artefact and is a protrusion from the adjacent myelin sheath. There are several collagen pockets in nearby Remak fibres (arrowheads). The use of methanolic uranyl acetate in staining has resulted in very pale collagen fibrils.This sural nerve biopsy is from a 73-year-old woman with CIDP. Bar = 1 µm. (b) This is an endoneurial mast cell. The characteristic inclusions can be seen (arrows), as can the pseudopodia that differentiate this type of cell from all others in the endoneurium. Collagen fibrils are well stained in this preparation. This biopsy is from a 42-year-old man with a paraproteinaemic neuropathy superimposed on neuropathy related to PMP22 duplication (CMT1A). Bar = 1 µm.

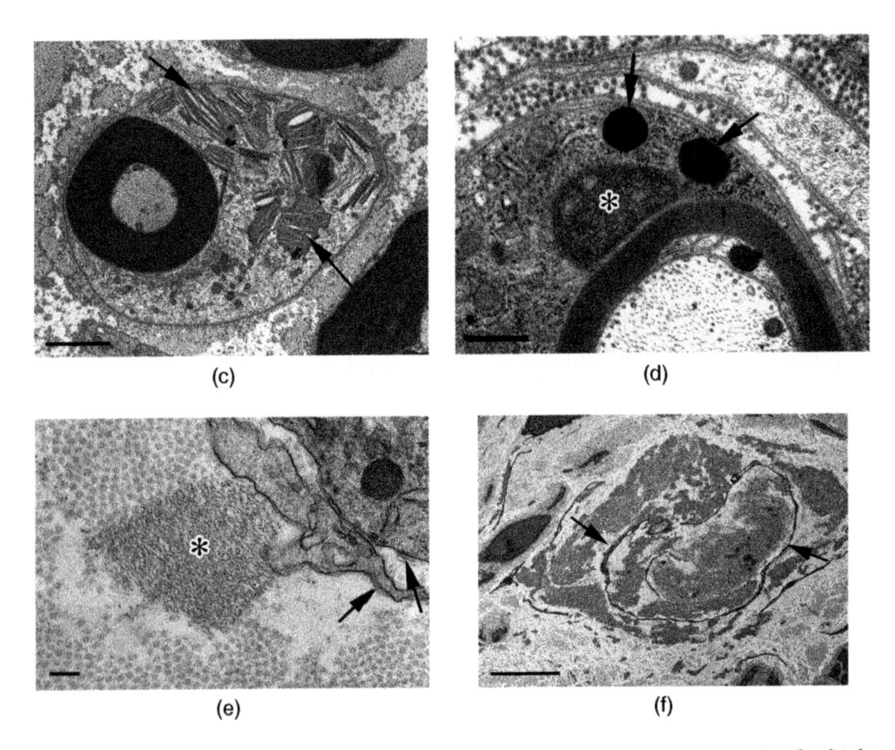

Figure 4.1 (c) This small myelinated fibre is abnormal with an inappropriately thick myelin sheath and dense axoplasm. The SC cytoplasm is rather dilated and pale; there are numerous Reich granules (arrows). The biopsy was from a 45-year-old man with vasculitis. Bar = 2 μm. (d) There are two lysosomes (arrows) in the cytoplasm of this myelinating SC suggesting a metabolic abnormality. The other dense body (*) is a glancing section across the SC nucleus. This is the same specimen as in (b). Bar = 0.5 μm. (e) Oxytalan fibrils (asterisk) are a nonspecific abnormality. They are larger than amyloid fibrils, not straight and less defined in cross-section. Arrows indicate extracellular dense coating. This sural nerve is from a 33-year-old woman with CMT1A. Bar = 0.2 μm. (f) A Renaut body may be reminiscent of amyloid deposits, but the fibrils are slightly larger and slightly more electron-dense; there are fibroblast processes in this typical circular arrangement (arrows). Sural nerve from an 81-year-old female with a chronic axonal neuropathy. Bar = 5 μm.

4.3.4 Node of Ranvier

This region between two adjacent SCs is bare of myelin and is ~1 μm long, a figure independent of axon diameter (reduced at the node in large fibres). TEM shows a dense undercoating of the nodal axolemma. The SCs from both sides extend to form the nodal processes, microvillus-like structures that interdigitate and terminate closely apposed to the axolemma (Figure 4.9c). The SC basal laminae are continuous across the nodal gap and form a continuous tube from the neuron to the end organ. The space between the nodal processes and the basal lamina sheath is filled with a moderately electron-dense material, the 'gap substance'.

4.3.5 Paranode

This is the region of termination of myelin lamellae. In fibres less than 5 μm in diameter, the lamellae terminate consecutively on the axolemma, whereas on larger fibres many terminate before they reach the axolemma, forming a complex structure reminiscent of an ear of corn (Berthold *et al.*, 2005), thus reducing the length of the paranodal region.

As each myelin lamella approaches the node, the major dense line opens up to enclose a small pocket of cytoplasm. These pockets form a continuous spiral connecting the adaxonal (closest to the axon) SC cytoplasm with the abaxonal (furthest from the axon) cytoplasm. The inner pockets are attached to the axolemma by septate-like junctions containing neurofascin 155 and caspr–paranodin (Boyle *et al.*, 2001), creating a barrier to movement of ions along the axonal membrane from the nodal region containing sodium channels to the juxtaparanodal region where potassium channels are located (Sherman *et al.*, 2005).

SC mitochondria are mainly concentrated in paranodal cytoplasmic pockets on the outer surface of the myelin. In large fibres, these pockets are dented into the myelin sheath so that its profile in cross-section is crenated with three or four arms in a 'Maltese cross' arrangement. These longitudinal grooves increase in number and size with the fibre diameter (Williams and Landon, 1963).

Near the node, the lamellae are stabilised by localised desmosome-like structures containing gap junction proteins.

4.3.6 Juxtaparanode

The potassium channels are situated in the juxtaparanodal region immediately adjacent to the paranode. The paranodal and juxtaparanodal regions are more difficult to preserve well than the internode or the node itself.

4.3.7 Internode

In the largest fibres, the internode may be 1.5 mm long, but only in the nuclear region and paranodes is it completely covered by cytoplasm.

Along the rest of the internode, the SC cytoplasm forms restricted irregular columns called canaliculi (Mi *et al.*, 1995, 1996) or Cajal bands (Court *et al.*, 2004), between which only the SC basal lamina separates the myelin from the extracellular space.

4.3.8 Schmidt–Lanterman Incisures

In addition to the complete interruption of myelin continuity at the nodes, there are also less complete disruptions at incisures of Schmidt–Lanterman. These can be visualised in living tissue as oblique dark lines crossing the pale myelin sheath and dividing it into cylindrico-conical segments (Hall and Williams, 1970). Consecutive lamellae open up to enclose pockets of cytoplasm forming a continuous helical pathway across the compact myelin lamellae. This allows transport of proteins from the abaxonal SC cytoplasm to the adaxonal cytoplasm and possibly also permits movement of the fibre without damage to the sheath. Connexin 32 gap junctions may be found in the outer part of Schmidt–Lanterman incisures and in localised patches on the paranode and outer mesaxon (Balice-Gordon, Bone and Scherer, 1998).

4.3.9 Remak Fibres

This is a convenient shorthand term for a complex of SCs and unmyelinated axons. In human nerves, between one and four small axons, less than ~2 μm in diameter, are associated with each SC process. The individual SCs have a very complex shape, and the axons move independently from one cell to an adjacent one. The details have been shown by three-dimensional reconstruction (Aguayo *et al.*, 1976). This complexity has implications in regeneration (see Section 4.4.2). Not only is regeneration more difficult, but also the new axons can be attracted to nearby regenerating myelinated fibres and directed to an inappropriate end organ (King and Thomas, 1971).

4.3.10 Fibroblasts

Peripheral nerve fascicles contain a small proportion of cells other than SCs. In particular, there are fibroblasts (probably more correctly called fibrocytes) whose function is the production of connective tissue, mainly collagen fibrils from tropocollagen monomers of collagens I and III. The result is a banded filament with a repeat of about 240 nm and a diameter of 60–80 nm in the endoneurium and perineurium, slightly larger in the epineurium. Fibroblasts have no basal laminae but often possess small dense patches of similar appearance on the external surface of the plasma membrane (Figure 4.1e); this differentiates them from lymphocytes.

4.3.11 Renaut Bodies

These are prominent fusiform structures 200–300 μm in diameter and rather greater in length. They are found most commonly under the perineurium but occasionally deeper in the endoneurium. They are normal components with no pathological significance. They are conspicuous by light microscopy (LM) and are often found in regions susceptible to damage. suggesting a function as shock absorbers. TEM shows that they are composed of connective tissue fibrils and fibroblast processes. Glancing sections through Renaut bodies lying deeper in the endoneurium may suggest amyloid deposits, but TEM examination will distinguish their oxytalan fibrils from amyloid fibrils.

Detailed descriptions of nerve structure are given in several textbooks, for example *Peripheral Neuropathy* (Berthold *et al.*, 2005) and *The Axon* (Waxman, Kocsis and Stys, 1995).

4.4 PATHOLOGICAL ULTRASTRUCTURAL FEATURES

4.4.1 Axonal Degeneration

In the early stages of axonal degeneration when the axon structure breaks down, myelin lamellae flow across the gaps to form a chain of large myelin ovoids within the tube of continuous SC basal lamina. These ovoids further break down and are either recycled within the SCs or removed by macrophages (Stoll *et al.*, 1989).

When the axon is damaged, it dies back to the last unaffected node of Ranvier. The undamaged part of the fibre can regenerate by forming between one and several axon sprouts from the last intact node. Ultrastructural examination is needed to distinguish between an axonal growth cone, which is a large structure filled with a variety of organelles, and an unmyelinated fibre in the earliest stages of degeneration. The latter are often clearly part of a Remak fibre, and there is a higher proportion of mitochondria (Thomas and King, 1974).

The complex of old basal lamina plus new SC processes and axon sprouts is often known as a band of Büngner.

The very earliest pathological changes are visible only by TEM. The SC at the node extends cytoplasmic processes that separate the terminal loops from the axolemma (Ballin and Thomas, 1969a). Experimental demyelination (Ballin and Thomas, 1969b), however, starts in the same way, so terminal loops lifted from the axolemma do not differentiate between demyelination and degeneration. The initial degeneration of filaments and tubules can be mimicked by fixation artefact, and a more convincing marker is the finding of abnormally large groups of mitochondria in the axon due to cessation of axonal flow (Figure 4.2a) as encountered in experimental studies of Wallerian degeneration (Ballin and Thomas, 1969a).

Toxic neuropathies, in particular glue sniffers' neuropathy due to the solvent n-hexane, can produce very dense, enlarged axons with inappropriately thin myelin (Oh and Kim, 1976). The axonal enlargement results from large numbers of closely packed neurofilaments lacking side arms. In giant axonal neuropathy (GAN), a hereditary disease of children, a much higher proportion of axons is affected, often accompanied by secondary demyelination. Patients with GAN usually also have tightly curled hair.

Scattered giant axons also occur in CMT2E, an autosomal, dominantly inherited disease related to a mutation in the gene for the light component of the neurofilament protein (*NEFL*) (Fabrizi *et al.*, 2004); in this disease, the hair is normal.

4.4.2 Axonal Regeneration

SC reduplication starts about 3 days after injury, and axonal sprouting very soon afterwards. If the continuity of the tube of SC basal lamina has not been affected, it can act as a guide for regenerating axon sprouts. Initially, the axon sprouts are very small and lie within one SC (Figure 4.2b), but as they grow (at about 1–2 mm per day), they separate into their own individual SCs. As the regenerating axon sprouts grow, the new SCs move along them. When these axons become large enough, they can myelinate, but if they fail to form functional connections with an appropriate end organ, they degenerate.

Bands of Büngner are tubes of basal lamina containing SC processes only or cell processes plus axon sprouts. Within a band of Büngner, it is often difficult to differentiate SC processes from small axons. Examination at high power in well-fixed material shows subtle differences

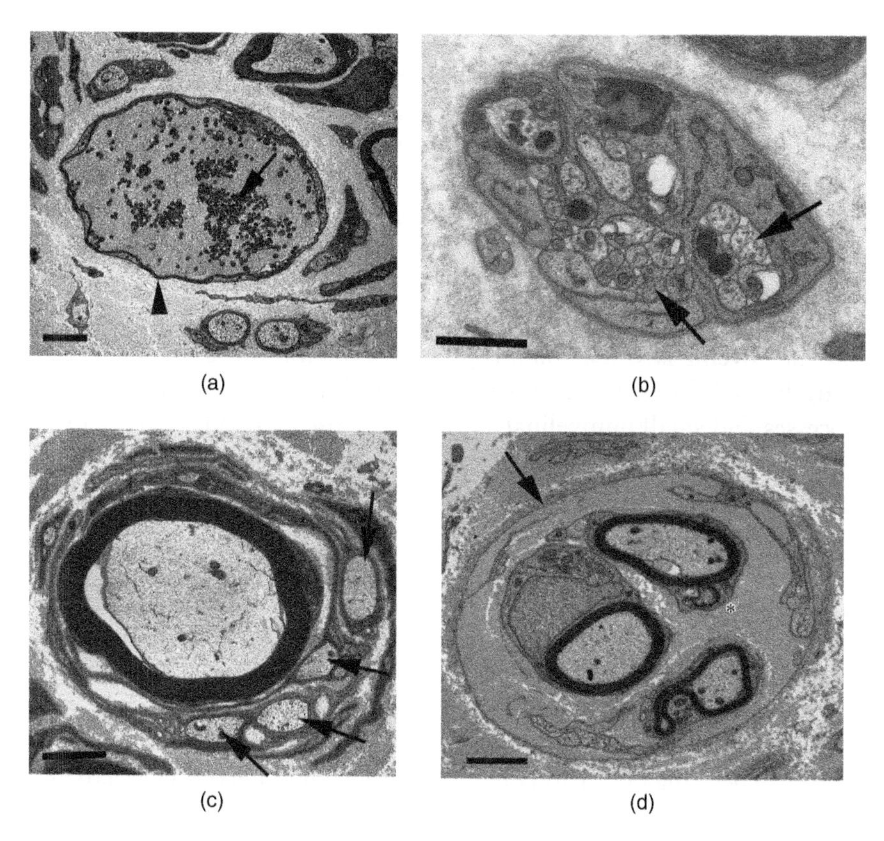

Figure 4.2 (a) This very enlarged axon is packed with closely spaced neurofilaments and large numbers of mitochondria (arrow). The myelin sheath consists of only two or three lamellae (arrowhead). This sural nerve is from an axonal neuropathy of uncertain cause in a 66-year-old woman. Bar = 2 μm.(b) Very early axonal regeneration. The axonal sprouts are very small and touch each other in a manner reminiscent of foetal nerve (arrows). This sural nerve is from a 55-year-old man with an inflammatory neuropathy. Bar = 0.5 μm. (c) This myelinated fibre might appear to be part of an onion bulb by LM, but it can be seen that there are several, smaller axons (arrows) close by that are probably regenerating sprouts from the same parent axon. This nerve from a 56-year-old woman had been damaged by vasculitis. Bar = 2 μm. (d) Three regenerating myelinated axon sprouts within an abnormally rounded and persistent SC basal lamina (arrow). There is extensive deposition of collagen fibrils within the tube of basal lamina (asterisk). This radial nerve biopsy is from a case of diabetic neuropathy in a 47-year-old man. Bar = 2 μm.

between the SC plasmalemma and the axolemma, the latter being denser and slightly irregular (Figures 4.2b and 4.5b). TEM is needed to identify small unmyelinated sprouts (Figure 4.2b,c). Differing growth rates result in regenerating clusters containing myelinated and unmyelinated axons (Figure 4.2c). With time, they increase in size and myelin thickness but usually remain smaller than the original fibre. The persistence of the original SC basal lamina is unusual except in diabetic neuropathy (Figure 4.2d).

Regenerating fibres have short internodes all of the same length, and the myelin is often inappropriately thin for the axon diameter. Remyelinated fibres may also have thin myelin, but the presence of SC processes and small unmyelinated axons in close proximity may help to differentiate them.

4.4.3 Remak Fibre Abnormalities

Bands of Büngner are composed of rounded SC processes (Figure 4.3a) and axonal profiles, whereas degeneration of unmyelinated axons results in bundles of flattened SC sheets (Figure 4.3b).

Remak fibre SC may enclose bundles of collagen fibrils as if they were unmyelinated axons. The resulting structures are known as collagen pockets. These may occasionally be seen in normal nerves, especially in older patients, but large numbers probably imply loss of small axons or an abnormality of the Remak fibres (Figure 4.3c). It is abnormal to find large numbers of axons (more than five or six) in a Remak fibre.

4.4.4 Polyglucosan Bodies

Large polyglucosan bodies (PGBs) are readily identified by LM, but occasionally TEM may reveal small collections of the characteristic filaments. PGBs are of no diagnostic significance unless they are found in very large quantities, in which case they may be related to adult PGB disease (Robitaille *et al.*, 1980).

4.4.5 Nonspecific Axonal Inclusions

A variety of abnormal axonal inclusions ranging from dense collections of glycogen granules to curved bodies (Figure 4.3d) and crystalline inclusions (Figure 4.3e) may be encountered. Most of these probably have no diagnostic significance beyond indicating an axonal neuropathy;

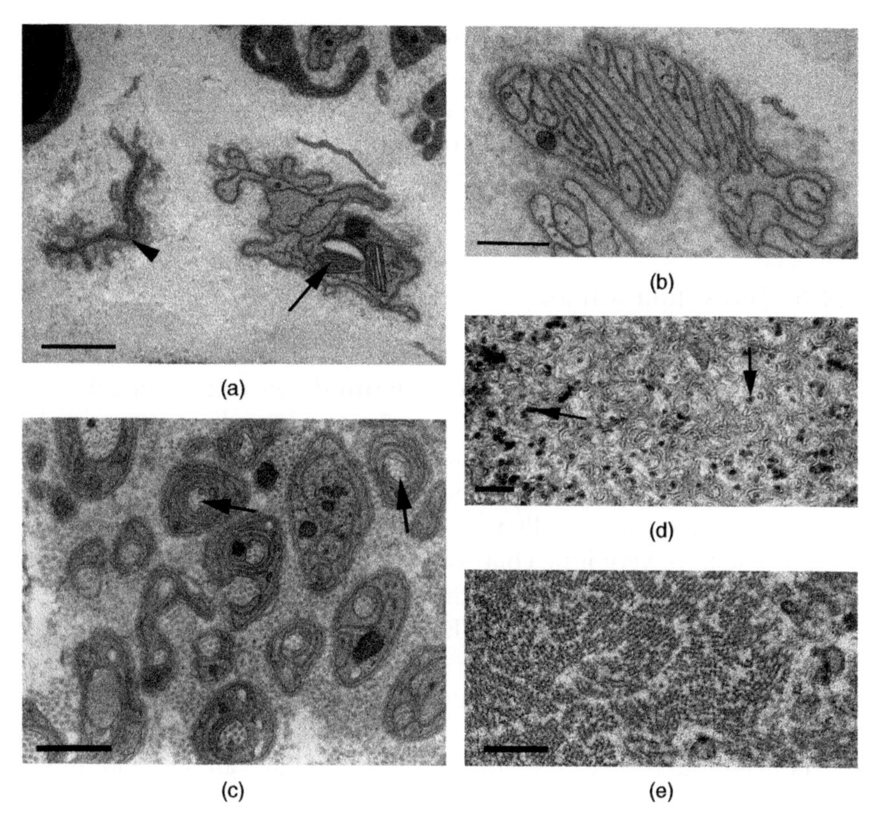

Figure 4.3 (a) Bands of Büngner remaining after myelinated fibre degeneration. The rounded processes are identified as having belonged to a myelinated fibre by the presence of Reich granules (arrow). The other SC process within a collapsed basal lamina is much more attenuated (arrowhead) and shows no sign of regenerating. Sural nerve from a 51-year-old female with a chronic active neuropathy. Bar = 2 μm. (b) Collections of flattened SC sheets indicate the loss of unmyelinated axons. Biopsy from an acute inflammatory neuropathy in a 47-year-old woman. Bar = 1 μm. (c) Collagen pockets; numerous small SC processes wrap around small groups of collagen fibrils as if they were axons (arrows). Sural nerve from a 36-year-old woman with a chronic axonal neuropathy. Bar = 1 μm. (d) Abnormal axonal inclusions; the whole axon was packed with these curved structures plus dense round bodies (arrows) that are probably glycogen granules. Sural nerve biopsy from a 72-year-old woman with a chronic axonal neuropathy. Bar = 0.2 μm. (e) These pseudo-crystalline structures lie in the axoplasm predominantly orientated along the axis of the axon, but some lie obliquely. This is a case of CIDP in a 52-year-old man. Bar = 0.2 μm.

the curved inclusions have been reported in ceroid lipofuscinosis, CMT 4D and experimental vitamin E deficiency (Lampert, 1967; Southam *et al.*, 1991; King *et al.*, 1999). Crystalline inclusions have been noted in a variety of diseases including leprosy, Friedreich ataxia, Fabry disease, CMT1 and GAN where they appear to be derived from closely packed abnormal neurofilaments (Donaghy *et al.*, 1988).

4.4.6 Demyelination and Remyelination

Primary segmental demyelination results from a failure of the SC to support its myelin sheath; the myelin sheath degenerates, and debris is found in the SC cytoplasm. TEM may show a demyelinated heminode (Figure 4.4a). If the myelin thickness on opposite sides is noticeably different, this implies segmental remyelination. Macrophages remove the debris, and inflammatory cells are not prominent. Axons with fewer than six lamellae have very little contrast on conventional LM, even in resin sections, and are invisible in paraffin sections. Suspected demyelination should always be confirmed by TEM as it may be mimicked by other abnormalities.

In a normal nerve, there is an intimate connection between the SC, myelin sheath and axon, so damage to any one may affect the others. The radial bonds between myelin lamellae can slip and allow some adaptation to changes in axonal diameter. If, however, there is excessive shrinkage or swelling, this is not possible, and the myelin sheath degenerates. In GAN, the axonal enlargement leads to extensive secondary demyelination and remyelination. Conversely, in slow axonal atrophy, as for example in some types of diabetic neuropathy, the myelin may adapt to a limited reduction in diameter; if this fails, however, there may be paranodal or internodal demyelination. Adjacent lamellae may extend to cover the defect (Figure 4.4d), or a new SC may produce a very short, intercalated internode.

In primary demyelination, remyelination may start anywhere along a bare internode and then spread to cover the whole defect (King, Pollard and Thomas, 1975). This frequently results in myelin segments of varied length and thickness; this contrasts with axonal regeneration, where all the new internodes will be of equal length and thickness. Teased fibre preparations are necessary to appreciate this.

Allergic demyelination is rather different, as demyelination is due to an autoimmune attack on the myelin sheath by activated macrophages. Cells may be found under the SC basal lamina before the myelin is attacked (Figure 4.4b). Next, macrophages insert processes which separate and remove the lamellae of the myelin sheath (Figure 4.4c). This is an important diagnostic feature in autoimmune neuropathies, as it differentiates demyelination in Guillain–Barré syndrome (GBS) and chronic inflammatory demyelinating polyneuropathy (CIDP) from primary demyelinating diseases due, for example, to diabetes, toxins or genetic abnormalities. Following demyelination, SCs usually multiply readily and attempt to repair the myelin defects. There may also be secondary axonal damage. In GBS, there are also numerous lymphocytes in the endoneurium. Finding macrophages on the inner surface of the

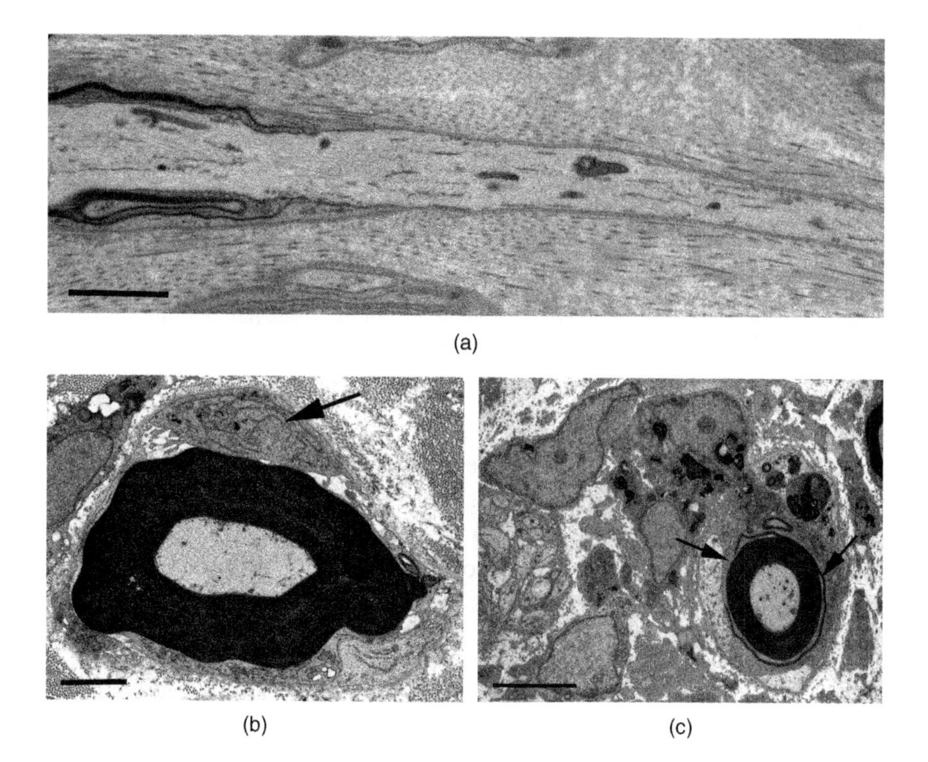

Figure 4.4 (a) In this demyelinated heminode, part of the axon is completely bare without SC covering. The myelinated side of the node has an abnormally thin myelin sheath, and there are no nodal processes. This nerve was from a 5-year-old boy with MLD. Bar = 2 μm. (b) Inflammatory cell within the SC basal lamina (arrow). The myelin sheath is rather thick for axon diameter, and axonal neurofilaments are densely packed. Sural nerve biopsy from a 29-year-old man with vasculitis or scleroderma. Bar = 2 μm. (c) In this example of myelin stripping, a debris-containing macrophage can be seen infiltrating the outer layers of the myelin sheath (arrows). This 60-year-old woman had an IgG paraproteinaemia and features of CIDP. Bar = 5 μm.

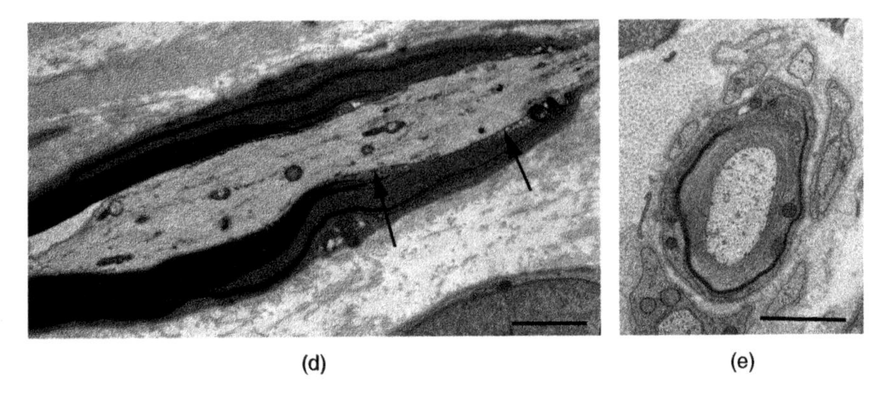

(d) (e)

Figure 4.4 (d) Here the myelin sheath appears paler in the paranode and juxtaparanode because the lamellae are abnormally widely spaced with a periodicity of 24 nm instead of the normal value of 14 nm. Myelin packing becomes normal further away from the node. There are two sets of myelin end loops (arrows) indicating local remyelination to repair a widened node. Sural nerve biopsy from a case of Waldenström's macroglobulinaemia in a 66-year-old man. Bar = 2 μm. (e) In this example of uncompacted myelin, the myelin in the inner part of the sheath is replaced by layers of SC plasma membrane. The extra SC processes and occasional mitochondria suggest that this section is through the paranode or juxtaparanode. Sural nerve biopsy from a 59-year-old woman with neurolymphomatosis. Bar = 1 μm.

myelin alongside the axon has been associated with attack on the axon rather than the myelin (Griffin *et al.*, 1995), but they were also reported in experimental demyelination due to ionomycin (a calcium-selective ionophore) (Smith and Hall, 1988).

Where the axon has never been myelinated, as in some hereditary demyelinating diseases, the bare axon is tightly covered by basal lamina (Figure 4.4a). In transverse section, this can be quite difficult to differentiate from a SC process. Loss of myelin, on the other hand, leaves a loose, crenated covering of basal lamina around the axon and SC.

4.4.6.1 Abnormal Myelin Periodicity

In the neuropathy associated with IgM kappa paraproteinaemia, ~50% of cases show a patchy increase in myelin periodicity from about 14 to 24 nm, due to an increased separation of the paired components of the less dense line (King and Thomas, 1984). This abnormality may be more pronounced in the paranode and is therefore better seen in longitudinal sections (Figure 4.4d).

The significance of other variations in periodicity occasionally encountered is uncertain (King and Thomas, 1984; Vital *et al.*, 1986). Local breaks in lamellae or disruptions in organisation are most likely to be created by poor handling or fixation rather than have any pathological significance.

4.4.6.2 Uncompacted Myelin

In this nonspecific alteration, no true myelin is formed (Figure 4.4e). It is difficult to distinguish from unusually extensive Schmidt–Lanterman incisures or abnormal paranodal compaction. Longitudinal sections are very useful for localising this abnormality.

4.4.6.3 Classical Onion Bulbs

Repeated episodes of demyelination and remyelination may result in superfluous SCs that lie in concentric whorls around the central fibre. The central fibre may be demyelinated or thinly remyelinated and also may undergo axonal degeneration and be replaced by a group of regenerative sprouts. Although onion bulbs are very conspicuous by LM, TEM is needed to identify the constituent cells in the whorls; this is especially important when only occasional onion bulbs are found (Figure 4.5b,c). The presence of very small, unmyelinated axons (Figure 4.5a) differentiates SCs from perineurial cells.

4.4.6.4 Pseudo-Onion Bulbs

Structures similar to onion bulbs may be found where the perineurial barrier is breached (e.g. in traumatic neuromas, perineuriosis and neurofibromatosis). The encircling cells are perineurial cells and not SCs (Figure 4.5b) (Thomas *et al.*, 1990, 2000). Perineurial cells have more conspicuous pinocytotic vesicles and are connected together by tight and gap junctions to form a local diffusion barrier replacing the damaged perineurium. Perineurial damage is most often seen in the perineuritis associated with systemic vasculitis.

There may be macrophages in the whorls of true onion bulbs, but occasionally in CIDP these may be sufficiently numerous to produce structures that TEM shows consist of lymphocytes and macrophages and not SC processes (Figure 4.5c).

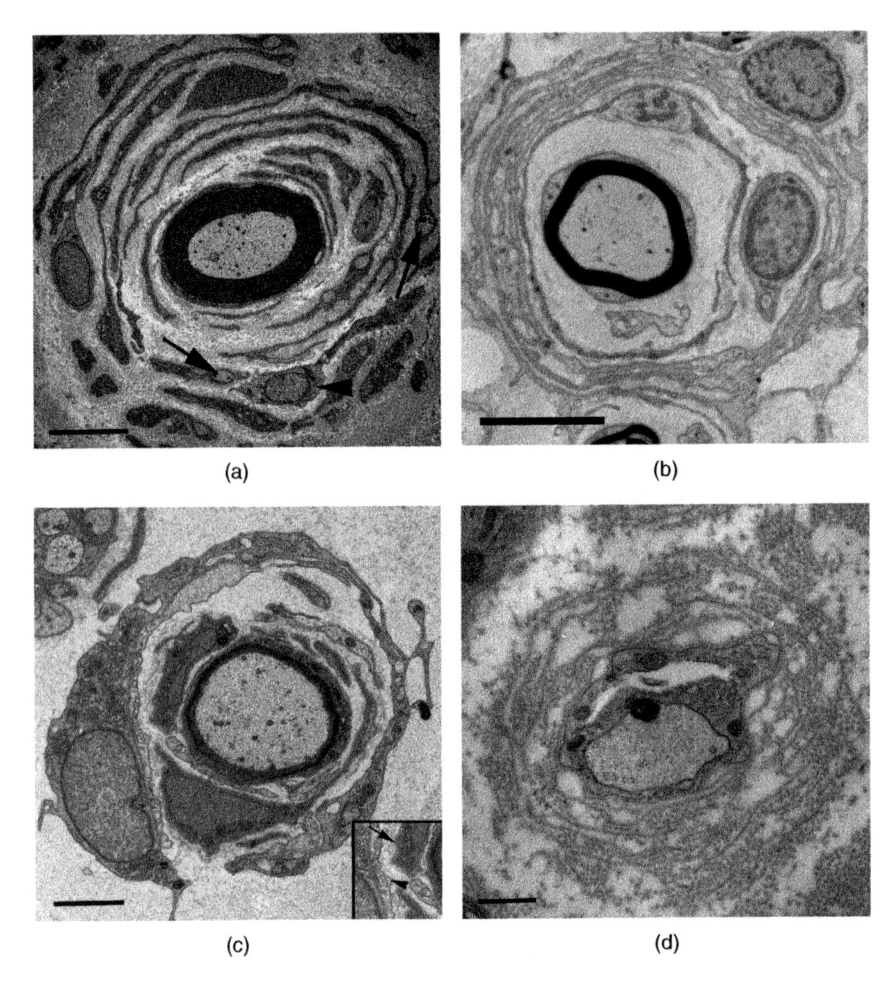

Figure 4.5 (a) In this classical onion bulb, the central myelinated fibre is encircled by SC processes. There are occasionally small unmyelinated axons in these processes (arrows). The cell process lacking a basal lamina is probably of a macrophage (arrowhead). Sural nerve biopsy from a case of CMT1A in a 40-year-old female. Bar = 5 μm. (b) On the edge of this ulnar nerve neuroma in a young man, there was a series of structures resembling onion bulbs. This small myelinated fibre has an inappropriately thin myelin sheath for axon diameter. There are no embedded axons in the encircling cells, which are perineurial cells and overlap to form a minifascicle. Bar = 5 μm. (c) The central fibre lacks myelin and probably remyelinating. The encircling cells are macrophage processes with very few SC processes. The cells can be distinguished by the presence (arrow) or absence (arrowhead) of basal lamina (insert). Sixty-eight-year-old man with an IgM kappa paraproteinaemia. Bar = 2 μm. (d) In this onion bulb, the central fibre lacks myelin, and the laminae consist only of remnants of basal lamina. Sixteen-year-old girl with CMT4C. Bar = 1 μm.

4.4.6.5 Basal Laminal Onion Bulbs

Classical onion bulbs are readily identified by LM, but sometimes the SC processes withdraw leaving their basal laminae behind. The 'onion bulb' then consists only of paired basal laminae with very little, if any, SC cytoplasm (Figure 4.5d). These structures are not resolvable by LM. Finding large numbers by TEM, especially in biopsies from children or infants, suggests a genetic cause such as point mutations in *PMP22* (Tyson *et al.*, 1997); rarer genetic causes are mutations in *EGR2*, *MPZ* or *SH3TC2* (KIAA1985). In the latter disease (CMT4C), the endoneurial abnormalities are very inhomogeneous with occasional thin myelin sheaths and onion bulbs with a central normally or thinly myelinated fibre surrounded by multiple layers of basal lamina and unusual Remak fibres with abnormally thin cell processes (Houlden *et al.*, 2009). If ultrastructural examination reveals this mixture of pathological abnormalities, DNA analysis should be performed to confirm the diagnosis.

4.4.7 Specific Schwann Cell Inclusions

4.4.7.1 Leprosy Bacteria

The perineurium forms a very efficient barrier, and the only bacteria that might be found in the endoneurium are *Mycobacterium leprae*, the causative organism of leprosy. These invade any type of cell but are particularly prevalent in Remak fibres (Figure 4.6a) (Rambukkana *et al.*, 2002). The pathological changes found in leprosy neuropathy are extraordinarily varied as they depend on the differing immune reactions of individuals. They are classified on a spectrum from lepromatous to tuberculoid (Parkash, 2009). Patients with tuberculoid leprosy (TL) have a strong immune reaction to the bacilli, which are killed or inhibited, whereas in patients with lepromatous leprosy (LL), there is little immune response and bacilli multiply unchecked in peripheral nerves. Once inside a cell (SC or macrophage), they are not detected by the host's immune system. The formation of foam cells (macrophages containing large quantities of lipid and numerous bacteria) makes LL easily diagnosed by LM, but the absence of bacteria and lack of cellular reaction make TL much more difficult to diagnose. In polar TL, the endoneurial contents may be entirely replaced by fibrous collagen in which few SCs survive. Finding bacilli using Gram-negative stains can be very difficult, and extensive searching by TEM may be additionally necessary to confirm the presence of *M. leprae*. The bacteria are usually very electron dense and have a pale lipid coat derived from the host (Cruz *et al.*, 2008). It is rare to find bacilli in axons, although they may be occasionally present in the adaxonal SC.

4.4.7.2 *Lipid Inclusions*

Many of the lipidoses affect SCs. These diseases most commonly present in early life and are usually diagnosed by quantification of leucocyte enzymes. They include metachromatic leucodystrophy (MLD) with so-called Tuffstein bodies that have a lipid periodicity of 5.6–5.8 nm (Thomas *et al.*, 1977), Farber's disease with curved inclusions (Vital *et al.*, 1976), Tangier disease with cholesteryl esters (Kocen *et al.*, 1973) and Krabbe's disease with galactocerebroside inclusions (Bischoff and Ulrich, 1969; Thomas *et al.*, 1984). The last three types of inclusions are all partially dissolved during processing for TEM. Very rarely, adult-onset variations may be encountered unexpectedly in a nerve biopsy. An example is a case of Krabbe's disease in a 33-year-old woman identified in the biopsy specimen by the universally thin myelin, distorted myelin sheaths and typical SC inclusions (Thomas *et al.*, 1984) (Figure 4.6b). The inclusions of MLD are less soluble and hence denser in TEM preparations. The largest quantities of these inclusions are found in the juvenile variant of MLD; the inclusions (Tuffstein bodies) found in myelinating SCs have pale regions giving them a 'moth-eaten' appearance (Figure 4.7c) differing from those found in Remak fibres that are evenly dense and resemble large lysosomes (Thomas *et al.*, 1977). The large collections of lipid in macrophages in MLD have a dramatic lamellar structure unlike any other inclusions.

4.4.8 Nonspecific Schwann Cell Inclusions

SCs do not normally divide in adult nerves; hence, finding centrioles or mitotic apparatus indicates abnormal activity. Similarly, pseudopodia are rarely seen and probably also imply abnormal movement of the cell.

4.4.8.1 *Reich Granules*

These are found only in myelinating SCs (Thomas *et al.*, 1993). When seen in Remak fibres or orphan SCs lacking an axon, their presence indicates that the original myelinated fibre has been lost. In this situation, they are therefore a marker of fibre degeneration (Figure 4.3a).

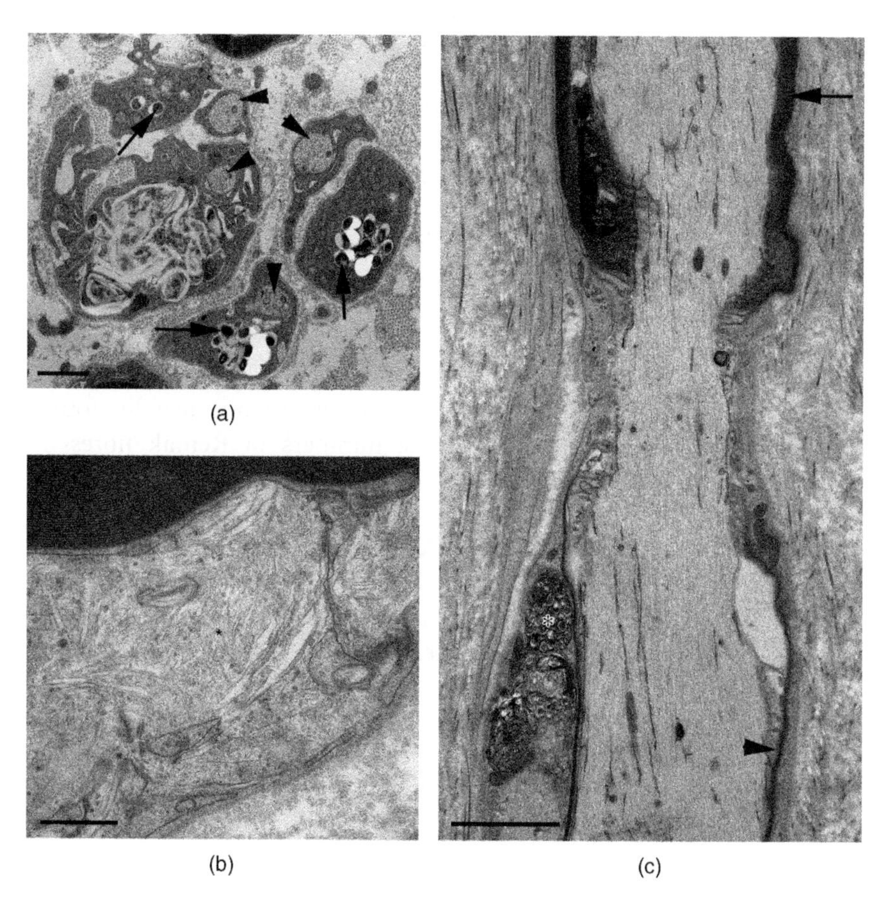

Figure 4.6 (a) Remak fibres containing *Mycobacterium leprae* (arrows). In the largest group, there is also extensive lipid deposition, but all bacilli are surrounded by an electron-lucent rim. The unmyelinated axons (arrowheads) appear normal.The density of the bacilli plus the surrounding lipid makes them very prone to tear out of the section, as seen here. This could be avoided by using coated grids. This biopsy is from a radial nerve from a 46-year-old man with borderline lepromatous leprosy. Bar = 0.5 μm. (b) SC cytoplasm adjacent to a myelinated fibre containing lipid crystals (asterisk) characteristic of Krabbe's disease that was unanticipated in this sural nerve biopsy from a 34-year-old female. Bar = 1.5 μm. (c) Longitudinal section through a node of Ranvier. On one side of the node, the myelin sheath is much thicker (arrow) than the other (arrowhead), indicating that that internode has undergone segmental demyelination and remyelination. The paranodal SC cytoplasm contains Tuffstein bodies typical of MLD (asterisk). Sural nerve from a 3-year-old boy found to have MLD on biopsy. Bar = 2 μm.

4.4.8.2 *Adaxonal Schwann Cell Inclusions*

Dense, pleomorphic inclusions may be found in the adaxonal SC cytoplasm of myelinated fibres. These probably suggest a primary defect in SC metabolism and may be used to differentiate demyelinating from axonal neuropathies, although this apparent correlation does not appear to have been investigated (Figure 4.7a). These can be readily distinguished by TEM from SC scavenging processes (see Section 4.4.8.7).

4.4.8.3 *Lysosomes*

As already noted in this chapter, lysosomes are not usually found in myelinated SCs (Figure 4.2d); large numbers in Remak fibres or other constituents of the endoneurium are probably also abnormal. In amiodarone toxicity, they are found in all types of endoneurial cells (Jacobs and Costa-Jussa, 1985), whereas in perhexilene maleate neuropathy, they are specific to SCs (Lhermitte *et al.*, 1976). It seems likely, however, that they indicate a disturbance of SC metabolism without any more specific significance.

4.4.8.4 *Lipofuscin*

This age-related breakdown product is probably also a nonspecific indicator of abnormality in SC metabolism.

→

Figure 4.7 (a) In this myelinated fibre, there is a dense inclusion in the adaxonal SC cytoplasm (asterisk). The axon and myelin appear normal. This sural nerve biopsy is from a 37-year-old man with hereditary neuropathy with liability to pressure palsies (HNPP). Bar = 2 μm. (b) This dense inclusion in Remak fibre cytoplasm has a crystalline fine structure. This sural nerve is from a 70-year-old man with both diabetes and an IgM kappa paraproteinaemia. Bar = 0.5 μm. (c) These round processes between the myelin and the axon are formed by SC networks scavenging debris from the axoplasm. This is the same case of CIDP as in Figure 4.3e. Bar = 1 μm. (d) This macrophage contains dense myelin debris and can be differentiated from a SC by the lack of basal lamina. Biopsy of the sural nerve from a 71-year-old female with inflammatory neuropathy. Bar = 2 μm. (e) SC process containing relatively pale myelin debris (asterisk). The cell processes are identified as SC by their basal laminae, although none contain axons. Biopsy from a 65-year-old woman with a systemic vasculitis. Bar = 2 μm. (f) In this longitudinal section through a demylinated fibre (asterisk), the associated SC contains numerous small myelin bodies, as does the adjacent macrophage (arrow). This sural nerve biopsy was taken from a normal 55-year-old woman. In normal nerves, there is an increasing incidence of small numbers of demyelinating fibres with increasing age from the middle 50s. Bar = 2 μm.

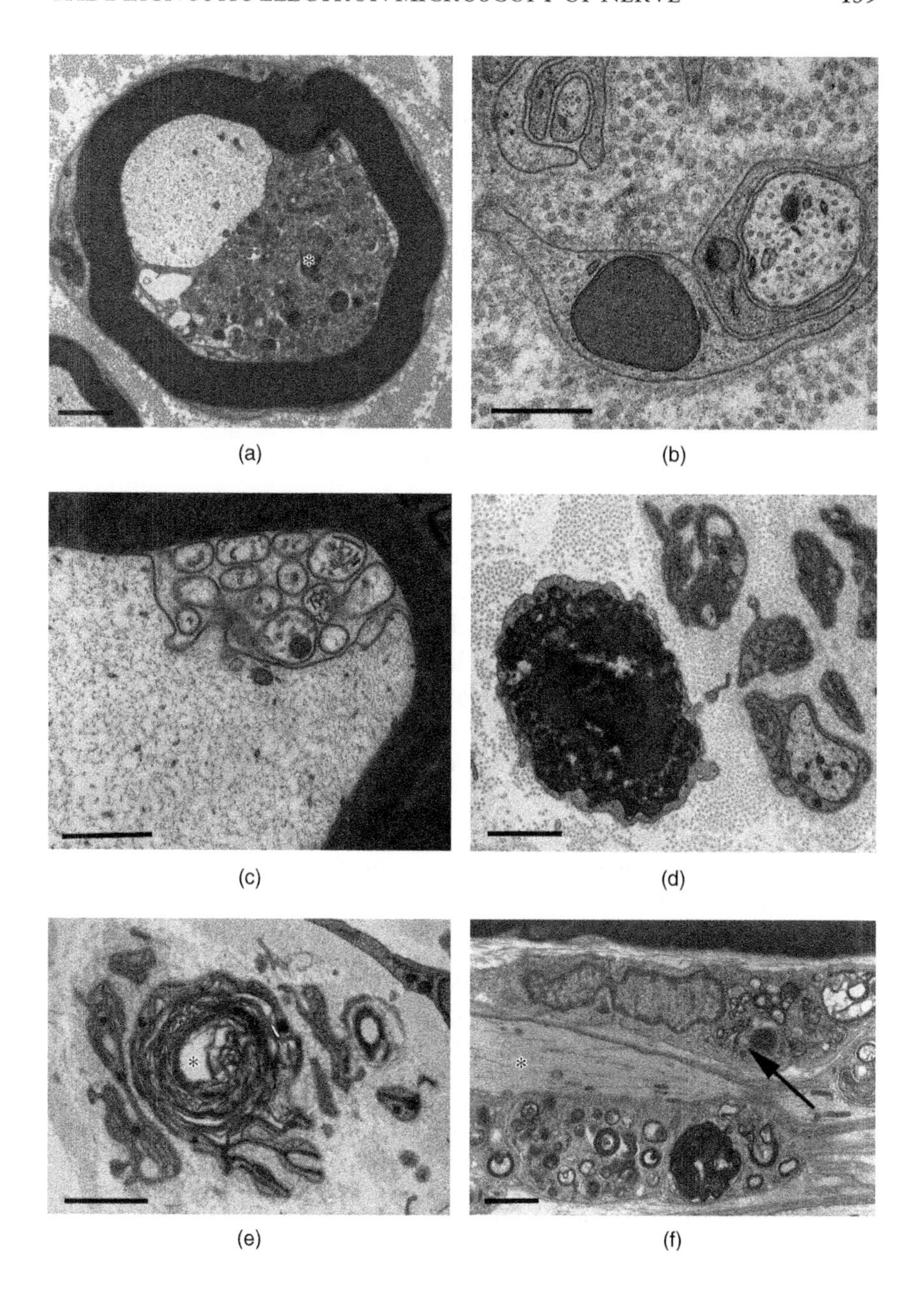
(a)
(b)
(c)
(d)
(e)
(f)

4.4.8.5 Hirano Bodies

These are eosinophilic, proteinaceous inclusions occasionally found in the adaxonal SC cytoplasm. They are moderately electron dense and have a pseudocrystalline fine structure, being composed of overlapping sheets of parallel filaments. It seems likely that they are derived from actin or actin-related filaments (Maselli *et al.*, 2003). They have been correlated with some degenerative diseases (Ogata, Budzilovich and Cravioto, 1972; Thomas, King and Sharma, 1980; Hirano, 1994). Extensive deposits of similar material were reported both in CMT4D and in the mouse model of this disease (King *et al.*, 1999, 2011).

4.4.8.6 Crystalline Inclusions

These cytoplasmic bodies may be found in the SC cytoplasm of Remak fibres. They are membrane bound with a predominantly pseudocrystalline ultrastructure (Figure 4.8b). There sometimes appear to be internal structures resembling mitochondrial cristae and also localised amorphous regions suggesting lipid. In longitudinal sections, they appear rod-like. Experimental studies suggest a relationship to cholesterol metabolism (Hedley-Whyte, 1973). Although initially reported in Refsum disease (Fardeau, Abelanet and Laudet, 1970), they are found

→

Figure 4.8 (a) In this micrograph, the long, stripy body is fibrous long-spacing collagen (FLSC) lying close to one of the cells in the perineurium (arrow). The adjacent collagen fibrils show the normal periodicity (arrowhead). This sural nerve biopsy was from a 44-year-old woman without any symptoms or signs of neuropathy. Bar = 0.2 μm. (b) This small endoneurial blood vessel has a layer of amyloid deposits around it mimicking the reduplicated basal lamina that is a common nonspecific abnormality but distinguishable at high power by the differing fine structure. There are also irregular small deposits (arrow). This sural nerve biopsy was from a 46-year-old man with a sporadic amyloid neuropathy unsuspected before examining the biopsy. Bar = 3 μm. (c) Straight, nonbranching amyloid fibrils can be seen to merge with the SC basal lamina (arrow), which is difficult to distinguish in places. This sural nerve biopsy was from a 30-year-old woman with familial hereditary amyloid polyneuropathy. Bar = 0.2 μm. (d) This abnormally large endoneurial blood vessel has numerous fenestrations where the vessel wall is reduced to plasma membrane only (insert). This is a case of perineuriosis, so the distinction between endoneurium and epineurium may be blurred. Sural nerve biopsy from a 33-year-old man. Bar = 5 μm. (e) Numerous small fibrin deposits (arrows) in the endoneurium. Most are aligned with the fascicle, but occasionally one lies transversely, so that the typical fine longitudinal periodicity is visible (inset). Sural nerve from a 15-year-old girl with an undiagnosed chronic neuropathy. Bar = 2 μm.

nonspecifically in abnormal nerves, particularly those in which there has been chronic axonal degeneration (Lyon and Evrard, 1970).

4.4.8.7 Schwann Cell and Axon Scavenging Networks

By extending processes into the axoplasm and engulfing debris, the SC is able to remove abnormal organelles from the axon. These convoluted SC processes are usually found in the juxtaparanode, particularly in large myelinated fibres (Figure 4.7c). Although occasionally they may be encountered in normal material, they are most common in metabolic

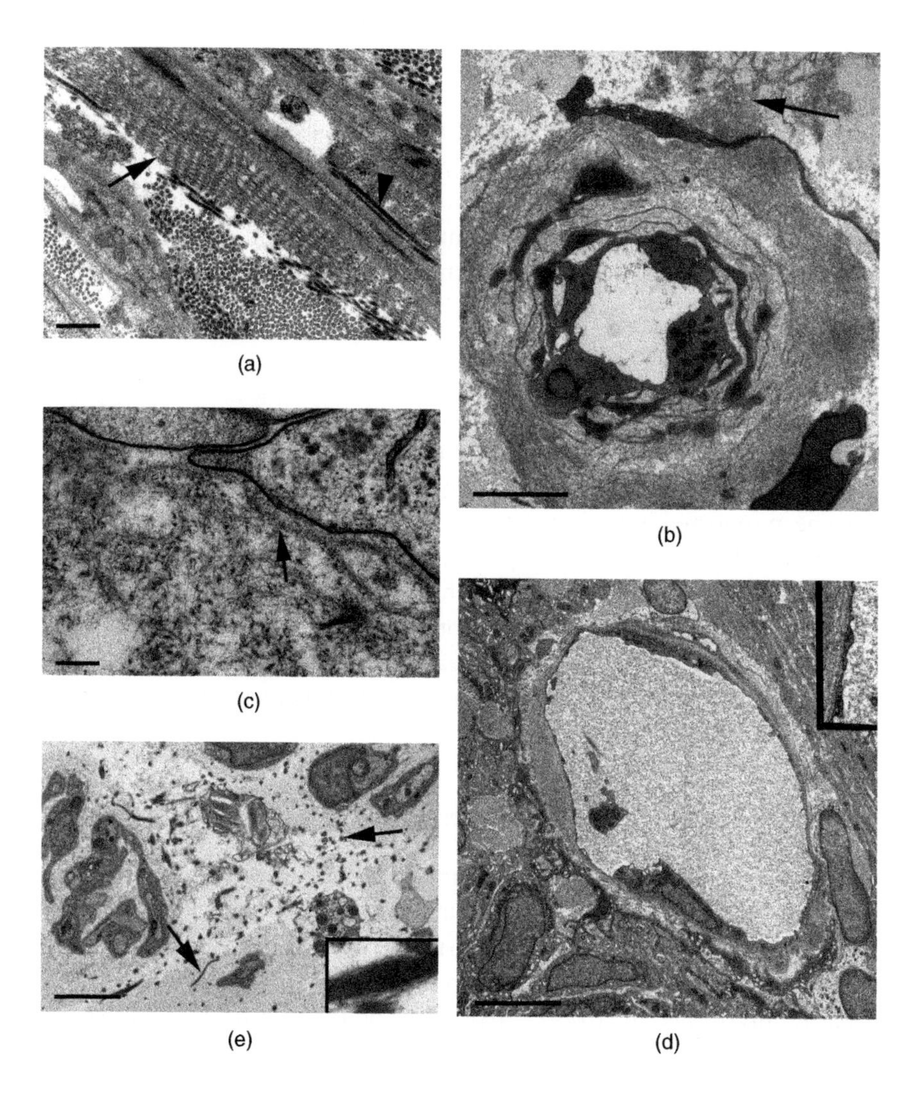

and toxic axonopathies of a variety of aetiologies (Midroni, Bilbao and Cohen, 1995).

4.4.8.8 Myelin Figures

Myelin debris seen in regenerating SCs in bands of Büngner originates from the original myelinated fibre that has been lost. Debris-laden SCs and macrophages can often appear very similar by LM, and TEM is required to distinguish them (Figure 4.7d) using the presence or absence of a basal lamina as the criterion. In the initial stages of myelinated fibre degeneration, the myelin debris forms an electron-dense inclusion. This becomes less electron-dense (Figure 4.7e) as it degenerates further into simpler lipids and fats and is removed by macrophages. During demyelination, the myelin sheath usually disintegrates into small, round myelin figures within the SC cytoplasm. Large myelin bodies can therefore be taken to suggest an axonopathy rather than a demyelinating process. Conversely, the presence of numerous, very small myelin figures is more likely to be associated with demyelination (Figure 4.7f), although this differentiation is not totally reliable.

4.4.9 Fibroblasts

The cytoplasm may contain lipid inclusions, thus superficially resembling macrophages. Fibroblasts containing large indentations and vacuoles have been reported in various experimental and hereditary neuropathies such as hereditary sensory neuropathy, hypertrophic neuropathy (Asbury, Cox and Baringer, 1971) and neurofibromatosis (Thomas *et al.*, 1990) and GBS (King, 1999). Vacuolated fibroblasts are normal in Renaut bodies (see Section 4.3.11). When seen adjacent to the perineurium, abnormal fibroblasts are probably related to the transition between fibroblast and perineurial cells.

4.4.10 Perineurial Abnormalities

Abnormal perineurial cells may be unusually thin, have very pale cytoplasm or, instead of forming continuous laminae, not be connected to adjacent cells. The basal lamina, particularly in the central laminae, is quite thick, lacks a lamina lucida and is often abnormal in diabetic neuropathy (Bradley *et al.*, 1994). Particularly on the endoneurial side, there may be layers of cells indeterminate in structure between fibroblasts and perineurial cells. This is probably a compensating process when there is a reduction in the volume of the endoneurial contents. Macrophages and myelin debris may be found between the perineurial laminae in many neuropathies. The extracellular material may be calcified, particularly in diabetic neuropathy (King *et al.*, 1988). The perineurium may also be affected in some cases of vasculitis and in LL. In these diseases, there are inflammatory cells in the perineurium and the laminae are disrupted.

4.4.10.1 Fabry Disease

This is the only neuropathy characterised by specific inclusions in the perineurium. Abnormal deposits of dense lipid bodies are found in perineurial cells and endoneurial capillaries. By TEM, these are found to have a specific periodicity of 5 nm (Kocen and Thomas, 1970).

4.4.11 Cellular Infiltration

Inflammatory cells are better investigated by immunocytochemistry than by TEM as many of the functionally different lymphocytes have no distinguishing morphological features. Orphan SCs (without an associated axon) and lymphocytes may both appear as small, round cells but they can be differentiated by the presence or absence of a basal lamina (Figures 4.5c and 4.7f).

4.4.12 Endoneurial Oedema

Proteinaceous material is often found in the endoneurium of abnormal nerves. This may be particularly prominent on the endoneurial face of the perineurium in cryoglobulinaemia or paraproteinaemia. It is rare for there to be a definable fine structure, but this does occasionally occur (Vallat *et al.*, 1980). Subperineurial oedema is not exclusively associated with inflammatory disease as it may be found in non-inflammatory neuropathies such as those with a genetic cause, but extensive, dense deposits are predominantly found in paraproteinaemic neuropathy.

4.4.13 Connective Tissue Abnormalities

4.4.13.1 Fibrous Long-Spacing Collagen

Probably the most dramatic connective tissue abnormality is the occurrence of fibrous long-spacing collagen (FLSC). This may been seen as irregular patches of varying sizes adjacent to the perineurial basal lamina (Figure 4.8a) or may form true Luse bodies which are larger and more clearly defined (Luse, Zopf and Cox, 1963). In these, the normal banding on collagen fibrils is altered from ~67 nm to 100–150 nm periodicity. In neurofibromatosis, the lengthened periodicity seems to be slightly different from that seen in CIDP or CMT1.

FLSC probably contains dermatan sulphate proteoglycan and is formed when alterations in the endoneurial pH or viscosity affect polymerisation of the monomer of tropocollagen (Dingemans and Teeling, 1994). Although commonest in the perineurium, FLSC may also be found adjacent to SCs in the endoneurium.

4.4.13.2 Basal Lamina Abnormalities

When myelinated fibres degenerate, the SC basal lamina remains (Figure 4.2a), but the precise timing of its breakdown after SC loss is unknown. Extracellular matrix glycation in diabetic neuropathy renders the tubes of basal lamina stiffer and abnormally resistant to proteolysis. They persist in the nerve as abnormally circular structures often filled with collagen (King *et al.*, 1989) (Figure 4.3d). Additionally, diabetic neuropathy is often characterised by extensive axonal sprouting that forms large regenerative clusters within them. These must be temporary as the axonal loss is progressive.

4.4.13.3 Amyloid

Amyloid deposition may be found in any part of the nerve trunk and may be identified by specific LM stains or immunostaining. It is preferable to examine suspected regions by TEM for confident diagnosis. Amyloid may also be found around the small endoneurial blood vessels and perineurium where it is not identifiable by LM (Figure 4.8b).

Amyloid fibrils may merge with the SC basal laminae, particularly in Remak fibres (Figure 4.8c). The fine structure is the same regardless of the type of amyloid and takes the form of straight, unbranching tubules with a diameter of 9 nm (Figure 4.8c). The length varies considerably in different specimens. In transverse section, they appear as dense annulae. They differ from oxytalan fibrils found in Renaut bodies (Figure 4.8f,g) which have a slightly larger diameter of 10 nm and are not so precisely straight.

4.4.14 Endoneurial Blood Vessels

The most conspicuous alteration is extensive thickening and reduplication of the endothelial basal lamina. This is found in diabetes, paraproteinaemic neuropathy and normal old age, but its identity should be confirmed by TEM.

Very occasionally, attenuation of the endoneurial blood vessel wall may be found that reduces the vessel wall to the thickness of the plasma membrane (Figure 4.8d). Experimental work correlates these regions with vessel leakage, but their occurrence in human nerve biopsies has rarely been noted and still less correlated with any specific pathology.

Endoneurial blood vessels often differ from true capillaries in their possession of associated pericytes that form a complete or partial layer. Occasionally, particularly in diabetic neuropathy (Bradley *et al.*, 1990), these pericytes appear effete with pale watery cytoplasm.

Deposits of fibrin may be found in the endoneurium, indicating vessel leakage (Figure 4.8e). These are recognised by their electron density and their fine periodicity when longitudinally sectioned. Finding such deposits is quite rare and suggests a vasculitic process, although correlation with abnormal blood vessels is often lacking.

4.4.15 Mast Cells

The only report of a pathological change in mast cells was a case of Chédiak–Higashi syndrome that was found to have mast cells with abnormally large granules (Misra *et al.*, 1991).

4.5 ARTEFACT

The abnormalities produced by clumsy handling of fresh nerve are, unsurprisingly, similar to those reported in experimental studies of the early changes found after nerve damage. The largest fibres are most susceptible to traumatic damage, and small myelinated fibres are often unaffected.

Any prolonged delay in fixation will, of course, produce post-mortem artefact, as can be seen in Figure 4.9a where none of the ultrastructural details are adequately preserved. Insufficient time in primary fixative can result in myelin lipids being dissolved into the axon and then being fixed there by the secondary osmium fixation (Figure 4.9b). Poorly preserved lipids will be removed during dehydration, producing disruption of the

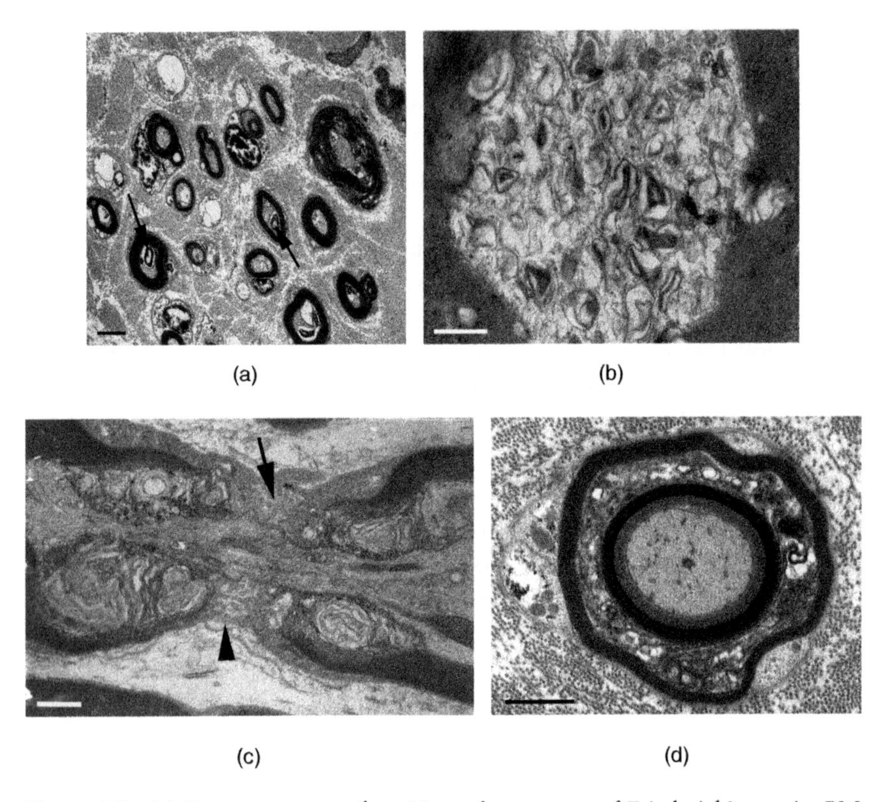

Figure 4.9 (a) Post-mortem artefact. Nerve from a case of Friedreich's ataxia. PM delay 2 days. Fixed first in paraformaldehyde for 24 hours, then in glutaraldehyde for 3 days. The nuclear chromatin is reduced to a dense circle.Many of the unmyelinated axons are empty and distended, and the SCs contain large empty vacuoles. There are myelin figures resulting from the delay in fixation (arrow). The myelin on small myelinated fibres is surprisingly well preserved, especially compared with the poorly preserved myelin on the larger fibres. Bar = 2 μm. (b) Experimental processing; fixation in cacodylate buffered glutaraldehyde for 3 hours, overnight in cacodylate buffer, fixed in osmium tetroxide for 4 hours and dehydration times are: 50% ethanol for 10 minutes, 70% ethanol for 10 minutes and absolute alcohol for 20 minutes. Glutaraldehyde stabilises but does not cross-link, so during the overnight wash in buffer, myelin lipids leached out into the axoplasm where they were fixed by osmium tetroxide. Bar = 2 μm. (c) Experimental processing, 2.5% glutaraldehyde–PIPES buffer (no added sucrose) for 4 hours; this buffer has a low osmolality, and the paranodal myelin is poorly preserved. Nodal processes are clearly seen (arrow) as is the SC basal lamina that continues from one SC across the node to the next cell (arrowhead). Bar = 1 μm. (d) This ulnar nerve neuroma was fixed in glutaraldehyde followed by 4 hours in osmium tetroxide plus potassium ferricyanide. This myelinated fibre in the centre of the neuroma is inadequately osmicated, as the time was insufficient to permit penetration throughout the whole thickness of the myelin sheath resulting in the inner layers being less well stained. The Schmidt–Lanterman incisure is poorly preserved, and the lamellae are altered to artefactual myelin figures. Bar = 1 μm.

myelin lamellae, particularly in the paranodes and Schmidt–Lanterman incisures (Figure 4.9c). Inadequate osmication results in the outer layers of the myelin sheath being more densely stained than the inner ones (Figure 4.9d).

Abnormal structures in the paranodes and Schmidt–Lanterman incisures are more likely to be artefactual than pathological; the earliest changes in degeneration and demyelination involve changes to the SC in the paranodal region before myelin abnormalities are seen. Experimentally, raised intracellular calcium levels produced vesicular demyelination in the paranodes and Schmidt–Lanterman incisures (Smith, Hall and Schauf, 1985).

4.6 CONCLUSIONS

The reluctance to perform diagnostic nerve biopsies nowadays must involve occasionally missing the correct diagnosis, as in rare cases unexpected ultrastructural findings may point to a totally unanticipated diagnosis. Where an autoimmune disease is suspected, finding myelin stripping is diagnostic. When demyelination is suspected, this must always be confirmed by TEM as LM does not distinguish a bare axon from enlarged SC processes or dilated unmyelinated fibres. Similarly, when the occasional onion bulb is seen, this should be examined further by TEM. The age of the patient is also important as occasional demyelinated fibres occur in normal nerves after the age of about 55 years. When widely spaced myelin is found by TEM, the chance of this neuropathy not being related to a paraprotein is extremely small.

Ultrastructural examination may reveal small deposits of amyloid that were not suspected by LM. It is particularly easy to miss if located in the perineurium or adjacent to endoneurial blood vessels. Leprosy bacteria are rare in nerves at the tuberculoid end of the spectrum, and TEM is the most efficient way of finding them.

Neuropathy due to a loss of one copy of the gene for PMP22 (hereditary neuropathy related to pressure palsy (HNPP)) is relatively common but can have a nonspecific presentation, and there is often no family history. This neuropathy is associated with abnormalities of folding of the myelin sheath (tomacula), especially in the paranode. Small as well as large fibres may be affected, but the abnormality will be resolved only by TEM. The proportion of affected fibres is often small, although a better idea of numbers will be seen if individual nerve fibres are teased out for examination. Finding tomaculous changes together with nonspecific demyelination and axonal degeneration should suggest DNA analysis. Finding sporadic onion bulbs consisting only of basal laminae suggests the much rarer disease, CMT4C.

In an ideal world, ultrastructural examination should be undertaken on all nerve biopsies. It will provide information as to the existence of early axonal regeneration and remyelination and hence whether recovery is underway. It also should confirm (or deny) the presence of a demyelinating component and disclose any significant inclusions in axons or SCs. Although this is a time-consuming procedure, it has been improved by the ability to record images digitally.

REFERENCES

Aguayo, A.J., Bray, G.M., Terry, L.C. *et al.* (1976) Three dimensional analysis of unmyelinated fibers in normal and pathologic autonomic nerves. *Journal of Neuropathology and Experimental Neurology*, 35, 136–151.

Asbury, A.K., Cox, S.C. and Baringer, J.R. (1971) The significance of giant vacuolation of endoneurial fibroblasts. *Acta Neuropathologica*, **18**, 123–131.

Balice-Gordon, R.J., Bone, L.J. and Scherer, S.S. (1998) Functional gap junctions in the Schwann cell myelin sheath. *The Journal of Cell Biology*, **142**, 1095–1104.

Ballin, R.H.M. and Thomas, P.K. (1969a) Changes at the nodes of Ranvier during Wallerian degeneration: an electron microscope study. *Acta Neuropathologica (Berlin)*, **14**, 237–249.

Ballin, R.H.M. and Thomas, P.K. (1969b) Electron microscope observations on demyelination and remyelination in experimental allergic neuritis. Part 1: demyelination. *Journal of the Neurological Sciences*, **8**, 1–18.

Baur, P.S. and Stacey, T.R. (1977) The use of PIPES buffer in the fixation of mammalian and marine tissues for electron microscopy. *Journal of Microscopy*, **109**, 315–327.

Berthold, C.H., Fraher, J.P., King, R.H.M. *et al.* (2005) Microscopic anatomy of the peripheral nervous system, in *Peripheral Neuropathy*, 4th edn, vol. 1 (eds P.J. Dyck and P.K. Thomas), Elsevier, Philadelphia, PA, pp. 35–91.

Berthold, C.H. and Rydmark, M. (1995) Morphology of normal peripheral axons, in *The Axon; Structure, Function and Pathophysiology* (eds S.G. Waxman, J.D. Kocsis and P.K. Stys), Oxford University Press, Oxford, pp. 13–48.

Bischoff, A. and Ulrich, J. (1969) Peripheral neuropathy in globoid cell leukodystrophy (Krabbe's disease). Ultrastructural and histochemical findings. *Brain*, **92**, 861–870.

Boyle, M.E., Berglund, E.O., Murai, K.K. *et al.* (2001) Contactin orchestrates assembly of the septate-like junctions at the paranode in myelinated peripheral nerve. *Neuron*, **30**, 385–397.

Bradley, J., Thomas, P.K., King, R.H.M. *et al.* (1990) Morphometry of endoneurial capillaries in diabetic sensory and autonomic neuropathy. *Diabetologia*, **33**, 611–618.

Bradley, J.L., Thomas, P.K., King, R.H.M. *et al.* (1994) A comparison of perineurial and vascular basal laminal changes in diabetic neuropathy. *Acta Neuropathologica*, **88**, 426–432.

Court, F.A., Sherman, D.L., Pratt, T. *et al.* (2004) Restricted growth of Schwann cells lacking Cajal bands slows conduction in myelinated nerves. *Nature*, **431**, 191–195.

Cruz, D., Watson, A.D., Miller, C.S. *et al.* (2008) Host-derived oxidized phospholipids and HDL regulate innate immunity in human leprosy. *Journal of Clinical Investigation*, **118**, 2917–2928.

Dingemans, K.P. and Teeling, P. (1994) Long-spacing collagen and proteoglycans in pathologic tissues. *Ultrastrucural Pathololog*y, **18**, 539–547.

Donaghy, M., King, R.H.M., Thomas, P.K. *et al.* (1988) Abnormalities of the axonal cytoskeleton in giant axonal neuropathy. *Journal of Neurocytology*, **17**, 197–208.

Fabrizi, G.M., Cavallaro, T., Angiari, C. *et al.* (2004) Giant axon and neurofilament accumulation in Charcot-Marie-Tooth disease type 2E. *Neurology*, **62**, 1429–1431.

Fardeau, M., Abelanet, R. and Laudet, P. (1970) Maladie de Refsum. Etude histologique, ultrastructurale et biochemique d'une biopsie de nerf périphérique. *Revue Neurologique*, **122**, 185–196.

Griffin, J.W., Li, C.Y., Ho, T.W. *et al.* (1995) Guillain-Barre syndrome in northern China. The spectrum of neuropathological changes in clinically defined cases. *Brain*, **118**, 577–595.

Hall, S.M. and Williams, P.L. (1970) Studies on the 'incisures' of Schmidt and Lanterman. *Journal of Cell Science*, **6**, 767–792.

Hedley-Whyte, E.T. (1973) Myelination of rat sciatic nerve: comparison of undernutrition and cholesterol biosynthesis inhibition. *Journal of Neuropathology and Experimental Neurology*, **32**, 284–302.

Hirano, A. (1994) Hirano bodies and related neuronal inclusions. *Neuropathology and Applied Neurobiology*, **20**, 3–11.

Hirokawa, N. (1982) Cross-linker system between neurofilaments, microtubules, and membranous organelles in frog axons revealed by the quick-freeze, deep-etching method. *The Journal of Cell Biology*, **94**, 129–142.

Houlden, H., Laura, M., Ginsberg, L. *et al.* (2009) The phenotype of Charcot-Marie-Tooth disease type 4C due to SH3TC2 mutations and possible predisposition to an inflammatory neuropathy. *Neuromuscular Disorders*, **19**, 264–269.

Jacobs, J.M. and Costa-Jussa, F.R. (1985) The pathology of amiodarone neurotoxicity. II. Peripheral neuropathy in man. *Brain*, **108**, 753–769.

Jacobs, J.M. and Love, S. (1985) Qualitative and quantitative morphology of human sural nerve at different ages. *Brain*, **108**, 897–924.

King, R.H.M. (1999) *Atlas of Peripheral Nerve Pathology*, Arnold, London.

King, R.H., Chandler, D., Lopaticki, S. *et al.* (2011) Ndrg1 in development and maintenance of the myelin sheath. *Neurobiology of Disease*, **42**, 368–380.

King, R.H.M., Llewelyn, J.G., Thomas, P.K. *et al.* (1988) Perineurial calcification. *Neuropathology and Applied Neurobiology*, **14**, 105–123.

King, R.H.M., Llewelyn, J.G., Thomas, P.K. *et al.* (1989) Diabetic neuropathy: abnormalities of Schwann cell and perineurial basal laminae. Implications for diabetic vasculopathy. *Neuropathology and Applied Neurobiology*, **15**, 339–355.

King, R.H.M., Pollard, J.D. and Thomas, P.K. (1975) Aberrant remyelination in chronic relapsing experimental allergic neuritis. *Neuropathology and Applied Neurobiology*, **1**, 367–378.

King, R.H. and Thomas, P.K. (1971) Electron microscope observations on aberrant regeneration of unmyelinated axons in the vagus nerve of the rabbit. *Acta Neuropathologica*, **18**, 150–159.

King, R.H.M. and Thomas, P.K. (1984) The occurrence and significance of myelin with unusually large periodicity. *Acta Neuropathologica (Berlin)*, **63**, 319–329.

King, R.H.M., Tournev, I., Colomer, J. *et al.* (1999) Ultrastructural changes in peripheral nerve in hereditary motor and sensory neuropathy-Lom. *Neuropathology and Applied Neurobiology*, **25**, 306–312.

Kleitman, N. and Bunge, R. (1995) The Schwann cell: morphology and development, in *The Axon: Structure, Function and Pathophysiology* (eds S.G. Waxman, J.D. Kocsis and P.K. Stys), Oxford University Press, New York, pp. 97–115.

Kocen, R.S., King, R.H.M., Thomas, P.K. *et al.* (1973) Nerve biopsy in two cases of Tangier disease. *Acta Neuropathologica (Berlin)*, **26**, 317–327.

Kocen, R.S. and Thomas, P.K. (1970) Peripheral nerve involvement in Fabry's disease. *Archives of Neurology (Chicago)*, **22**, 81–88.

Lampert, P.W. (1967) A comparative electron microscopic study of reactive, degenerating, regenerating and dystrophic axons. *Journal of Neuropathology and Experimental Neurology*, **26**, 345–368.

Lhermitte, F., Fardeau, M., Chedru, F. *et al.* (1976) Polyneuropathy after perhexilene maleate therapy. *British Medical Journal*, **1** (6020), 1256.

Luse, S.A., Zopf, D. and Cox, J.W. (1963) An electron microscopic study of *in vitro* and *in vivo* long-spacing collagen. *The Anatomical Record*, **145**, 254–255.

Lyon, G. and Evrard, P. (1970) Sur la présence d'inclusions cristallines dans les cellules de Schwann dans diverses neuropathies périphériques. *Comptes Rendues Hebdominaires des Sceances de l'Academie des Sciences (Paris). D: Sciences Naturelles*, **271**, 1000–1002.

Maselli, A., Furukawa, R., Thomson, S.A. *et al.* (2003) Formation of Hirano bodies induced by expression of an actin cross-linking protein with a gain-of-function mutation. *Eukaryotic Cell*, **2**, 778–787.

Mi, H., Deerinck, T.J., Ellisman, M.H. *et al.* (1995) Differential distribution of closely related potassium channels in rat Schwann cells. *Journal of Neuroscience*, **15**, 3761–3774.

Mi, H., Deerinck, T.J., Jones, M. *et al.* (1996) Inwardly rectifying K+ channels that may participate in K+ buffering are localized in microvilli of Schwann cells. *Journal of Neuroscience*, **16**, 2421–2429.

Midroni, G., Bilbao, J.M. and Cohen, S.M. (1995) The axon: normal structure and pathological alterations, in *Biopsy Diagnosis of Peripheral Neuropathy* (eds G. Midroni and J.M. Bilbao), Butterworth-Heinemann, Boston, pp. 45–74.

Misra, V.P., King, R.H.M., Harding, A.E. *et al.* (1991) Peripheral neuropathy in the Chédiak-Higashi syndrome. *Acta Neuropathologica (Berlin)*, **81**, 354–358.

Ogata, J., Budzilovich, G.N. and Cravioto, H. (1972) A study of rod-like structures (Hirano bodies) in 240 normal and pathological brains. *Acta Neuropathologica*, **21**, 61–67.

Oh, S.J. and Kim, J.M. (1976) Giant axonal swelling in "huffer's" neuropathy. *Archives of Neurology*, **33**, 583–586.

Ohnishi, A., O'Brien, P.C. and Dyck, P.J. (1976a) Studies to improve fixation of human nerves. IV. Effect of time elapsed between death and glutaraldehyde fixation on density of microtubules and neurofilaments. *Journal of Neuropathology and Experimental Neurology*, **35**, 26–29.

Ohnishi, A., O'Brien, P.C. and Dyck, P.J. (1976b) Studies to improve fixation of human nerves. V. Effect of temperature, fixative and $CaCl^2$ on density of microtubules and neurofilaments. *Journal of Neuropathology and Experimental Neurology*, **35**, 167–179.

Parkash, O. (2009) Classification of leprosy into multibacillary and paucibacillary groups: an analysis. *FEMS Immunoogy and Medical Microbiology*, **55**, 1–5.

Quarles, R.H. (2007) Myelin-associated glycoprotein (MAG): past, present and beyond. *Journal of Neurochemistry*, **100**, 1431–1448.

Rambukkana, A., Zanazzi, G., Tapinos, N. *et al.* (2002) Contact-dependent demyelination by *Mycobacterium leprae* in the absence of immune cells. *Science*, **296**, 927–931.

Robitaille, Y., Carpenter, S., Karpati, G. *et al.* (1980) A distinct form of adult polyglucosan body disease with massive involvement of central and peripheral neuronal processes and astrocytes: a report of four cases and a review of the occurrence of polyglucosan bodies in other conditions such as Lafora's disease and normal ageing. *Brain*, **103**, 315–336.

Shanthaveerappa, T.R. and Bourne, G.H. (1962) 'The perineurial epithelium', a metabolically active, continuous, protoplasmic cell barrier surrounding peripheral nerve fasciculi. *Journal of Anatomy*, **96**, 527–537.

Sherman, D.L., Tait, S., Melrose, S. *et al.* (2005) Neurofascins are required to establish axonal domains for saltatory conduction. *Neuron*, **48**, 737–742.

Smith, K.J. and Hall, S.M. (1988) Peripheral demyelination and remyelination initiated by the calcium-selective ionophore ionomycin: *in vivo* observations. *Journal of Neurological Science*, **83**, 37–53.

Smith, K.J., Hall, S.M. and Schauf, C.L. (1985) Vesicular demyelination induced by raised intracellular calcium. *Journal of Neurological Science*, **71**, 19–37.

Southam, E., Thomas, P.K., King, R.H.M. *et al.* (1991) Experimental vitamin E deficiency in rats. Morphological and functional evidence of abnormal axonal transport secondary to free radical damage. *Brain*, **114**, 915–936.

Stoll, G., Griffin, J.W., Li, C.Y. *et al.* (1989) Wallerian degeneration in the peripheral nervous system: participation of both Schwann cells and macrophages in myelin degradation. *Journal of Neurocytology*, **18**, 671–683.

Thomas, P.K., Halpern, J.P., King, R.H.M. *et al.* (1984) Galactosylceramide lipidosis: novel presentation as a slowly progressive spinocerebellar degeneration. *Annals of Neurology*, **16**, 618–620.

Thomas, P.K. and King, R.H. (1974) The degeneration of unmyelinated axons following nerve section: an ultrastructural study. *Journal of Neurocytology*, **3** (4), 497–512.

Thomas, P.K., King, R.H., Chiang, T.R. *et al.* (1990) Neurofibromatous neuropathy. *Muscle and Nerve*, **13**, 93–101.

Thomas, P.K., King, R.H.M., Kocen, R.S. *et al.* (1977) Comparative ultrastructural observations on peripheral nerve abnormalities in the late infantile, juvenile and late onset forms of metachromatic leukodystrophy. *Acta Neuropathologica (Berlin)*, **39**, 237–245.

Thomas, P.K., King, R.H.M. and Sharma, A.K. (1980) Changes with age in the peripheral nerves of the rat. An ultrastructural study. *Acta Neuropathologica (Berlin)*, 52, 1–6.

Thomas, P.K., King, R.H., Workman, J.M. *et al.* (2000) Hypertrophic perineurial dysplasia in multifocal and generalized peripheral neuropathies. *Neuropathology and Applied Neurobiology*, **26**, 536–543.

Thomas, P.K., Ochoa, J.L., Berthold, C.H. *et al.* (1993) Microscopic anatomy of the peripheral nervous system, in *Peripheral Neuropathy*, 3rd edn, vol. 1 (eds P.J. Dyck and P.K. Thomas), W.B. Saunders, Philadelphia, PA, pp. 28–92.

Tyson, J., Ellis, D., Fairbrother, U. *et al.* (1997) Hereditary demyelinating neuropathy of infancy. A genetically complex syndrome. *Brain*, **120**, 47–63.

Vallat, J.M., Desproges-Gotteron, R., Leboutet, M.J. *et al.* (1980) Cryoglobulinemic neuropathy: a pathological study. *Annals of Neurology*, **8**, 179–185.

Vital, C., Battin, J., Rivel, J. *et al.* (1976) Aspects ultrastructuraux des lésions du nerf périphérique dans un cas de maladie de Farber. *Revue Neurologique (Paris)*, **132**, 419–423.

Vital, C., Dumas, P., Latinville, D. *et al.* (1986) Relapsing inflammatory demyelinating polyneuropathy in a diabetic patient. *Acta Neuropathologica*, **71**, 94–99.

Waxman, S.G., Kocsis, J.D. and Stys, P.K. (1995) *The Axon: Structure, Function and Pathophysiology*, Oxford University Press, Oxford.

Williams, P.L. and Landon, D.N. (1963) Paranodal apparatus of peripheral nerve fibres of mammals. *Nature (London)*, **198**, 670–673.

5

The Diagnostic Electron Microscopy of Tumours

Brian Eyden

Department of Histopathology, Christie NHS Foundation Trust, Manchester, United Kingdom

5.1 INTRODUCTION

Transmission electron microscopy (TEM) has been used for the diagnosis of problematical tumours – those which challenge the pathologist working mainly at the light microscopy level – since the late 1960s. The technique enjoyed widespread popularity throughout the 1970s and 1980s especially, transforming our understanding of cells generally and tumour cells in particular, but thereafter declined in importance as immunohistochemistry (IHC) became a dominant ancillary technique; it was considered to give the same answers as TEM but on the basis of different criteria, as well as being more convenient and less expensive. Now, some 30 years on from the first diagnostic applications of IHC, there are far fewer TEM units attached to pathology departments. However, in spite of these negative consequences for TEM, there are still operational units in most developed countries, in which the ultrastructural diagnosis of tumours continues to be practised. There is at least one major reason for this – the earlier supposed specificity of

Diagnostic Electron Microscopy: A Practical Guide to Interpretation and Technique, First Edition. Edited by John W. Stirling, Alan Curry and Brian Eyden.

antibodies used in IHC has been shown to be fallacious; indeed, there are very few completely tumour-cell-specific antibodies. As a result, large panels of immunostains may be needed to define a tumoural cell differentiation in a problematical case, and this increases the time and cost of IHC. Other limitations of IHC which compromise confidence of interpretation include the lack of immunoreactivity of some tumours (e.g. fibroblastic lesions) to large panels of antibodies, and the obtaining of different results in different laboratories for the same antibody and from changing techniques and instrumentation. Consequently, even today and despite the mature development of IHC, tumours continue to be encountered where there are discrepancies between clinical, histological and immunohistochemical findings. In such circumstances, as well as in the case of tumours where diagnostic confidence may be low as a result of their rarity, TEM can be diagnostically contributory. Finally, outside the diagnostic field, ultrastructural analysis can be important in research by contributing to our understanding of the cells which have a role in cancer progression, such as the myofibroblast and macrophage (Eyden, 2005a,b; De Wever *et al.*, 2008).

5.2 PRINCIPLES AND PROCEDURES FOR DIAGNOSING TUMOURS BY ELECTRON MICROSCOPY

5.2.1 The Objective of Tumour Diagnosis

The purpose of diagnosing a tumour is to give it a name which an oncologist can recognise and for which an appropriate treatment can be selected. Pathologists have done their best to use names that reflect the biology or nature (cell differentiation) of the lesional cells – for example 'myomelanocytic tumour' for neoplasms combining features of smooth-muscle and melanocytic differentiation (Barnard and Lajoie, 2001; Tazelaar, Batts and Srigley, 2001). However, the ideal of basing terminology exclusively on cell differentiation is far from achieved, and terminological idiosyncrasies and misnomers exist. For example, myofibroblastoma is not, as its name might suggest, a myofibroblastic proliferation but one showing a smooth-muscle ultrastructure (Eyden *et al.*, 1999; Eyden and Chorneyko, 2001). In these circumstances, TEM has a role in developing more appropriate terminologies for tumours.

5.2.2 The Intellectual Requirements for Tumour Diagnosis by Electron Microscopy

The intellectual requirements for practising tumour diagnosis by TEM include, first and foremost, knowledge of the ultrastructural features of named tumours. It is therefore essential for the practitioner to spend time making his or her own observations on named tumours. The advantage of this approach is that one always feels a greater conviction from one's own observations than those of others, even if the latter have appeared in peer-reviewed journals. Reading the literature is, however, essential for rare tumour entities, which may not fall within the experience of a given practitioner, even in the compass of an entire career. Consultation of some particular journals – *Cancer, American Journal of Surgical Pathology, Human Pathology, Virchows Archiv* and *Histopathology* – is recommended, especially the years 1970–1990, since these are full of useful, generously illustrated tumour case reports, while *Ultrastructural Pathology* harbours some of the definitive ultrastructural descriptions of tumours. Several monographs have also contributed to our knowledge of tumour ultrastructure (Henderson *et al.*, 1986; Erlandson, 1994; Eyden, 1996; Dickersin, 2000; Eyden *et al.*, 2012).

Having knowledge of the light and electron microscopy features of normal cells and tissues is also important because tumours often recapitulate the features of the so-called normal putative cellular counterpart from which they are commonly (but erroneously) supposed to derive. For example, squamous cell carcinomas show the dense tonofibrils, desmosomes and basal lamina with anchoring fibrils which are seen in normal keratinocytes, malignant melanoma cells possess the melanosomes of the normal melanocyte and Schwannoma have the lamina-coated processes regarded as tumoural counterparts of the lamina and myelin sheath in the normal cell. A histology or cell biology textbook, with an ultrastructural emphasis, is therefore valuable, and Fawcett (1986), Rhodin (1975), Pollard and Earnshaw (2002) and Pavelka and Roth (2005) can be recommended.

Finally, and related to the preceding point, one needs to have knowledge of the features that collectively make up a given cell differentiation, for example squamous-cell, glandular, neuroendocrine, steroidogenic, smooth-muscle, striated-muscle, endothelial and all the other cell differentiations that are expressed in tumours. The value of ultrastructure in defining these cell differentiations is founded on the fact that each cell has a distinctive ultrastructure, based on a collection of organelles or cell structures. There are admittedly few completely specific single organelles for a given cell type (the Weibel–Palade body and the Langerhans cell granule come to mind), but the overall aggregate of ultrastructural findings is often strongly indicative of a given cell differentiation and a given tumour. Table 5.1 gives a list of the main cell structures and organelles characterising the common cell types and their tumours, while Table 5.2 lists the characteristic ultrastructural features of named tumours. It should be remembered that neoplastic cells may harbour variant or scarce organelles compared with the putative normal cell counterpart, and that completely new features may appear (in the process of metaplasia (transdifferentiation)) (Banerjee and Eyden, 2008). Inevitably, some tumours will be encountered which are so poorly differentiated as to have no identifiable marker organelles.

5.2.3 Technical Considerations

Most tumour tissue for diagnostic TEM is retrieved from histological formalin at the cut-up stage. Glutaraldehyde on fresh tissue gives better structural preservation, but this requires staff and is not possible in all departments. For formalin fixation, small pieces of tumour tissue (usually selected by a pathologist or trained practitioner) of approximately 1 mm side ('millimetre cubes') are processed by conventional procedures into Epoxy resin. It is good practice to keep some tissue in formalin aside for *all* specimens until the haematoxylin-and-eosin (H&E) sections reveal whether or not there is a diagnostic problem; the choice can then be made to process or discard the tissue.

If wet aldehyde-fixed tissue is not available, tissue for TEM can be obtained by deparaffinisation of the wax block (Johannessen, 1977). It is convenient to excise a small fragment using a scalpel blade from an area of the wax block based on matching it with the H&E section and to leave it in xylene overnight. A condensed procedure in which osmium tetroxide is dissolved in the deparaffinising xylene has been found to be successful (Van Den Bergh Weerman and Dingemans, 1984). Not surprisingly, given that the tissue has been exposed to hot xylene and hot wax, structural preservation is usually compromised. The cell and matrix constituents that appear to be relatively well preserved mostly contain a significant amount of protein (for example heterochromatin, desmosomal plaques, tonofibrils, melanosomes and Langerhans cell granules); those that are not well retained tend to have a substantially lipidic composition (smooth endoplasmic reticulum (sER)), mitochondria (variably

Table 5.1 Cell structures defining differentiation in common tumours

Organelle	Differentiation
Round or epithelioid cell tumours	
Rough endoplasmic reticulum (rER)	Plasma cell
Melanosomes	Melanocytic
Neuroendocrine granules	Neuroendocrine Neuronal
Mucigen and serous granules	Glandular epithelial
Desmosomes	Epithelial
Tonofibrils	Epithelial
Basal lamina	Epithelial
Lumina and microvilli	Glandular epithelial
Long, slender, smooth microvilli	Mesothelial
Cilia	Epithelial or ependymal
Mitochondria	Oncocytic (glandular epithelial) Steroidogenic
Sarcomeres	Rhabdomyoblastic
Lysosomes	Histiocytic (macrophage)
Langerhans cell ('Birbeck') granules	Langerhans cell
Glycogen	Nonspecific but distinctive for Ewing sarcoma
Lipid	Adipocytic Steroidogenic
Smooth endoplasmic reticulum (sER)	Steroidogenic
Cell processes	Glandular epithelial (microvilli) Mesothelial (microvilli) Histiocytic
Spindle cell tumours	
Desmosomes	Epithelial[a]
Tonofibrils	Epithelial[a]
(Basal) lamina	Epithelial[a] Schwannian Perineurial Myogenic Endothelial

(*continued overleaf*)

Table 5.1 (*continued*)

Organelle	Differentiation
Processes coated in lamina	Schwannian Perineurial cell
rER[b]	Fibroblastic Myofibroblastic
Smooth-muscle myofilaments with focal densities	Smooth muscle (e.g. leiomyoma and leiomyosarcoma) Myofibroblastic Pericytic
Fibronexus junctions	Myofibroblastic

[a]For example, spindle-cell carcinoma.
[b]rER is prominent also in chondroblastic and osteoblastic tumours, which typically do not have a spindled morphology.
Modified from Eyden (2005) and Eyden and Banerjee (2008).

preserved), lipid droplets and the small clear vesicles of chromophobe renal cell carcinoma (Bonsib, Bray and Timmerman, 1993). For further details on technique, see Chapter 14.

5.2.4 Identifying Good Preservation

Interpretation is facilitated by good preservation. To a certain extent, good and poor preservation can be identified in the semithin toluidine-blue-stained section, and appropriate blocks should be selected on this basis for ultramicrotomy. Sharply delineated ('contrasty') nuclei with clear contents are often indicative of poor preservation (even though pathologists may like this appearance because it reminds them of the sharply defined nuclei in H&E sections). By contrast, pale-staining cells, where perhaps only strongly stained nucleoli mark the position of nuclei, are well preserved (see Eyden, 2002, for further details). These light microscopy images have ultrastructural correlates. Nuclei with prominent peripheral heterochromatin (except for normal or reactive haematolymphoid cells), clear intranuclear spaces and an expanded, clear perinuclear cisterna are to be regarded as showing poor preservation; well-preserved nuclei have a fine uniform granular texture to the nuclear interior. Similarly, in the cytoplasm, masses of intermediate filaments which have been shown by correlated IHC to be vimentin may artefactually condense to look like tonofibrils, and generally any

Table 5.2 Selected uncommon tumours with distinctive organelles

Adrenocortical carcinoma	Lipid droplets Smooth endoplasmic reticulum (sER) Mitochondria with tubular cristae Basal lamina No desmosomes
Alveolar soft-part sarcoma	Rhomboidal crystals Lamina
Aggressive angiomyxoma	Intermediate filaments
Angiomyofibroblastoma	Lamina Attachment plaques Plasmalemmal caveolae No fibronexus junctions
Angiosarcoma	Lumina without microvilli and bordered by pinocytotic vesicles Intermediate filaments Lamina
Astrocytoma	Processes containing intermediate (glial) filaments Lysosomes Rosenthal fibres
Biphasic synovial sarcoma	Lumina and microvilli Junctional complexes Desmosomes Basal lamina
Chromophobe renal-cell carcinoma	Small clear vesicles
Embryonal sarcoma of liver	Dense cytoplasmic granules
Ependymoma	Primitive junctions (not desmosomes) Lumina lined by microvilli and cilia Basal lamina
Epithelioid sarcoma	Abundant intermediate filaments Tonofibrils No desmosomes No lamina Peripheral smooth-muscle type myofilaments Fibronexus junctions
Ewing's sarcoma and peripheral neuro-ectodermal tumour	Glycogen Cell processes Intermediate filaments and microtubules Neuroendocrine granules

(*continued overleaf*)

Table 5.2 (*continued*)

Follicular dendritic reticulum cell-sarcoma	Desmosomes
Gastrointestinal stromal tumour and gastrointestinal autonomic nerve tumour (GANT)	Cell processes Intermediate filaments Mitochondria sER Neuroendocrine granules (GANT) Microtubules in processes (GANT) Skeinoid fibres
Granular-cell tumour	Cell processes coated in lamina Secondary lysosomes
Granulosa-cell tumour	Desmosomes Processes Call–Exner bodies lined by basal lamina
Hibernoma	Mitochondria
Interdigitating reticulum-cell sarcoma	Interdigitating cell processes
Juxtaglomerular-cell tumour	Rhomboidal crystals
Kaposi's sarcoma	Tubuloreticular inclusions Siderosomes
Langerhans cell granulomatosis	Birbeck granules Interdigitating cell processes
Leydig cell tumour	Lipid droplets sER Mitochondria with tubular cristae Reinke crystals
Malignant rhabdoid tumour	Paranuclear whorls of intermediate filaments
Merkel cell tumour	Paranuclear whorls of intermediate filaments Spinous processes Neuroendocrine granules
Mesothelioma	Desmosomes Tonofibrils Basal lamina Lumina Long, slender, smooth microvilli (length: diameter exceeding 15) Collagen–microvilli associations Amorphous hyaluronic acid–like deposits associated with microvilli
Myoepithelioma	Desmosomes Tonofibrils Lamina Smooth-muscle-type myofilaments

Table 5.2 (*continued*)

Myofibroblastoma	Lamina Attachment plaques Plasmalemmal caveolae No fibronexus junctions
Neuroblastoma	Microtubules in cell processes Neuroendocrine granules
Oligodendroglioma	Short processes Microtubules Lysosomes
Oncocytoma	Mitochondria
Perivascular epithelioid cell tumour (PEComa)	Melanosomes
	Lamina Attachment plaques Smooth-muscle-type myofilaments
Seminoma	Rope-like nucleoli Glycogen Small desmosomes
Sertoli cell tumour	Lumina lined by microvilli Junctional complexes Desmosomes Basal lamina
Signet-ring cell carcinoma	Mucigen granules
Signet-ring cell lymphoma	Vacuoles rER
Signet-ring cell melanoma	Intermediate filaments
Spindle-cell carcinoma	Tonofibrils Desmosomes Smooth-muscle-type myofilaments Fibronexus junctions (rare)
Yolk sac tumour	Glandular lumina Junctional complexes Desmosomes Basal lamina Glycogen Dense ('hyaline') inclusions

clear structureless space (unless it is an unambiguous membrane-bound vacuole) suggests artefactual loss of cytoplasmic constituents.

5.2.5 Distinguishing Reactive from Neoplastic Cells

An important caveat in the diagnostic TEM of tumours is to avoid interpreting reactive cells as neoplastic. In the course of examining an ultrathin section, well-defined organelles are being sought (e.g. a lamina to confirm leiomyosarcoma or a lumen to confirm adenocarcinoma), and care needs to be taken not to interpret the structures of reactive elements as organelles of a neoplastic cell. The important reactive cells commonly found in tumours are those of vessels (where lamina, lumina and Weibel–Palade bodies may be seen), reactive myofibroblasts, rhabdomyoblasts and a variety of haematolymphoid cells. One image which can help in identifying vessels with a relatively closed lumen which are juxtaposed to tumour cells is the *mirror-image polarity* of lamina-cell cytoplasm–lumen-cell cytoplasm–lamina (see Eyden, 1996).

5.3 ORGANELLES AND GROUPS OF CELL STRUCTURES DEFINING CELLULAR DIFFERENTIATION

This section illustrates some of the principal cell structures which have diagnostic value, and it includes some pertinent comments and caveats on their use. Space allows for only an illustrative treatment, and the following monographs – Henderson *et al.* (1986), Erlandson (1994), Eyden (1996), Dickersin (2000) and Eyden *et al.* (2012) – provide additional information.

5.3.1 Rough Endoplasmic Reticulum

Rough endoplasmic reticulum (rER), mitochondria and lysosomes are almost universal cell constituents, and therefore they lack cell specificity. For these organelles, therefore, diagnostic importance arises only when they are found in large quantities or large numbers (e.g. mitochondria in oncocytoma) or when there are distinctive features (e.g. tubular cristae in steroidogenic tumours).

- In spindled cells (Figure 5.1a), abundant rER suggests fibroblastic or myofibroblastic differentiation, while in rounded cells, it can suggest

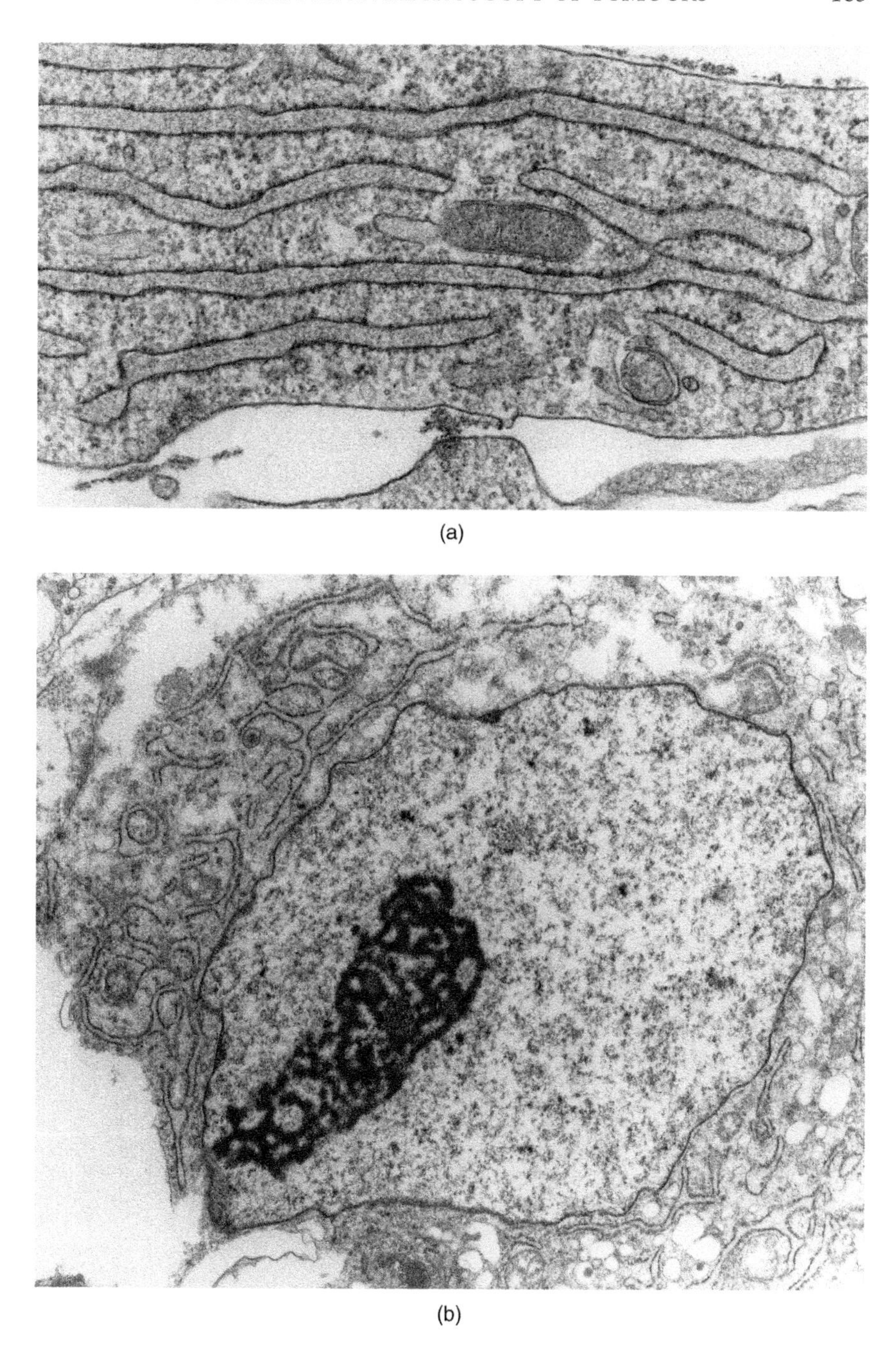

Figure 5.1 Rough endoplasmic reticulum (rER). (a) rER in a spindled-cell profile from a cultured lung fibroblast. The cisternae have a finely textured internal material representing newly synthesised protein not yet transported to the Golgi apparatus for subsequent secretion. (b) rER in a round-cell context in sclerosing epithelioid fibrosarcoma.

plasma cell differentiation, but also, for example, sclerosing epithelioid fibrosarcoma (Figure 5.1b) (Eyden *et al.*, 1998). This illustrates the importance of *context* and how a single organelle can have different diagnostic implications depending on the context of other features.

Lysosomes also assume diagnostic significance when present in large numbers, mostly as a marker of macrophage (histiocytic) differentiation.

- When well preserved, both primary and secondary lysosomes have a narrow uniform space beneath the limiting membrane (Figure 5.2a,b), which helps to distinguish them from neuroendocrine granules with which they can be confused (see Section 5.3.7).

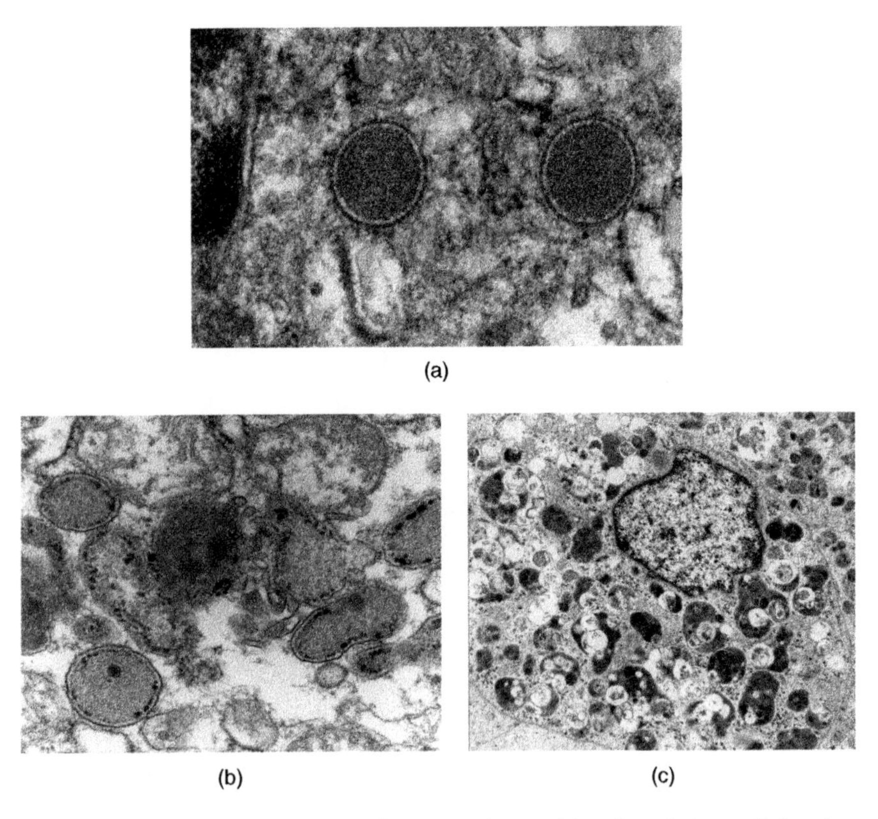

Figure 5.2 Lysosomes. (a) Primary lysosomes in a multinucleated giant cell showing the characteristic fine and uniform space under the limiting membrane. (b) A similar space in late secondary lysosomes in a macrophage from a malignant melanoma. The small dense granules associated with this space are melanin. (c) Abundant secondary lysosomes in a granular-cell tumour (Schwannoma).

- Secondary lysosomes are readily identified by their heterogeneous content, which suggests the digestion of cellular components or ingested extracellular materials. Secondary lysosomes account for the cytoplasmic granularity of granular-cell tumours, the majority of which are Schwannian (mostly benign Schwannoma) (Figure 5.2c). However, the granular-cell appearance is recognised as a phenotype, applicable to several types of cell differentiation and tumours, the main ones being angiosarcoma (McWilliam and Harris, 1985), astrocytoma (Ohta *et al.*, 2010) and leiomyoma and leiomyosarcoma (Abenoza and Sibley, 1987; LeBoit *et al.*, 1991). For a more comprehensive list, see Eyden *et al.* (2012).

5.3.2 Melanosomes

The melanosome is the ultrastructural marker of melanocytic differentiation. In the context of appropriate clinical and light microscopical features, the melanosome can provide confirmation of malignant melanoma in cases of uncertain immunostaining; some malignant melanomas, for example, lack the most reliable melanocytic markers, S100 protein and HMB45 (Bishop *et al.*, 1993; Argenyi *et al.*, 1994).

- Type II (unmelanised) melanosomes are the most readily identifiable (Figure 5.3a). Type III melanosomes are partially melanised and are also diagnostically helpful (Figure 5.3b). However, type IV melanosomes are fully pigmented and can be confused with lysosomes and other pigment granules, and so these are less useful.
- Lattice-deficient melanosomes (with a rather structureless interior) can be confirmed as melanosomes by immunoelectron microscopy using gold-labelled anti-HMB45 antibodies (Eyden *et al.*, 2005a).
- Compound melanosomes are collections of melanosomes enclosed within lysosomes and are not markers of melanocytic differentiation, being found in keratinocytes, macrophages, fibroblasts and myofibroblasts (Figure 5.3c).
- Compound melanosomes are also found in malignant melanoma where they have no diagnostic significance. Biologically, they reflect crinophagy, the intracellular digestion of organelles destined for exteriorisation outside the cell.
- Some tumours, which are not acceptable terminologically as malignant melanomas, contain free type II melanosomes and therefore show melanocytic differentiation but co-express other differentiations. Examples include smooth-muscle differentiation (in myomelanocytic

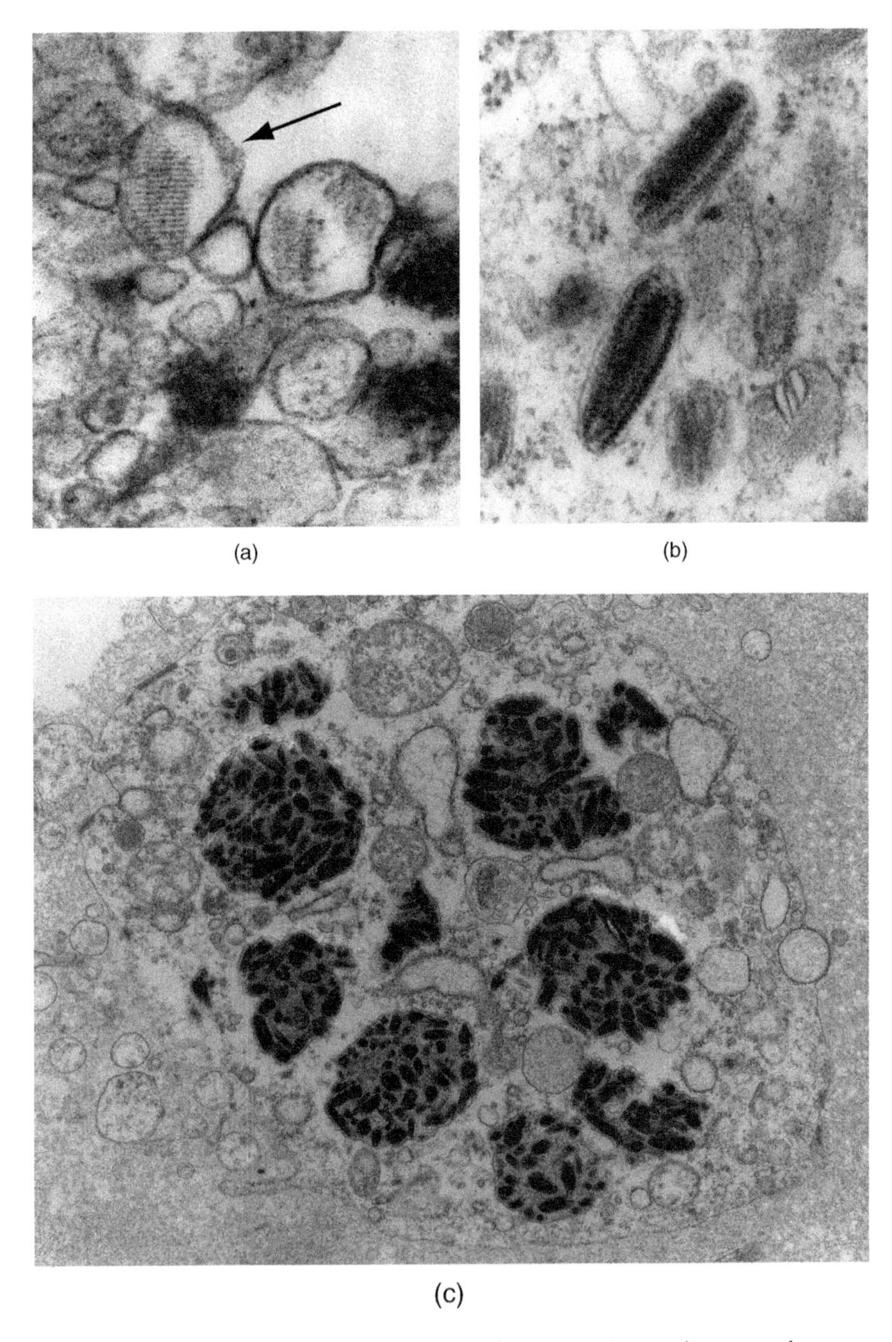

Figure 5.3 Melanosomes. (a) Two type II melanosomes in a malignant melanoma; one (arrow) shows the distinctive internal 6–7 nm periodicity. Neither contains electron-dense melanin. (b) Partially melanised type III melanosomes in a malignant melanoma. (c) Compound melanosomes (secondary lysosomes containing partially degraded melanosomes) in a macrophage in a basal cell carcinoma.

tumours (perivascular epithelioid cell tumours)) (Barnard and Lajoie, 2001; Tazelaar, Batts and Srigley, 2001), neuroendocrine differentiation (Gould *et al.*, 1981) and Schwannian differentiation (Carney, 1990).

5.3.3 Desmosomes

The desmosome is responsible for the cell cohesiveness associated with epithelial tissues and carcinomas. It is one of the most important markers for epithelial differentiation, and is identified by a combination of features – the 'intermediate line' (the result of the imaging of traversing filaments) in the space between the areas of apposed cell membrane, dense plaques under the apposed membranes and tonofibril tails (Figure 5.4a).

- Finding a desmosome in a malignant tumour does not make it a carcinoma. As mentioned in this chapter, organelles need to be interpreted in the context of overall findings. For example, desmosomes are found in tumours which are not referred to as carcinoma, such as peripheral primitive neuroectodermal tumour, granulosa cell tumour, Sertoli cell tumour, cardiac rhabdomyoma, meningioma and dendritic reticulum cell sarcoma. The presence of desmosomes in these tumours simply gives evidence of **an element of epithelial differentiation** to their overall cell differentiation.
- Apart from epithelial neoplasms, most intercellular junctions found in tumours are not desmosomes. The majority of them are primitive structures of no diagnostic importance, even though many have been referred to, somewhat confusingly, as 'desmosome-like' junctions (Figure 5.4b).

5.3.4 Tonofibrils

Tonofibrils are bundles of individual cytokeratin intermediate filaments that are responsible for the cytokeratin immunoreactivity of epithelial tumours. They are less conspicuous in adenocarcinomas than in squamous carcinomas.

- In glandular epithelial tumours, they appear loosely organised ('pale-staining'), and often individual intermediate filaments within the fibril can be discerned.

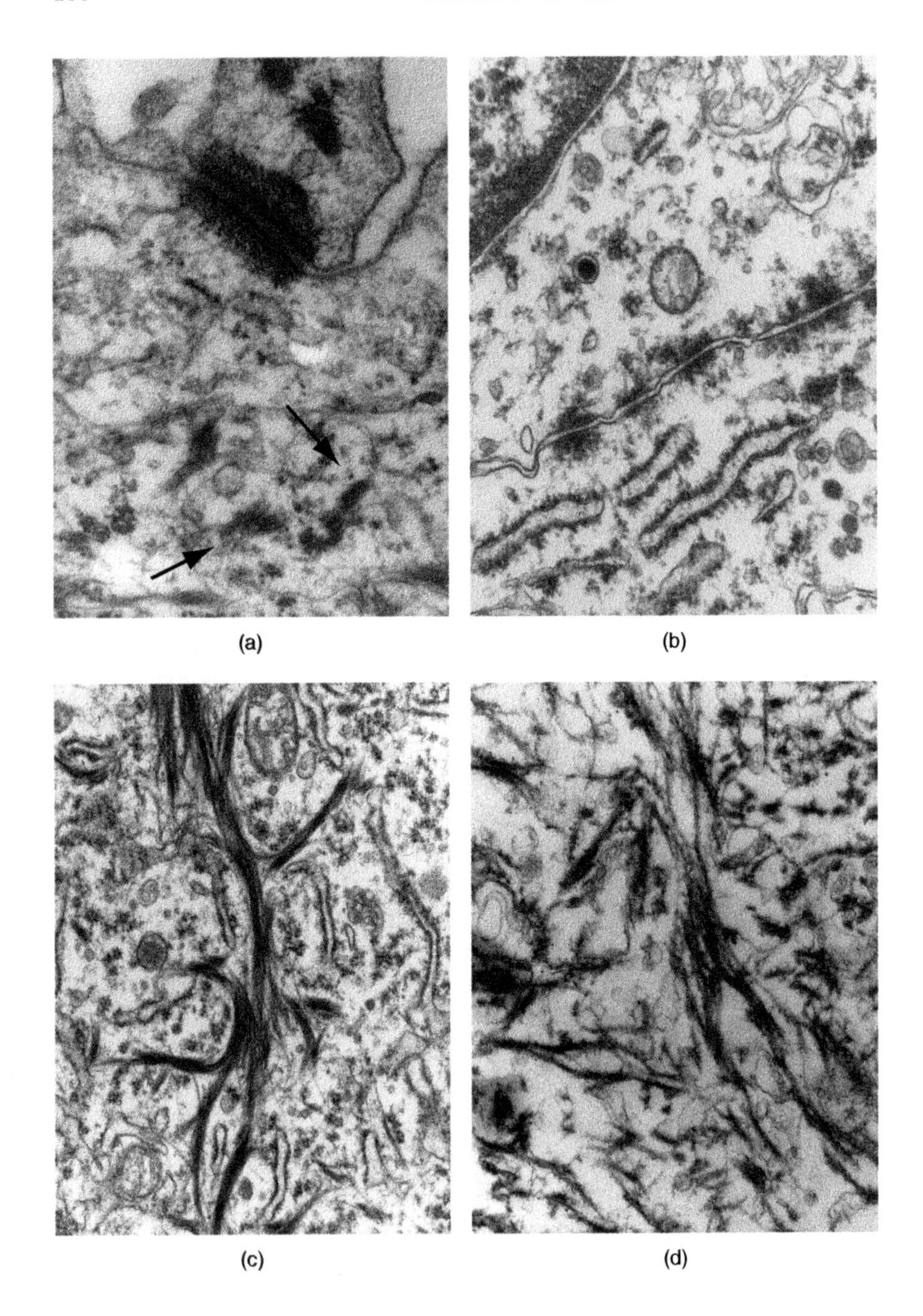
(a)
(b)
(c)
(d)

- In squamous neoplasms, the tonofibrils are more densely staining, and sometimes individual filaments cannot be discerned, appearing as if they have almost fused together (Figure 5.4c). This may reflect suboptimal preservation.
- When tonofibrils are poorly developed or lying in an inappropriate orientation within the section, they may be difficult to identify and may appear as little more than dense granules.
- A diagnostic caveat to note is that in tumours which are immunoreactive solely for vimentin, tonofibril-like structures can be present owing to the artefactual condensation of vimentin (Figure 5.4d). One can be alerted to this by other signs of poor preservation, such as excessively condensed heterochromatin and clear nuclear interiors.

5.3.5 Basal Lamina

Basal lamina consists of a plate of proteinaceous material marking the boundary between basal epithelial cells and underlying stroma (dermis, in the case of skin); a similar structure covers muscle and nerve-sheath cells (Figure 5.5a). In epithelium, it adheres to the basal epithelial cell surfaces by hemidesmosomes and to the stroma by anchoring fibrils. It is regarded as a barrier, the destruction of which is considered to be a prerequisite for tumoural invasion of stroma.

- The basal lamina can be seen in carcinomas, although it can be either less well organised or excessively and aberrantly developed in comparison with the normal putative cell counterpart – for example, as multiple layering, reticulation or an 'amorphous, non-laminate' structure (Eyden, 2005a, b) (Figure 5.5b).
- There is some terminological difficulty with basal lamina. As mentioned in this chapter, a similar structure is found on the surfaces

Figure 5.4 Junctions and tonofibrils. (a) Fully developed desmosome in a squamous cell carcinoma showing intermediate line, plaques and tonofibril tails. Away from the desmosomes are transversely and obliquely sectioned fibrils (arrows), not obviously recognisable as the tonofibrils that they are. (b) Primitive junctions in a gastrointestinal autonomic nerve tumour. They have no diagnostic significance, being common in all kinds of tumour. (c) Well-defined dense tonofibrils more in keeping with squamous-cell differentiation in an ovarian adenocarcinoma. (d) Artefactually condensed vimentin mimicking tonofibrils in a cytokeratin-negative malignant melanoma.

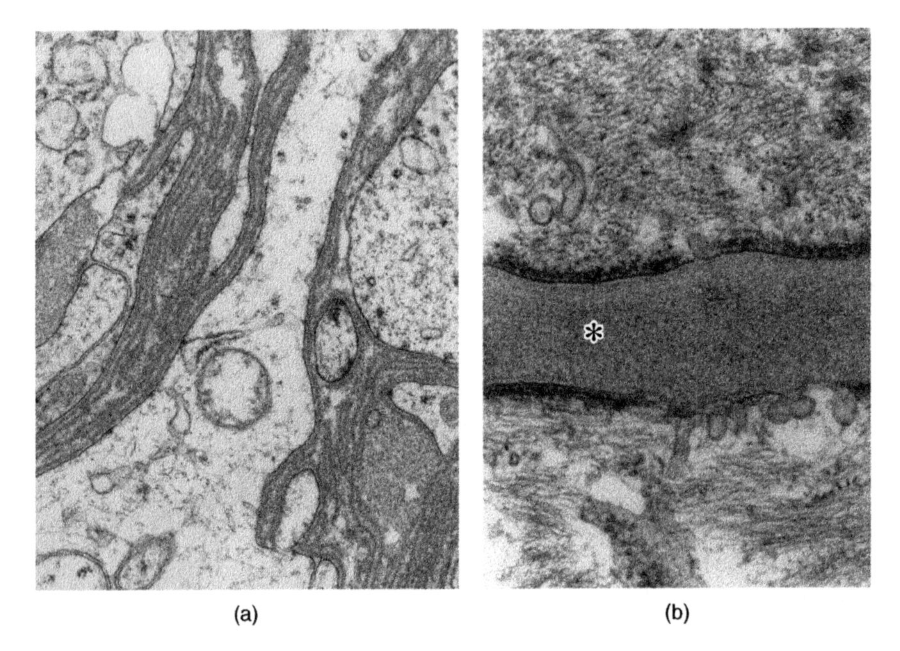

(a) (b)

Figure 5.5 Lamina. (a) Lamina over cell bodies and processes in a Schwannoma. (From Eyden (2005) with permission from Wiley-Blackwell.) (b) A variant of lamina having an amorphous, 'nonlaminate' structure (*) associated with a normal smooth-muscle cell. Reproduced from B. Eyden, Electron Miscroscopy in pathology, The Science, of Laboratory diagnosis, 2nd ed., D Burnett and J Crocker (Eds.), pp43-60, 2005, with permission of John Wiley & Sons.

of Schwann cells and muscle cells. However, since these cells lack a baso-apical polarity, it is inappropriate to refer to this lamina as 'basal' lamina. The term external lamina has been used, but since all laminae are external, this seems illogical. The author has preferred the simple term 'lamina' and restricts 'basal lamina' to where there is a true baso-apical polarity or to tumours which are unambiguously epithelial on the basis of other findings.

- The lamina around endothelium is unique in that endothelial cells are mesenchymal in origin, but the vessel itself has some undoubtedly epithelial features (notably, lumen bearing processes, tight junctions and some cytokeratin immunoreactivity) so that the lamina has a glandular context and 'basal' lamina is permissible.
- Lamina distinguishes epithelium, myoepithelium, endothelium, Schwann and perineurial cells and muscle cells (which have lamina) from neurons, glial cells, fibroblasts, macrophages (and other haematolymphoid cells) and myofibroblasts (which do not).

- While myofibroblasts are often stated in the literature as having focal lamina ('basement membrane'), there is evidence that this is a misinterpretation of the fibronectin fibril (Eyden, 2005a, b).
- It is important to avoid misinterpreting the lamina of reactive cells for the lamina of neoplastic tumour cells.

5.3.6 Glandular Epithelial Differentiation and Cell Processes

While desmosomes, tonofibrils and lamina are pan-epithelial markers, lumina bearing microvillous processes (Figure 5.6a), mucigen granules, serous (zymogen) granules and type II pneumocyte surfactant bodies provide evidence of glandular epithelial differentiation. Microvillous processes which are often long, smooth and without actin filament cores and rootlets suggest mesothelial differentiation and mesothelioma (Coleman, Henderson and Mukherjee, 1989) (Figure 5.6b).

5.3.7 Neuroendocrine Granules

Neuroendocrine granules constitute the ultrastructural markers of neuroendocrine differentiation in tumours. These granules have a classical or archetypal size and appearance – a single limiting membrane enclosing a rounded, homogeneously dense granular core and a halo (a clear space) beneath the membrane. Although varying in size, they are typically 200–300 nm in diameter (Figure 5.7a).

- Neuroendocrine granules need to be distinguished from primary lysosomes, with which they can share similar features. Distinguishing them partly depends on context (Figure 5.7b), certain ultrastructural features and specialised TEM procedures such as the uranaffin reaction, which stains neuroendocrine granules but not lysosomes (Figure 5.7c) (Payne, 1993).
- Neuroendocrine granules can show variations from the archetypal appearance in different tumours. The norepinephrine granules of phaeochromocytomas have an eccentric core in an expanded vacuolar space (Figure 5.7d) (Gómez *et al.*, 1991), insulinomas have a crystalloid core (Feiner, 1978), pituitary adenomas can show large and pleomorphic granules (Horvath and Kovacs, 1992) and neuroendocrine tumours of the lower gastrointestinal and reproductive tracts can show a dimorphic population of rounded and rod-shaped granules (Eyden, 2005).

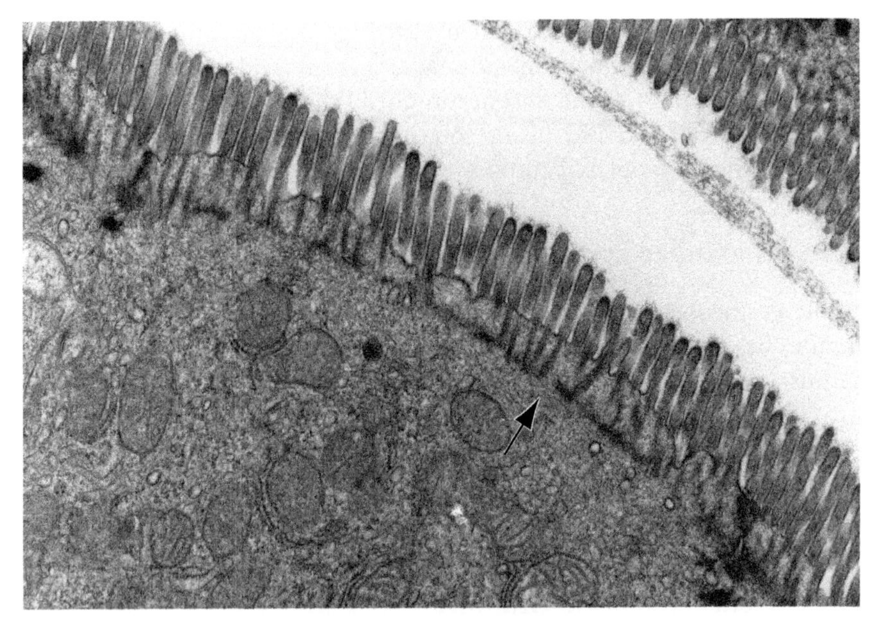

(a)

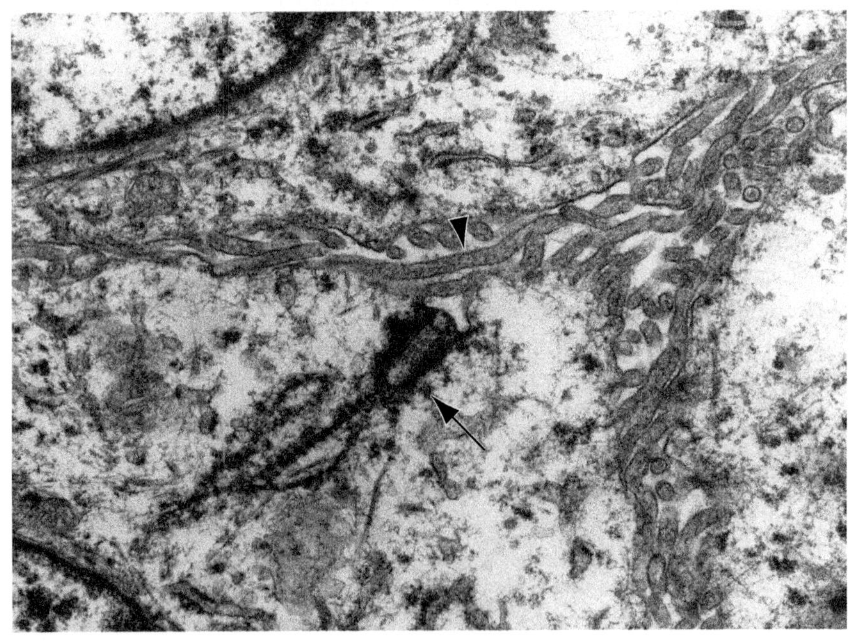

(b)

- Neuroendocrine granules may be found in rare examples of malignant melanoma (Eyden, Pandit and Banerjee, 2005b), extraskeletal myxoid chondrosarcoma (Harris *et al.*, 2000) and alveolar rhabdomyosarcoma (Houreih *et al.*, 2009).

5.3.8 Smooth-Muscle Myofilaments

Smooth-muscle myofilaments with focal densities can indicate smooth-muscle, myofibroblastic, pericytic or epithelial differentiation, depending on the context of other TEM, light microscopy and immunohistochemical features, as follows.

- Lamina, attachment plaques, plasmalemmal caveolae and desmin immunostaining indicate smooth-muscle differentiation, as in leiomyoma and leiomyosarcoma (Figure 5.8a) (Dickersin, Selig and Park, 1997).
- Lamina, attachment plaques, plasmalemmal caveolae and negative desmin immunostaining can indicate pericytic differentiation, although it is fair to say that true pericytic differentiation in tumoural cells using TEM has not been easy to demonstrate because in ultrastructural terms, the normal pericyte is tantamount to a poorly differentiated smooth-muscle cell.
- Fibronexus junctions, abundant rER and fibronectin immunostaining and negative desmin immunostaining indicate myofibroblastic differentiation, as in the pseudotumoural myofibroblastic lesions such as nodular fasciitis and the malignancy known as myofibrosarcoma (Figure 5.8b) (Eyden, 2005a, b).
- Desmosomes, tonofibrils and cytokeratin staining indicate epithelial differentiation, as in spindle-cell carcinoma (Balercia, Bhan and Dickersin, 1995).

Figure 5.6 Microvilli. (a) Glandular epithelial microvilli (normal small intestine) with a fine glycocalyx and rootlets anchored into a terminal web (arrow). Such structures are recapitulated in some adenocarcinomas. (b) Smooth, tortuous microvilli, some seen to be long (arrowhead) in an intercellular channel in a mesothelioma. They lack glycocalyces and actin filament rootlets. The centriole of an oligocilium (sensory cilium) with attached rootlet is also present (arrow).

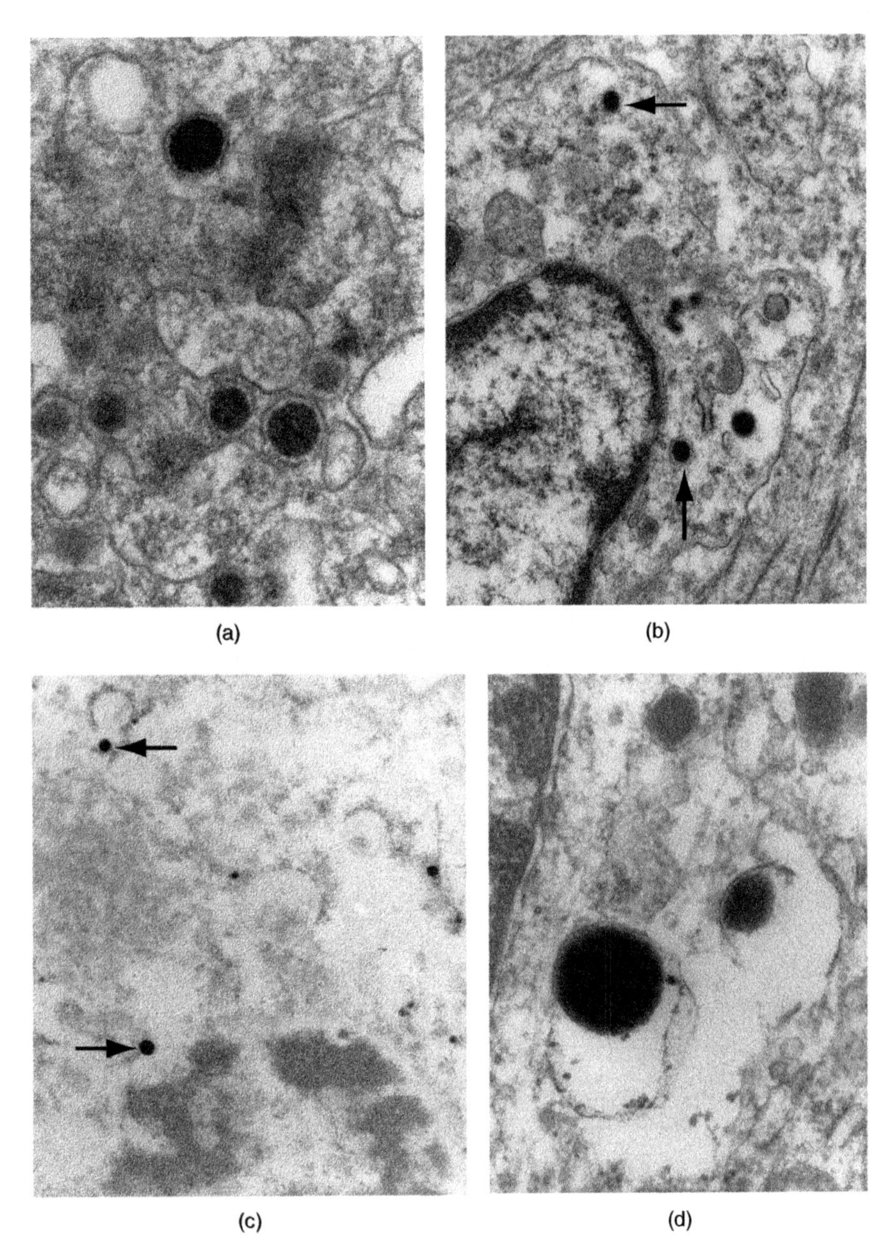

Figure 5.7 Neuroendocrine granules. (a) Archetypal neuroendocrine granules 250 nm in diameter in a liver carcinoid. (b) Although at a lower magnification, the granules in this figure (arrows) bear some similarity to the granules in 'a', but are primary lysosomes in a reactive lymphocyte, showing a degree of suboptimal preservation. (c) Granules, about 100 nm in diameter, stained positively in a gastrointestinal autonomic nerve tumour by the uranaffin procedure (arrows). (d) Norepinephrine granules in a phaeochromocytoma showing the eccentrically placed core and vacuolar expansion.

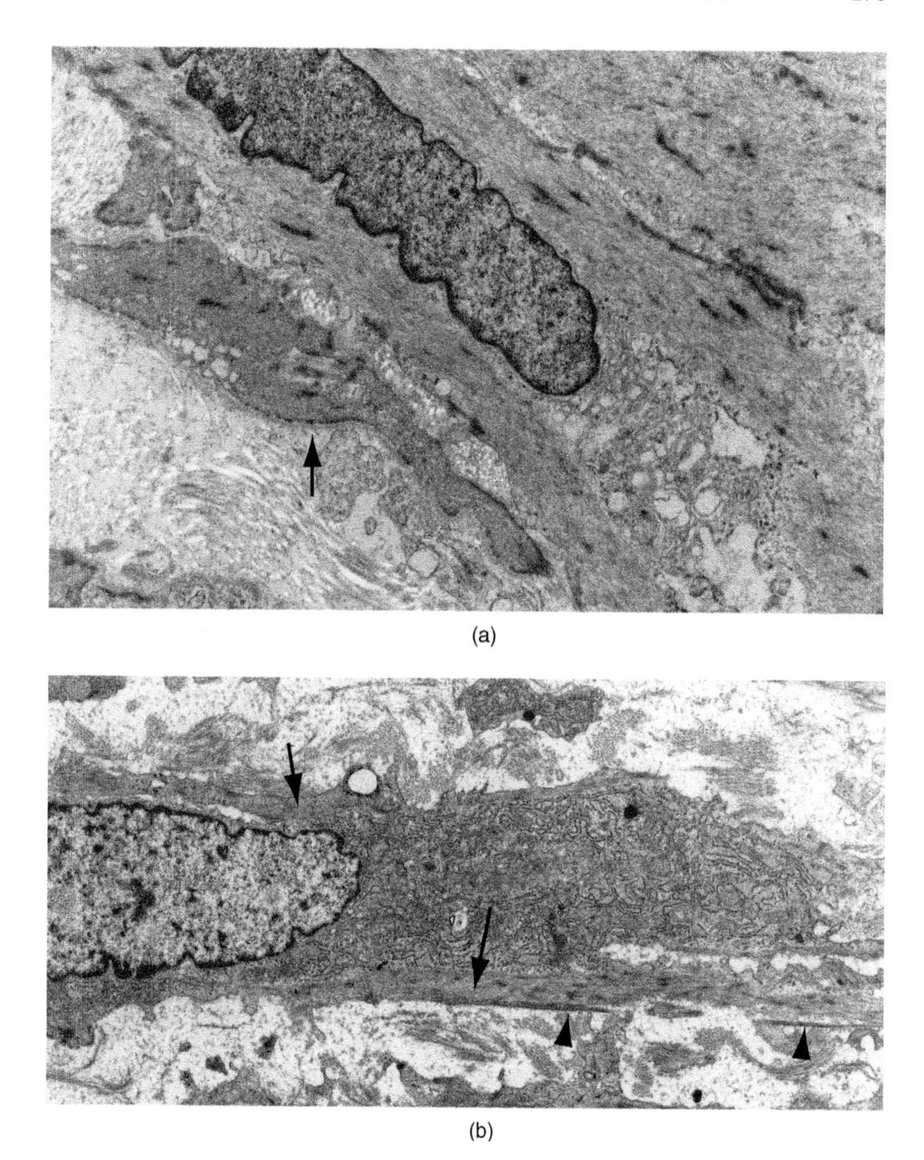

Figure 5.8 Smooth-muscle myofilaments. (a) Uterine leiomyoma showing abundant smooth-muscle myofilaments with focal densities, attachment plaques and lamina (arrow) – the full ultrastructural complement of organelles for smooth-muscle differentiation. (b) Smooth-muscle myofilaments with focal densities in more limited numbers and confined to subplasmalemmal regions of a myofibroblast (arrows). The cell has the distinctively dense and straight fibronectin fibril anchored onto the surface (arrowheads).

5.3.9 Sarcomeric Myofilaments (Thick-and-Thin Filaments with Z-Disks)

Sarcomeric myofilaments are the major ultrastructural feature for defining striated-muscle differentiation (Figure 5.9a), which in the context of a malignancy indicates rhabdomyosarcoma.

- The clinical and light microscopical context also dictates whether this malignancy is a rhabdomyosarcoma or another tumour, for example one showing divergent rhabdomyoblastic differentiation (such as malignant melanoma) (Banerjee and Eyden, 2008). Neoplastic rhabdomyoblasts are also found in Triton tumour, which is essentially a Schwannian tumour (Azzopardi *et al.*, 1983).
- Z-disks can be poorly developed, but they can be identified with some confidence when embedded within bundles of sarcomeric filaments (Figure 5.9b).

In summary, TEM can contribute to diagnostic precision in cases of tumours where there are inconsistencies between clinical and light microscopy features; such cases continue to be encountered, in many cases as a result of the limitations of IHC. Whether or not an ultrastructural analysis is pursued depends on several factors – the confidence with which the main diagnostic contender is held, whether treatment is compromised by the diagnostic uncertainty which ultrastructural analysis might subsequently reduce or eliminate, the size of the surgical specimen (in that it is common practice for small specimens less than about 10 mm across to be embedded in wax *in toto* and therefore be less available for TEM), the academic orientation of the investigator (who might wish to pursue publication) and cost. These factors lie within the judgement of the pathologist, a reflection of the fact that the technique of electron microscopy applied to tumour diagnosis is most effective when integrated with other disciplines and their staff.

→

Figure 5.9 Sarcomeric filaments. (a) Normal striated-muscle cell showing the high level of sarcomeric organisation which can be found in rhabdomyomas and well-differentiated rhabdomyosarcomas. (b) Poorly developed sarcomeres represented by aligned bundles of thick-and-thin filaments with interspersed and structurally poorly defined densities (Z-disks – arrows) in a pleomorphic cytokeratin-positive rhabdomyosarcoma where TEM was undertaken to exclude a carcinoma.

(a)

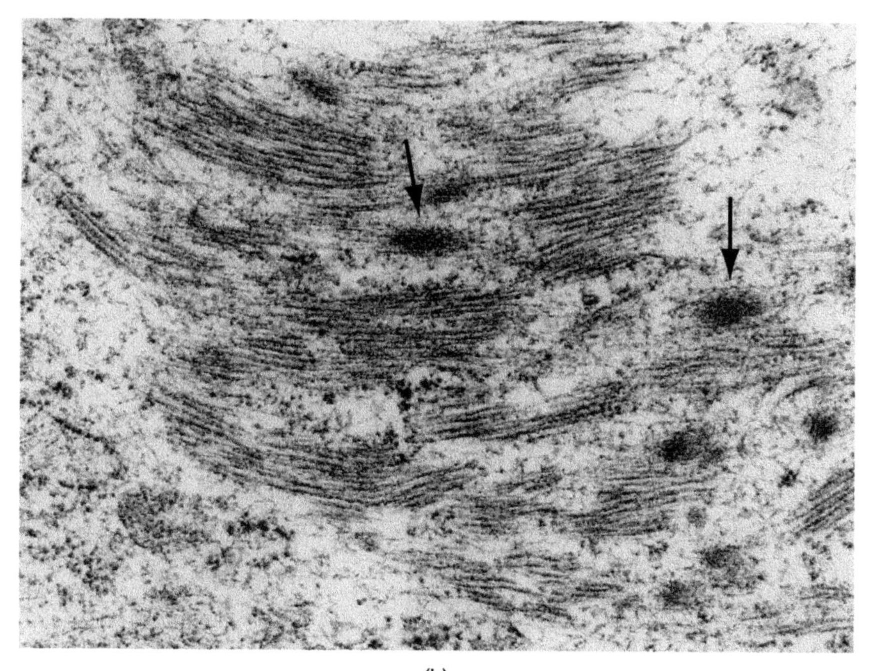

(b)

REFERENCES

Abenoza, P. and Sibley, R.K. (1987) Granular cell myoma and Schwannoma: fine structural and immunohistochemical study. *Ultrastructural Pathology*, **11** (1), 19–28.

Argenyi, Z.B., Cain, C., Bromley, C. *et al.* (1994) S-100 protein-negative malignant melanoma: fact or fiction? A light-microscopic and immunohistochemical study. *American Journal of Surgical Pathology*, **16** (3), 233–240.

Azzopardi, J.G., Eusebi, V., Tison, V. and Betts, C.M. (1983) Neurofibroma with rhabdomyomatous differentiation: benign 'Triton' tumour of the vagina. *Histopathology*, 7 (4), 561–572.

Balercia, G., Bhan, A.K. and Dickersin, G.R. (1995) Sarcomatoid carcinoma: an ultrastructural study with light microscopic and immunohistochemical correlation of 10 cases. *Ultrastructural Pathology*, **19** (4), 249–263.

Banerjee, S.S. and Eyden, B. (2008) Divergent differentiation in malignant melanoma: a review. *Histopathology*, **52** (2), 119–129.

Barnard, M. and Lajoie, G. (2001) Angiomyolipoma: immunohistochemical and ultrastructural study of 14 cases. *Ultrastructural Pathology*, **25** (1), 21–29.

Bishop, P.W., Menasce, L.P., Yates, A.J. *et al.* (1993) An immunophenotypic survey of malignant melanomas. *Histopathology*, **23** (2), 159–166.

Bonsib, S.M., Bray, C. and Timmerman, T.G. (1993) Renal chromophobe cell carcinoma: limitations of paraffin-embedded tissue. *Ultrastructural Pathology*, **17** (5), 529–536.

Carney, J.A. (1990) Psammomatous melanotic schwannoma. A distinctive heritable tumor with special associations, including cardiac myxoma and the Cushing syndrome. *American Journal of Surgical Pathology*, **14** (3), 206–222.

Coleman, M., Henderson, D.W. and Mukherjee, T.M. (1989) The ultrastructural pathology of malignant pleural mesothelioma. *Pathology Annual*, **24** (Pt 1), 303–353.

De Wever, O., Demetter, P., Mareel, M. and Bracke, M. (2008) Stromal myofibroblasts are drivers of invasive cancer growth. *International Journal of Cancer*, **123** (10), 2229–2238.

Dickersin, G.R. (2000) *Diagnostic Electron Microscopy. A Text/Atlas*, 2nd edn, Springer-Verlag, New York.

Dickersin, G.R., Selig, M.K. and Park, Y.N. (1997) The many faces of smooth muscle neoplasms in a gynecological sampling: an ultrastructural study. *Ultrastructural Pathology*, **21** (2), 109–134.

Erlandson, R.A. (1994) *Diagnostic Transmission Electron Microscopy of Tumors with Clinicopathological, Immunohistochemical, and Cytogenetic Correlations*, Raven Press, New York.

Eyden, B. (1996) *Organelles in Tumor Diagnosis: An Ultrastructural Atlas*, Igaku-Shoin Medical Publishers, New York.

Eyden, B. (2002) Diagnostic electron microscopy of tumours. *Current Diagnostic Pathology*, 8 (4), 216–224.

Eyden, B. (2005) Electron microscopy in pathology, in *The Science of Laboratory Diagnosis*, 2nd edn (eds D. Burnett and J. Crocker), John Wiley & Sons, Inc., Hoboken, NJ, pp. 43–60.

Eyden, B. (2005a) The myofibroblast: a study of normal, reactive and neoplastic tissues, with an emphasis on ultrastructure. Part 1 – normal and reactive cells. *Journal of Submicroscopic Cytology and Pathology*, 37 (2), 109–204.

Eyden, B. (2005b) The myofibroblast: a study of normal, reactive and neoplastic tissues, with an emphasis on ultrastructure. Part 2 – tumour and tumour-like lesions. *Journal of Submicroscopic Cytology and Pathology*, 37 (3–4), 231–296.

Eyden, B.P. and Banerjee, S.S. (2008) Origins, contemporary applications, and future of diagnostic electron microscopy applied to soft tissue tumors. *Pathology Case Reviews*, 13 (2), 51–56.

Eyden, B., Banerjee, S.S., Ru, Y-X. and Liberski, P. (2012) *The Ultrastructure of Human Tumours: Applications in Diagnosis and Research*, Zhejiang University Press-Springer Verlag, Hangzhou, in press.

Eyden, B. and Chorneyko, K.A. (2001) Intranodal myofibroblastoma: study of a case suggesting smooth-muscle differentiation. *Journal of Submicroscopic Cytology and Pathology*, 33 (1–2), 157–163.

Eyden, B.P., Manson, C., Banerjee, S.S. *et al.* (1998) Sclerosing epithelioid fibrosarcoma: a study of five cases emphasizing diagnostic criteria. *Histopathology*, 33 (3), 354–360.

Eyden, B., Moss, J., Shore, I. and Banerjee, S.S. (2005a) Metastatic small cell malignant melanoma: a case requiring immunoelectronmicroscopy for the demonstration of lattice-deficient melanosomes. *Ultrastructural Pathology*, 29 (1), 71–78.

Eyden, B., Pandit, D. and Banerjee, S.S. (2005b) Malignant melanoma with neuroendocrine differentiation: clinical, histological, immunohistochemical and ultrastructural features of three cases. *Histopathology*, 47 (4), 402–409.

Eyden, B.P., Shanks, J.H., Ioachim, E. *et al.* (1999) Myofibroblastoma of breast: evidence favoring smooth-muscle rather than myofibroblastic differentiation. *Ultrastructural Pathology*, 23 (4), 249–257.

Fawcett, D.W. (1986) *Bloom and Fawcett: A Textbook of Histology*, 11th edn, W.B. Saunders Company, Philadelphia, PA.

Feiner, H. (1978) Electron microscopy of neoplasms of pancreatic islet cells. *Journal of Dermatology and Surgical Oncology*, 4 (10), 751–757.

Gómez, R.R., Osborne, B.M., Ordoñez, N.G. and Mackay, B. (1991) Pheochromocytoma. *Ultrastructural Pathology*, 15 (4–5), 557–562.

Gould, V.E., Memoli, V.A., Dardi, L.E. *et al.* (1981) Neuroendocrine carcinomas with multiple immunoreactive peptides and melanin production. *Ultrastructural Pathology*, 2 (3), 199–217.

Harris, M., Coyne, J., Tariq, M. *et al.* (2000) Extraskeletal myxoid chondrosarcoma with neuroendocrine differentiation. A pathologic, cytogenetic, and molecular study of a case with a novel translocation t(9;17)(q22;q11.2). *American Journal of Surgical Pathology*, 24 (7), 1020–1026.

Henderson, D.W., Papadimitriou, J.M. and Coleman, M. (1986) *Ultrastructural Appearances of Tumours. Diagnosis and Classification of Human Neoplasia by Electron Microscopy*, 2nd edn, Churchill Livingstone, Edinburgh.

Horvath, E. and Kovacs, K. (1992) Ultrastructural diagnosis of human pituitary adenomas. *Microscopy Research and Technique*, 20 (2), 107–135.

Houreih, M.A., Lin, A.Y., Eyden, B. *et al.* (2009) Alveolar rhabdomyosarcoma with neuroendocrine/neuronal differentiation: report of 3 cases. *International Journal of Surgical Pathology*, 17 (2), 135–141.

Johannessen, J.V. (1977) Use of paraffin material for electron microscopy. *Pathology Annual*, **12** (Pt 2), 189–224.
LeBoit, P.E., Barr, R.J., Burall, S. *et al.* (1991) Primitive polypoid granular-cell tumor and other cutaneous granular-cell neoplasms of apparent nonneural origin. *American Journal of Surgical Pathology*, **15** (1), 48–58.
McWilliam, L.J. and Harris, M. (1985) Granular cell angiosarcoma of the skin: histology, electron microscopy and immunohistochemistry of a newly recognised tumour. *Histopathology*, **9** (11), 1205–1216.
Ohta, T., Yachi, K., Ogino, A. *et al.* (2010) Pleomorphic granular cell astrocytoma in the pineal gland: case report. *Neuropathology*, **30** (6), 615–620.
Pavelka, M. and Roth, J. (2005) *Functional Ultrastructure*, Springer-Verlag, Vienna.
Payne, C.M. (1993) Use of the uranaffin reaction in the identification of neuroendocrine granules. *Ultrastructural Pathology*, **17** (1), 49–82.
Pollard, T.D. and Earnshaw, W.C. (2002) *Cell Biology*, Saunders, Philadelphia, PA.
Rhodin, J.A.G. (1975) *An Atlas of Histology*, Oxford University Press, Oxford.
Tazelaar, H.D., Batts, K.P. and Srigley, J.R. (2001) Primary extrapulmonary sugar tumor (PEST): a report of four cases. *Modern Pathology*, **14** (6), 615–622.
Van Den Bergh Weerman, M.A. and Dingemans, K.P. (1984) Rapid deparaffinization for electron microscopy. *Ultrastructural Pathology*, 7 (1), 55–57.

6

Microbial Ultrastructure

Alan Curry

Health Protection Agency, Clinical Sciences Building, Manchester Royal Infirmary, Manchester, United Kingdom

6.1 INTRODUCTION

Microbiology covers several groups of organisms – bacteria, fungi, viruses and protozoa (but single-celled algae causing the very rare condition called chlorellosis will not be covered here). Electron microscopy (EM) is not normally used in the diagnosis of bacterial or fungal infections, but it has been used, until comparatively recently, in the investigation of viral gastroenteritis and viral infections associated with skin lesions. Before the advent of antiretroviral treatment for human immunodeficiency virus (HIV) infection, some specialist EM facilities were used in the investigation of opportunistic infections associated with this disease. Examples of opportunistic organisms include *Cryptosporidium* and *Isospora*, which are protozoan, and microsporidia and *Pneumocystis*, which were classified as protozoa but, because of molecular studies, are now regarded as belonging to the fungal lineage. All of these groups of microorganisms have distinctive ultrastructural appearances allowing identification, sometimes to the species level, and this chapter shows examples of these organisms from diagnostic cases (Fuller and Lovelock, 1976; Wills, 1992).

Diagnostic Electron Microscopy: A Practical Guide to Interpretation and Technique, First Edition. Edited by John W. Stirling, Alan Curry and Brian Eyden.

6.2 PRACTICAL GUIDANCE

Even if the investigator is familiar with the spectrum of ultrastructure shown by microorganisms, identifying them in cells and tissues often requires patience and diligence unless the organism is present in significant numbers. If a microorganism is suspected of being present in a biopsy, the sections should be screened at a magnification that is a compromise between being high enough to allow recognition of the expected organism but low enough to allow a sufficient area of the section or grid to be scanned.

The fixative of choice is glutaraldehyde, but when this is unavailable, histological formalin can be used. In addition, tissue samples embedded in wax may be all that is available for reprocessing into resin blocks. Such material can show relatively poor ultrastructural preservation, and therefore, examination of such samples requires more cautious interpretation. However, the resistant spores of, for example, microsporidia can reveal important features that can aid specific identification.

All microscopic examination should be 'open-minded' as unexpected organisms can occasionally be encountered even when other, more appropriate diagnostic tests are apparently negative. EM can sometimes detect parasitic infections (protozoan, microsporidial or the bacterial rickettsiae) not initially suspected, particularly if there is any immunodeficiency involved.

Bacteria are not infrequently encountered during biopsy examination, but these may be contaminants rather than an infection. This usually occurs because of inappropriate handling prior to embedding. In some centres, for instance, biopsies are transported in buffer to the laboratory for fixation. If the buffer supports bacterial growth, these organisms can contaminate the biopsy. This may involve only the biopsy surfaces, but in renal biopsies, be aware that bacteria can penetrate into the tubules and, given enough time, into the urinary space of the glomerulus. In addition, some parasites may appear outside their traditional geographical area of distribution as a result of climate change (i.e. global warming), and some infections as a result of increased intercontinental travel. Such acquired infections may be unfamiliar to medical diagnosticians who encounter the returned traveller in his or her country of origin, making the diagnosis a more prolonged process or one that requires referral to a specialist centre.

6.3 VIRUSES

For historical applications of EM to diagnostic virology, see Curry, Appleton and Dowsett (2006), and for texts on virus morphology, see Doane and Anderson (1987), Madeley and Field (1988), Madeley (1992) and Hsiung, Fong and Landry (1994).

As EM is not dependent on the presence of nucleic acid or antigenicity, it can be used as an independent investigative technique for testing for viruses, particularly if the very specific polymerase chain reaction (PCR) test fails to detect one or more pathogens.

Viruses can be classified according to morphological features seen after negative staining under the EM. Traditionally, two EM methods have been used to study or diagnose virus infections, namely, negative staining of whole-virus particles or resin-embedded sections of virus-infected cells or tissues. Examination of thin sections of virus-infected cells gives information about how viruses replicate and assemble within infected cells. For example, adenoviruses and polyomavirus capsids assemble within the nucleus of virus-infected cells (Figures 6.1 and 6.2), whereas sapoviruses (sometimes called caliciviruses), rotaviruses and poxviruses assemble in the cytoplasm of infected cells (Figure 6.3). The ribonucleoproteins of paramyxoviruses (e.g. measles or mumps) assemble in the cytoplasm and/or the nucleus depending on type. The mumps paramyxovirus virus assembles ribonucleoproteins in the cytoplasm before budding to produce mature particles, whereas measles shows assembly in both the nucleus and cytoplasm before budding. Infected cells show evidence of replication, and careful examination can reveal assemblages of virus capsids. Other viruses, such as influenza (an orthomyxovirus) and HIV (a retrovirus), show no obvious internal signs of replication in infected cells, but the viruses can be detected budding from the plasma membrane of infected cells. After budding from the surface, HIV undergoes a maturation process externally. However, ultrastructural studies on the pandemic influenza A (H1N1) 2009 virus isolate have shown dense tubules within the nuclei of some infected cells (see Goldsmith *et al.*, 2011).

Until relatively recently, viral gastroenteritis (see Table 6.1), outbreaks of winter vomiting disease, viruses excreted in the urine of immunocompromised patients and skin lesions (see Table 6.2) were investigated by negative-staining EM. Viruses were concentrated (e.g. from stool,

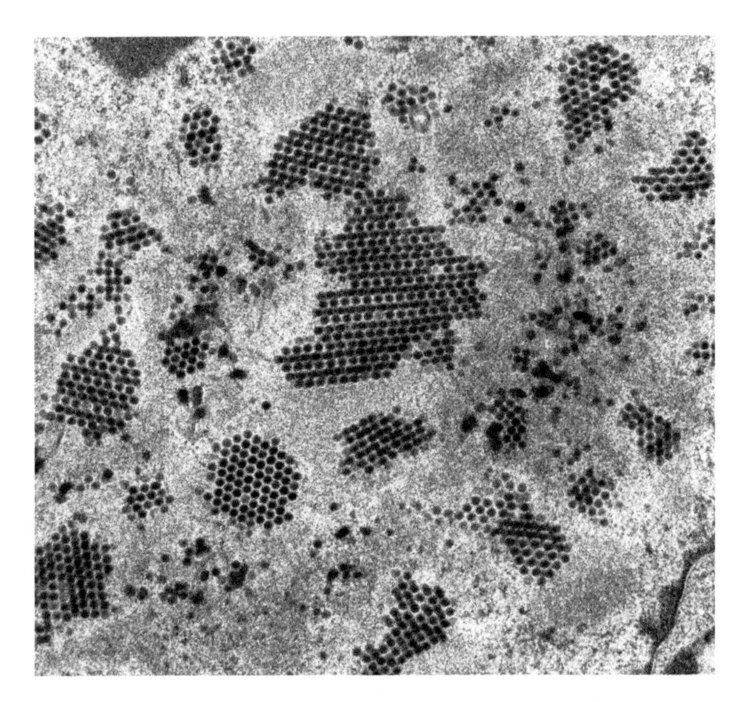

Figure 6.1 Adenovirus assembly in a cultured cell. Virus capsids assemble within the nucleus where they form crystalline arrays. Individual capsids measure about 70–75 nm in diameter.

vomitus or early morning urine samples) and partially purified before adsorption onto a thin plastic (or carbon) support film covering the grid. The concentration of virus required for detection under the EM is thought to be about 10^6 or 10^7 virus particles per ml of sample. After staining with a heavy metal stain, such as phosphotungstic acid, the grids could be examined at an appropriate magnification for evidence of the presence of virus particles. Experience is needed to identify some of the smaller viruses associated with gastroenteritis (astroviruses, sapoviruses and noroviruses). The specific morphological features associated with these small round viruses may not be obvious, and careful examination is required (see Caul and Appleton, 1982).

Skin lesions could be prepared and examined quickly (within about 30 minutes of receipt in the laboratory), but diarrhoeal samples requiring ultracentrifugation take longer to prepare. An advantage of EM in virology is that it is a 'catch-all' system and can potentially find double or multiple infections that may not have been clinically expected. Even though EM and virology should be thought of as inseparable, it is a fact that in many centres EM is no longer required in virus

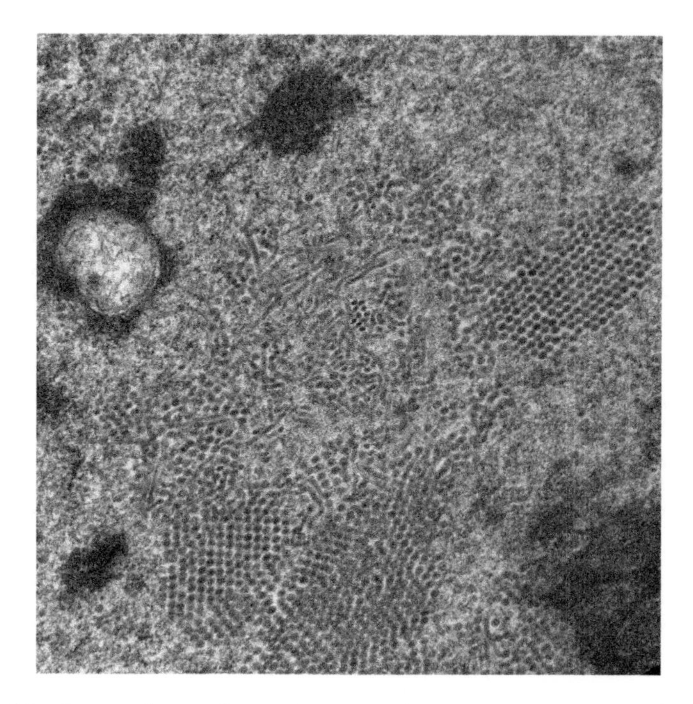

Figure 6.2 Intranuclear polyomavirus capsids (JC virus) in the nucleus of an oligodendrocyte from a brain biopsy of a patient with progressive multifocal leucoencephalopathy (PML). Individual virus capsids are about 40–45 nm in diameter, but note tubular forms. These tubules, which are unusual, can form when excess capsid protein is produced. (Referred case from Dr Andrew Dean.)

diagnosis because serological and molecular tests have superseded EM investigations.

6.4 CURRENT USE OF EM IN VIROLOGY

Although the use of diagnostic EM has lessened in importance in recent years, EM in specialised laboratories is used to support the development of new tests. In particular, recombinant virus capsid proteins are being used in the making of monoclonal antibodies and development of vaccines. EM can be used to show that the viral proteins are assembling into capsid-like particles, where they may be more antigenic than the individual unassembled proteins, for example papilloma viruses such as HPV 16 and HPV 18.

With the re-emergence of bioterrorism as a threat to society, EM could play a role in differentiating between poxviruses and herpesviruses in

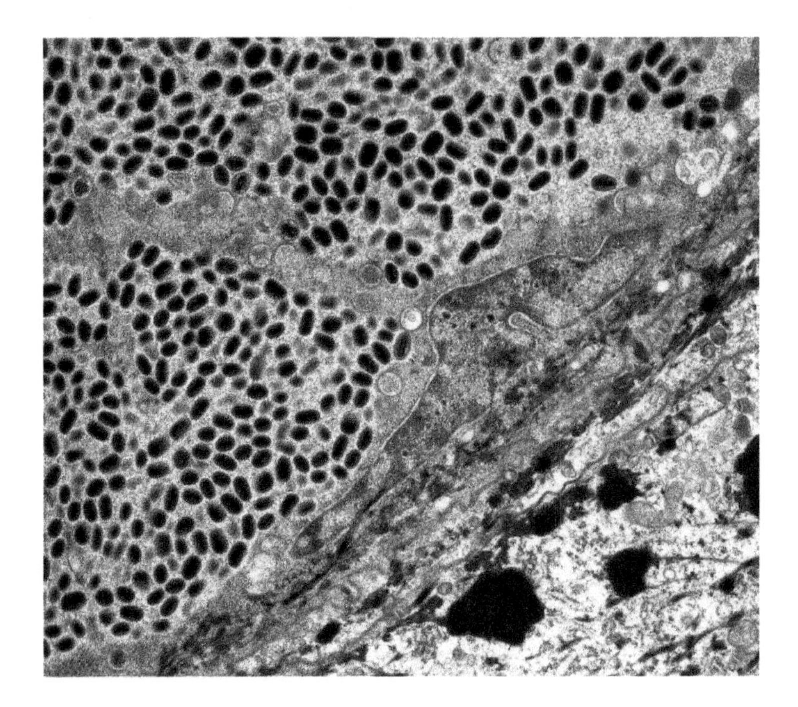

Figure 6.3 *Molluscum contagiosum*, a poxvirus, in a skin lesion. These complex viruses assemble in the cell cytoplasm and can push the nucleus out to the periphery of the infected cell. Individual virus particles measure 250 × 200 nm.

skin lesions (see Table 6.2), which have a similar clinical appearance in the early stages of infection (Hazelton and Gelderblom, 2003; Miller, 2003). The main advantage of EM is that it is a rapid diagnostic technique. Suspicious samples should be handled only in laboratories with high levels of containment. However, any orthopoxvirus found should be properly investigated to ensure that there is no possibility that it is smallpox. EM also has a use in the characterisation of new infectious agents. The severe acute respiratory syndrome (SARS) outbreak was not bioterrorist related, but EM played an important role in the identification and characterisation of this new coronavirus (Falsey and Walsh, 2003; Ksiazek *et al.*, 2003).

6.5 VIRUSES IN THIN SECTIONS OF CELLS OR TISSUES

Although a specialised area, requests are sometimes made to examine tissue biopsies or cultured cells for evidence of virus replication and to

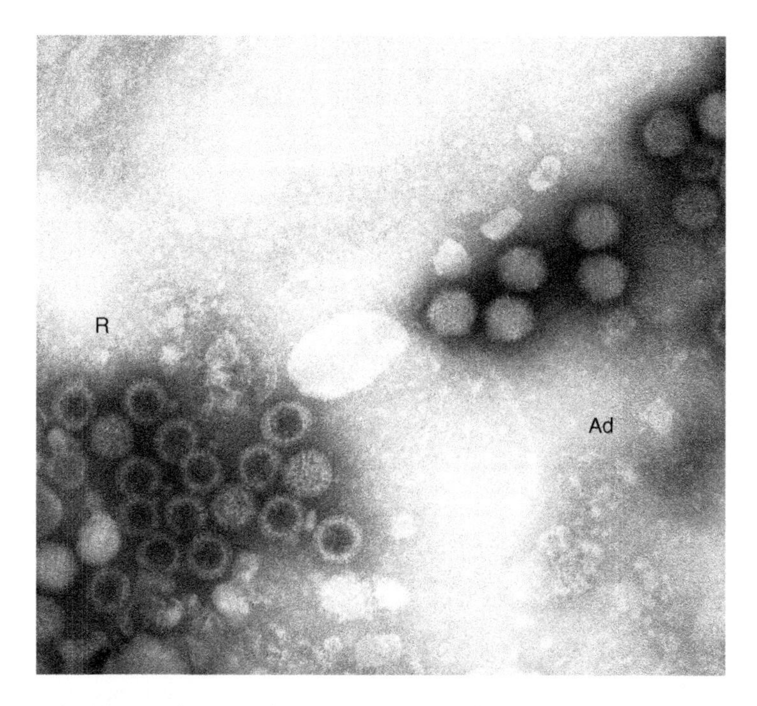

Figure 6.4 Negatively stained rotaviruses (R) and adenoviruses (Ad) from a case of infantile gastroenteritis. Such dual infections of the intestine are not uncommon. Note the double-shelled appearance of the intact rotaviruses and the icosahedral shape of adenoviruses. Intact double-shelled rotaviruses are about 70 nm in diameter, and adenoviruses 70–75 nm in diameter.

find significant ultrastructural features to aid specific identification of the virus involved or to elucidate aspects of virus assembly. An example of an orthopox virus factory found in an infected cell line is shown in Figure 6.8.

Papovaviruses (papillomaviruses and polyomaviruses) have neoplastic potential and are common infections that can be found in specific tissues. For example, polyomaviruses such as BK and JC can be found in some brain biopsies (Williams-Gray *et al.*, 2007) (Figure 6.2) or in the tubular epithelial cells of some renal transplant biopsies. Polyomaviruses in renal transplants can give rise to symptoms mimicking rejection, and although there are more appropriate methods of diagnosis, EM can confirm infection. Papillomaviruses are known to cause cancer, particularly cervical cancer and oropharyngeal cancer (Stanbridge *et al.*, 1981; Ramqvist and Dalianis, 2010), and viruses can be found by EM in biopsies.

Table 6.1 Viruses associated with gastroenteritis

Virus name	Size	Morphological details	Comments
Rotavirus (see Figure 6.4)	70–75 nm	Icosahedral – appears round and 'wheel-like', but three forms are known. Intact particles appear spherical and are double shelled and have a smooth outer layer. Spokes link inner and outer shells. The outer layer is often absent in preparations ('rough rotaviruses'), and rarely a tubular form may be present that forms from excess structural protein.	Normally associated with diarrhoea in young children (or young animals). Seasonal (winter months) in temperate regions. Rotaviruses (particularly 'rough rotaviruses') can be confused with reoviruses that also occur in some faecal samples.
Adenovirus (see Figure 6.4)	70–75 nm	Classical icosahedral shape	Enteric adenoviruses belong to Group F (Types 40 and 41). Infection occurs throughout the year in temperate regions, but there is usually a winter increase. Affects all age groups.
Astrovirus	27–28 nm	Icosahedral – small round viruses with a five or six-pointed surface star (not seen on all particles). Surface spikes inhibit close packing of particles.	Normally associated with diarrhoea in young children. Seasonal (winter months) in temperate regions.
Sapovirus (Calicivirus)	31–35 nm	Icosahedral – Small round viruses with distinctive surface hollows that fill with negative stain. Appearance of hollows varies depending on 5 : 3 : 2 symmetry plane.	Normally associated with diarrhoea in young children. Seasonal (winter months) in temperate regions.

(*continued overleaf*)

Table 6.1 (*continued*)

Virus name	Size	Morphological details	Comments
Norovirus (Norwalk-like virus or small round structured virus – SRSV)	30–35 nm	Icosahedral – small round viruses with a 'fuzzy' surface, sometimes described as 'a ragged-edge morphology'	Often associated with outbreaks of gastroenteritis ('winter vomiting disease'). All age groups can be infected. Spontaneous vomiting is a major symptom, and viruses in vomit are a major cause of transmission from person to person.

Note: It is not uncommon for two or more of the viruses shown in this table to be found in the same sample, particularly if from young children (Figure 6.4). Also, other viruses have been found in diarrhoeal samples (e.g. parvoviruses, enteric coronaviruses and reoviruses), but, perhaps controversially, these are not thought to be a significant cause of symptoms.

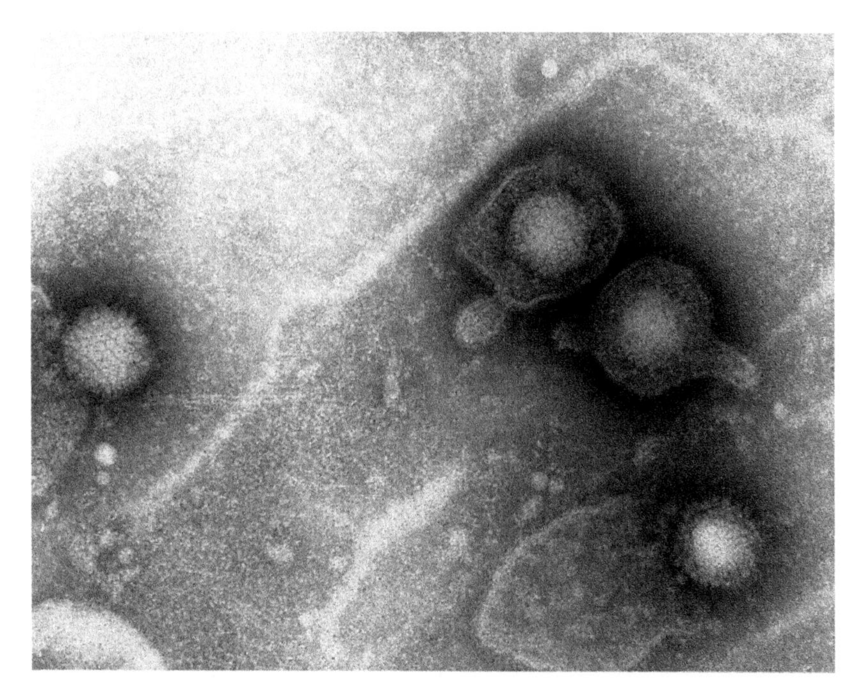

Figure 6.5 Negatively stained herpesviruses from a skin lesion. Enveloped and non-enveloped viruses are shown. Note tubular capsomeres that make up the virus capsid. Individual capsids of herpesviruses measure about 100 nm in diameter.

Table 6.2 Viruses associated with skin lesions

Virus name	Size	Morphology	Comments
Herpesviruses, for example *Herpes simplex*; *Varicella zoster* (Figure 6.5)	Capsid about 100 nm in diameter	Infective capsid surrounded by a membrane envelope, which has no particular size or shape. Capsid appears spherical (icosahedral) with protruding tubular capsomeres.	Sometimes found without an outer membrane in EM preparations. Occasionally two capsids can be found within a single enclosing membrane ('double yolker').
Poxviruses: orthopoxviruses (e.g. cowpox; smallpox) (Figure 6.6)	250 × 200 nm	'Brick-shaped' viruses with thread-like structures running over the surface.	All poxviruses are slightly pleomorphic. These are viruses showing complex symmetry. Two forms are seen in preparations – Mulberry form (mature virus) and capsulate form (immature virus).
Parapoxviruses (e.g. Orf; pseudocowpox) (Figure 6.7)	250 × 150 nm	'Sausage-shaped' viruses with a surface thread that appears to be spiral in form.	–
Molluscipoxviruses (*Molluscum contagiosum*) (Figure 6.3)	250 × 200 nm	Brick shaped but with rounded ends. Surface threads are more clearly seen than in orthopoxviruses but can be difficult to differentiate from orthopoxviruses.	–
Papovaviruses (papillo-maviruses – Wart viruses)	45–55 nm	Capsid spherical with prominent capsomeres. Virus is icosahedral but shows skewed symmetry.	–

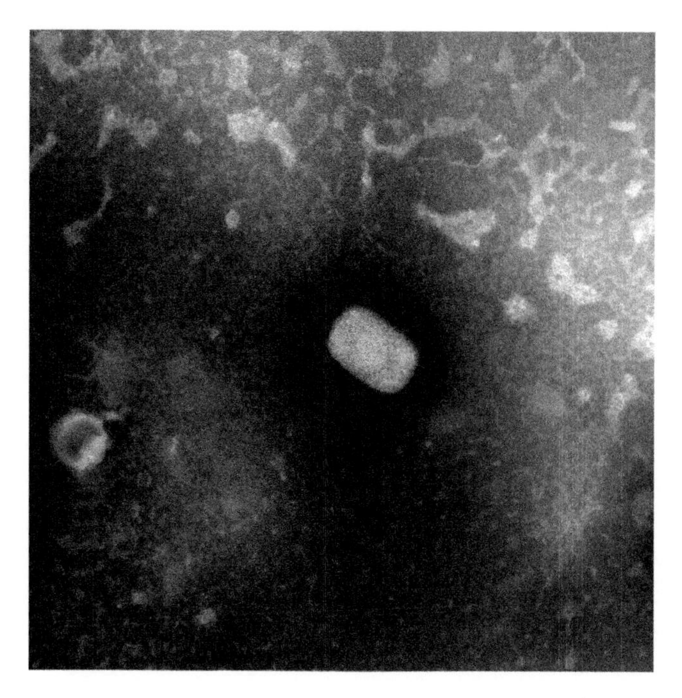

Figure 6.6 Negatively stained orthopoxvirus (cowpox). Note that this virus is more brick shaped than parapoxviruses and that the surface details are less distinct. Individual virus particles measure 250 × 200 nm.

A potential problem in identifying viruses in cells, particularly in cell nuclei, is that chromatin or material from the nucleolus can aggregate in such a way that they mimic virus capsids. Perichromatin granules, which are spherical and often of viral size, are well-recognised 'interpretational' artefacts (Doane and Anderson, 1987; Ghadially, 1988).

6.6 BACTERIA

Unlike viruses, most bacteria are capable of an independent existence. They show little internal compartmentalisation (prokaryotic organisation) and contain a non-enveloped filamentous nucleoid surrounded by bacterial ribosomes (Murray, 1978). The surrounding wall structure can indicate whether the organism is Gram-positive or Gram-negative (Figure 6.9), and surface features such as flagella (Figure 6.9), pili and fimbriae and capsular material can indicate features of the biology of the organism (such as, motility, adhesion and resistance). Some bacteria,

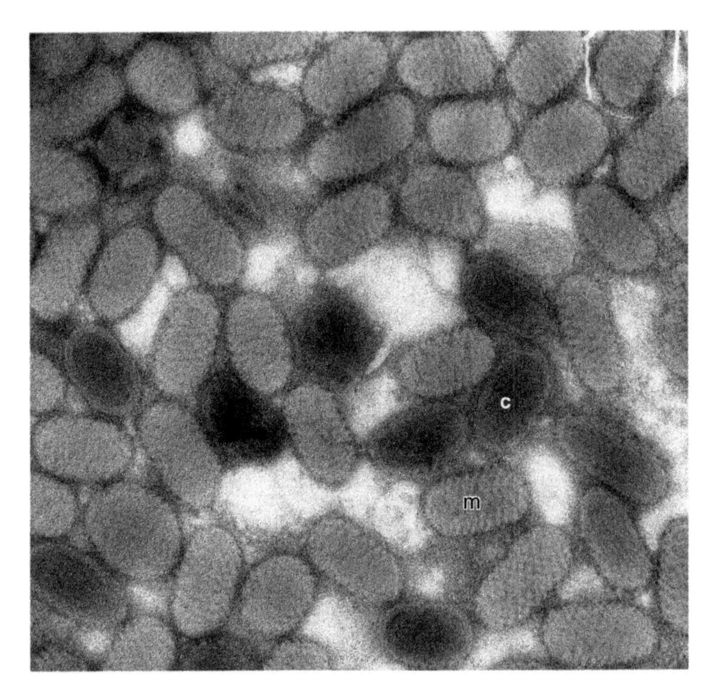

Figure 6.7 Negatively stained parapoxvirus (Orf virus) from a finger lesion of a sheep farmer. Although slightly pleomorphic, these viruses are sausage shaped, are about 250 × 150 nm in size and have a prominent surface filament or tubule that appears to wind around the virus particle giving a criss-cross appearance. Mature (Mulberry form: m) and immature (capsulate form: c) virus particles are shown.

such as *Bacillus* and *Clostridium*, have stages (spores) that can resist extreme environmental conditions. Spores show multiple external layers surrounding the internal genetic material and, as such, have a characteristic appearance. Although comparatively simple, bacteria show a range of morphology that can help in identification, particularly if they are found in samples being examined by EM. Many bacteria are either coccoid (e.g. *Staphylococcus* and *Neisseria*) or rod-shaped (e.g. *Salmonella, Escherichia* and *Bacillus*), but others can be spiral (e.g. *Campylobacter* and *Helicobacter*). Another distinctive group are the spirochaetes (e.g. *Treponema* and *Borrelia*). These flexible, motile bacteria show a unique helical morphology with internal axial filaments responsible for movement (Johnson, 1977). The internal axial filaments are structurally similar to the external flagella found on other types of bacteria. Mycoplasmas also have a distinctive appearance, and these small organisms that lack peptidoglycan in the wall are not infrequently found contaminating cultured cells (Figure 6.10).

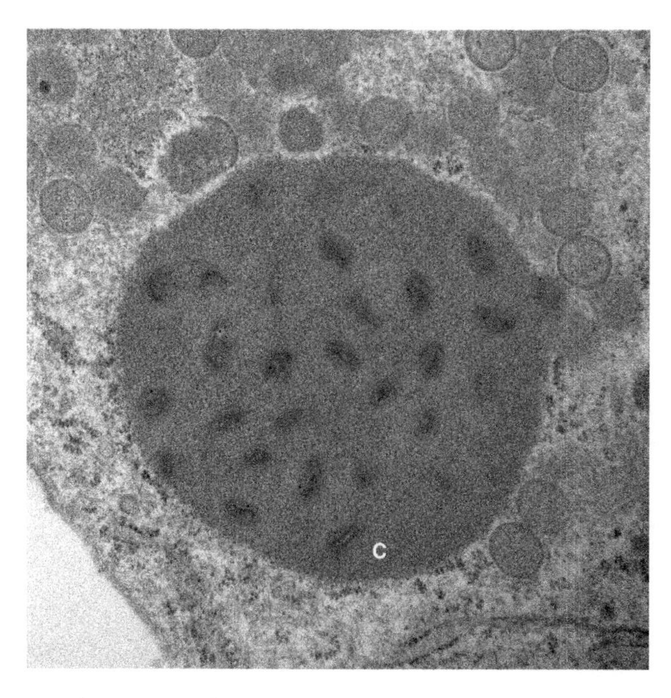

Figure 6.8 Orthropoxvirus factory (cowpox) in the cytoplasm of a cultured cell line (MRC cells). Note characteristic dumbbell appearance of the virus cores (C) within the factory. Virus cores measure about 200 nm in length.

One of the best understood forms of bacterial mobility is swimming that involves external flagellar filaments. The number and distribution of flagella on the bacterial cell surface vary between species but are usually constant in any one species. Some Gram-negative genera have sheathed flagella (Figure 6.9) where the external filament is tightly covered by an extension of the outer membrane-like envelope (e.g. *Vibrio* and *Helicobacter*).

Motility and chemotaxis are important biological attributes. Although several types of bacterial movement are known, flagella are particularly interesting as they are totally dissimilar to the flagellar apparatus found in eukaryotic cells (see De Pamphilis and Adler, 1971; Silverman and Simon, 1977).

Some significant bacterial pathogens develop within cells (e.g. *Chlamydia, Coxiella* and *Rickettsia*) from which they derive their particular nutritional and developmental requirements.

Some human bacterial infections can look similar under the EM but are, in fact, caused by different groups of bacteria. *Trophoderma*

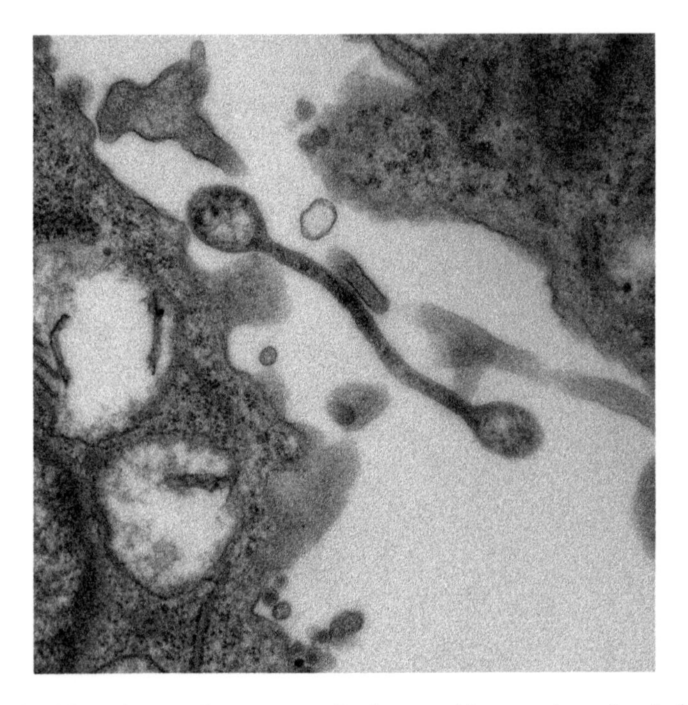

Figure 6.9 Mycoplasma: these are small, pleomorphic organisms that lack pepdidoglycan in the cell wall. Such organisms are often (but not exclusively) associated with contamination of cell cultures. Mycoplasmas measure up to about 0.5 μm.

whippleii (the cause of Whipple's disease) and *Mycobacterium avium-intracellulare* complex, which is often associated with HIV infection, have similar appearances in the lamina propria of the small intestine. The bacteria are extracellular but are frequently phagocytosed. Bacterial outer membranes accumulate within phagocytes as membrane whorls.

Like viruses, bacteria can be examined by both negative staining and thin sectioning techniques. Negative staining is the method of choice for studying external appendages and shape, but significant features can still be found by careful examination of sectioned material, for example the sheathed flagellar filaments of *Helicobacter pylori.*

6.7 FUNGAL ORGANISMS

Filamentous fungi are not normally identified by EM, but they may occasionally be encountered during biopsy examination. However, there are several groups of microorganisms that were previously considered to be protozoan that are now classified as fungi following molecular

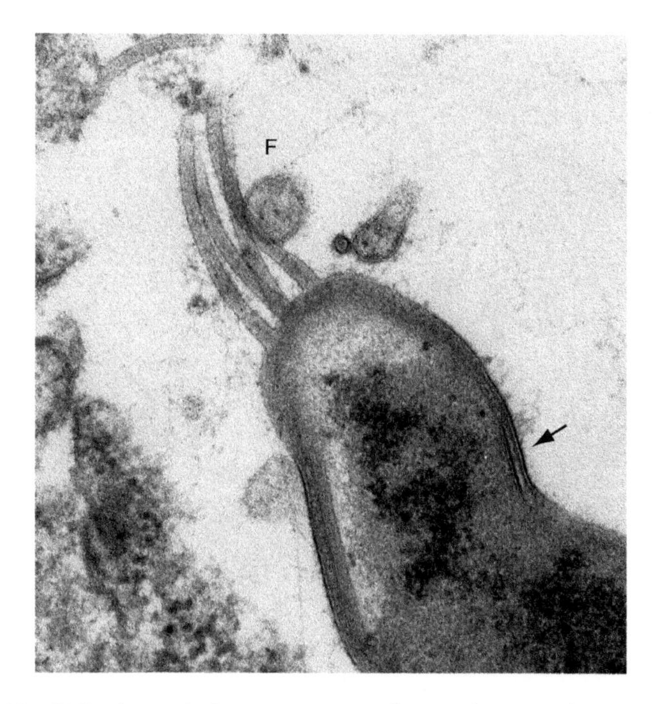

Figure 6.10 *Helicobacter heilmanii* in a gastric biopsy showing a bunch of sheathed flagellar filaments (F) at one end of the organism. Each sheathed flagellum is 30 nm in diameter. Note the Gram-negative wall structure (internal plasma membrane and outer membrane-like structure) (arrow).

studies. In particular, *Pneumocystis* and microsporidia are now thought to be fungi that diverged from the main fungal lineage at an early stage of the evolution. Fungal organisms are all eukaryotic (i.e. have a nucleus enclosed in a membrane envelope).

Pneumocystis jiroveci (previously called *Pneumocystis carinii*) was the organism that first alerted the world to the existence of AIDS in the early 1980s. Diagnosis at that time often required a needle biopsy into an infected lung, followed by light microscopy (LM) and sometimes EM. This organism is known to be able to spread widely in immunodeficient individuals. *Pneumocystis* develops extracellularly, normally in lung alveolar spaces. It occurs in two forms, and the terminology used reflects that they were once thought to be protozoan. A trophozoite stage and a cystic stage are the two main forms of this organism. The trophozoites, often seen in clusters, vary in shape, are about 1–2 μm in diameter and are surrounded by a thin pellicle comprising a membrane and an outer electron-dense layer (Figure 6.11). Often short surface microvilli are present. Internally, there is an enveloped nucleus surrounded by

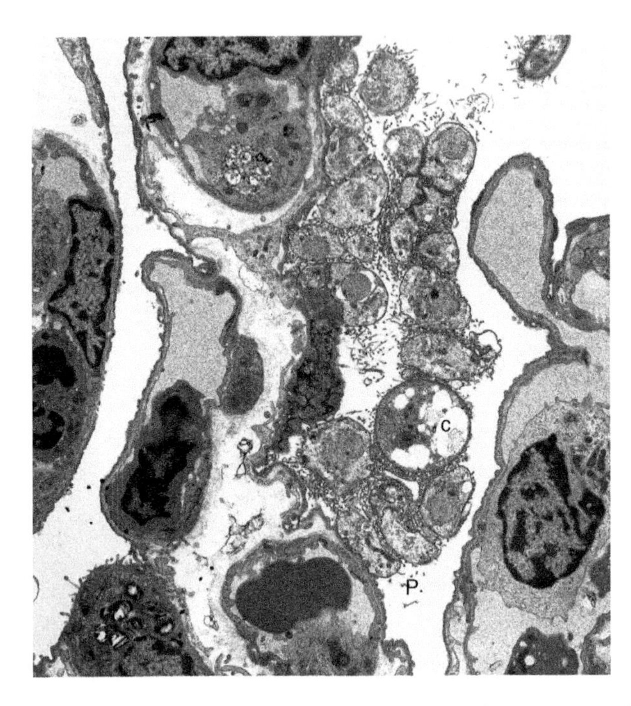

Figure 6.11 *Pneumocystis jiroveci* (P) trophozoites and cyst (C) in the alveolar space of a lung (from a needle biopsy) of an immunocompromised patient. Individual cysts measure 4–6 μm in diameter.

cytoplasm containing ribosomes, poorly developed mitochondria and endoplasmic reticulum. The cystic stage is generally spherical, 4–6 μm in diameter and surrounded by a thickened layered pellicle. Mature cysts contain intracystic bodies that are similar to trophozoites. Occasionally, cysts can be found releasing intracystic bodies, and collapsed empty cysts (which are crescent shaped) can also occasionally be found. It is thought that *Pneumocystis* infection is caught by inhalation of cysts.

6.8 MICROSPORIDIA

Microsporidia are all small obligate intracellular parasites that have a unique mode of infecting cells via a polar tube (sometimes called the polar filament) located in a highly resistant spore (Canning and Lom, 1986; Wittner and Weiss, 1999). The first human cases were ophthalmic, but human microsporidial infection came into prominence only when it was found commonly in AIDS patients (Curry, 2005). Most human microsporidial infections associated with AIDS were enteric, causing

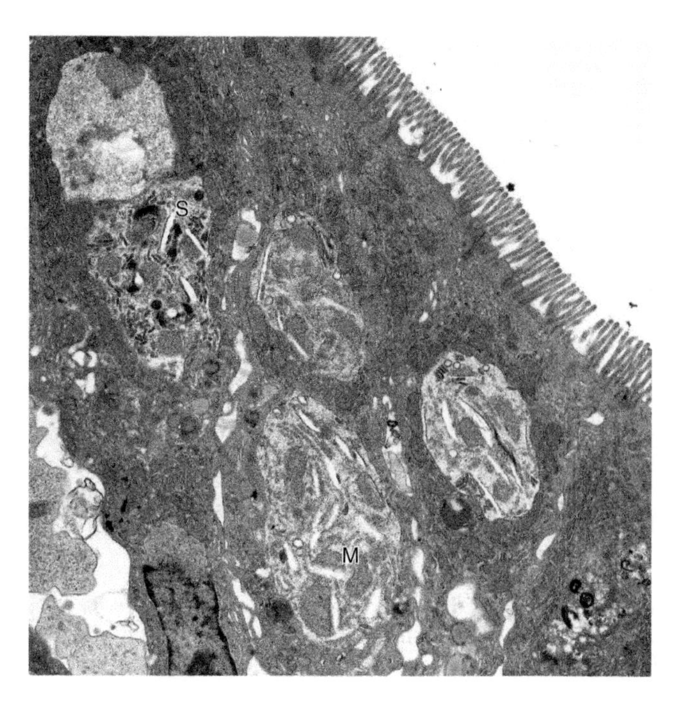

Figure 6.12 *Enterocytozoon bieneusi* in small intestinal enterocytes of an AIDS patient. Two stages are shown, several multinucleated merogonic stages (M) and a sporogonial stage (containing disc-shaped organelles that form the polar tube) (S). All stages develop in direct contact with the enterocyte cytoplasm. Uninucleate merogonic stages measure about 1.5 µm in diameter, whereas the multinucleated merogonic stages measure up to about 5 µm in diameter.

chronic watery diarrhoea, and were often associated with very low CD4 lymphocyte counts. Diagnosis of enteric microsporidial infections is possible by staining spores in stool samples. Commonly used methods include a modified Trichrome stain (Weber *et al.*, 1992) or an optical brightening agent such as Calcofluor (Van Gool *et al.*, 1993). The latter reagent requires the use of a fluorescence microscope and a 330–380 nm excitation filter. Once spores are detected, confirmation and speciation should be undertaken using either EM or molecular methods. Confirmation of species is important as some microsporidia, such as *Enterocytozoon bieneusi*, can be refractory to some treatments (e.g. Albendazole) and excretion can be prolonged (in some cases, several years).

At least 18 microsporidian species have now been described from humans (see Table 6.3), and it is likely that more will be found. With the advent of antiretroviral treatment, few microsporidia are now found in

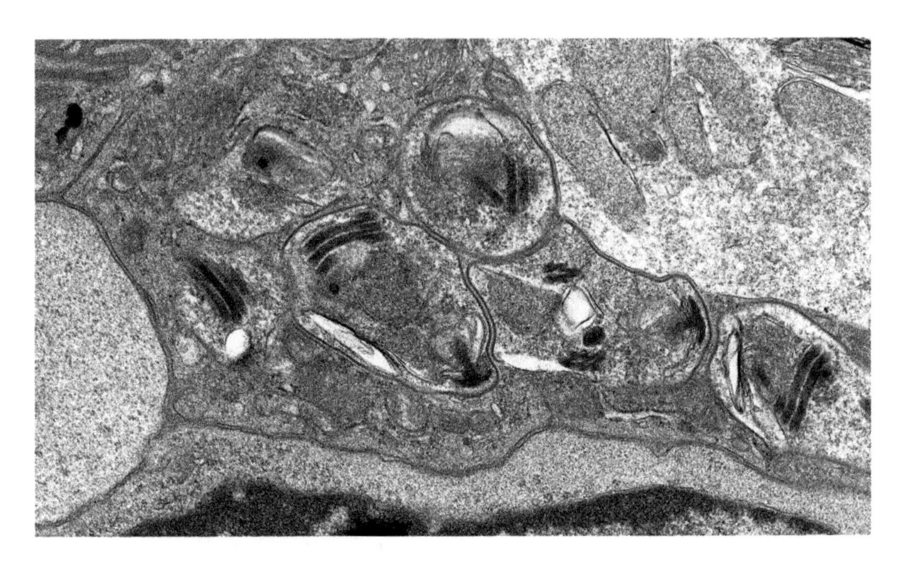

Figure 6.13 Sporoblasts of *Enterocytozoon bieneusi* within an infected small intestinal enterocyte. This stage is rarely seen, as division of the preceding sporogonial stage is a rapid process. Note the single nucleus, polar tube and anterior anchoring disc. Individual sporoblasts measure about 1.5 μm in length.

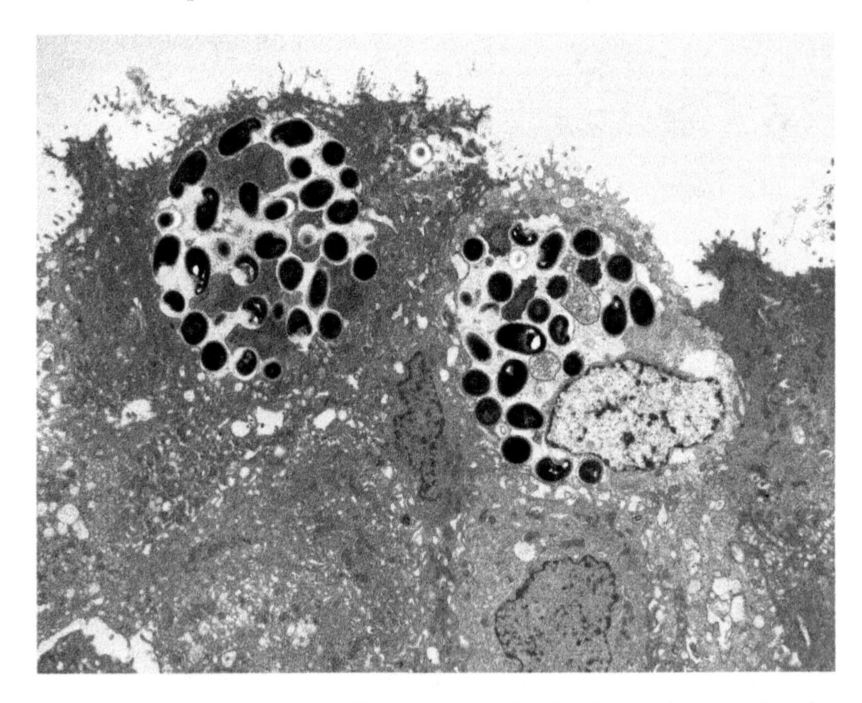

Figure 6.14 *Encephalitozoon hellem* in epithelial cells of a nasal polyp taken from an AIDS patient. Note the development of spores within a vacuole. Spores measure 2–2.5 × 1–1.5 μm.

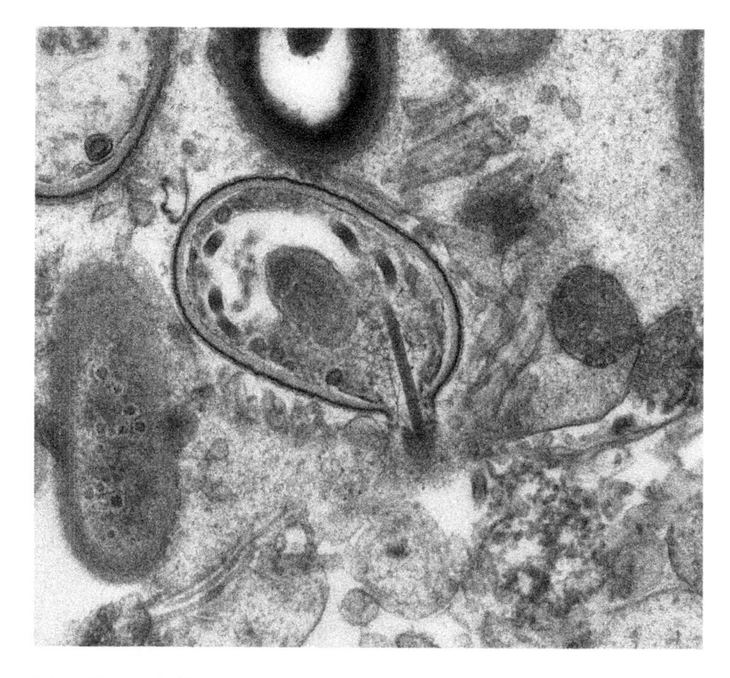

Figure 6.15 *Encephalitozoon cuniculi* spore found in the urine sediment of an AIDS patient with renal symptoms. Note that the polar tube is partially discharged. Spores measure 2.5–3.2 × 1.2–1.6 μm.

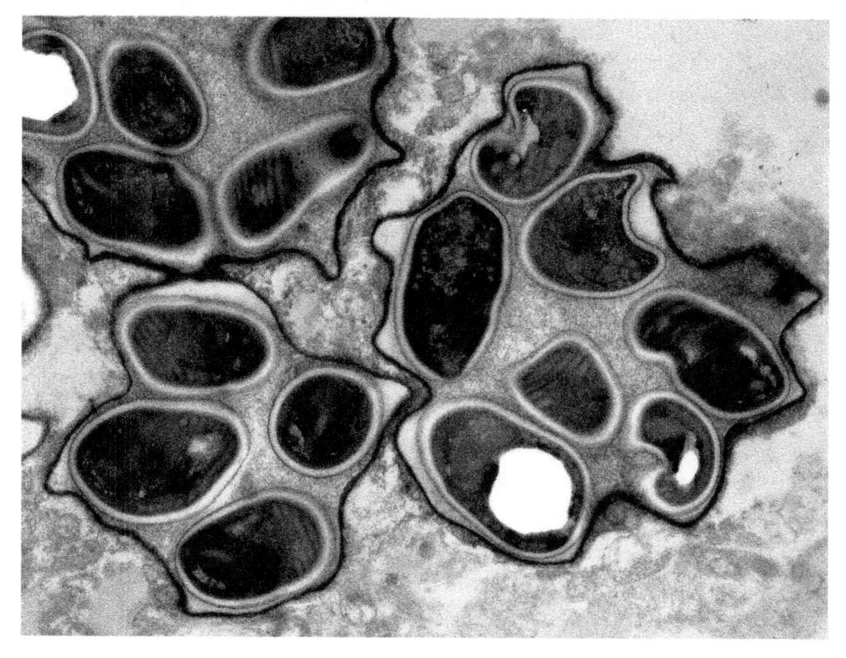

Figure 6.16 *Trachipleistophora hominis* sporophorous vesicles containing spores in the skeletal muscle of an AIDS patient. Spores measure 5.2 × 2.4 μm.

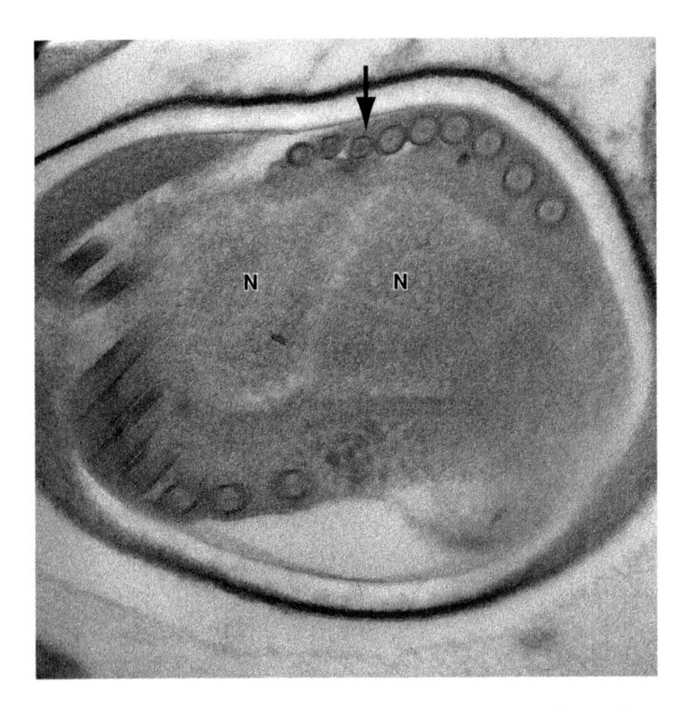

Figure 6.17 Diplokaryon (double nucleus) (N) in a spore of *Vittaforma corneae* found in the cornea of a non-immunocompromised patient with stromal keratitis. Spores measure $3.8 \times 1.2\,\mu m$. Note the anisofilar polar tube arrangement (posterior coils are of a smaller diameter than more anterior coils) (arrow). Although far from ideal, this sample was reprocessed into resin for EM from a wax block and shows that diagnostically significant detail can be found in such samples with careful searching.

AIDS patients, but occasionally microsporidia are found in individuals with other immunodeficiencies (e.g. renal transplant patients) or in ophthalmic samples from immunocompetent individuals. EM remains an important technique (the 'gold standard') in identifying microsporidia, which, because of their small size, are often difficult to recognise by LM.

Microsporidia lack mitochondria, and ribosomes are similar to those found in prokaryotes. The nuclear envelope remains intact throughout the developmental cycle. During nuclear division, plaques are formed on the nuclear envelope, and these are involved in spindle formation. A single nucleus or double nucleus (diplokaryon) may be present, depending on the species. The highly resistant spore is the most easily recognised stage, and even if tissues are poorly preserved or reprocessed from wax, this stage can be identified – hence the emphasis on spore morphology shown in Table 6.3. The spore wall consists of a dark

Table 6.3 Microsporidia currently identified in humans

Species	Main tissues infected	References	Spore size and morphology	Some significant ultrastructural features of developmental stages
Enterocytozoon bieneusi[a] (Figures 6.12 and 6.13)	Mainly small intestinal epithelium but also bile duct epithelium, nasal polyps and bronchial epithelium	Desportes *et al.* (1985)	Spores very small, 1.5 × 0.5 μm; single nucleus; 4–6 coils of PT in two rows	Uninucleate; all developmental stages lie in direct contact with the host cell cytoplasm; in multinucleated sporogonial plasmodia, polar tube precursors form as electron-dense discs; a unique characteristic of this genus is the complete differentiation of the polar tube within the sporogonial plasmodium prior to fission.
Encephalitozoon intestinalis[a] (synonym for *Septata intestinalis*)	Intestinal tract, kidney and gall bladder	Cali, Kotler and Orenstein (1993), Canning *et al.* (1994) and Hartskeerl *et al.* (1995)	2.2 × 1.2 μm; single nucleus; 5–7 coils of PT in single row	Uninucleate; development occurs in parasitophorous vacuoles.
E. hellem[a] (Figure 6.14)	Eyes, kidney, nasal polyps and lung	Didier *et al.* (1991), Canning *et al.* (1992) and Hollister *et al.* (1993)	2–2.5 × 1–1.5 μm; single nucleus; 6–8 coils of PT in single row	Uninucleate; development occurs in parasitophorous vacuoles.

(*continued overleaf*)

Table 6.3 (*continued*)

Species	Main tissues infected	References	Spore size and morphology	Some significant ultrastructural features of developmental stages
E. cuniculi[a] (Figure 6.15)	Liver, peritoneum and kidney	Canning and Lom (1986)	2.5–3.2 × 1.2–1.6 μm; single nucleus; 4–6 coils of PT in single row	Uninucleate; development occurs in parasitophorous vacuoles.
Trachipleistophora hominis (Figure 6.16)	Skeletal muscle, pectoral muscle myocardium, nasal sinus and epithelium and stroma of the eye	Hollister *et al.* (1996), Rauz *et al.* (2004) and Curry *et al.* (2005)	Spores in packets (e.g. SPOV); 5.2 × 2.4 μm; single nucleus; 10–13 coils of PT in single row; PT is anisofilar (8–11 + 1–3).	Unpaired nuclei throughout life cycle; meronts form as multinucleate plasmodia that possess a thick, amorphous, electron-dense surface coat external to the plasma membrane; surface coat becomes wall of SPOV.
T. anthropophthera	Brain, myocardium, kidney, liver, thyroid, parathyroid, bone marrow, lymph nodes, spleen and cornea	Yachnis *et al.* (1996), Vavra *et al.* (1998), Juarez *et al.* (2005) and Pariyakanok and Jongwutiwes (2005)	Spores in SPOVs of two types; type I SPOV contains usually eight spores measuring 3.7 × 2 um with nine coils of the PT; anisofilar (7 + 2); type II SPOV contains two spores measuring 2.5 × 1.4 um with 4–5 coils of the PT.	Unpaired nuclei throughout life cycle; meronts form as multinucleate plasmodia that possess a thick, amorphous, electron-dense surface coat external to the plasma membrane; surface coat becomes wall of SPOV.

Pleistophora ronneafiei	Skeletal muscle	Ledford *et al.* (1985) and Cali and Takvorian (2003)	Spores in SPOVs; spores about 3.3–4 μm × 2–2.8 μm with 9–11 coils of the PT.	Unpaired nuclei throughout the life cycle; meronts form as multinucleate plasmodia that possess a thick, amorphous, electron-dense surface coat external to the plasma membrane; surface coat becomes wall of SPOV.
Pleistophora sp.	Skeletal muscle	Ledford *et al.* (1985), Chupp *et al.* (1993) and Grau *et al.* (1996)	Spores in SPOVs; spores about 4 × 2 μm with 9–12 coils of the PT (Chupp *et al.*, 1993) or spores about 3.2–3.4 × 2.8 μm with 11 coils of the PT	Unpaired nuclei throughout life cycle; meronts form as multinucleate plasmodia that possess a thick, amorphous, electron-dense surface coat external to the plasma membrane; surface coat becomes wall of SPOV.
Anncaliia vesicularum	Skeletal muscle	Franzen *et al.* (2006)	2.5 × 2 μm; diplokaryotic; 7–10 coils of PT arranged in up to three rows (usually two); anisofilar with 2–3 narrow posterior coils	Diplokaryotic; developmental stages show strands of vesicular structures penetrating into the cytoplasm of the host cell.
A. connori (synonym for *Nosema connori*)	Myocardium and diaphragm but widely disseminated	Sprague (1974) and Shadduck, Kelsoe and Helmke (1979)	4–4.5 × 2–2.5 μm; diplokaryotic; 10–12 coils of PT arranged in a single row; not known if anisofilar	Diplokaryotic; developmental stages show strands of vesicular structures penetrating into the cytoplasm of the host cell.

(*continued overleaf*)

Table 6.3 (*continued*)

Species	Main tissues infected	References	Spore size and morphology	Some significant ultrastructural features of developmental stages
A. algerae (synonym for *Brachiola algerae* and *Nosema algerae*)	Eye, skin, muscle and vocal cords	Visvesvara *et al.* (1999), Coyle *et al.* (2004) and Cali *et al.* (2010)	3.7–5.4 × 2.3–3.9 μm; diplokaryotic; 8–11 coils of PT arranged in a single row; anisofilar with 0–3 narrow posterior coils	Developmental stages show strands of vesicular structures penetrating into the cytoplasm of the host cell.
Nosema ocularum	Eye	Cali *et al.* (1991)	3 × 5 μm; diplokaryotic; 9–12 coils of PT in a single row	Parasites are in direct contact with host cell cytoplasm; diplokaryotic throughout the life cycle.
Nosema sp.[b]	Eye	Curry *et al.* (2007)	2.5–3.5 μm in length; diplokaryotic; 6–7 coils of PT in single row	Parasites are in direct contact with host cell cytoplasm; diplokaryotic throughout the life cycle.
Tubulinosema sp.	Muscle (tongue)	Choudhary *et al.* (2011)	1.4–2.4 μm in length; diplokaryotic; 11 coils of PT in a single row; anisofilar (eight wide and three narrow)	Developmental stages not seen.
Vittaforma corneae (Figure 6.17)	Eye	Silveira and Canning (1995) and Rauz *et al.* (2004)	3.8 × 1.2 μm; diplokaryotic; 5–7 coils of PT in a single row; PT isofilar in mature spores,	All developmental stages are individually enveloped by cisternae of host cell rER.

			but immature spores show an anisofilar arrangement (5 + 2 narrow or 4 + 3 narrow); host rER surrounds spores.	
Vittaforma-like species	Intestinal tract	Sulaiman *et al.* (2003)	1.8–3.6 × 1–1.4 µm	NA
'*Microsporidium*'[c] *ceylonensis*	Eye	Ashton and Wirasinha (1973) and Canning *et al.* (1998)	3.5 × 1.5 µm; single nucleus; about 8–13 coils of PT; PT anisofilar (6–10 wide and 2–3 narrow)	Sporoblasts and spores are seen in macrophages of wax sections processed into resin for EM.
'*Microsporidium*'[c] *africanum*	Eye	Pinnolis *et al.* (1981)	4.5–5 × 2.5–3 µm; single nucleus; 11–13 coils of PT; PT anisofilar	Develop in macrophages.

[a]Commonest species found in humans.
[b]'*Nosema* group or type' are mainly insect pathogens that will generate more human infections as this group is investigated by molecular methods. Many of the *Nosema* types we now know of will be separated into new genera and species.
[c]True genus undetermined – '*Microsporidium*' is a convenient collective group but given a specific name.
Anisofilar PT = polar tube has two different diameters.
Isofilar PT = polar tube all of the same diameter.
PT: polar tube; SPOV: sporophorous vesicle; rER: rough endoplasmic reticulum and NA: not available.

proteinaceous exospore layer and a lucent chitinous layer, which is lined internally with the plasma membrane. Internally, the spore contains a diagnostically significant coiled filament (the polar tube) that originates at the anterior of the spore (the manubrium) and is coiled towards the posterior of the spore. Around the anterior straight portion of the tube is the polaroplast, an organelle consisting of many membranous lamellae. A posterior vacuole is present in intact spores, but this normally collapses during processing for EM and consequently distorts the posterior of the spore.

Although it may seem that microsporidia will infect only a specific organ, it should be noted that the range of tissues reportedly infected by these organisms is expanding and that dissemination should be expected, particularly if a species of *Encephalitozoon* or *Anncaliia* is involved. Equally, in the investigation of enteric microsporidial infection, a rectal biopsy may be considered because of the relative ease of providing biopsy material. However, microsporidia such as *Enterocytozoon bieneusi* do not normally colonise this terminal segment of the gut, and if suspected, a small intestinal biopsy should be considered instead.

In optimally fixed material, details of developmental stages can be found. However, microsporidia can show a complex and wide spectrum of ultrastructural features – see articles by Larsson (1986) and Vavra and Larsson (1999) – some of which can aid specific identification. These articles show mainly microsporidia infecting various animals (particularly insects and fish) rather than humans, but as these organisms are opportunistic, particularly in the immunocompromised individual, some may be able to infect humans, and this seems to be the case from the published literature.

6.9 PARASITIC PROTOZOA

Protozoa are a diverse group of eukaryotic single-cell organisms, some of which have developed a parasitic existence. Parasitic protozoa are some of the most important pathogens of animals and humans (Levine, 1985). Four main groups can be found infecting humans: flagellates (e.g. *Giardia, Trichomonas, Leishmania* and *Trypanosoma*), apicomplexans (e.g. *Plasmodium, Cryptosporidium, Isospora* and *Toxoplasma*), amoebae (e.g. *Acanthamoeba* and *Entamoeba*) and ciliates (only *Balantidium coli* in humans). About 80–100 protozoan parasites have been described from humans (Ashford and Crewe, 1998). All groups show ultrastructural features that can help to identify the protozoan organism involved and, sometimes, can identify the particular species (Aikawa and Sterling, 1974; Scholtyseck, 1979; Warton, 1992).

There are too many parasitic protozoan infections for this section of the chapter to cover adequately, but other published works (see this chapter's references) can be consulted for a more comprehensive coverage of protozoan parasite ultrastructure. The organisms covered here are examples of what can be found by EM, with pertinent ultrastructural features shown. Most are intracellular parasitic protozoa that may be found in the examination of tissues. Normally, most of these organisms would be detected by other diagnostic methods if they were suspected. As EM is potentially a catch-all method, examination can sometimes reveal unexpected infectious agents. If an organism is detected, characteristic ultrastructural features can assist in identification. Coccidia (phylum Apicomplexa) are normally elongated cells that share some common morphological features, particularly the apical complex seen in some developmental stages. However, characteristic ultrastructural features, such as those associated with the apical complex, might not be present in all life cycle stages. In this chapter, particular emphasis is placed on enteric protozoan infections – see Table 6.4. If an enteric coccidian is found in a tissue section or is suspected, a stool sample should be examined for oocysts that can confirm infection and identify the enteric species involved.

6.9.1 Cryptosporidium

Cryptosporidium infection causes diarrhoea in humans which can be of relatively short duration in normal individuals, but in the immunocompromised (such as those with AIDS), it can cause chronic, life threatening, watery diarrhoea (Chalmers and Davies, 2010). EM is still occasionally requested on intestinal, rectal or bile duct biopsies (e.g. the ampulla of Vater) to detect these organisms from individuals who are immunocompromised for reasons other than HIV infection. *Cryptosporidium* undergoes a complex developmental cycle (asexual, sexual and sporogonic) in the gut, and many of the life cycle stages are distinctive and recognisable ultrastructurally (see Current and Reese, 1986). Although *Cryptosporidium* is a genuinely intracellular parasite for most of its developmental cycle, it superficially appears to be an extracellular infection in intestinal or rectal biopsies. The organism 'appears' to be attached to the surface of intestinal enterocytes because of its luminal existence, but the plasma membrane of the infected enterocytes actually encloses

Table 6.4 Some enteric protozoan parasites

Organism	Site of infection	Size and ultrastructural features	Characteristics of oocysts or cystic stages found in stool specimens	Comments
Cryptosporidium parvum and *C. hominis* (Figure 6.18)	Occurs throughout the gastrointestinal tract	Type I meronts (containing 6–8 merozoites) about 4.5–6 × 4.5–6 μm; type II meronts (containing four merozoites) about 3.5–5 μm × 3.5–5 μm; merozoites show typical apical complex features; macrogamont 4–5.5 μm × 4 μm × 5.5 μm; microgamont 3.5–5 μm × 3.5–5 μm	Oocysts[a] are spherical, measure about 4.5 × 5.5 μm and contain four naked sporozoites.	Developmental stages are intracellular but extracytoplasmic (i.e. appear to be surface parasites); complex life cycle; microgametes are nonflagellated and with an elliptical nucleus – see Current and Reese (1986).
Isospora belli (Figure 6.19)	Mainly small intestinal epithelium and lamina propria, but it affects the oesophagus, mesenteric and mediastinal lymph nodes, liver, gallbladder and spleen in AIDS.	Ultrastructural features typically apicomplexan, and resting stage (hypnozoite) shows crystalloid and fibrous surrounding (Figure 6.19). Merozoites measure approximately 4–11 μm × 3–4 μm; tissue cysts measure about 8 μm × 3.5–4 μm.	Oocysts[a] are elliptical, measure 20–33 × 10–19 μm and contain four sporozoites in each of two sporocysts.	Rare, but seen in immunocompromised patients and especially those with AIDS in temperate regions; complex, mainly intracellular, life cycle.

Cyclospora cayetanensis	Enterocytes of the upper small intestine	Ultrastructural features typically apicomplexan; merozoites about 5.5 μm × 3.6 μm; microgametocyte about 5.6 μm × 3.6 μm	Oocysts[a] are spherical, measure 8–10 μm and contain two sporozoites in each of two sporocysts.	Complex, mainly intracellular, life cycle; microgametes are flagellated.
Giardia intestinalis (synonym for *G. lamblia* and *G. duodenalis*)	Small intestine	Trophozoites measure 10–20 × 5–10 μm	Cysts are oval in shape and measure 8–14 × 6–10 μm	Lives in the lumen of the small intestine; lining cells sometimes show microvillous shortening; not an intracellular parasite. Trophozoite has four pairs of flagella, two nuclei, median bodies and an attachment disc.

[a]Oocyst examination in stool samples can determine the species involved. i.e. the oocysts of *Cryptosporidium*, *Isospora* and *Cyclospora* all differ in size and morphology.

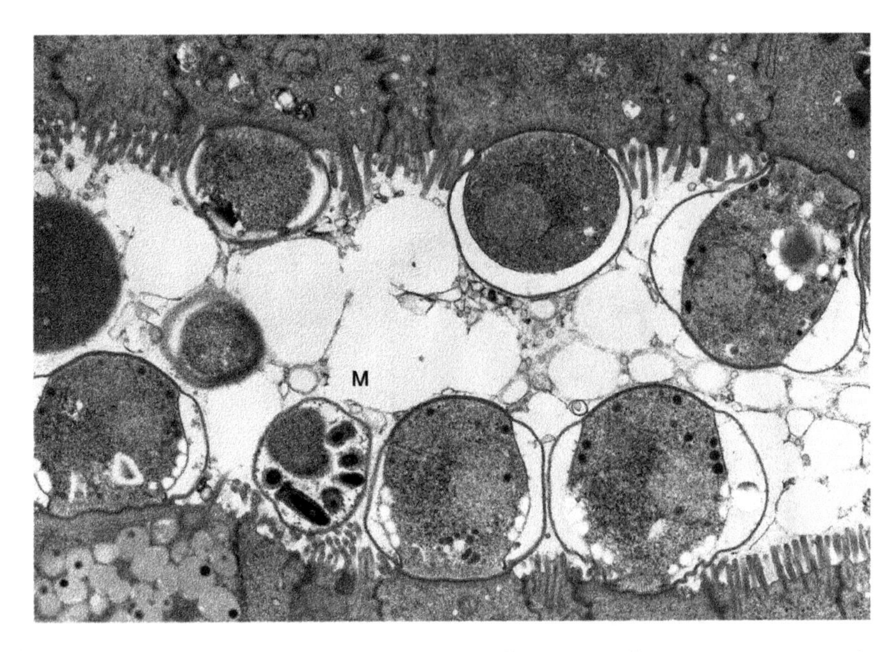

Figure 6.18 *Cryptosporidium sp.* in the small intestine of an AIDS patient with chronic diarrhoea. Although an intracellular parasite, *Cryptosporidium* appears to be adherent to the surface of infected cells. This parasite has a complex life cycle, and several of these life cycle stages are shown in this view, including a microgametocyte (m). Microgametocytes measure about 5 μm in diameter.

the *Cryptosporidium* parasite (Figure 6.18). Sometimes, intestinal enterocytes can be infected by more than one *Cryptosporidium* parasite, and infected enterocytes often have elongated microvilli. From published literature, it seems that *Cryptosporidium* is capable of infecting cells other than those in the gastrointestinal tract. There are reports of tracheo-bronchial infection and sinusitis in severely immunocompromised individuals (Clavel *et al.*, 1996; Dunand *et al.*, 1997). In addition, the genus *Cryptosporidium* is diverse in both host range and described species. The most common human infections involve *C. parvum* or *C. hominis*, and these are thought to be morphologically identical. However, other species, such as *C. muris* (Chalmers *et al.*, 1994), which has also been reported from humans, differs in some significant ultrastructural details, which can aid identification. Many species and cryptic species have been identified by molecular methods, but few of these have equivalent ultrastructural descriptions. *Cryptosporidium* is easily missed by LM but can be readily identified under EM. If this significant parasite is detected in a biopsy, it may only be possible to say that a species of

Cryptosporidium has been identified; if speciation is required, refer the sample for molecular analysis.

6.9.2 Isospora belli

In the pre-AIDS era, according to published literature, *Isospora belli* was regarded as a very rare enteric infection with most cases being reported from the tropics. In addition, this parasite is thought to have no animal reservoir, with infections being entirely human. A second species of *Isospora, I. natalensis*, has been reported from South Africa (Elson-Dew, 1953), but human infection with this species has not been reported subsequently (Dubey, 1993). The not-infrequent finding of *I. belli* in AIDS patients (and other immunocompromised patients) in temperate locations is at odds with the previously published view of its rarity. An animal reservoir still has not been identified, but it is possible that human infection is much commoner than was previously thought. Symptomatic infection may be transient or non-existent in immunocompetent individuals and may be followed by latency. Immunosuppression, by whatever cause, might allow symptomatic reactivation of this protozoan parasite. Infection by *I. belli* should normally be diagnosed by the finding of characteristic oval oocysts in diarrhoeal samples or the detection of intracellular stages developing in conspicuous intracellular vacuoles in intestinal biopsies by LM. However, careful EM examination of intestinal biopsy specimens can sometimes reveal this parasite, particularly in cases where faecal oocysts have been missed or where the intracellular vacuolar membrane is tightly adherent to the parasite pellicle, making detection difficult except by EM (Figure 6.19).

I. belli belongs to a group of parasitic protozoa called apicomplexans (as do *Cryptosporidium* and *Cyclospora*), so-called because of the presence in some life cycle stages of the aptly named apical complex that consists of a conoid, polar rings, rhoptries and micronemes (see Figure 6.19). These structures facilitate penetration and infection of new cells. *I. belli* can also be found as a resting stage in the blood vessel pericytes in the lamina propria. These intracellular sporozoite resting stages are covered in a thick fibrous layer (the cyst wall) and internally contain a prominent crystalloid structure. *I. belli* can also disseminate away from the intestine, as it has been found in mesenteric, mediastinal and tracheo-bronchial lymph nodes, liver and spleen (Restrepo, Macher and Radany, 1987; Michiels *et al.*, 1994).

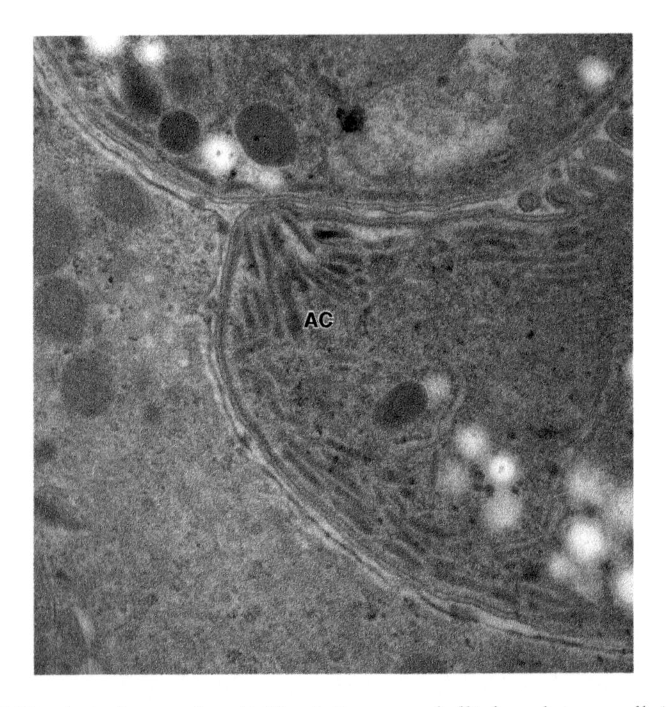

Figure 6.19 Apical complex (AC) of *Isospora belli* found in small intestinal enterocytes of an AIDS patient with diarrhoea. The apical complex consists of a conoid, polar rings, rhoptries and micronemes. At its widest, the diameter of the conoid is about 400 nm. All protozoa of the apicomplexan group possess these structures associated with cell entry at some stage of their life cycle. Note that in this case, the vacuolar membrane surrounding the organisms is closely adherent.

6.10 EXAMPLES OF NON-ENTERIC PROTOZOA

Kinetoplastid flagellates, such as *Leishmania* and *Trypanosoma*, show a characteristic cytoplasmic structure that readily identifies this group of protozoa. This organelle, the kinetoplast, is a modified area of the mitochondrion, containing a dense band of DNA (Figure 6.20). In addition, these parasites show conspicuous sub-pellicular microtubules.

Leishmania exists in two forms – a nonflagellated (amastigote) rounded form in humans or other vertebrates (note that a flagellum is indeed present in the amastigote form, but it is shortened and contained in a flagellar pocket) and an elongated flagellated (promastigote) form in the sandfly vector. The kinetoplast is situated anteriorly between the nucleus and the basal body found associated with the flagellar apparatus. Although all species show a similar morphology, they are grouped

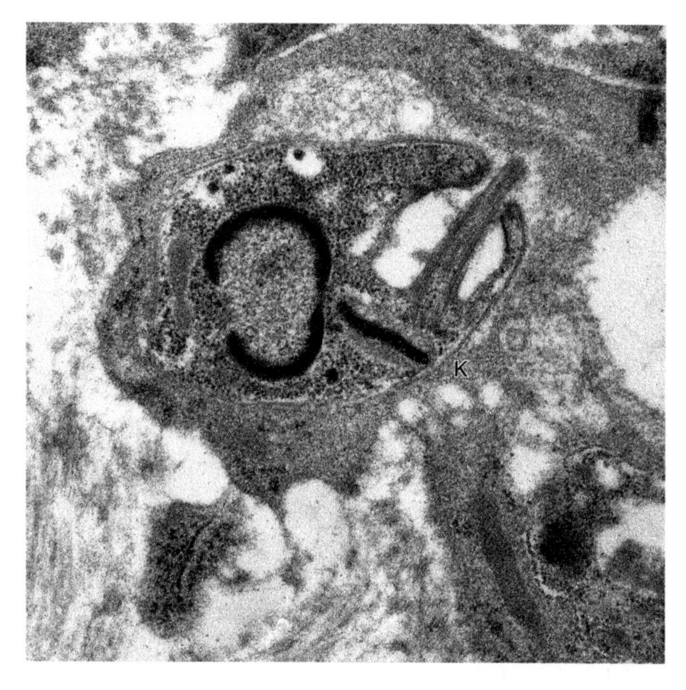

Figure 6.20 *Leishmania* from the tongue of an immunocompromised patient showing the kinetoplast (K) at the base of the short flagellum located within a pocket (amastigote form). The kinetoplast measures about 1 μm in length.

as species-complexes (e.g. the *Leishmania donovani* complex). *Leishmania* parasites cause several diseases in humans depending on the species complex involved (e.g. cutaneous leishmaniasis, visceral leishmaniasis and mucocutaneous leishmaniasis).

6.11 PARASITIC AMOEBAE

Amoebae can be difficult to identify in tissues because of their pleomorphic form, but careful examination can reveal some of the characteristic ultrastructural features of these parasitic cells (Martinez *et al.*, 1971, 1976). *Entamoeba* shows no mitochondria, endoplasmic reticulum or Golgi apparatus in the cytoplasm (Martinez-Palomo *et al.*, 1976). Mitochondria are present in *Acanthamoeba*, and these show tubular cristae (Bowers and Korn, 1968). Rough endoplasmic reticulum and smooth endoplasmic reticulum are also present. The cysts of *Acanthamoeba* in the cornea are thick-walled and show a characteristic lobed or stellate appearance (Bowers and Korn, 1969).

6.12 CONCLUSIONS

EM remains an important diagnostic technique in the investigation of both microsporidial and parasitic protozoan infections in biopsies. In addition, the technique still has a valuable role in research into the biological aspects of many infectious diseases, particularly interactions between microorganism and cells and/or intracellular development of the agent.

ACKNOWLEDGEMENTS

Sincere thanks are due to Dr Andrew Dean, Dr Rachel Chalmers, Maureen Metcalfe, Dr John Kennaugh and Mrs Hilary Cotterill for contributing information and for critical comments on the manuscript.

REFERENCES AND ADDITIONAL READING

General

Fuller, R. and Lovelock, D.W. (1976) *Microbial Ultrastructure. The Use of the Electron Microscope*, Society for Applied Bacteriology, Technical Series, Vol. 10, Academic Press, London.

Warton, A. (1992) Diagnostic ultrastructure of human parasites, in *Diagnostic Ultrastructure of Non-Neoplastic Diseases* (eds J.M. Papadimitriou, D.W. Henderson and D.V. Spagnolo). Churchill Livingstone, Edinburgh, pp. 145–169.

Wills, E.J. (1992) Infectious agents: fungi, bacteria and viruses, in *Diagnostic Ultrastructure of Non-Neoplastic Diseases*, Chapter 9 (eds J.M. Papadimitriou, D.W. Henderson and D.V. Spagnolo), Churchill Livingstone, Edinburgh, pp. 212–233.

Viruses

Caul, E.O. and Appleton, H. (1982) The electron microscopical and physical characteristics of small round human faecal viruses: an interim scheme for classification. *Journal of Medical Virology*, **9**, 257–265.

Curry, A., Appleton, H. and Dowsett, B. (2006) Application of transmission electron microscopy to the clinical study of viral and bacterial infections: present and future. *Micron*, 37, 91–106.

Doane, F.W. and Anderson, N. (1987) *Electron Microscopy in Diagnostic Virology. A Practical Guide and Atlas*, Cambridge University Press, Cambridge.

Falsey, A.R. and Walsh, E.E. (2003) Novel coronavirus and severe acute respiratory syndrome. *Lancet*, **361**, 1312–1313.

Ghadially, F.N. (1988) *Ultrastructural Pathology of the Cell and Matrix. A Text and Atlas of Physiological and Pathological Alterations in the Fine Structure of Cellular and Extracellular Components*, Butterworths, Oxford.

Goldsmith, C.S., Metcalfe, M.G., Rollin, D.C. *et al.* (2011) Ultrastructural characterization of pandemic (H1N1) 2009 virus. *Emerging Infectious Diseases*, **17**, 2056–2059.

Hazelton, P.R. and Gelderblom, H.R. (2003) Electron microscopy for rapid diagnosis of infectious agents in emergent situations. *Emerging Infectious Diseases*, **9**, 294–303.

Hsiung, G.D., Fong, C.K.Y. and Landry, M.L. (1994) *Hsiung's Diagnostic Virology as Illustrated by Light and Electron Microscopy*, 4th edn, Yale University Press, New Haven, CT.

Ksiazek, T.G., Erdman, D., Goldsmith, C.S. *et al.* The SARS Working Group (2003) A novel coronavirus associated with severe acute respiratory syndrome. *New England Journal of Medicine*, **348**, 1953–1966.

Madeley, C.R. (1992) The search for viruses by negative contrast, in *Diagnostic Ultrastructure of Non-Neoplastic Diseases*, Chapter 8 (eds J.M. Papadimitriou, D.W. Henderson and D.V. Spagnolo), Churchill Livingstone, Edinburgh, pp. 194–211.

Madeley, C.R. and Field, A.M. (1988) *Virus Morphology*, 2nd edn, Churchill Livingstone, Edinburgh.

Miller, S.E. (2003) Bioterrorism and electron microscopic differentiation of poxviruses from herpesviruses: dos and don'ts. *Ultrastructural Pathology*, **27**, 133–140.

Ramqvist, T. and Dalianis, T. (2010) Oropharyngeal cancer epidemic and human papillomavirus. *Emerging Infectious Diseases*, **16**, 1671–1677.

Stanbridge, C.M., Mather, J., Curry, A. and Butler, E.B. (1981) Demonstration of papilloma virus particles in cervical and vaginal scrape material: a report of 10 cases. *Journal of Clinical Pathology*, **34**, 524–531.

Williams-Gray, C.H., Aliyu, S.H., Lever, A.M.L. *et al.* (2007) Reversible parkinsonism in a patient with progressive multifocal leucoencephalopathy. *Journal of Neurology, Neurosurgery, and Psychiatry*, **78**, 408–410.

Bacteria

De Pamphilis, M.L. and Adler, J. (1971) Fine structure and isolation of the hook-basal body complex of flagella from *Escherichia coli* and *Bacillus subtilis*. *Journal of Bacteriology*, **105**, 384–395.

Johnson, R.C. (1977) The spirochaetes. *Annual Review of Microbiology*, **31**, 89–106.

Murray, R.G.E. (1978) Form and function – 1. Bacteria, in *Essays in Microbiology* (eds J.R. Norris and M.H. Richmond), John Wiley & Sons, Inc., Hoboken, NJ, pp. 1–32.

Silverman, M. and Simon, M.I. (1977) Bacterial flagella. *Annual Review of Microbiology*, **31**, 397–419.

Fungal Organisms

Ashton, N. and Wirasinha, P.A. (1973) Encephalitozoonosis (nosematosis) of the cornea. *The British Journal of Ophthalmology*, **57**, 669–674.

Cali, A., Kotler, D.P. and Orenstein, J.M. (1993) *Septata intestinalis* n.g. n.sp., an intestinal microsporidian associated with chronic diarrhea and dissemination in AIDS patients. *The Journal of Eukaryotic Microbiology*, **40**, 101–112.

Cali, A., Meisler, D.M., Rutherford, I. *et al.* (1991) Corneal microsporidiosis in a patient with AIDS. *The American Journal of Tropical Medicine and Hygiene*, **44**, 463–468.

Cali, A., Neafie, R., Weiss, L.M. *et al.* (2010) Human vocal cord infection with the microsporidium *Anncaliia algerae*. *The Journal of Eukaryotic Microbiology*, **57**, 562–567.

Cali, A. and Takvorian, P.M. (2003) Ultrastructure and development of *Pleistophora ronneafiei* n. sp., a microsporidium (protista) in the skeletal muscle of an immune-compromised individual. *The Journal of Eukaryotic Microbiology*, **50**, 77–85.

Canning, E.U., Curry, A., Lacey, C.J.N. and Fenwick, J.D. (1992) Ultrastructure of *Encephalitozoon* sp. infecting the conjunctival, corneal and nasal epithelia of a patient with AIDS. *European Journal of Protistology*, **28**, 226–237.

Canning, E.U., Curry, A., Vavra, J. and Bonshek, R.E. (1998) Some ultrastructural data on *Microsporidium ceylonensis*, a cause of corneal microsporidiosis. *Parasite*, **5**, 247–254.

Canning, E.U., Field, A.S., Hing, M.C. and Marriott, D.J. (1994) Further observations on the ultrastructure of *Septata intestinalis* Cali, Kotler and Orenstein, 1993. *European Journal of Protistology*, **30**, 414–422.

Canning, E.U. and Lom, J. (1986) *The Microsporidia of Vertebrates*, Academic Press, London.

Chupp, G.L., Alroy, J., Adelvan, L.S. *et al.* (1993) Myositis due to *Pleistophora* (microsporidia) in a patient with AIDS. *Clinical Infectious Diseases*, **16**, 15–21.

Coyle, C., Weiss, L.M., Rhodes, L.V. *et al.* (2004) Fatal myositis due to the microsporidian *Brachiola algerae*, a mosquito pathogen. *The New England Journal of Medicine*, **351**, 42–47.

Curry, A. (2005) Microsporidiosis, in *Topley & Wilson's Microbiology & Microbial Infections*, 10th edn, Parasitology Volume (eds F.E.G. Cox, D. Wakelin, S.H. Gillespie and D.D. Despommier), Hodder Arnold, London, pp. 171–195.

Curry, A., Beeching, N.J., Gilbert, J.D. *et al.* (2005) *Trachipleistophora hominis* infection in the myocardium and skeletal muscle of a patient with AIDS. *Journal of Infection*, **51**, e139–e144.

Curry, A., Mudhar, H.S., Dewan, S. *et al.* (2007) A case of bilateral microsporidial keratitis from Bangladesh – infection by an insect parasite from the genus *Nosema*. *Journal of Medical Microbiology*, **56**, 1250–1252.

Desportes, I., Le Charpentier, Y., Galian, A. *et al.* (1985) Occurrence of a new microsporidian: *Enterocytozoon bieneusi* n.g., n.sp. in the enterocytes of a human patient with AIDS. *The Journal of Protozoology*, **32**, 250–254.

Didier, E.S., Didier, P.J., Friedberg, D.N. *et al.* (1991) Isolation and characterization of a new human microspridian *Encephalitozoon hellem* (n. sp.), from three

AIDS patients with keratoconjunctivitis. *The Journal of Infectious Diseases*, **163**, 617–621.

Franzen, C., Nassonova, E.S., Scholmerich, J. and Issi, I.V. (2006) Transfer of the members of the genus *Brachiola* (Microsporidia) to the genus *Anncaliia* based on ultrastructural and molecular data. *The Journal of Eukaryotic Microbiology*, **53**, 26–35.

Grau, A., Valls, M.E., Williams, J.E. *et al.* (1996) Miositis por *Pleistophora* en un paciente consida. *The Medical Clinics*, **107**, 779–781.

Hartskeerl, R.A., Van Gool, T., Schuitema, A.R. *et al.* (1995) Genetic and immunological characterization of the microsporidian *Septata intestinalis* Cali, Kotler and Orenstein, 1993: reclassification to *Encephalitozoon intestinalis*. *Parasitology*, **110**, 277–285.

Hollister, W.S., Canning, E.U., Colbourn, N.I. *et al.* (1993) Characterization of *Encephalitozoon hellem* (Microspora) isolated from the nasal mucosa of a patient with AIDS. *Parasitology*, **107**, 351–358.

Hollister, W.S., Canning, E.U., Weidner, E. *et al.* (1996) Development and ultrastructure of *Trachipleistophora hominis* n.g., n.sp. after in vitro isolation from an AIDS patient and inoculation into athymic mice. *Parasitology*, **112**, 143–154.

Juarez, S.I., Putaporntip, C., Jongwutiwes, S. *et al.* (2005) In vitro cultivation and electron microscopy characterization of *Trachipleistophora anthropophthera* isolated from the cornea of an AIDS patient. *The Journal of Eukaryotic Microbiology*, **52**, 179–190.

Larsson, R. (1986) Ultrastructure, function and classification of Microsporidia. *Progress in Protozoology*, **1**, 325–390.

Ledford, D.K., Overman, M.D., Gonzalvo, A. *et al.* (1985) Microsporidiosis myositis in a patient with acquired immunodeficiency syndrome. *Annals of Internal Medicine*, **102**, 628–630.

Choudhary, M.M., Metcalfe, M.G., Arrambide, K. *et al.* (2011) *Tubulinosema sp.* Microsporidian myositis in immunosuppressed patient. *Emerging Infectious Diseases*, **17**, 1727–1730.

Pariyakanok, L. and Jongwutiwes, S. (2005) Keratitis caused by *Trachipleistophora anthropopthera*. *The Journal of Infection*, **51**, 325–328.

Pinnolis, M., Egbert, P.R., Font, R.L. and Winter, F.C. (1981) Nosematosis of the cornea. Case report, including electron microscopic studies. *Archives of Ophthalmology*, **99**, 1044–1047.

Rauz, S., Tuft, S., Dart, J.K.G. *et al.* (2004) Ultrastructural examination of two cases of stromal microsporidial keratitis. *Journal of Medical Microbiology*, **53**, 775–781.

Silveira, H. and Canning, E.U. (1995) *Vittaforma corneae* N. Comb. for the human microsporidium *Nosema corneum* Shadduck, Meccoli, Davis & Font, 1990, based on its ultrastructure in the liver of experimentally infected athymic mice. *The Journal of Eukaryotic Microbiology*, **42**, 158–165.

Shadduck, J.A., Kelsoe, G. and Helmke, J. (1979) A microsporidian contaminant of a non-human primate cell culture: ultrastructural comparison with *Nosema connori*. *Journal of Parasitology*, **65**, 185–188.

Sprague, V. (1974) *Nosema connori* n. sp., a microsporidian parasite of man. *Transactions of the American Microscopical Society*, **93**, 400–403.
Sulaiman, I.M., Matos, O., Lobo, M.L. and Xiao, L. (2003) Identification of a new microsporidian parasite related to *Vittaforma corneae* in HIV-positive and HIV-negative patients from Portugal. *The Journal of Eukaryotic Microbiology*, **50**, 586–590.
Van Gool, T., Snijders, F., Eeftinck Schattenkerk, J.K.M. *et al.* (1993) Diagnosis of intestinal and disseminated microsporidial infections in HIV-infected individuals with a new rapid fluorescent technique. *Journal of Clinical Pathology*, **46**, 694–699.
Vavra, J. and Larsson, J.I.R. (1999) Structure of microsporidia, in *The Microsporidia and Microsporidiosis*, Chapter 2 (eds M. Wittner and L.M. Weiss), ASM Press, Washington, DC, pp. 7–84.
Vavra, J., Yachnis, A.T., Shadduck, J.A. and Orenstein, J.M. (1998) Microsporidia of the genus *Trachipleistophora*-causative agents of human microsporidiosis: description of *Trachipleistophora anthropophthera* n.sp. (Protozoa: Microsporidia). *The Journal of Eukaryotic Microbiology*, **45**, 273–283.
Visvesvara, G.S., Belloso, M., Moura, H. *et al.* (1999) Isolation of *Nosema algerae* from the cornea of an immunocompetent patient. *The Journal of Eukaryotic Microbiology*, **46** (Suppl.), 10S.
Weber, R., Bryan, R.T., Owen, R.L. *et al.* (1992) Improved light-microscopical detection of microsporidia spores in stool and duodenal aspirates. *The New England Journal of Medicine*, **326**, 161–166.
Wittner, M. and Weiss, L.M. (1999) *The Microsporidia and Microsporidiosis*, ASM Press, Washington, DC.
Yachnis, A.T., Berg, J., Martinez-Salazar, A. *et al.* (1996) Disseminated microsporidiosis especially infecting the brain, heart, and kidneys. Report of a newly recognized pansporoblastic species in two symptomatic AIDS patients. *American Journal of Clinical Pathology*, **106**, 535–543.

Parasitic Protozoa

Aikawa, M. and Sterling, C.R. (1974) *Intracellular Parasitic Protozoa*, Academic Press, London.
Ashford, R.W. and Crewe, W. (1998) *The Parasites of Homo sapiens. An Annotated Checklist of the Protozoa, Helminths and Arthropods for Which We Are Home*, Liverpool School of Tropical Medicine, Liverpool.
Bowers, B. and Korn, E.D. (1968) The fine structure of *Acanthamoeba castellanii*. I. The trophozoite. *The Journal of Cell Biology*, **39**, 95–111.
Bowers, B. and Korn, E.D. (1969) The fine structure of *Acanthamoeba castellanii* (Neff strain). II. Encystment. *The Journal of Cell Biology*, **41**, 786–805.
Chalmers, R.M. and Davies, A.P. (2010) Minireview: clinical cryptosporidiosis. *Experimental Parasitology*, **124**, 138–146.
Chalmers, R.M., Sturdee, A.P., Casemore, D.P. *et al.* (1994) *Cryptosporldium muris* in wild house mice (*Mus musculus*): first report in the UK. *European Journal of Protistology*, **30**, 151–155.
Clavel, A., Arnal, A.C., Sanchez, E.C. *et al.* (1996) Respiratory cryptosporidiosis: case series and review of literature. *Infection*, **24**, 341–346.

Current, W.L. and Reese, N.C. (1986) A comparison of endogenous development of three isolates of *Cryptosporidium* in suckling mice. *The Journal of Protozoology*, 33, 98–108.

Dubey, J.P. (1993) *Toxoplasma*, *Neospora*, *Sarcocystis*, and other tissue cyst-forming coccidia of humans and animals, in *Parasitic Protozoa*, 2nd edn, vol. 6 (ed. J.P. Kreier), Academic Press, London, pp. 1–158.

Dunand, V.A., Hammer, S.M., Rossi, R. *et al.* (1997) Parasitic sinusitis and otitis in patients infected with human immunodeficiency virus: report of five cases and review. *Clinical Infectious Diseases*, **25**, 267–272.

Elson-Dew, R. (1953) *Isospora natalensis* (sp. nov.) in man. *The American Journal of Tropical Medicine and Hygiene*, **56**, 149–150.

Levine, N.D. (1985) *Veterinary Protozoology*, Iowa State University Press, Iowa City, IA.

Martinez, A.J., Fultz, D.G. and Armstrong, C.E. (1976) The value of electron microscopy in the differential diagnosis of amebic meningoencephalitis, in *34th Annual Conference of Electron Microscopy Society of America*, Vol. 26 (ed. GW Bailey), Miami Beach, FL, pp. 226–227.

Martinez, A.J., Nelson, E.C., Jones, M.M. *et al.* (1971) Experimental *Naegleria* meningoencephalitis in mice. An electron microscope study. *Laboratory Investigation*, **25**, 465–475.

Martinez-Palomo, A., Gonzalez-Robles, A. and Chavez de Ramirez, B. (1976) Ultrastructural study of various *Entamoeba* strains, in *Proceedings of the International Conference on Amebiasis* (eds B. Sepulveda and L.S. Diamond), IMSS, Mexico, pp. 226–237.

Michiels, J.F., Hofman, P., Bernard, E. *et al.* (1994) Intestinal and extraintestinal *Isospora belli* infection in an AIDS patient. *Pathology, Research and Practice*, **190**, 1089–1093.

Restrepo, C., Macher, A.M. and Radany, E.H. (1987) Disseminated extraintestinal isosporiasis in a patient with acquired immune deficiency syndrome. *American Journal of Clinical Pathology*, **87**, 536–542.

Scholtyseck, E. (1979) *Fine Structure of Parasitic Protozoa: An Atlas of Micrographs, Drawings and Diagrams*, Springer-Verlag, Berlin.

Warton, A. (1992) Diagnostic ultrastructure of human parasites, in *Diagnostic Ultrastructure of Non-Neoplastic Diseases* (eds J.M. Papadimitriou, D.W. Henderson and D.V. Spagnolo). Churchill Livingstone, Edinburgh.

7

The Contemporary Use of Electron Microscopy in the Diagnosis of Ciliary Disorders and Sperm Centriolar Abnormalities

P. Yiallouros,[1] M. Nearchou,[2] A. Hadjisavvas[2] and K. Kyriacou[2]

[1] *Cyprus International Institute, Cyprus University of Technology, Limassol, Cyprus*

[2] *Department of Electron Microscopy/Molecular Pathology, The Cyprus Institute of Neurology and Genetics, Nicosia, Cyprus*

7.1 INTRODUCTION

Cilia and flagella are hair-like structures which extend from the cell membrane and are involved in a variety of biological functions in eukaryotes. Cilia are found in a wide range of human tissues and cell types including: the epithelium of the upper and lower respiratory tracts, the epithelium of the kidney, bile and pancreatic ducts, the embryonic node, and the ependymal cells in the brain and in the oviduct epithelium. Flagella, which are structurally the same as cilia, form the tail of sperm (Lee, 2011). In line with their wide distribution in different specialised cells, cilia are involved in a variety of important functions and are best

Diagnostic Electron Microscopy: A Practical Guide to Interpretation and Technique, First Edition. Edited by John W. Stirling, Alan Curry and Brian Eyden.

known for the transport of mucus and other fluids as well as for cell motility. Less obvious functions are the emerging roles of cilia which enable organisms and cells to experience and communicate with the extracellular environment (Pan, Wang and Snell, 2005).

The name 'cilium' originates from the Latin word for 'eyelash'; cilia can be subdivided into two subtypes, based on their function and structure, which are: (i) motile cilia, which are structurally identical to flagella and (ii) primary (sensory) cilia (Chodhari, Mitchison and Meeks, 2004). Both subtypes consist of a basal body located below the surface of the cell and a plasmalemma-covered cytoskeletal element, the axoneme that extends away from the cell which comprises more than 250 proteins that assemble into a specific arrangement of microtubules, outer and inner rows of dynein arms and radial spokes. In motile cilia, the axoneme has a '9 + 2' arrangement of microtubules (nine peripheral doublets and two central separate microtubules), whereas sensory cilia have a '9 + 0' arrangement and are immotile. Sensory cilia exist as single projections extending from the surface of many cell types, and defects in their structure or function have been implicated in an expanding list of diseases, including for example polycystic kidney disease, Bardet–Biedl syndrome and Meikel–Graber syndrome (Bisgrove and Yost, 2006). Dysfunction of nodal cilia during embryogenesis causes randomisation of left-right body asymmetry; this is the mechanism through which approximately 50% of patients with primary ciliary dyskinesia (PCD) have situs inversus at random.

Mutations in genes that are essential for the structure, function and assembly of motile and sensory cilia cause a diverse spectrum of ciliary disorders, also referred to as ciliopathies. Consequently, ciliopathies are a heterogeneous group of disorders; they have a broad range of phenotypes and encompass a number of different autosomal recessive and, rarely, X-linked syndromes. Among the ciliopathies, the most prominent congenital autosomal recessive syndrome affecting the movement and function of motile cilia is PCD (Chodhari, Mitchison and Meeks, 2004). This is a rare disorder with an estimated prevalence ranging from 1:10 000 to 1:30 000 in different populations (Kuehni *et al.*, 2010). This condition was first described in 1900 in patients with bronchiectasis and situs inversus; and a triad of symptoms – sinusitis, bronchiectasis and situs inversus – was reported by Kartagener in four patients (Kartagener, 1933). He recognised this as a clinical syndrome which was subsequently named after him, as Kartagener syndrome. In the 1970s, Afzelius and colleagues were the first to describe the association of sinusitis, bronchiectasis and situs inversus with an absence of dynein arms in the tails of the immotile spermatozoa in infertile men (Afzelius *et al.*, 1975). Subsequently, Eliasson *et al.* (1977) coined the term immotile cilia syndrome to describe all congenital ciliary defects causing impaired mucociliary clearance and male infertility. The term primary ciliary dyskinesia was proposed by Sleigh (1981) to describe the congenital ultrastructural ciliary alterations that are the main feature of these disorders. The breakthrough discovery of Afzelius was followed by a flurry of publications highlighting the crucial role of electron microscopy (EM) in the diagnosis of these disorders (Sturgess and Turner, 1984; Bush *et al.*, 1998; Roomans *et al.*, 2006; Papon *et al.*, 2010). These studies proposed ultrastructural investigation as the 'gold standard' for accurate diagnosis of PCD.

Currently, the diagnosis of PCD is still challenging and requires good clinical work-up together with a number of laboratory investigations that include ultrastructural examination of cilia, assessment of ciliary beat frequency (CBF) and beat pattern and, where possible, immunofluorescent microscopy, nasal nitric oxide (NO) measurements as well as molecular genetics. Transmission electron microscopy (TEM) remains essential for the accurate diagnosis of PCD. However, as 30% of PCD patients have normal ciliary ultrastructure (Zariwala, Knowles and Omran, 2007), the results must be interpreted with caution and always in line with the other findings. TEM reveals phenotypic structural abnormalities, most commonly a reduced number in outer or inner dynein arms (IDAs), or both. In the last 5 years, great progress has been achieved in our understanding of the function(s) as well as the role of these organelles in normal mammalian development. An important aspect has been the cloning and identification of genes that encode structural and/or functional components of primary cilia (Gherman, Davis and Katsanis, 2006). Mutational analysis shows that nearly 40% of PCD patients carry mutations in the dynein genes DNAI1 and DNAH5, although the underlying genetics of these disorders are complex and the list of implicated genes is rapidly expanding (Lee, 2011). In view of recent developments in molecular genetics and the emerging applications of other tools for PCD diagnosis, such as nasal NO measurements and immunofluorescent microscopy, the aim of this chapter is to highlight the still important role of EM in the diagnosis of ciliary disorders such as PCD.

7.2 ULTRASTRUCTURE OF MOTILE CILIA

The landmark observation by Afzelius *et al.* (1975) that immotile spermatozoa are deficient in dynein arms established the ultrastructural evaluation of respiratory cilia as an essential diagnostic test for PCD. Although currently the accurate diagnosis of ciliopathies including PCD requires additional diagnostic modalities, such as assessment of ciliary beat pattern and frequency (Leigh *et al.*, 2009), TEM is still considered the cornerstone for diagnosis. Cilia and flagella are evolutionally ancient organelles whose structure and function have been faithfully conserved. Each normal cilium contains an array of longitudinal microtubules, consisting of nine doublets arranged in an outer circle around a central pair, the canonical 9+2 pattern. Dynein arms are large protein complexes which consist of heavy, intermediate and light chains. These are attached to the nine microtubule doublets and assemble as distinct IDAs and outer

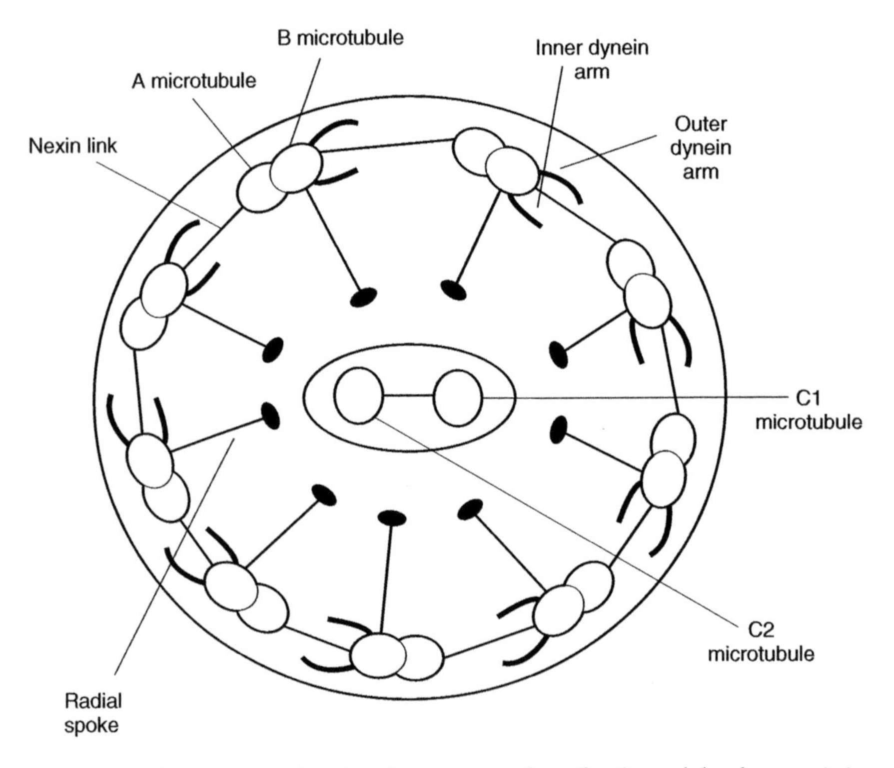

Figure 7.1 Cross-section showing the axoneme of motile cilia and the characteristic 9+2 pattern of microtubules A, B and C1, C2.

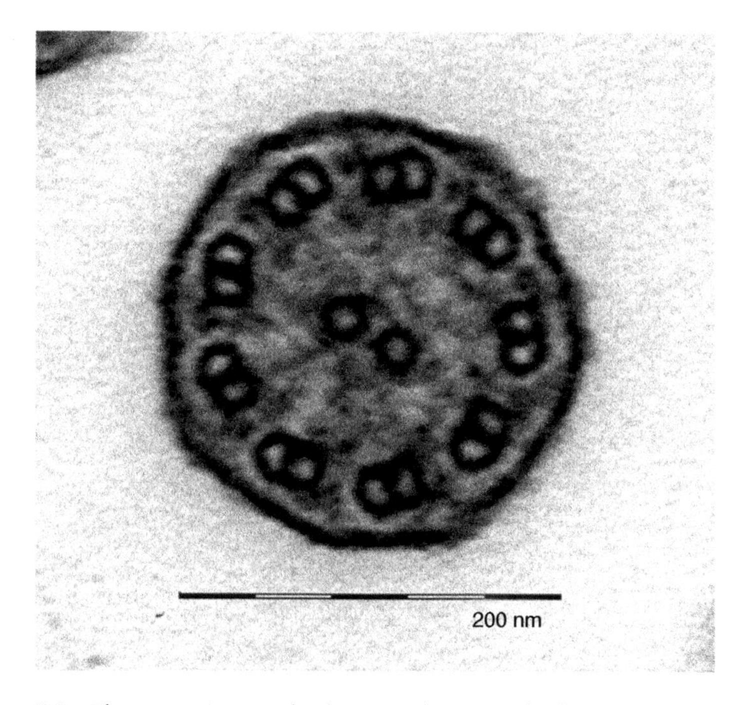

Figure 7.2 Electron micrograph showing the normal ultrastructure of motile cilia featuring the 9+2 arrangement of microtubules and the presence of IDAs and ODAs.

dynein arms (ODAs), and they are best visualised by examining cilia in cross-section. The 9+2 arrangement of microtubules is maintained by a network of microtubular links and accessory proteins, appearing as radial spokes and nexin links (see Figures 7.1 and 7.2). Normal cilia beat at a frequency between 8 and 20 Hz and in a synchronous manner as reflected by their orientation in electron micrographs (Figure 7.3). Several defects in the ultrastructure of cilia have been described, especially in patients with PCD. However, the most frequent abnormalities involve the absence or shortening of ODAs or IDAs (Table 7.1 and Figures 7.4 and 7.5), lack of radial spoke or loss of the central pair of microtubules (Figure 7.6), with transposition of a peripheral doublet to the centre (Bush *et al.*, 1998). These pathognomonic features can only be visualised by EM, which remains the only tool that can distinguish morphological ciliary phenotypes and direct investigation into corresponding genotypes.

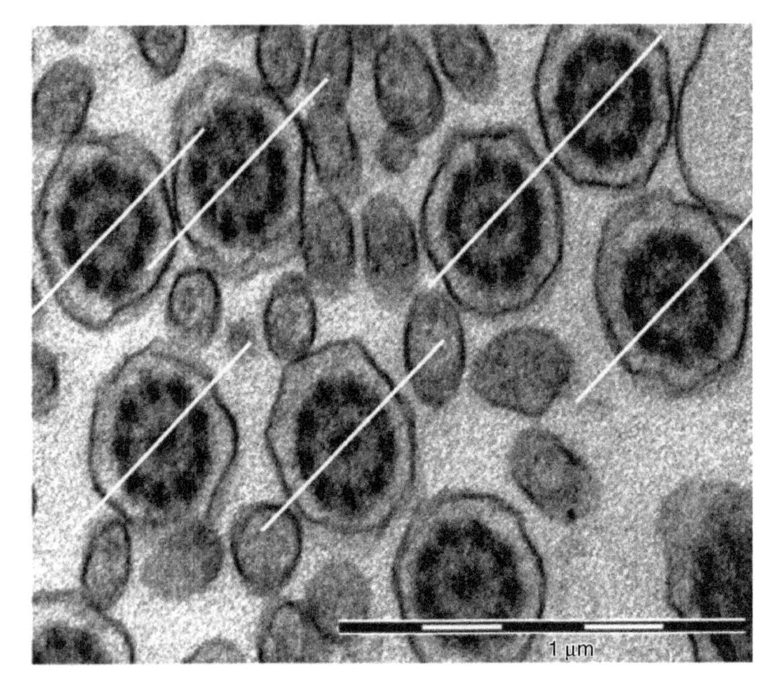

Figure 7.3 Electron micrograph showing well-orientated normal cilia, as revealed by the white parallel lines drawn through the central pair of microtubules.

Table 7.1 Ultrastructural changes frequently associated with PCD

Outer dynein arm defect	Shortened or absent
Inner dynein arm defect	Shortened or absent
Inner and outer dynein arm defect	Shortened or absent
Radial spoke defect	Missing radial spokes
Transposition defect	Central pair missing
	Defective 9+2 arrangement

7.3 GENETICS OF PCD

PCD is an autosomal recessive disorder, but rare cases of dominant or X-linked transmission have also been documented. PCD is genetically heterogeneous and is caused by mutations in a number of different genes (Blouin *et al.*, 2000; Geremek and Witt, 2004). It is estimated that several hundred proteins are needed to construct the axoneme within a single cilium, each encoded by a separate gene, and many other genes participate in the assembly and functional regulation of

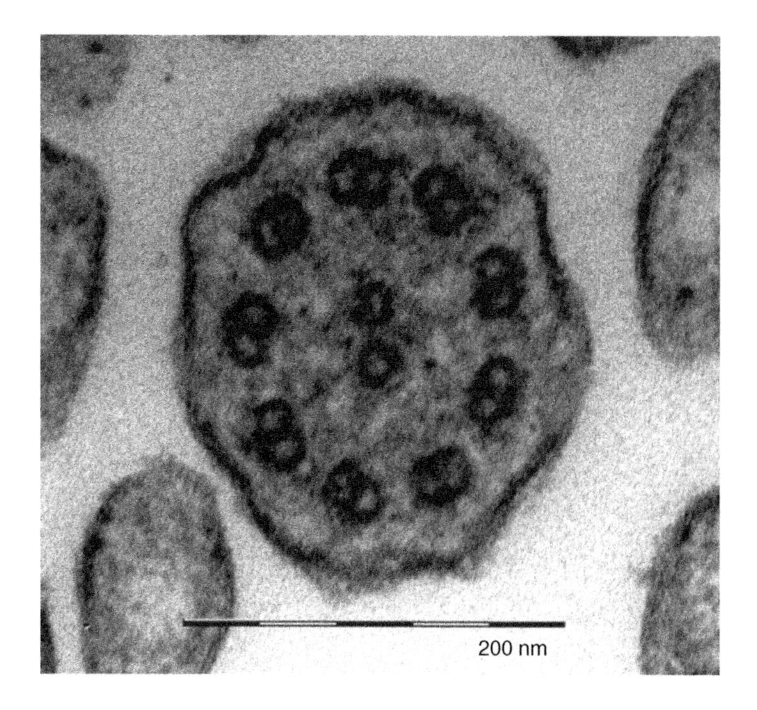

Figure 7.4 Electron micrograph showing an absence of both IDAs and ODAs.

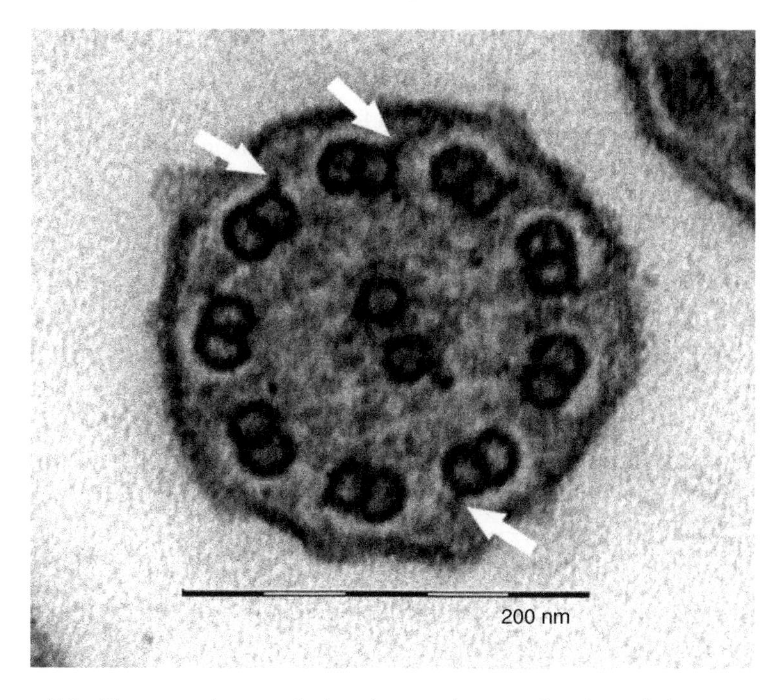

Figure 7.5 Electron micrograph showing an absence of IDAs and the presence of shortened ODAs (arrows).

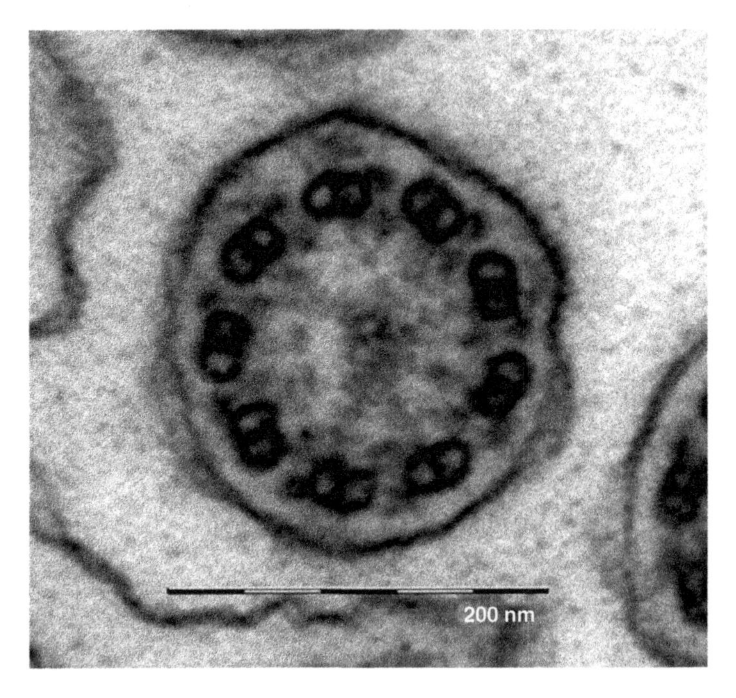

Figure 7.6 Electron micrograph showing an absence of the central pair of microtubules. IDAs and ODAs are intact.

cilia. Although the majority of these genes remain to be discovered, a number of candidate genes have been identified (Leigh *et al.*, 2009). The first gene in which a mutation was found to be a cause of PCD was DNAI1, which encodes an ODA intermediate chain (Pennarun *et al.*, 1999). Currently, mutations in several other genes have been shown to cause PCD, including DNAH5 and DNAH11, which encode ODA heavy chains (Bartoloni *et al.*, 2002; Olbrich *et al.*, 2002). The number of genes implicated in PCD is increasing (Lee, 2011), and so far more than 20 different genes, which are associated with PCD in humans and mouse models, have been identified. Table 7.2 presents eight genes in which pathogenic mutations have been detected in PCD in humans.

7.4 CURRENT DIAGNOSTIC MODALITIES

Accurate diagnosis of ciliopathies, including PCD, requires specialised investigations and expertise that are available in only a few dedicated research centres. The diagnosis requires detailed supporting clinical data,

Table 7.2 Major genes implicated in PCD in humans

Gene/protein	Chromosome locus	Exons	Location	Ultrastructural defect
DNAI1	9p13-21	20	Dynein arms	Lack of ODA
DNAI2	17q25	14	Dynein arms	Lack of ODA
DNAH5	5p15-p14	80	Dynein arms	Lack of ODA
DNAH11	7p15	82	Dynein arms	None
RSPH4A	6q22.1	6	Radial spoke	Absent CP
RSPH9	6p21.1	5	Radial spoke	Absent CP
KTU	14q21.3	3	Cytoplasmic (dynein arms assembly)	Lack of IDA
TXNDC3	7p14.1	18	Ciliary function unknown	Lack of ODA

Note: ODA: outer dynein arm; IDA: inner dynein arm; and CP: central pair.

including the presence of characteristic clinical phenotypic features, information on CBF and pattern as well as morphological details. In addition, new imaging techniques such as high-resolution digital high-speed video allow the precise beat pattern of cilia to be examined. Cilia beat in a forward and backward planar motion without a sideways recovery sweep, and observation of the ciliary beat pattern is important, as this is associated with specific ultrastructural defects (Chilvers, Rutman and O'Callaghan, 2003). In addition to the beat pattern, electron micrographs can be used to detect any other anomalies in the orientation of cilia, as shown in Figure 7.3. In dedicated centres, additional modalities, such as nasal NO measurement, molecular genetic testing, cell cultures of ciliated epithelium as well as immunofluorescent microscopy of cilia, may be available, and these contribute to the diagnosis of difficult cases (Bush *et al.*, 2007).

7.5 CLINICAL FEATURES

Ciliopathies including PCD are phenotypically and genetically heterogeneous disorders which are associated with a variety of ultrastructural and functional abnormalities of cilia. Clinical manifestations of PCD include recurrent respiratory infections, chronic rhinitis and sinusitis, bronchiectasis and subfertility. Patients are often misdiagnosed and poorly managed, and suffer from significant long-term morbidity and increased mortality. A recent survey found that Cyprus has a prevalence of 1 : 10 000, which is the highest reported in Europe (Kuehni *et al.*, 2010). This may partly be due to the excellent national network already organised between clinicians and scientists as a result of a funded national project for diagnosing these disorders. Usually, patients are diagnosed late in life, following many years of chronic respiratory symptoms with established and frequently severe lung disease (Ellerman and Bisgaard, 1997). Patients may undergo a battery of invasive and unnecessary diagnostic procedures in an attempt to delineate their underlying problem (Bush and O'Callaghan, 2002). Delayed diagnosis is primarily due to lack of appropriate diagnostic facilities as well as to inaccessibility to tertiary diagnostic centres.

7.6 PROCUREMENT AND ASSESSMENT OF CILIATED SPECIMENS

In our centre, ciliated samples are procured by brushing the inferior nasal turbinate, without local anaesthetic. Nasal brushings are collected in culture medium 199, which contains 100 units per ml of antibiotic solution (streptomycin and penicillin; Gibco). Each sample is evaluated by measuring CBF and by observing beat pattern as described in this chapter (Chilvers and O'Callaghan, 2000).

Briefly, wedges of ciliated epithelium are observed at 37 °C using a ×100 oil immersion objective lens in a ZEISS brightfield light microscope. Beating ciliated edges, free of mucus, are recorded using a high-speed video camera (FASTCAM, Super 10k; Photron) at 500 frames per second onto a Sony video tape. Around 15–25 frames are recorded, each of 5.4 seconds duration, from at least 10 different ciliated wedges per biopsy. Recorded video sequences are played back at reduced frame rates, in order to count the number of frames required to complete 10 cycles. The average of 50 different cycles is converted to CBF by using the formula: CBF = (500/number of frames for 10 beats) × 10. The mean CBF, 95% CIs and range are calculated and then compared with the normal CBF value, which ranges from 8 to 20 Hz (Leigh *et al.*, 2009). In addition to the calculation of CBF, the ciliary beat pattern is observed and recorded in each case. Following the CBF measurements, the remaining ciliated sample is centrifuged, and the supernatant culture medium is removed and replaced by a 2.5% glutaraldehyde solution in 0.1 M phosphate buffer for primary fixation. Subsequently, the samples are stabilised with agar, post-fixed in osmium tetroxide, dehydrated and embedded in an Epon–Araldite resin mixture.

Polymerised blocks are sectioned to produce survey semithin sections of ciliated epithelium, and these areas are selected for cutting ultrathin sections of silver to gold interference colour. Sections are then stained with saturated uranyl acetate and lead citrate and in our centre have been examined in a JEOL JEM 1010 transmission electron microscope. In each case, at least 50 cilia are examined, and ultrastructural defects, as described in Table 7.1, are recorded; for statistical purposes, the mean and 95% CI are calculated for each sample. Ciliated specimens showing ultrastructural abnormalities in more than 10% of cilia examined are deemed as abnormal. It may be necessary also to resort to tilting during the examination in the EM in order to obtain and examine good-quality cross-sectional profiles through cilia.

7.7 CENTRIOLAR SPERM ABNORMALITIES

Sperm motility is considered the 'best' predictor of fertility, despite the fact that in normal sperm it ranges from 53 to 62% and shows a wide variation in repeat samples up to 20% (Natali and Turek, 2011). As with the formation of cilia, sperm tails arise from centrioles which function as a site of nucleation for polymerisation and formation of the nine doublet microtubules of the axoneme (Fawcett, 1981). As already discussed, PCD patients are subfertile, and in males this is due to immotile sperm; in females, it may be caused by the inability of cilia to move the female egg towards the uterus. It is becoming increasingly recognised that many cases of male infertility in humans are caused by genetic abnormalities associated with mutations in genes encoding proteins involved in axoneme structure, flagellar transport, motor proteins and transcription factors (Turner, 2003). Indeed, because the complex interactions between the structural and functional components of the axoneme are just beginning to be unravelled (Inaba, 2007), it is recommended that in order to detect abnormalities in sperm motility and flagellar transport, which are essential for fertility, semen analysis should include the TEM examination of sperm. Many such studies reveal a faulty development of microtubules from the sperm-derived centrosome defects, which not only affect sperm motility but also prevent syngamy between the oocyte and sperm DNA. Furthermore, the mother's centrosome is dissolved in the mature human oocyte, and defects in the male sperm centriole (Figures 7.7 and 7.8) prevent the formation of a mitotic spindle and cell division (Rawe *et al.*, 2002).

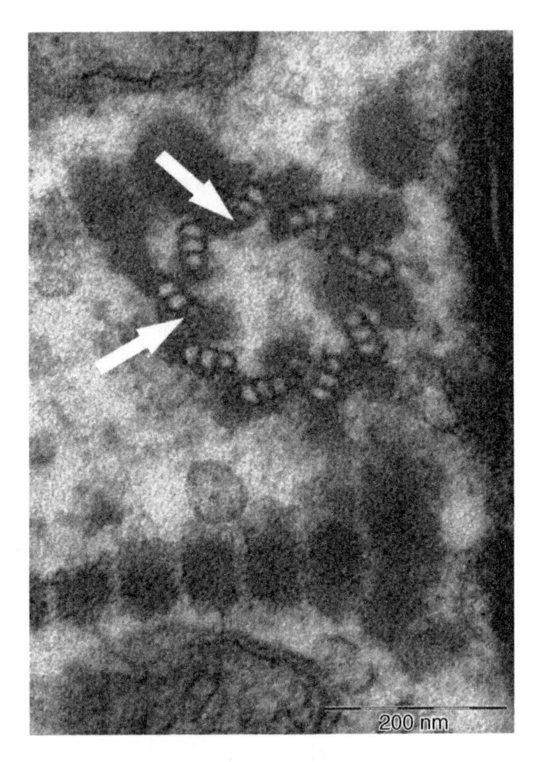

Figure 7.7 Electron micrograph of sperm centriole showing defects (arrows) in the microtubular triplet arrangement causing zygote division arrest following intracytoplasmic sperm injection. (Figure kindly donated by Dr J.A. Schroeder, EM Department of Pathology, University Hospital Regensburg, Germany).

7.8 DISCUSSION

Early and accurate diagnosis of PCD is essential as the vast majority of PCD patients have symptoms from birth (Greenstone *et al.*, 1988) or early infancy. However, the diagnosis is often made later, frequently resulting in severe morbidity and a reduced lifespan of these patients. Despite the great advances achieved in recent years, the diagnosis of PCD is still difficult and challenging because it requires specialised expertise and laboratory tests that are available in only a few dedicated research centres.

Current ciliary ultrastructural diagnostic criteria include the absence or structural modification of IDAs, ODAs or both. Cilia with defective radial spokes and transposition of a peripheral pair of microtubules to the centre have been associated with PCD, but such cases are rare. Some

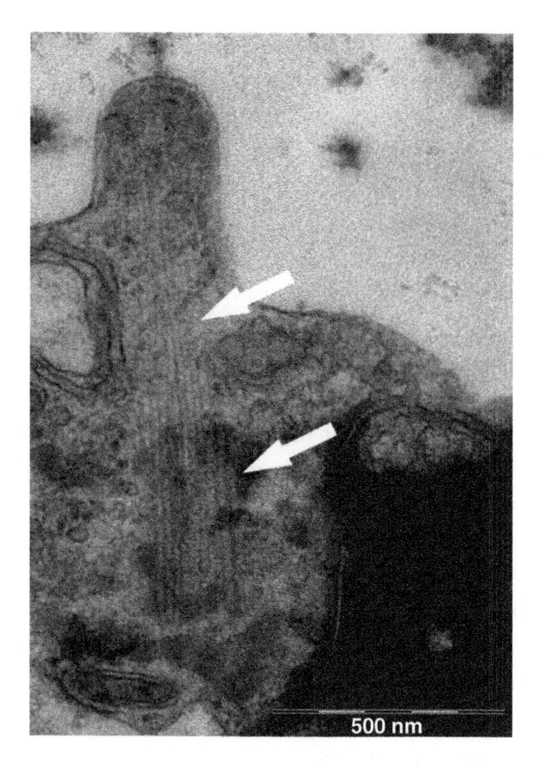

Figure 7.8 Electron micrograph of sperm centriole showing abnormally long microtubules of the centrosome extending laterally to the sperm neck (arrows). (Figure kindly donated by Dr J.A. Schroeder, EM Department of Pathology, University Hospital Regensburg, Germany).

reports also noted other ultrastructural defects, including the presence of compound cilia, disorganised axonemes, alterations in the number of the peripheral microtubules and ciliary disorientation (Pizzi *et al.*, 2003); most probably, these are caused by infection or inflammation. These secondary abnormalities may be distinguished from the gold standard pathognomonic features of PCD, as typically they are not present in all cilia examined. This makes it mandatory to use a semi-quantitative approach to evaluate ciliary defects, assessing at least 50 cilia cross-sections per case and recording any ultrastructural ciliary abnormalities, in an effort to increase the accuracy of PCD diagnosis. It should be mentioned that even in healthy individuals, up to 5% of cilia examined are defective; thus, any abnormalities should be present in at least 10% of cilia examined in order to be of diagnostic significance. Although examination of cilia in the EM is still vitally important for diagnosis, recent data demonstrate that some cases of PCD may harbour mutations in ciliary genes and have compromised motility, but nevertheless display normal cilia ultrastructure (Schwabe *et al.*, 2008).

In addition, new modalities have emerged that aim to contribute to a more accurate diagnosis of PCD and ciliary disorders, which include measurement of nasal NO, immunofluorescent microscopy and molecular genetic analysis. Nasal NO is altered in several airway disorders, and in PCD it is always extremely low and has a proven good sensitivity and specificity in the diagnosis of PCD (Stehling *et al.*, 2006), so it is now used routinely in specialised centres. Immunofluorescent microscopy, using specific antibodies to axonemal structural proteins, increases the diagnostic accuracy of PCD, but this is not widely used. Molecular genetics is important for accurate diagnosis, but currently its application is limited as these methods are expensive, cumbersome and time-consuming, since many of the genes implicated in PCD are very large, spanning more than 80 exons. In conclusion, we are currently witnessing an exponential expansion of our knowledge related to ciliary structure and function. However, despite the application of an increasing range of auxiliary methods for diagnosing ciliary disorders, EM is still and will remain an essential tool for their accurate diagnosis.

ACKNOWLEDGEMENTS

We would like to thank the Cyprus Research Promotion Foundation for funding the project ENTAX/0506/24, Mrs S. Aristodemou and T. Michael for helping with electron microscopy and the Cyprus Institute of Neurology and Genetics for supporting this work.

REFERENCES

Afzelius, B.A., Eliasson, R., Johnsen, O. and Lindholmer, C. (1975) Lack of dynein arms in immotile human spermatozoa. *The Journal of Cell Biology*, **66** (2), 225–232.

Bartoloni, L., Blouin, J.L., Pan, Y. *et al.* (2002) Mutations in the DNAH11 (axonemal heavy chain dynein type 11) gene cause one form of situs inversus totalis and most likely primary ciliary dyskinesia. *Proceedings of the National Academy of Sciences of the United States of America*, **99** (16), 10282–10286.

Bisgrove, B.W. and Yost, H.J. (2006) The roles of cilia in developmental disorders and disease. *Development*, **133** (21), 4131–4143.

Blouin, J.L., Meeks, M., Radhakrishna, U. *et al.* (2000) Primary ciliary dyskinesia: a genome-wide linkage analysis reveals extensive locus heterogeneity. *European Journal of Human Genetics: EJHG*, **8** (2), 109–118.

Bush, A., Chodhari, R., Collins, N. *et al.* (2007) Primary ciliary dyskinesia: current state of the art. *Archives of Disease in Childhood*, **92** (12), 1136–1140.

Bush, A., Cole, P., Hariri, M. *et al.* (1998) Primary ciliary dyskinesia: diagnosis and standards of care. *The European Respiratory Journal: Official Journal of the European Society for Clinical Respiratory Physiology*, **12** (4), 982–988.

Bush, A. and O'Callaghan, C. (2002) Primary ciliary dyskinesia. *Archives of Disease in Childhood*, **87** (5), 363–365; discussion 363–365.

Chilvers, M.A. and O'Callaghan, C. (2000) Analysis of ciliary beat pattern and beat frequency using digital high speed imaging: comparison with the photomultiplier and photodiode methods. *Thorax*, **55** (4), 314–317.

Chilvers, M.A., Rutman, A. and O'Callaghan, C. (2003) Ciliary beat pattern is associated with specific ultrastructural defects in primary ciliary dyskinesia. *The Journal of Allergy and Clinical Immunology*, **112** (3), 518–524.

Chodhari, R., Mitchison, H.M. and Meeks, M. (2004) Cilia, primary ciliary dyskinesia and molecular genetics. *Paediatric Respiratory Reviews*, **5** (1), 69–76.

Eliasson, R., Mossberg, B., Camner, P. and Afzelius, B.A. (1977) The immotile-cilia syndrome. A congenital ciliary abnormality as an etiologic factor in chronic airway infections and male sterility. *The New England Journal of Medicine*, **297** (1), 1–6.

Ellerman, A. and Bisgaard, H. (1997) Longitudinal study of lung function in a cohort of primary ciliary dyskinesia. *The European Respiratory Journal: Official Journal of the European Society for Clinical Respiratory Physiology*, **10** (10), 2376–2379.

Fawcett, D.W. (ed.) (1981) *The Cell*, W.B. Saunders, Philadelphia, PA.

Geremek, M. and Witt, M. (2004) Primary ciliary dyskinesia: genes, candidate genes and chromosomal regions. *Journal of Applied Genetics*, **45** (3), 347–361.

Gherman, A., Davis, E.E. and Katsanis, N. (2006) The ciliary proteome database: an integrated community resource for the genetic and functional dissection of cilia. *Nature Genetics*, **38** (9), 961–962.

Greenstone, M., Rutman, A., Dewar, A. *et al.* (1988) Primary ciliary dyskinesia: cytological and clinical features. *The Quarterly Journal of Medicine*, **67** (253), 405–423.

Inaba, K. (2007) Molecular basis of sperm flagellar axonemes: structural and evolutionary aspects. *Annals of the New York Academy of Sciences*, **1101**, 506–526.

Kartagener, M. (1933) Zur Pathogenese der Bronchiektasien: Bronchiektasien bei Situs viscerum inversus. *Beitr Klin Tuberk*, **83**, 489–501.

Kuehni, C.E., Frischer, T., Strippoli, M.P. *et al.* (2010) Factors influencing age at diagnosis of primary ciliary dyskinesia in European children. *The European Respiratory Journal: Official Journal of the European Society for Clinical Respiratory Physiology*, **36** (6), 1248–1258.

Lee, L. (2011) Mechanisms of mammalian ciliary motility: insights from primary ciliary dyskinesia genetics. *Gene*, **473** (2), 57–66.

Leigh, M.W., Pittman, J.E., Carson, J.L. *et al.* (2009) Clinical and genetic aspects of primary ciliary dyskinesia/Kartagener syndrome. *Genetics in Medicine*, **11** (7), 473–487.

Natali, A. and Turek, P.J. (2011) An assessment of new sperm tests for male infertility. *Urology*, **77** (5), 1027–1034.

Olbrich, H., Haffner, K., Kispert, A. *et al.* (2002) Mutations in DNAH5 cause primary ciliary dyskinesia and randomization of left-right asymmetry. *Nature Genetics*, **30** (2), 143–144.

Pan, J., Wang, Q. and Snell, W.J. (2005) Cilium-generated signaling and cilia-related disorders. *Laboratory Investigation*, **85** (4), 452–463.

Papon, J.F., Coste, A., Roudot-Thoraval, F. *et al.* (2010) A 20-year experience of electron microscopy in the diagnosis of primary ciliary dyskinesia. *The European Respiratory Journal: Official Journal of the European Society for Clinical Respiratory Physiology*, **35** (5), 1057–1063.

Pennarun, G., Escudier, E., Chapelin, C. *et al.* (1999) Loss-of-function mutations in a human gene related to *Chlamydomonas reinhardtii* dynein IC78 result in primary ciliary dyskinesia. *American Journal of Human Genetics*, **65** (6), 1508–1519.

Pizzi, S., Cazzato, S., Bernardi, F. *et al.* (2003) Clinico-pathological evaluation of ciliary dyskinesia: diagnostic role of electron microscopy. *Ultrastructural Pathology*, **27** (4), 243–252.

Rawe, V.Y., Terada, Y., Nakamura, S. *et al.* (2002) A pathology of the sperm centriole responsible for defective sperm aster formation, syngamy and cleavage. *Human Reproduction*, **17** (9), 2344–2349.

Roomans, G.M., Ivanovs, A., Shebani, E.B. and Johannesson, M. (2006) Transmission electron microscopy in the diagnosis of primary ciliary dyskinesia. *Uppsala Journal of Medical Sciences*, **111** (1), 155–168.

Schwabe, G.C., Hoffmann, K., Loges, N.T. *et al.* (2008) Primary ciliary dyskinesia associated with normal axoneme ultrastructure is caused by DNAH11 mutations. *Human Mutation*, **29** (2), 289–298.

Sleigh, M.A. (1981) Primary ciliary dyskinesia. *Lancet*, **2** (8244), 476.

Stehling, F., Roll, C., Ratjen, F. and Grasemann, H. (2006) Nasal nitric oxide to diagnose primary ciliary dyskinesia in newborns. *Archives of Disease in Childhood: Fetal Neonatal Edition*, **91** (3), F233.

Sturgess, J.M. and Turner, J.A. (1984) Ultrastructural pathology of cilia in the immotile cilia syndrome. *Perspectives in Pediatric Pathology*, **8** (2), 133–161.

Turner, R.M. (2003) Tales from the tail: what do we really know about sperm motility? *Journal of Andrology*, **24** (6), 790–803.

Zariwala, M.A., Knowles, M.R. and Omran, H. (2007) Genetic defects in ciliary structure and function. *Annual Review of Physiology*, **69**, 423–450.

8

Electron Microscopy as a Useful Tool in the Diagnosis of Lysosomal Storage Diseases

Joseph Alroy,[1,2] Rolf Pfannl[2] and Angelo A. Ucci[2]

[1]*Department of Pathology, Tufts University Cumming's School of Veterinary Medicine, Grafton, Massachusetts, United States and Department of Pathology and Laboratory Medicine, Tufts Medical Center and Tufts University School of Medicine, Boston, Massachusetts, United States*
[2]*Department of Pathology and Laboratory Medicine, Tufts Medical Center and Tufts University School of Medicine, Boston, Massachusetts, United States*

8.1 INTRODUCTION

Lysosomes are the cell's digestive and recycling centre. More than 70 hydrolytic enzymes participate in the turnover of cellular and extracellular macromolecules. These enzymes degrade proteins, glycoproteins, complex carbohydrates, glycosaminoglycans (GAGs), nucleic acids and glycolipids and provide amino acids, fatty acids, nucleic acids and sugars for the synthesis of cellular and extracellular molecules. In addition, lysosomes are involved in other cellular processes such as post-translational maturation of proteins and the presentation of exogenous antigens to the immune system, and they play a role in the invasion of neoplastic cells. Currently, 12 different mechanisms are known to cause lysosomal storage disorders (LSDs) (Table 8.1) (Alroy, Ucci, and Pfannl, 2006),

Diagnostic Electron Microscopy: A Practical Guide to Interpretation and Technique, First Edition. Edited by John W. Stirling, Alan Curry and Brian Eyden.

Table 8.1 Classification of lysosomal storage diseases according to the mechanisms that cause an intralysosomal deficiency of hydrolase activity or abnormal storage

Mechanisms	Examples
Disorders in which no immunologically detectable enzyme is synthesized; includes conditions with grossly abnormal structural genes	Infantile forms of glycoprotein, glycolipid, glycogen or lipid storage diseases and mucopolysaccharides
Disorders in which a catalytically inactive polypeptide is co-synthesized; the mutation may also affect the stability or transport of the polypeptide.	Juvenile or adult forms of glycoprotein, glycolipid, glycogen, or lipid storage diseases and mucopolysaccharides
Disorders in which a catalytically active enzyme is synthesized but the enzyme is not segregated into lysosomes	I-cell disease, pseudo-Hurler polydystrophy
Disorders in which a catalytically active enzyme is synthesized; however, the enzyme is unstable in pre-lysosomal or lysosomal compartments.	Galactosialidosis
Disorders in which activator proteins (saposins) of lipid degrading hydrolases are missing	AB variant of G_{M2}-gangliosidosis, variants of Gaucher and Farber disease and juvenile variant of metachromatic leukodystrophy
Disorders in which the structural gene of the hydrolase is normal but there is a mutation of the gene(s) that code for post-translational modification of the hydrolase	Multiple sulfatase deficiencies
Disorders due to abnormal transport to the lysosomes	Mucolipidosis IV
Disorder of lack of fusion between endosome and lysosome due to mutation of the gene that codes for lysosomal membrane protein-2	Danon disease
Disorders due to a decrease in transport of degradation end products (i.e. free sialic acid and free cysteine) out of the lysosomes	Salla disease, infantile sialidosis and cystinosis
Disorders due to oversupply of substrate	Some cases of chronic myelocytic leukemia, sickle cell anemia and thalassemia that are associated with Gaucher-like cells (i.e. sea-blue histiocytes)
Disorders in which deficiencies in activity of lysosomal enzymes result from intoxication with neutral or synthetic inhibitors of lysosomal enzymes	Cationic amphophilic drugs (i.e. amiodarone and chloroquine)
Disorders due to loss of chloride channel 7 (CIC-7) of late endosomes and lysosomes	LSD and osteopetrosis

most of which are caused by inherited defects, while some are acquired, such as those caused by cationic amphophilic drugs such as amiodarone (Hruban, 1984) and chloroquine (Albay *et al.*, 2005). Rarely, the injection of vaccines containing aluminium hydroxide may mimic lysosomal storage (Gherardi *et al.*, 2001).

To date, 58 genetically inherited LSDs have been described, most of them with multiple identified pathogenic mutations. Table 8.2 lists specific LSDs resulting from deficient and/or missing enzymes, activator proteins or lysosomal membrane proteins, the number of known mutations in each disorder and the principal stored product and the primary tissues involved.

The age of onset of clinical manifestations of these diseases depends on the nature of the mutated gene, the type of mutation and its interaction with other, unknown factors. Often, the clinical changes are not unique for specific LSDs. The disease onset may be seen in infants, juveniles or adults, and the phenotype may range from neurological disorders, hepatosplenomegaly, cloudy corneas, skeletal abnormalities (Gelb, Bromme, and Desnick, 2001) and cardiomyopathy (Chabas, Duque, and Gort, 2007) to a combination of any of these. In the majority of lysosomal storage diseases, the specific diagnosis can be established by demonstrating a deficient activity of the lysosomal enzyme in leukocytes or cultured fibroblasts, or by identifying the mutated gene. Since deficient enzymatic activity occurs in all cell types of affected individuals, the abnormal substrate accumulation in affected cells is determined by the amount of available substrate in these cells. The composition of the cell's plasma membrane, the rate of membrane recycling, cellular catabolism, the lifespan of the cell, the phagocytic ability of the affected cells as well as the nature of the phagocytosed substrate influence the substrate load and nature.

Since the early 1970s, electron microscopic examinations of skin (Belcher, 1972), liver, lymph node, leukocyte pellets (Isenberg and Sharp, 1976), conjunctiva (Merin *et al.*, 1975), rectum (Ushiyama *et al.*, 1985), kidneys (Castagnaro *et al.*, 1987), mast cells (Hammel *et al.*, 1993), pancreatic exocrine and endocrine cells (Hammel and Alroy, 1995) and chorionic villi (Fowler *et al.*, 2007) have been reported. It is quite challenging to suggest a specific diagnosis by electron microscopy (EM) only, given the large number of lysosomal storage diseases and the similarity of their stored products as well as overlapping symptoms with other developmental and neurological disorders. However, EM examination can exclude nonlysosomal disorders, and in most cases, the ultrastructure of the accumulated material indicates the nature of the compound (i.e. glycolipids, glycoproteins or mucopolysaccharides) and guides further specific testing. It is noteworthy that EM examination of

Table 8.2 Lysosomal diseases

Disease	Deficient enzyme membrane proteins	Gene locus	Number of known mutations	Primary storage products	Major organs involved
Mucopolysaccharidosis (MPS) MPS I Hurler–Scheie syndrome	α-Iduronidase	4p16.3	>100 (Bertola *et al.*, 2011)	Dermatan sulfate Heparan sulfate	Central nervous system (CNS), connective tissue, heart, skeleton and cornea
MPS II Hunter syndrome	Iduronate sulfatase	Xq 27–28	>370 (Lualdi *et al.*, 2010)	Dermatan sulfate Heparan sulfate	CNS, connective tissue, heart and skeleton
MPS III Sanfillipo syndrome MPS III A	Heparan-*N*-sulfatase	17q25.3	>75 (Meyer *et al.*, 2008)	Heparan sulfate	CNS
MPS III B	α-*N*-acetylglucosa-minidase	17q21.1	114 (Beesley *et al.*, 2005)	Heparan sulfate	CNS
MPSIII C	Acetyl coenzyme A	8p11-q13	50 (Feldhammer *et al.*, 2009)	Heparan sulfate	CNS
MPSIII D	*N*-acetylglucos-amine	12q14	29 (Valstar *et al.*, 2010)	Heparan sulfate	CNS
MPS IVA Morquio syndrome	Galactose 6-sulfatase	16q24.3	150 (Montano *et al.*, 2007)	Chondroitin-4-sulfate	Skeleton and cornea

MPS IVB	β-galactosidase	3p21-3pter	16 (Gucev *et al.*, 2008)	Keratan sulfate	Skeleton and cornea
MPS VI Maroteaux–Lamy syndrome	Arylsulfatase B	5q11-13	83 (Karageorgos *et al.*, 2007)	Dermatan sulfate	Skeleton and cornea
MPS VII	β-glucuronidase	7q21	58 (Tomatsu *et al.*, 2009)	Dermatan sulfate Heparan sulfate	CNS, connective tissue and skeleton
MPS IX	Hyaluronidase	3p21.3	2 (Triggs-Raine *et al.*, 1999)	Hyaluronan	Connective tissue and joints
Mucolipidosis (ML) ML I (sialidosis)	α-neuraminidase	6p21.3	5 (Pattison *et al.*, 2004)	Sialyl–oligosaccharides sialylated glycolipids	CNS, liver, spleen, skeleton and connective tissue
ML II (I cell disease)	N-acetyl-glucosaminyl-1 phosphotransferase	12q23.2	40 (Tappino *et al.*, 2008)	Glycosaminoglycan glycolipids	CNS, connective tissue, skeleton, heartand kidney
ML III (pseudo-Hurler polydystrophy)	N-acetyl-glucosaminyl-1 phosphotransferase	12q23.2	16 (Tappino *et al.*, 2008)	Glycolipids Glycosaminoglycan	Connective tissue and predominant joint
ML IV	Mucolipin and transient calcium channel	19p13.2–13.3	>15 (Zeevi, Frumkin, and Bach, 2007)	Glycolipids Glycosaminoglycan	CNS and connective tissue
G_{M1}–gangliosidosis	β-galactosidase	3p21.33	102 (Santamaria *et al.*, 2007)	Ganglioside G_{M1} sialo G_{M1} Oligosaccharides keratan sulfate	CNS, liver, spleen, eye and skeleton

(*continued overleaf*)

Table 8.2 (*continued*)

G_{M2}–gangliosidosis Tay–Sachs (variant B)	β-hexosaminidase α-subunit (Hex A)	15q23-q24	>100 (Mark *et al.*, 2003)	Ganglioside G_{M2}	CNS and eye
Sandhoff	β-subunit (Hex A+B)	5q13	30 (Mark *et al.*, 2003)	Ganglioside G_{M2} oligosaccharides	CNS, liver, spleen, eye and skeleton
G_{M2} activator deficiency (variant AB)	G_{M2}-activator protein Saposin B	5q32-q31	5 (Mark *et al.*, 2003)	Ganglioside G_{M2}	CNS and eye
Gaucher disease Type I, II and III	β-glucocerebrosiase	1q21-q31	300 (Sun *et al.*, 2006)	Glucosylceramide	Reticuloendothelial cells, CNS, spleen, liver and bone marrow
Activator deficiency	Saposin C	10q21	3 (Tylki-Szymanska *et al.*, 2007)	Glucosylceramide	Reticuloendothelial cells, spleen, liver and bone marrow
Activator deficiency	Saposin D	10q21	1 (Diaz-Font *et al.*, 2005a)	Glucosylceramide	Reticuloendothelial cells, spleen, liver and bone marrow
Prosaposin	Saposin A, B, C and D	10q21	1 (Kuchar *et al.*, 2009)	Glucosylceramide	CNS
Galactosialidosis	Protective protein–cathepsin A	20q13.1	20 (Malvagia *et al.*, 2004)	Glycolipids	CNS, spleen, liver and skeleton
Farber's lipogranulomatosis	Acid ceramidase	8p21.3–22	17 (Park and Schuchman, 2006)	Ceramide	Subcutaneous nodules, larynx, liver, lung, heart and joints

Ceramidase activator protein deficiency	Saposin D	10q21	N/A[b]	Ceramide	Subcutaneous nodules, larynx, liver, lung, heart and joints
Wolman disease	Acid lipase	10q24-25	14 (Lohse *et al.*, 2000)	Cholesteryl and ester triglycerides	Liver, spleen and heart
Fabry disease	α-galactosidase	Xq21.33-p22	>400 (Benjamin *et al.*, 2009)	Glycosphingolipids with terminal α-galactosyl	Blood vessels, CNS, kidney and heart
Schindler	α-*N*-acetylgalacto-saminidase	1q13-13.2	5 (Bakker *et al.*, 2001)	Glycosphingolipids and oligosaccharides with terminal α-galactosyl	CNS and endothelial cell
Metachromatic leukodystrophy	Arylsulfatase A	22q13	126 (Cesani *et al.*, 2009)	Galactosylsulfatide	CNS, liver, kidney and gall bladder
Cerebroside–sulfate	Saposin B	10q21	5 (Kuchar *et al.*, 2009)	Galactosylsulfatide	CNS, liver, kidney and gall bladder
Multiple sulfatase	Arylsulfatase ABC	3 p26	12 (Diaz-Font *et al.*, 2005b)	Sulfated lipids and sulfated glycosaminoglycans	CNS, liver, kidney and gall bladder
Niemann–Pick disease (NP)	Sphingomyelinase	11p15.1–15.4	>100 (Rodriguez-Pascau *et al.*, 2009)	Sphingomyelin cholesterol	CNS, spleen, bone marrow, liver, kidney and eye

(*continued overleaf*)

Table 8.2 (*continued*)

NPC1	Transmemberane protein of late endosome, a cholesterol transporter	18q11-12	200 (Fernandez-Valero *et al.*, 2005)	Sphingomyelin, glycolipids and phospholipids	CNS, spleen, bone marrow, liver, kidney and eye
NPC2	Lysosomal protein, a cholesterol transporter	14q24.3	15 (Verot *et al.*, 2007)	Sphingomyelin, glycolipids and phospholipids	CNS, spleen, bone marrow, liver, kidney and eye
Krabbe disease Globoid cell leukodystrophy	Galactosylceraminidase	4q24.3–32	<100 (Tappino *et al.*, 2008)	Galactosylsphingosine (psychosin)	CNS and peripheral nervous system (PNS)
Galactosylceramide activator protein	Saposin A	10q21	1 (Spiegel *et al.*, 2005)	Galactosylsphingosine (psychosine)	CNS and PNS
Fucosidosis	α-fucosidase	2q24-23	27 (Lin *et al.*, 2007)	Oligosaccharides glycolipids	CNS, spleen and liver
A-mannosidosis	α-mannosidase	19p12-q12	125 (Riise Stensland *et al.*, 2012)	Oligosaccharides	CNS, spleen, liver and skeleton
B-mannosidosis	β-mannosidase	4q21-25	14 (Riise Stensland *et al.*, 2008)	Oligosaccharides	CNS and spleen
N-aspartylglu-cosaminuria	Acid aspartyl–glucosaminidase	4q34-35	20 (Aronson, 1999)	Oligosaccharides	CNS, connective tissue and bone marrow
Pompe disease Glycogen type II	α-glucosidase (acid maltase)	17q25.2–25.3	289 (Park *et al.*, 2006)	Glycogen	CNS, muscle and heart

Danon disease	Lysosomal-associated membrane-protein 2	Xq24	>15 (Tunon *et al.*, 2008)	Glycogen	CNS, muscle and heart
Neuronal ceroid lipofuscinosis (CLN), infantile (CLN1)	Palmitoyl protein thioesterase	1p32	49 (Autti *et al.*, 2011)	Saposins A and D	CNS and retina
Late infantile (CLN2)	Tripeptydyl–peptidase	11p15	<70 (Chang *et al.*, 2011)	Late stage of protein catabolism	CNS and retina
Juvenile (CLN3)	Transmembrane protein	16p12	>40 (Aberg *et al.*, 2011)	Lipoproteins SCMAS[a]	CNS and retina
Adult onset (CLN4)	Unknown	Unknown	(Boustany *et al.*, 2011)	SCMAS[a]	CNS
Late infantile (CLN5) (Finnish variant)	Transmembrane protein	13q31-32	27 (Aberg *et al.*, 2011)	SCMAS[a] saposins A and D	CNS and retina
Late infantile (CLN6) (Indian variant)	Transmembrane protein endoplasmic reticulum	15q21-23	41 (Alroy *et al.*, 2011)	Glycolipidcathepsin D	CNS and retina
Late infantile (CLN7) (Turkish variant)	Integral membrane	4q28.1–28.2	22 (Elleder *et al.*, 2011)	SCMAS[a] protein	CNS and retina

(*continued overleaf*)

Table 8.2 (*continued*)

Progressive epilepsy with mental retardation (CLN8)	Membrane protein in endoplasmic reticulum and Golgi apparatus	8p23	16 (Aiello *et al.*, 2011)	SCMAS[a] saposins	CNS and retina
Juvenile onset (CLN9)	Unknown	Unknown	Boustany *et al.* (2011)	SCMAS[a]	CNS and retina
Congenital (CLN10)	Cathepsin D	11p15.5	4 (Cooper *et al.*, 2011)	Gangliosides monoacylglycerol	CNS and retina
Free sialic acid storage disease (Salla disease)	Sialin membrane protein	6q14-15	20 (Aula *et al.*, 2002)	Free sialic acid	CNS, liver and kidney
Cystinosis	Cystinosin membrane protein	17p113	80 (Bendavid *et al.*, 2004)	Cystine	Kidney
Pycnodysostosis	Cathepsin K	1q21	30 (Khan, Ahmed, and Ahmad, 2010)	Collagen	Skeleton
Lysosomal storage with osteopetrosis	Chloride channel 7 (CCL-7)	NA[b]	2 (Pangrazio *et al.*, 2010)	NA[b]	CNS, eye and skeleton

[a]SCMAS: subunit C mitochondria.
[b]NA: not available.

Table 8.3 Cell types present in skin

Keratinocytes	Unmyelinated axons
Apocrine cells	Myelinated axons
Eccrine cells	Schwann cells
Sebaceous cells	Macrophages
Fibroblasts	Langerhans cells
Endothelial cells	Mast cells
Pericytes	Smooth muscle cells
Melanocytes	Merkel cells

skin biopsies is less invasive than biopsies from other sites and is the most cost-effective first-line diagnostic tool (Prasad, Kaye, and Alroy, 1996). Skin biopsies have been reported to be useful in diagnosing conditions such as heritable connective tissue disorders (Holbrook and Byers, 1989), Guamanian neurodegenerative disease and sporadic amyotrophic lateral sclerosis (ALS) (Ono *et al.*, 1997) and cerebral autosomal dominant arteriopathy with subcortical infarcts and leukoencephalopathy (Ruchoux and Maurage, 1998). Skin biopsies are the most suitable tissue because they contain an array of cell types of ectodermal, endodermal and mesodermal origin including neuronal elements. Table 8.3 lists the various cell types present in skin.

8.2 MORPHOLOGICAL FINDINGS

Light microscopic examination of resin-embedded, 1 μm thick, toluidine-stained sections often suggests a diagnosis of lysosomal storage disease and the possible specific class of the storage substrate. The presence of clear vacuoles may suggest the storage of oligosaccharides or mucopolysaccharides, while the presence of dark staining granules may indicate the storage of glycolipids. The ultrastructural appearance of each type of stored material is distinctive and contributes to the diagnosis. For instance, cases in which the lysosomes of all the various skin cell types have electron-lucent or fine fibrillar material are indicative of lysosomal storage of oligosaccharides or free sialic acids, as is seen in α-mannosidosis (Figure 8.1), β-mannosidosis, aspartyl glucosaminuria (Figure 8.2), free sialic acid (Figure 8.3) and mucopolysaccharidosis (MPS) IX (Figure 8.4). Lysosomal storage of glycolipids, however, is characterized ultrastructurally by the presence of lysosomes that contain lamellated membrane structures, that is, fingerprints or zebra bodies as seen in Tay–Sachs, a B variant of G_{M2}-gangliosidosis, in deficiencies

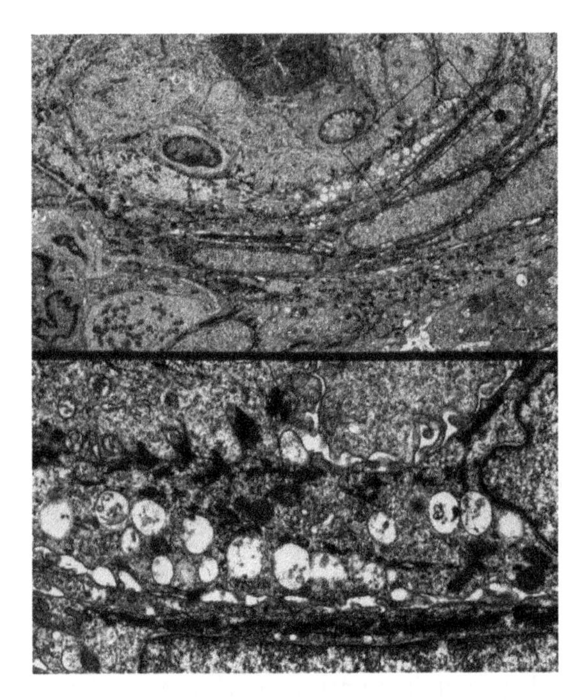

Figure 8.1 Low-magnification electron micrograph of a skin biopsy from a 2.5-year-old boy with α-mannosidosis; the majority of the cells contain few electron-lucent lysosomes. However, some of the basal cells of the eccrine gland contain numerous lysosomes. Higher magnification demonstrates numerous large electron-lucent lysosomes.

of G_{M2} activator (variant AB), in Fabry disease and in mucolipidosis IV (MLIV) (Figure 8.5). In the various types of Niemann–Pick disease, the lysosomes contain coarse lamellated membrane structures (Figure 8.6). A distinctive ultrastructural feature of Fabry disease as opposed to other glycolipid storage diseases is the presence of multiple layers of basal lamina around the blood vessels (Dvorak *et al.*, 1981) (Figure 8.7). Such changes are similar to those seen in diabetic vasculopathy, which is thought to reflect an increased turnover of endothelial cells and pericytes (Vracko and Benditt, 1974). It is noteworthy that some lysosomal enzymes remove the same terminal residue from both glycolipids and glycoproteins resulting in the accumulation of lamellated membrane structures in cells that recycle mostly glycolipids, while those cells that recycle primarily glycoproteins accumulate fibrillar material or electron-lucent lysosomes. The latter phenomenon is seen in fucosidosis (Figure 8.8), G_{M1} gangliosidosis (Figure 8.9), Sandhoff disease, which is a G_{M2} gangliosidosis (variant O), and sialidosis where endothelial cells,

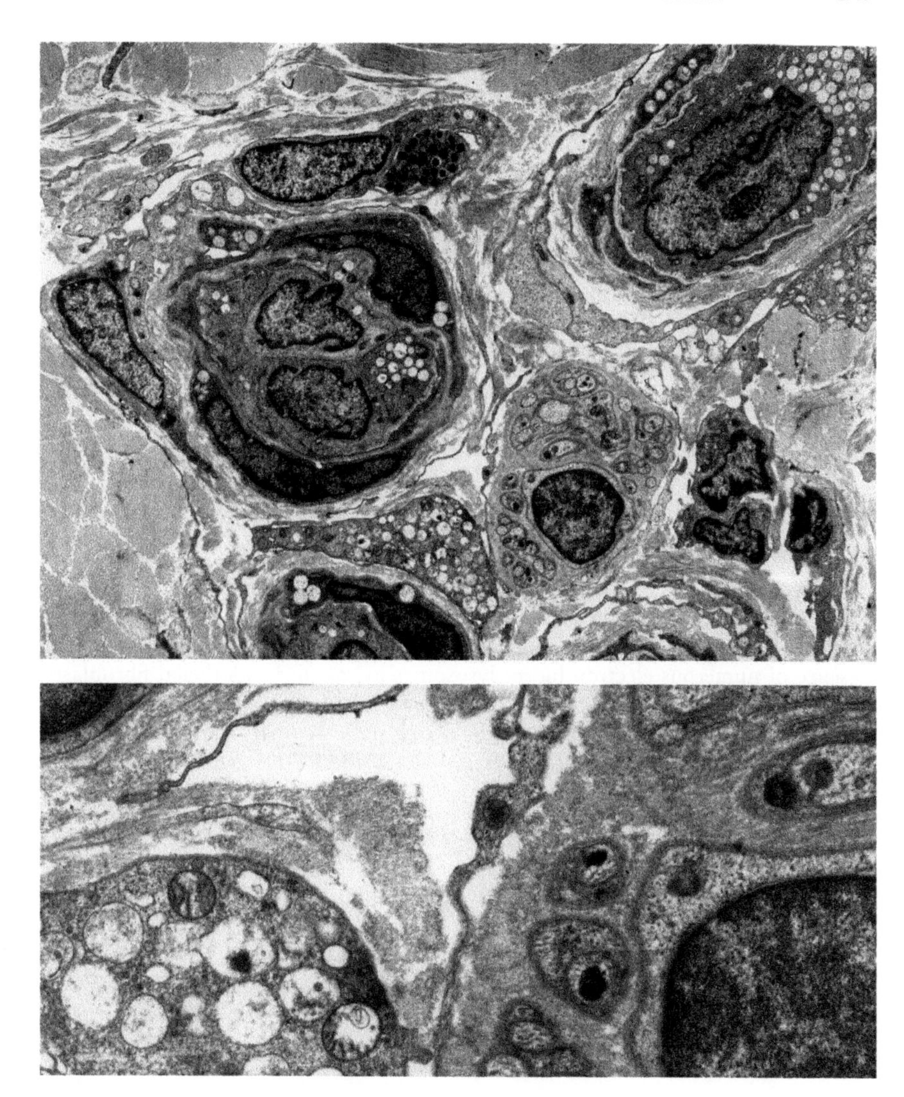

Figure 8.2 Low-magnification electron micrograph of a skin biopsy from a patient with aspartylglucosaminuria revealing the presence of numerous electron-lucent lysosomes in vascular endothelium, pericytes and fibroblasts.

fibroblasts and smooth-muscle cells have electron-lucent lysosomes or contain fine fibrillar material, while Schwann cells may contain lamellated membrane structures, that is, fingerprints or zebra bodies (Alroy and Ucci, 2006). The presence of clusters of mononuclear and multinuclear globoid cells in the white matter characterizes Krabbe's disease, that is, globoid cell leukodystrophy. Schwann cells and sweat glands

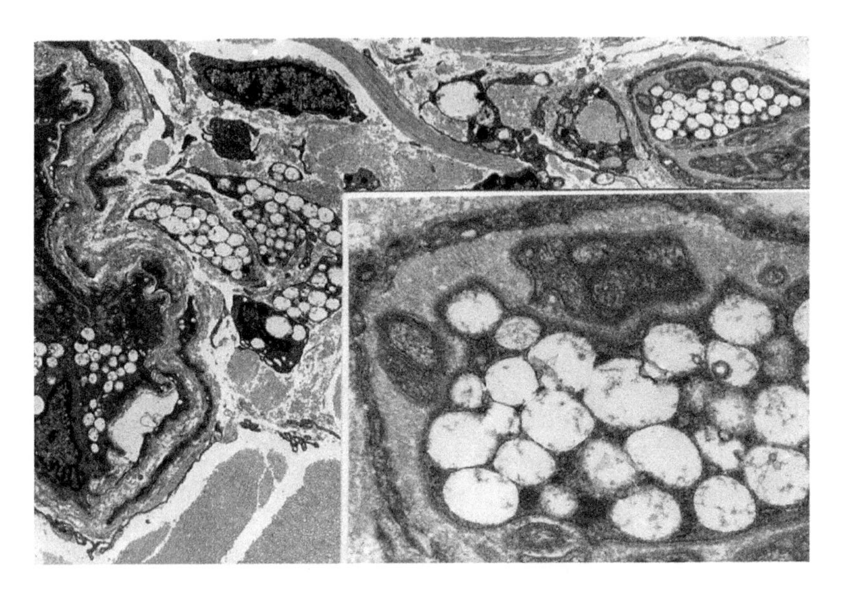

Figure 8.3 Low- and high-magnification electron micrographs of skin biopsy from an 18-month-old girl with storage of free sialic acid. They demonstrate the presence of numerous electron-lucent lysosomes in vascular endothelia, fibroblasts and Schwann cells. Reproduced from R. Kleta, R. P. Morse, J. Alroy, et al., Clinical, biochemical, and molecular diagnosis of a free sialic acid storage disease patient of moderate severity, Mol Gent & Metab 82:137–143 2005, with permission of Elsevier.

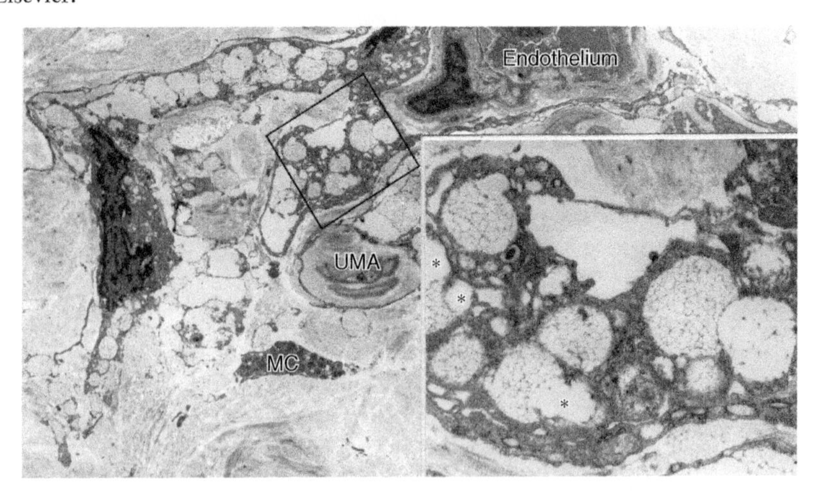

Figure 8.4 Low- and high-magnification electron micrographs of skin biopsy from an 8-year-old girl with MPS IX, illustrating vascular endothelium, pericytes, mast cells (MCs), large fibroblasts and a peripheral nerve with unmyelinated axons (UMA). The fibroblasts contain large lysosomes ($_*$) that are filled with very fine fibrils.

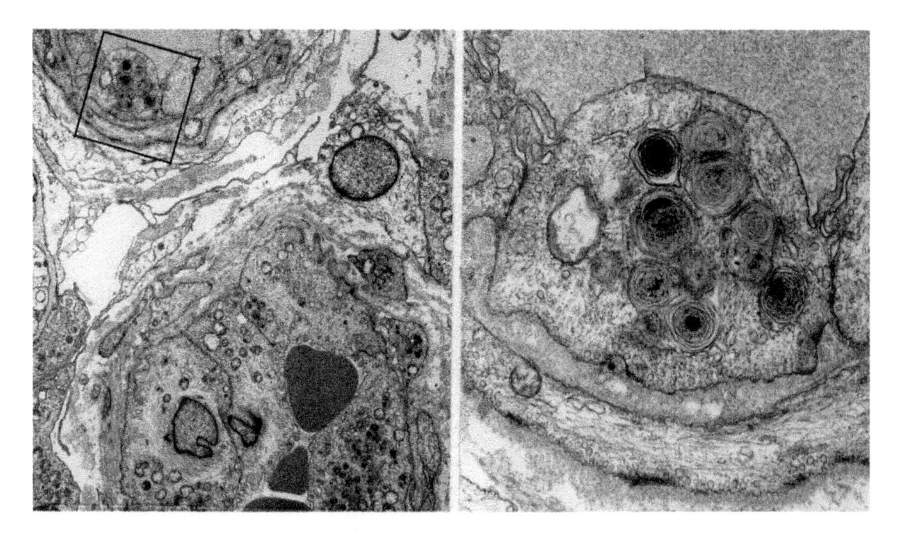

Figure 8.5 Low- and high-magnification electron micrographs of skin biopsy from an 11-month-old boy with ML IV. They reveal an endothelial cell containing lysosomes packed with lamellated membrane structures.

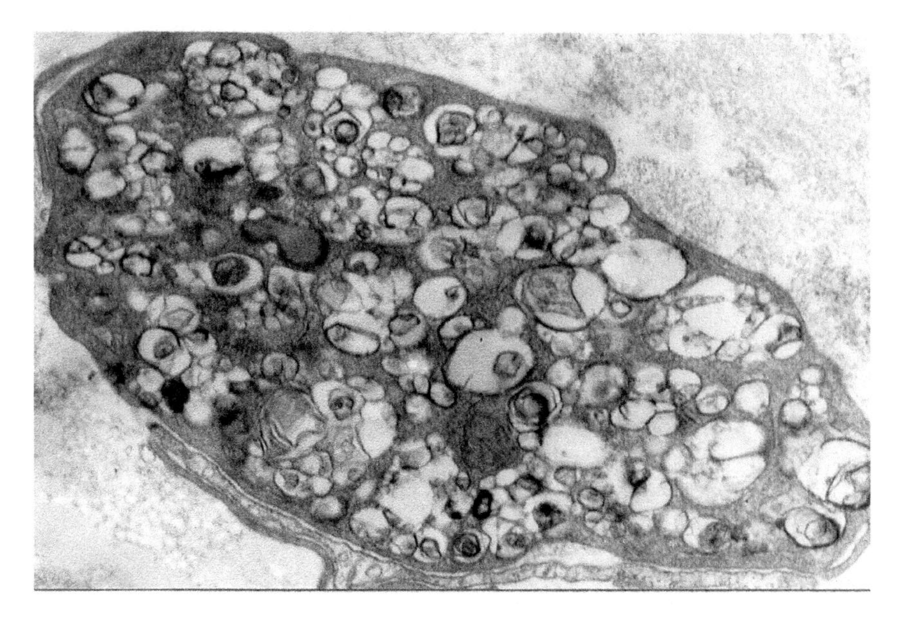

Figure 8.6 High-magnification electron micrograph of skin biopsy from a 34-year-old male with Niemann Pick B, that is, adult onset. It highlights the presence of numerous lysosomes packed with coarse lamellated membrane structures. Reproduced from J. Alroy, Skin Biopsy: A Useful Tool in the Diagnosis of Lysosomal Storage Diseases, Ultrastruc Pathol 30: 489–503, 2007 with permission of Informa Healthcare.

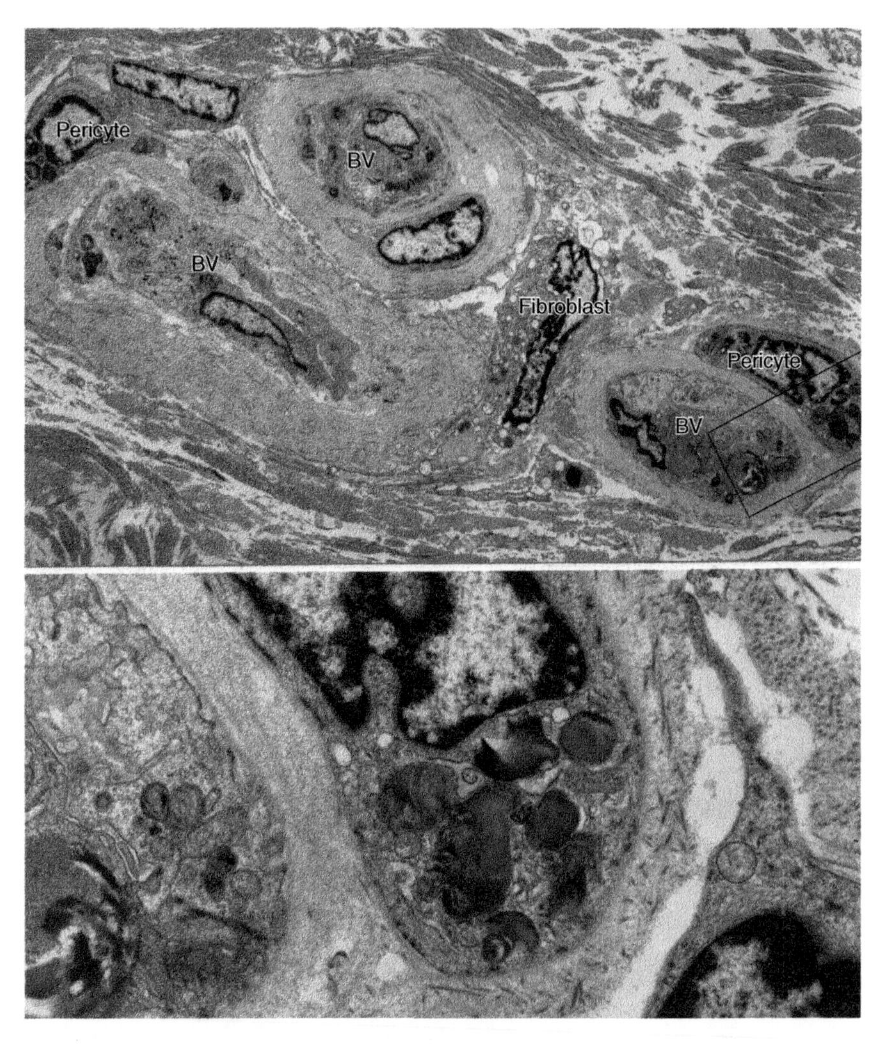

Figure 8.7 Low-magnification electron micrograph of a skin biopsy from a 53-year-old male with Fabry disease. It illustrates the presence of lysosomes that contain lamellated membrane structures in endothelial cells and pericytes. The blood vessels are surrounded by multiple layers of basal lamina.

may have large lysosomes containing slightly tubular, curved or cleft-like structures (Figure 8.10) (Alroy and Ucci, 2006). Similar structures are seen in macrophages, spleen, liver and bone marrow from patients with Gaucher's disease (Figure 8.11). Deficiency of acid lipase, as seen in Wolman disease, results in the accumulation of cholesterol esters and triglycerides (Assmann and Seedorf, 2001), characterized by the presence of drop-like and cleft-like lysosomes (Figure 8.12).

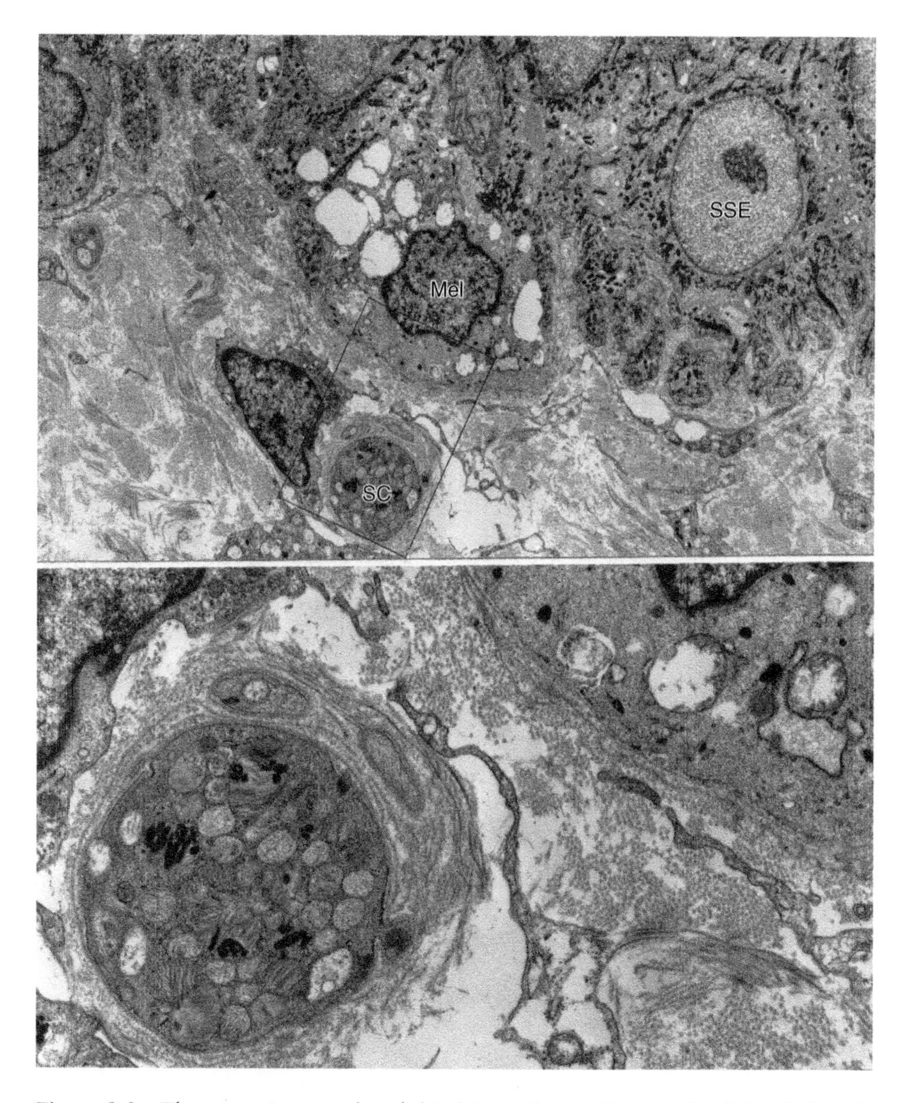

Figure 8.8 Electron micrographs of skin biopsy from a 9-year-old girl with fucosidosis. The low magnification shows stratified squamous epithelium (SSE), fibroblasts, a melanocyte (Mel), unmyelinated axons, perineural cells and a Schwann cell (SC). The basal layer of skin, the fibroblasts and the melanocyte contain large electron-lucent lysosomes. The higher magnification of the nerve with unmyelinated axons shows a Schwann cell containing lysosomes packed with lamellated membrane structures, that is, zebra bodies.

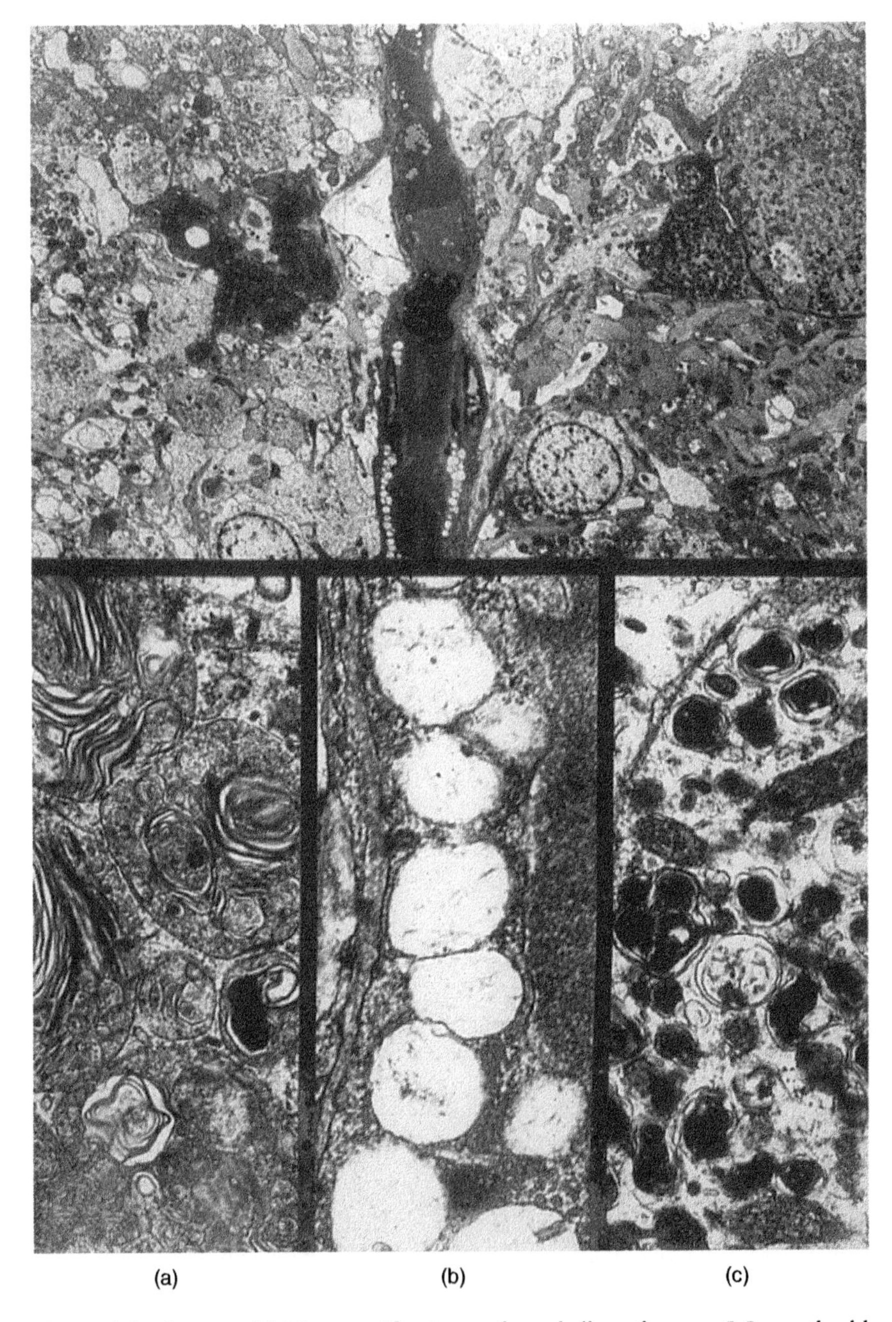

(a) (b) (c)

Figure 8.9 Low and high magnifications of cerebellum from a 5.5-month-old Portuguese water dog with G_{M1} gangliosidosis. There are accumulated lamellated membrane structures in the lysosomes of the neuronal soma (a), oligosaccharides in lysosomes of endothelial cells (b) as well as spheroids in axons (c).

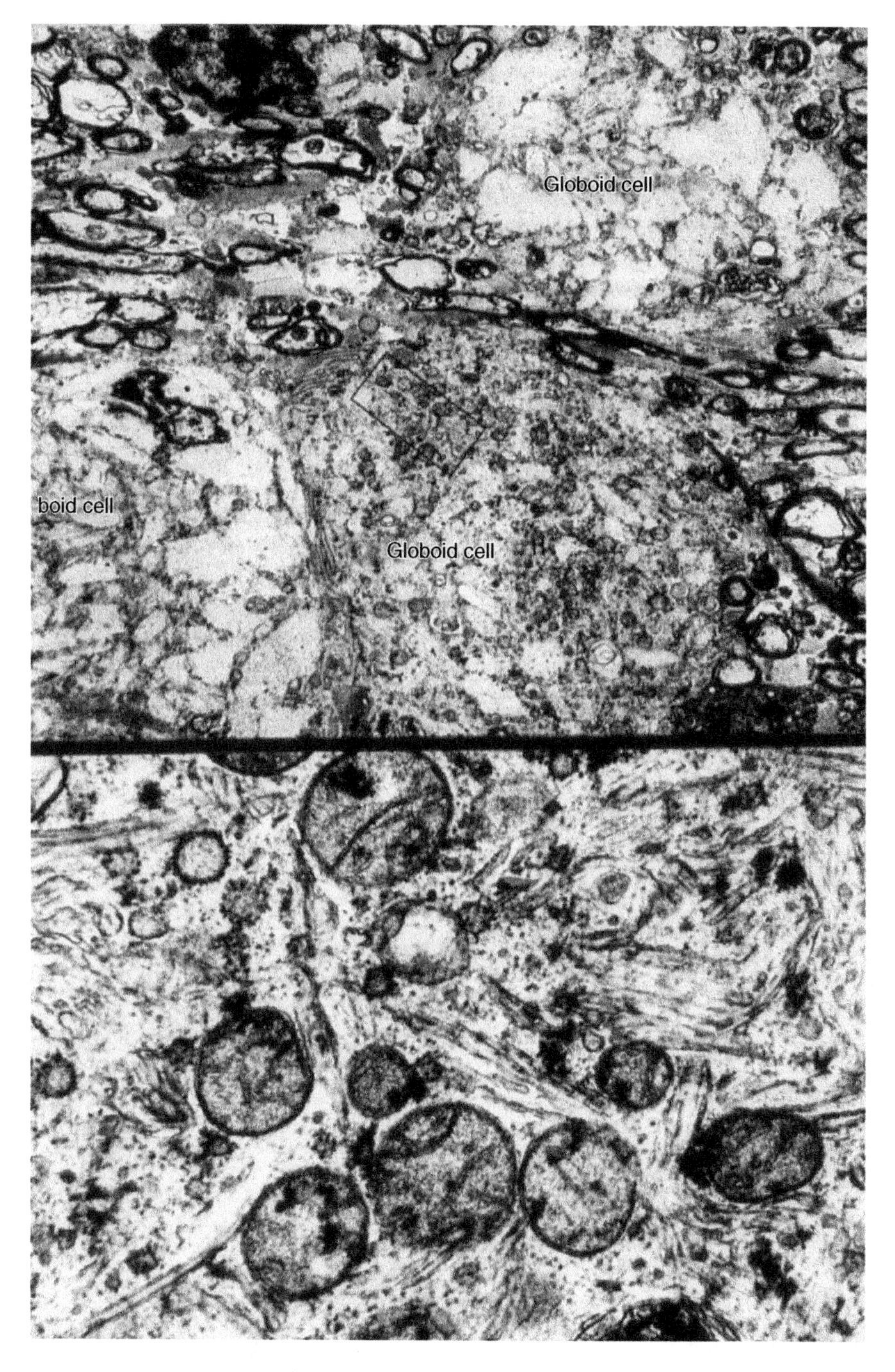

Figure 8.10 Low-magnification electron micrograph of white matter in Rhesus monkey with Krabbe disease illustrates myelinated axons and macrophages. High-magnification view shows macrophages containing tubular, curved or cleft-like structures.

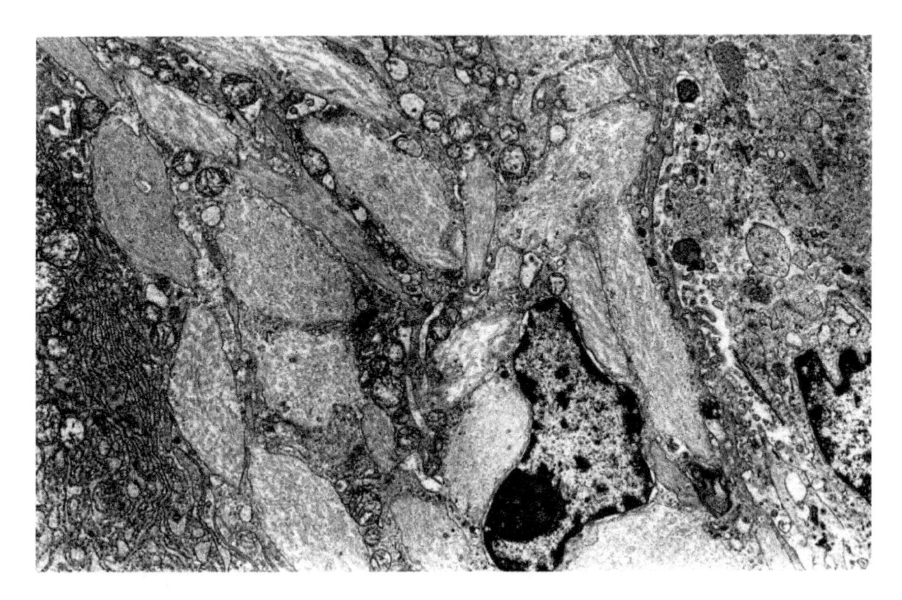

Figure 8.11 Electron micrograph of macrophage obtained from the bone marrow of a patient with Gaucher disease. It contains lysosomes with numerous bundles of twisted tubules. (Courtesy of Theodore Iancu.).

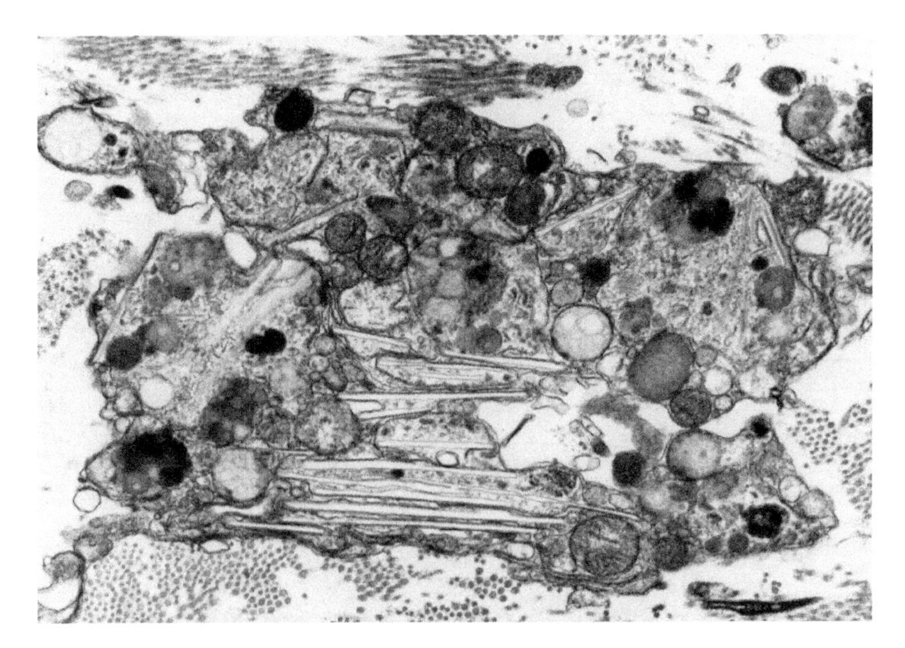

Figure 8.12 High-magnification electron micrograph of a skin biopsy from a 12-year-old boy with Wolman disease revealing a macrophage containing cleft-like cholesterol and lysosomes with lipid-like material.

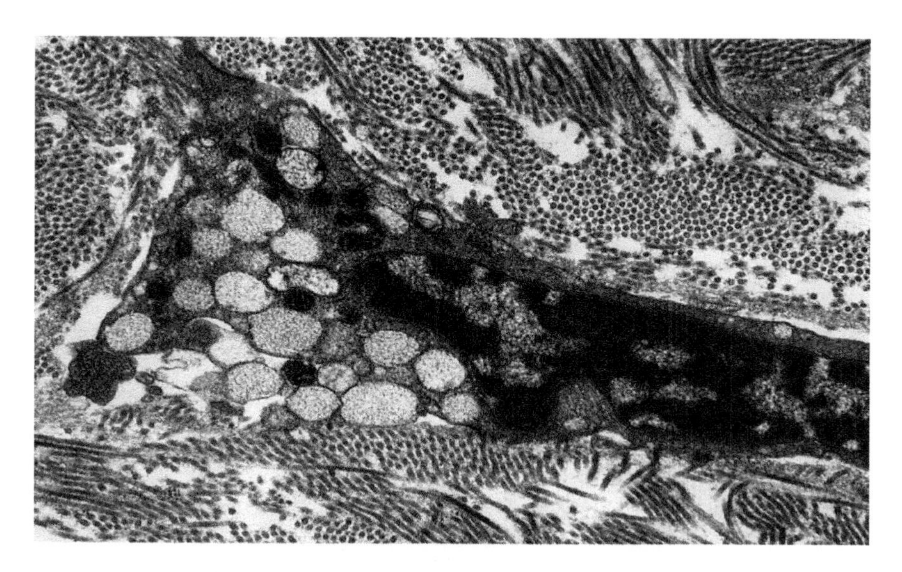

Figure 8.13 High-magnification electron micrograph of a skin biopsy from a 5-year-old boy with MPS II (Hunter disease) illustrating a fibroblast with numerous lysosomes, which contain fine fibrillar material.

In MPS, the major storage materials are undegraded GAGs. These appear in electron micrographs as fine fibrillar material (Figure 8.13). In addition, there is also accumulation of other substrates such as glycolipids, which appear as lamellated membrane structures in affected cells (Alroy and Ucci, 2006). Thus, MPS lysosomes are characterized by contents of fine fibrillar material admixed with lamellated membrane structures (Figure 8.14). The reason for the accumulation of mixed substrates within the same lysosome is thought to be due to the secondary inhibition of over 27 different lysosomal hydrolases by the accumulation of GAGs (Avila and Convit, 1975).

Pompe disease is a glycogen storage disease type II, and is characterized by lysosomal storage of glycogen in different cell types (Hirschhorn and Reuser, 2001). Unlike Pompe disease, in Danon disease, also called lysosomal glycogen storage disease with normal acid maltase, the autophagic vacuoles contain electron-dense, granular particles and some fingerprint-like structures (Alroy *et al.*, 2010). In cystinosis, there is lysosomal accumulation of hexagonal cystine crystals (Figure 8.15) observed in different cell types (Gahl, Thoene, and Schneider, 2001).

Different types of neuronal ceroid lipofuscinosis (NCL) demonstrate diverse morphology of the stored material. In infantile NCL, known

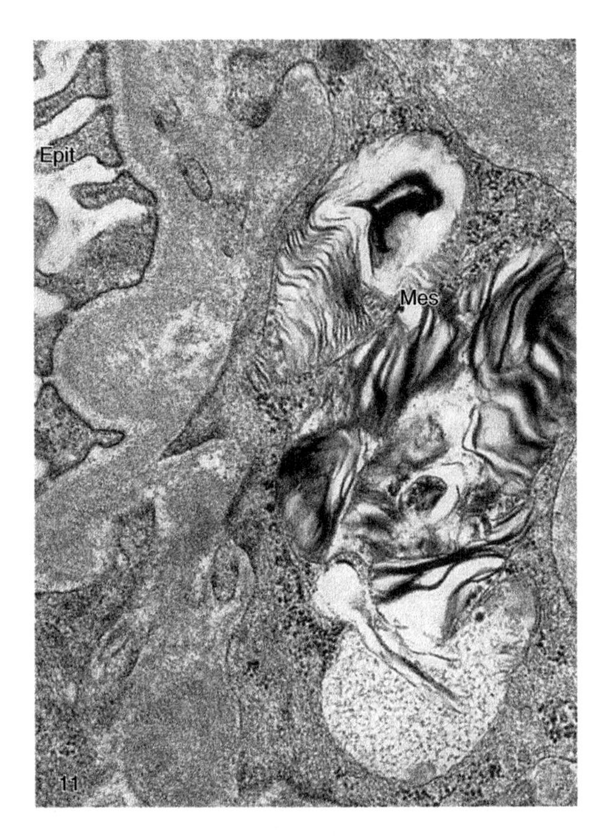

Figure 8.14 High-magnification electron micrograph of kidney from a cat with mucopolysaccharidosis VI (Maroteaux–Lamy) illustrating a mesangial cell containing enlarged lysosomes with a mixture of lamellated membrane structures and fine fibrillar material. Reproduced from M. Castagnaro, J. Alroy, A.A. Ucci, et al., Lectin histochemistry and ultrastructure of feline kidneys from six different storage diseases, Virchows Archive B 54: 16–26, 1988, with permission from Springer.

as ceroid lipofuscinosis, neuronal 1 (CLN1), and in ceroid lipofuscinosis, neuronal 10 (CLN10), a congenital late-infantile form (Jalanko and Braulke, 2009), there is storage of granular osmiophilic deposits (GRAD) (Figure 8.16). In late-infantile neuronal ceroid lipofuscinosis (CLN2) and in ceroid-lipofuscinosis, neuronal 8 (CLN8, congenital late-infantile northern epilepsy), there is storage of curvilinear profiles (CLPs). CLN3 and CLN4 reveal storage of a fingerprint pattern (FPP). In ceroid-lipofuscinosis, neuronal 5 (CLN5, the late-infantile Finnish variant), in ceroid-lipofuscinosis, neuronal 6 (CLN6, another late-infantile variant) and in ceroid-lipofuscinosis, neuronal 7 (CLN7, the late-infantile Turkish variant), there is storage of mixed CLP, FPP and rectilinear

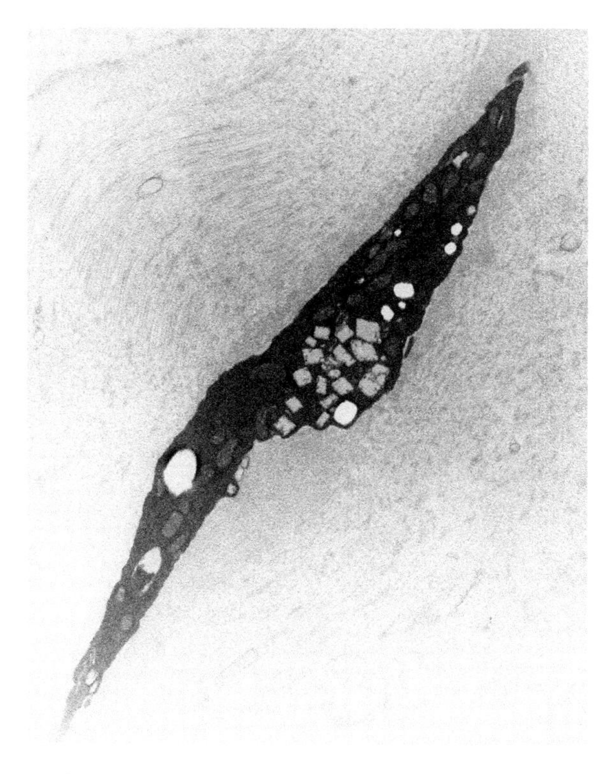

Figure 8.15 High-magnification electron micrograph of cornea from a 29-year-old female with cystinosis revealing a keratocyte containing hexagonal and irregular cystine crystals within lysosomes. Reproduced from J. Alroy, Skin Biopsy: A Useful Tool in the Diagnosis of Lysosomal Storage Diseases, Ultrastruc Pathol 30: 489–503, 2007 with permission of Informa Healthcare.

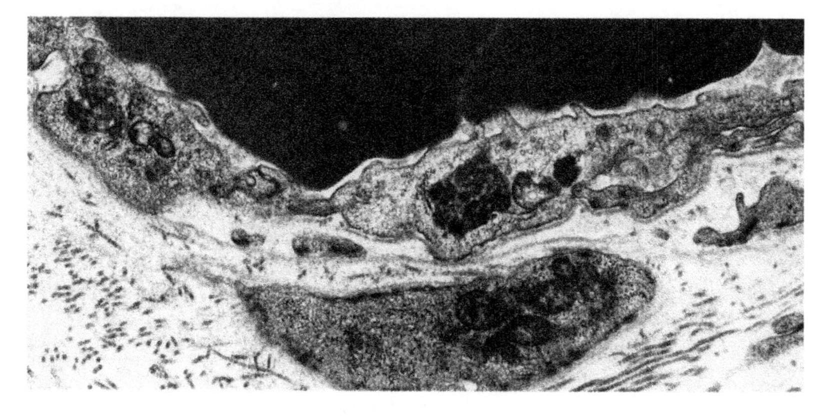

Figure 8.16 High-magnification electron micrograph of an endothelial cell and pericyte from a patient with the infantile form of NCL. Both cell types have secondary lysosomes filled with granular osmiophilic deposits (GRADs).

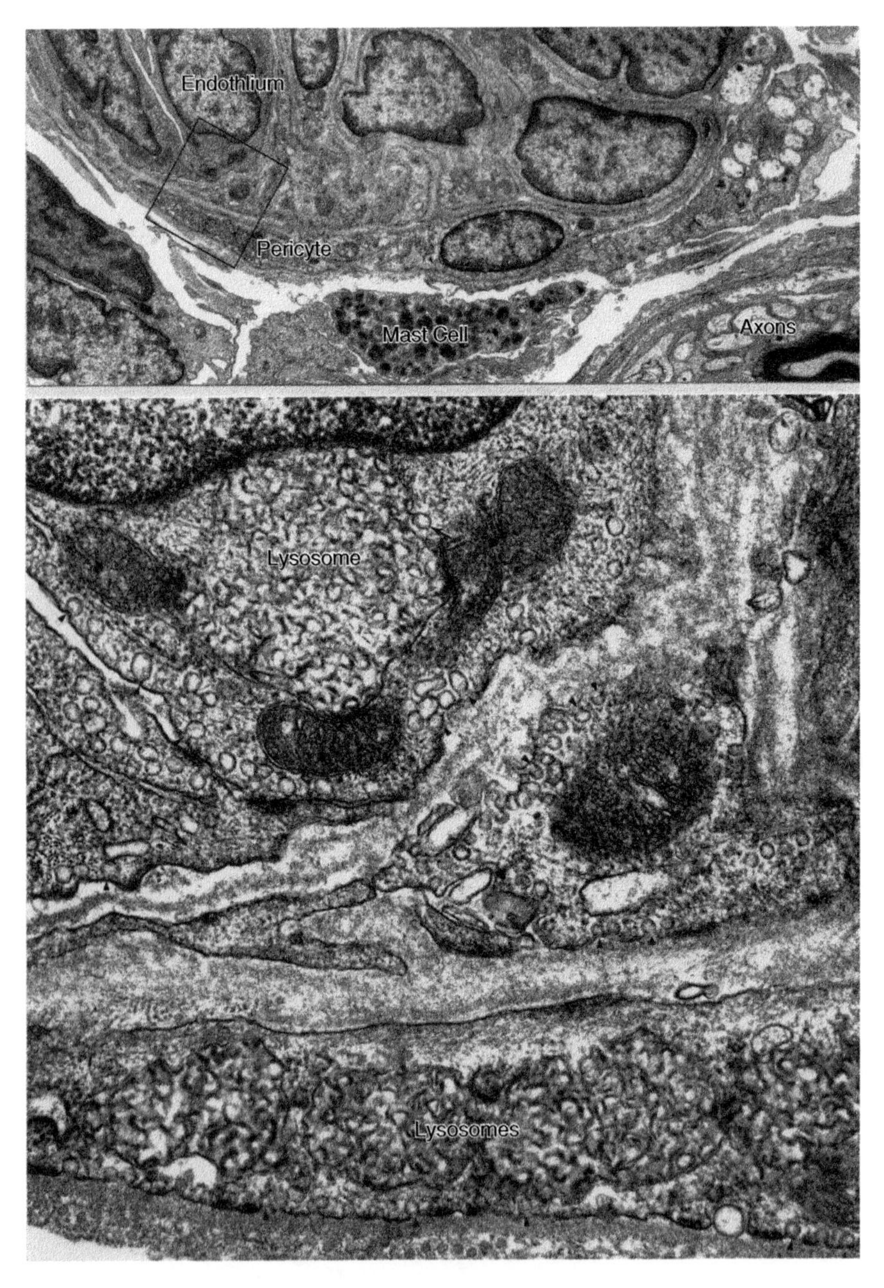

Figure 8.17 High magnification of skin biopsy from a 4.5-year-old boy with juvenile ceroid lipofuscinosis revealing curvilinear structures in an endothelial cell and pericyte.

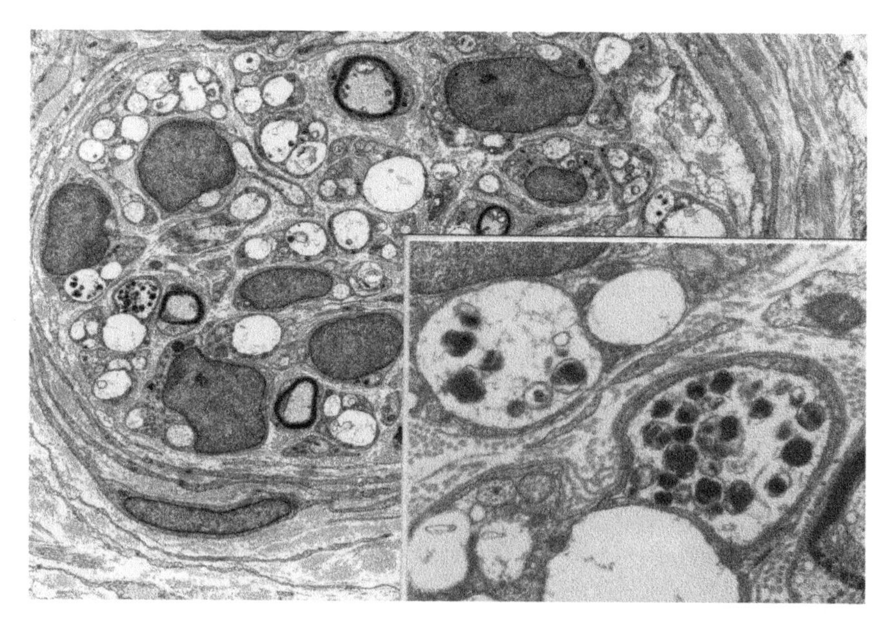

Figure 8.18 Low- and high-magnification electron micrographs of skin from a 1-month-old boy. They demonstrate a peripheral nerve with few myelinated axons with narrow myelin sheaths and many swollen unmyelinated axons, some containing numerous spheroids. Reproduced from J. Alroy, Electron microscopic findings in skin biopsies from patients with infantile osteopetrosis and neuronal storage disease, Ultrastruc Pathol 31: 333–338, 2008 with permission of Informa Healthcare.

profiles (RLPs) (Figure 8.17). In addition, ceroid-lipofuscinosis, neuronal 9 (CLN9, juvenile NCL), shows mixed storage of CLP, FPP and GRAD (Jalanko and Braulke, 2009).

In the brains of patients with infantile osteopetrosis and neuronal storage, there is lysosomal accumulation of granular osmiophilic, amorphous material (Takahashi *et al.*, 1990) or lamellated structures (Ambler *et al.*, 1983). Skin biopsies from patients with infantile osteopetrosis and neuronal storage show an accumulation of spheroids in unmyelinated axons (UMAs) and poor myelination of myelinated axons (Figure 8.18) (Alroy *et al.*, 2007).

8.3 CONCLUSION

Electron microscopic examination is a useful and relatively inexpensive method to distinguish lysosomal storage diseases from other diseases with similar clinical symptoms and sort lysosomal storage diseases into broad storage disease classes involving glycoproteins, complex carbohydrates and glycolipids.

REFERENCES

Aberg, L., Autti, T., Cooper, J.D. *et al.* (2011) CLN5, in *The Neuronal Ceroid Lipofuscinoses (Batten Disease)*, 2nd edn (eds S. Mole, R.E. Williams and H.H. Goebel), Oxford University Press, Oxford, pp. 140–158.

Aiello, C., Cannelli, N., Cooper, J.D. *et al.* (2011) CLN8, in *The Neuronal Ceroid Lipofuscinoses (Batten Disease)*, Oxford, 2nd edn (eds S. Mole, R.E. Williams, and H.H. Goebel), Oxford University Press, pp. 189–202.

Albay, D., Adler, S.G., Philipose, J. *et al.* (2005) Chloroquine-induced lipidosis mimicking Fabry disease. *Modern Pathology*, **18**, 733–738.

Alroy, J., Braulke, T., Cismondi, I.A. *et al.* (2011) CLN6, in *The Neuronal Ceroid Lipofuscinoses (Batten Disease)*, Oxford, 2nd edn (eds S. Mole, R.E. Williams, and H.H. Goebel), Oxford University Press, pp. 159–175.

Alroy, J., Pfannl, R., Slavov, D. and Taylor, M.R. (2010) Electron microscopic findings in skin biopsies from patients with Danon disease. *Ultrastructural Pathology*, **34**, 333–336.

Alroy, J., Pfannl, R., Ucci, A. *et al.* (2007) Electron microscopic findings in skin biopsies from patients with infantile osteopetrosis and neuronal storage disease. *Ultrastructural Pathology*, **31**, 333–338.

Alroy, J. and Ucci, A.A. (2006) Skin biopsy: a useful tool in the diagnosis of lysosomal storage diseases. *Ultrastructural Pathology*, **30**, 489–503.

Alroy, J., Ucci, A.A. and Pfannl, R. (2006) Why skin biopsy is useful for the diagnosis of lysosomal storage diseases. *Current Medical Literature*, **5**, 70–76.

Ambler, M.W., Trice, J., Grauerholz, J. and O'Shea, P.A. (1983) Infantile osteopetrosis and neuronal storage disease. *Neurology*, **33**, 437–441.

Aronson, N.N. Jr. (1999) Aspartylglycosaminuria: biochemistry and molecular biology. *Biochimica et Biophysica Acta*, **1455**, 139–154.

Assmann, G. and Seedorf, U. (2001) Acid lipase deficiency: Wolman disease and cholesteryl ester storage disease, in *The Metabolic and Molecular Basis of Inherited Disease*, 8th edn (eds C. Scriver, A.L. Beaudet, W.S. Sly and D. Valle), McGraw-Hill, New York, pp. 3551–3572.

Aula, N., Jalanko, A., Aula, P. and Peltonen, L. (2002) Unraveling the molecular pathogenesis of free sialic acid storage disorders: altered targeting of mutant sialin. *Molecular Genetics and Metabolism*, **77**, 99–107.

Autti, T., Cooper, J.D., Van Diggelen, O.P. *et al.* (2011) CLN1, in *The Neuronal Ceroid Lipofuscinoses (Batten Disease)*, 2nd edn (eds S. Mole, R.E. Williams and H.H. Goebel) Oxford University Press, Oxford, pp. 55–79.

Avila, J.L. and Convit, J. (1975) Inhibition of leucocytic lysosomal enzymes by glycosaminoglycans *in vitro*. *The Biochemical Journal*, **152**, 57–64.

Bakker, H.D., De Sonnaville, M.L., Vreken, P. *et al.* (2001) Human alpha-N-acetylgalactosaminidase (alpha-NAGA) deficiency: no association with neuroaxonal dystrophy? *European Journal of Human Genetics: EJHG*, **9**, 91–96.

Beesley, C.E., Jackson, M., Young, E.P. *et al.* (2005) Molecular defects in Sanfilippo syndrome type B (mucopolysaccharidosis IIIB). *Journal of Inherited Metabolic Disease*, **28**, 759–767.

Belcher, R.W. (1972) Ultrastructure of the skin in the genetic mucopolysaccharidoses. *Archives of Pathology*, **94**, 511–518.

Bendavid, C., Kleta, R., Long, R. *et al.* (2004) FISH diagnosis of the common 57-kb deletion in CTNS causing cystinosis. *Human Genetics*, **115**, 510–514.

Benjamin, E.R., Flanagan, J.J., Schilling, A. *et al.* (2009) The pharmacological chaperone 1-deoxygalactonojirimycin increases alpha-galactosidase A levels in Fabry patient cell lines. *Journal of Inherited Metabolic Disease*, **32**, 424–440.

Bertola, F., Filocamo, M., Casati, G. *et al.* (2011) IDUA mutational profiling of a cohort of 102 European patients with mucopolysaccharidosis type I: identification and characterization of 35 novel alpha-L-iduronidase (IDUA) alleles. *Human Mutation*, **32**, E2189–E2210.

Boustany, R.M., Ceuterick-De Groote, C., Goebel, H.H. *et al.* (2011) Genetically unassigned or unusual NCLs, in *The Neuronal Ceroid Lipofuscinoses (Batten Disease)*, 2nd edn (eds S. Mole, R.E. Williams and H.H. Goebel), Oxford University Press, Oxford, pp. 213–236.

Castagnaro, M., Alroy, J., Ucci, A.A. and Glew, R.H. (1987) Lectin histochemistry and ultrastructure of feline kidneys from six different storage diseases. *Virchows Archiv. B, Cell Pathology Including Molecular Pathology*, **54**, 16–26.

Cesani, M., Capotondo, A., Plati, T. *et al.* (2009) Characterization of new arylsulfatase A gene mutations reinforces genotype-phenotype correlation in metachromatic leukodystrophy. *Human Mutation*, **30**, E936–E945.

Chabas, A., Duque, J. and Gort, L. (2007) A new infantile case of alpha-N-acetylgalactosaminidase deficiency. Cardiomyopathy as a presenting symptom. *Journal of Inherited Metabolic Disease*, **30**, 108.

Chang, M., Cooper, J.D., Davidson, B.L. *et al.* (2011) CLN2, in *The Neuronal Ceroid Lipofuscinoses (Batten Disease)* (eds S. Mole, R.E. Williams and H.H. Goebel), Oxford University Press, Oxford, pp. 80–109.

Cooper, J.D., Partanen, S., Siintola, E. *et al.* (2011) CLN10, in *The Neuronal Ceroid Lipofuscinoses (Batten Disease)*, 2nd edn (eds S. Mole, R.E. Williams and H.H. Goebel), Oxford University Press, Oxford, pp. 203–212.

Diaz-Font, A., Cormand, B., Santamaria, R. *et al.* (2005a) A mutation within the saposin D domain in a Gaucher disease patient with normal glucocerebrosidase activity. *Human Genetics*, **117**, 275–277.

Diaz-Font, A., Santamaria, R., Cozar, M. *et al.* (2005b) Clinical and mutational characterization of three patients with multiple sulfatase deficiency: report of a new splicing mutation. *Molecular Genetics and Metabolism*, **86**, 206–211.

Dvorak, A.M., Cable, W.J., Osage, J.E. and Kolodny, E.H. (1981) Diagnostic electron microscopy. II. Fabry's disease: use of biopsies from uninvolved skin. Acute and chronic changes involving the microvasculature and small unmyelinated nerves. *Pathology Annual*, **16** (Pt 1), 139–158.

Elleder, M., Kousi, M., Lehesjoki, A.E. *et al.* (2011) CLN7, in *The Neuronal Ceroid Lipofuscinoses (Batten Disease)*, 2nd edn (eds S. Mole, R.E. Williams and H.H. Goebel), Oxford University Press, Oxford, pp. 176–188.

Feldhammer, M., Durand, S., Mrazova, L. *et al.* (2009) Sanfilippo syndrome type C: mutation spectrum in the heparan sulfate acetyl-CoA: alpha-glucosaminide N-acetyltransferase (HGSNAT) gene. *Human Mutation*, 30, 918–925.

Fernandez-Valero, E.M., Ballart, A., Iturriaga, C. *et al.* (2005) Identification of 25 new mutations in 40 unrelated Spanish Niemann-Pick type C patients: genotype-phenotype correlations. *Clinical Genetics*, **68**, 245–254.

Fowler, D.J., Anderson, G., Vellodi, A. *et al.* (2007) Electron microscopy of chorionic villus samples for prenatal diagnosis of lysosomal storage disorders. *Ultrastructural Pathology*, **31**, 15–21.

Gahl, W.A., Thoene, J.G. and Schneider, J.A. (2001) Cystinosis: a disorder of lysosomal membrane transport, in *The Metabolic and Molecular Basis of Inherited Disease*, 8th edn (eds C. Scriver, A.L. Beaudet, W.S. Sly and D. Valle), McGraw-Hill, New York, pp. 5085–5108.

Gelb, B.D., Bromme, D. and Desnick, R.J. (2001) Pycnodysostosis: cathepsin K deficiency, in *The Metabolic and Molecular Basis of Inherited Disease*, 8th edn (eds C. Scriver, A.L. Beaudet, W.S. Sly and D. Valle), McGraw-Hill, New York, pp. 4244–4261.

Gherardi, R.K., Coquet, M., Cherin, P. *et al.* (2001) Macrophagic myofasciitis lesions assess long-term persistence of vaccine-derived aluminium hydroxide in muscle. *Brain*, **124**, 1821–1831.

Gucev, Z.S., Tasic, V., Jancevska, A. *et al.* (2008) Novel beta-galactosidase gene mutation p.W273R in a woman with mucopolysaccharidosis type IVB (Morquio B) and lack of response to *in vitro* chaperone treatment of her skin fibroblasts. *American Journal of Medical Genetics. Part A*, **146A**, 1736–1740.

Hammel, I. and Alroy, J. (1995) The effect of lysosomal storage diseases on secretory cells: an ultrastructural study of pancreas as an example. *Journal of Submicroscopic Cytology and Pathology*, **27**, 143–160.

Hammel, I., Alroy, J., Goyal, V. and Galli, S.J. (1993) Ultrastructure of human dermal mast cells in 29 different lysosomal storage diseases. *Virchows Archiv. B, Cell Pathology Including Molecular Pathology*, **64**, 83–89.

Hirschhorn, R. and Reuser, A. (2001) Glycogen storage disease type II: acid α-glucosidase (acid maltase) deficiency, in *The Metabolic and Molecular Basis of Inherited Disease*, 8th edn (eds C. Scriver, A.L. Beaudet, W.S. Sly and D. Valle), McGraw-Hill, New York, pp. 3389–3420.

Holbrook, K.A. and Byers, P.H. (1989) Skin is a window on heritable disorders of connective tissue. *American Journal of Medical Genetics*, **34**, 105–121.

Hruban, Z. (1984) Pulmonary and generalized lysosomal storage induced by amphiphilic drugs. *Environmental Health Perspectives*, **55**, 53–76.

Isenberg, J.N. and Sharp, H.L. (1976) Aspartylglucosaminuria: unique biochemical and ultrastructural characteristics. *Human Pathology*, 7, 469–481.

Jalanko, A. and Braulke, T. (2009) Neuronal ceroid lipofuscinoses. *Biochimica et Biophysica Acta*, **1793**, 697–709.

Karageorgos, L., Brooks, D.A., Pollard, A. *et al.* (2007) Mutational analysis of 105 mucopolysaccharidosis type VI patients. *Human Mutation*, **28**, 897–903.

Khan, B., Ahmed, Z. and Ahmad, W. (2010) A novel missense mutation in cathepsin K (CTSK) gene in a consanguineous Pakistani family with pycnodysostosis. *Journal of Investigative Medicine*, **58**, 720–724.

Kuchar, L., Ledvinova, J., Hrebicek, M. *et al.* (2009) Prosaposin deficiency and saposin B deficiency (activator-deficient metachromatic leukodystrophy): report on two patients detected by analysis of urinary sphingolipids and carrying novel PSAP gene mutations. *American Journal of Medical Genetics, Part A*, **149A**, 613–621.

Lin, S.P., Chang, J.H., De La Cadena, M.P. *et al.* (2007) Mutation identification and characterization of a Taiwanese patient with fucosidosis. *Journal of Human Genetics*, **52**, 553–556.

Lohse, P., Maas, S., Elleder, M. *et al.* (2000) Compound heterozygosity for a Wolman mutation is frequent among patients with cholesteryl ester storage disease. *Journal of Lipid Research*, **41**, 23–31.

Lualdi, S., Tappino, B., Di Duca, M. *et al.* (2010) Enigmatic *in vivo* iduronate-2-sulfatase (IDS) mutant transcript correction to wild-type in Hunter syndrome. *Human Mutation*, **31**, E1261–E1285.

Malvagia, S., Morrone, A., Caciotti, A. *et al.* (2004) New mutations in the PPBG gene lead to loss of PPCA protein which affects the level of the beta-galactosidase/neuraminidase complex and the EBP-receptor. *Molecular Genetics and Metabolism*, **82**, 48–55.

Mark, B.L., Mahuran, D.J., Cherney, M.M. *et al.* (2003) Crystal structure of human beta-hexosaminidase B: understanding the molecular basis of Sandhoff and Tay-Sachs disease. *Journal of Molecular Biology*, **327**, 1093–1109.

Merin, S., Livni, N., Berman, E.R. and Yatziv, S. (1975) Mucolipidosis IV: ocular, systemic, and ultrastructural findings. *Investigative Ophthalmology*, **14**, 437–448.

Meyer, A., Kossow, K., Gal, A. *et al.* (2008) The mutation p.Ser298Pro in the sulphamidase gene (SGSH) is associated with a slowly progressive clinical phenotype in mucopolysaccharidosis type IIIA (Sanfilippo A syndrome). *Human Mutation*, **29**, 770.

Montano, A.M., Sukegawa, K., Kato, Z. *et al.* (2007) Effect of 'attenuated' mutations in mucopolysaccharidosis IVA on molecular phenotypes of N-acetylgalactosamine-6-sulfate sulfatase. *Journal of Inherited Metabolic Disease*, **30**, 758–767.

Ono, S., Waring, S.C., Kurland, L.L. *et al.* (1997) Guamanian neurodegenerative disease: ultrastructural studies of skin. *Journal of the Neurological Sciences*, **146**, 35–40.

Pangrazio, A., Pusch, M., Caldana, E. *et al.* (2010) Molecular and clinical heterogeneity in CLCN7-dependent osteopetrosis: report of 20 novel mutations. *Human Mutation*, **31**, E1071–E1080.

Park, Y.E., Park, K.H., Lee, C.H. *et al.* (2006) Two new missense mutations of GAA in late onset glycogen storage disease type II. *Journal of the Neurological Sciences*, **251**, 113–117.

Park, J.H. and Schuchman, E.H. (2006) Acid ceramidase and human disease. *Biochimica et Biophysica Acta*, **1758**, 2133–2138.

Pattison, S., Pankarican, M., Rupar, C.A. *et al.* (2004) Five novel mutations in the lysosomal sialidase gene (NEU1) in type II sialidosis patients and assessment of their impact on enzyme activity and intracellular targeting using adenovirus-mediated expression. *Human Mutation*, **23**, 32–39.

Prasad, A., Kaye, E.M. and Alroy, J. (1996) Electron microscopic examination of skin biopsy as a cost-effective tool in the diagnosis of lysosomal storage diseases. *Journal of Child Neurology*, **11**, 301–308.

Riise Stensland, H.M., Klenow, H.B., Van Nguyen, L. *et al.* (2012) Identification of 83 novel alpha-mannosidosis-associated sequence variants: functional analysis of MAN2B1 missense mutations. *Human Mutation*, **33**, 511–520.

Riise Stensland, H.M., Persichetti, E., Sorriso, C. *et al.* (2008) Identification of two novel beta-mannosidosis-associated sequence variants: biochemical analysis of beta-mannosidase (MANBA) missense mutations. *Molecular Genetics and Metabolism*, **94**, 476–480.

Rodriguez-Pascau, L., Gort, L., Schuchman, E.H. *et al.* (2009) Identification and characterization of SMPD1 mutations causing Niemann-Pick types A and B in Spanish patients. *Human Mutation*, **30**, 1117–1122.

Ruchoux, M.M. and Maurage, C.A. (1998) Endothelial changes in muscle and skin biopsies in patients with CADASIL. *Neuropathology and Applied Neurobiology*, **24**, 60–65.

Santamaria, R., Blanco, M., Chabas, A. *et al.* (2007) Identification of 14 novel GLB1 mutations, including five deletions, in 19 patients with GM1 gangliosidosis from South America. *Clinical Genetics*, **71**, 273–279.

Spiegel, R., Bach, G., Sury, V. *et al.* (2005) A mutation in the saposin A coding region of the prosaposin gene in an infant presenting as Krabbe disease: first report of saposin A deficiency in humans. *Molecular Genetics and Metabolism*, **84**, 160–166.

Sun, Y., Quinn, B., Xu, Y.H. *et al.* (2006) Conditional expression of human acid beta-glucosidase improves the visceral phenotype in a Gaucher disease mouse model. *Journal of Lipid Research*, **47**, 2161–2170.

Takahashi, K., Naito, M., Yamamura, F. *et al.* (1990) Infantile osteopetrosis complicating neuronal ceroid lipofuscinosis. *Pathology, Research and Practice*, **186**, 697–706.

Tappino, B., Regis, S., Corsolini, F. and Filocamo, M. (2008) An Alu insertion in compound heterozygosity with a microduplication in GNPTAB gene underlies mucolipidosis II. *Molecular Genetics and Metabolism*, **93**, 129–133.

Tomatsu, S., Montano, A.M., Dung, V.C. *et al.* (2009) Mutations and polymorphisms in GUSB gene in mucopolysaccharidosis VII (Sly syndrome). *Human Mutation*, **30**, 511–519.

Triggs-Raine, B., Salo, T.J., Zhang, H. *et al.* (1999) Mutations in HYAL1, a member of a tandemly distributed multigene family encoding disparate hyaluronidase activities, cause a newly described lysosomal disorder, mucopolysaccharidosis IX. *Proceedings of the National Academy of Sciences of the United States of America*, **96**, 6296–6300.

Tunon, T., Guerrero, D., Urchaga, A. *et al.* (2008) Danon disease: a novel LAMP-2 gene mutation in a family with four affected members. *Neuromuscular Disorders*, **18**, 167–174.

Tylki-Szymanska, A., Czartoryska, B., Vanier, M.T. *et al.* (2007) Non-neuronopathic Gaucher disease due to saposin C deficiency. *Clinical Genetics*, **72**, 538–542.

Ushiyama, M., Ikeda, S., Nakayama, J. *et al.* (1985) Type III (chronic) GM1-gangliosidosis. Histochemical and ultrastructural studies of rectal biopsy. *Journal of the Neurological Sciences*, **71**, 209–223.

Valstar, M.J., Bertoli-Avella, A.M., Wessels, M.W. *et al.* (2010) Mucopolysaccharidosis type IIID: 12 new patients and 15 novel mutations. *Human Mutation*, **31**, E1348–E1360.

Verot, L., Chikh, K., Freydiere, E. *et al.* (2007) Niemann-Pick C disease: functional characterization of three NPC2 mutations and clinical and molecular update on patients with NPC2. *Clinical Genetics*, **71**, 320–330.

Vracko, R. and Benditt, E.P. (1974) Manifestations of diabetes mellitus – their possible relationships to an underlying cell defect. A review. *American Journal of Pathology*, **75**, 204–224.

Zeevi, D.A., Frumkin, A. and Bach, G. (2007) TRPML and lysosomal function. *Biochimica et Biophysica Acta*, **1772**, 851–858.

9

Cerebral Autosomal Dominant Arteriopathy with Subcortical Infarcts and Leukoencephalopathy (CADASIL)

John W. Stirling

Centre for Ultrastructural Pathology, IMVS – SA Pathology, Adelaide, Australia

9.1 INTRODUCTION

Cerebral autosomal dominant arteriopathy with subcortical infarcts and leukoencephalopathy (CADASIL) has a highly variable clinical presentation and is the most common form of hereditary vascular dementia. *De novo* mutations also occur (Chabriat *et al.*, 2009; Tikka *et al.*, 2009). Significant symptoms (migraine and cerebrovascular disease) generally manifest at between 30 and 50 years of age, but childhood-onset cases have also been reported (Kalimo *et al.*, 2002; Hartley *et al.*, 2010). CADASIL was recognised as a distinct entity when mutations in the Notch3 gene on chromosome 19 were identified (Joutel *et al.*, 1996). Notch3 encodes a transmembrane receptor ('Notch3') with a 97 kDa intracellular domain and a 210 kDa extracellular domain containing

Diagnostic Electron Microscopy: A Practical Guide to Interpretation and Technique, First Edition. Edited by John W. Stirling, Alan Curry and Brian Eyden.

34 epidermal growth factor–like (EGF-like) repeats (Joutel *et al.*, 2000). At least 180 mutations have been identified in exons 2–24 of the Notch3 gene that code for the EGF-like repeats (Yamamoto *et al.*, 2011). The principal exons affected vary geographically, but most pathogenic mutations are clustered in exons 3–6, 8 and 11, with the majority occurring in exons 3 and 4 (exon 4 predominates) (Chabriat *et al.*, 2009; Tikka *et al.*, 2009; André, 2010; Hervé and Chabriat, 2010). All mutations are associated with the formation of discrete deposits around vascular smooth muscle cells (VSMCs) and pericytes. These deposits, which are composed of granular osmiophilic material (GOM) (Figure 9.1) containing the Notch3 receptor extracellular domain (Ishiko *et al.*, 2006), are pathognomonic of CADASIL and the target of ultrastructural screening. Similar deposits are not found in cerebral autosomal recessive arteriopathy with subcortical infarcts and leukoencephalopathy (CARASIL) (Yamamoto *et al.*, 2011). Mitochondrial dysfunction (mtDNA mutations) (Finsterer, 2007) and mitochondrial structural abnormalities such as paracrystalline inclusions (Malandrini *et al.*, 2002) have also been linked to CADASIL but have no current diagnostic application.

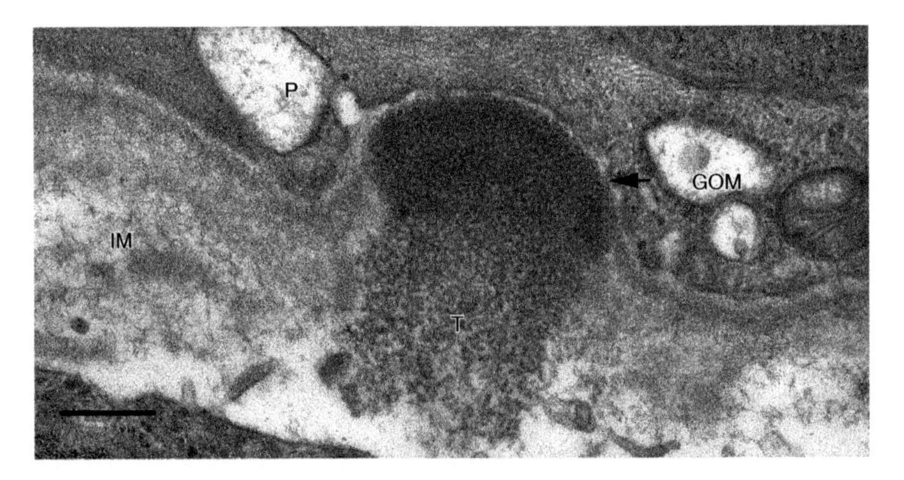

Figure 9.1 A typical deposit of dense granular osmiophilic material (GOM) which is pathognomonic for CADASIL. The GOM sits within an indentation of the pericyte (P) in close contact with the cell wall and with no intervening cell lamina. A characteristic tail of coarse granular material (T) extends out from the head of the deposit into the intercellular matrix (IM), giving the deposit an appearance similar to that of a mushroom or jellyfish. Bar = 250 nm.

9.2 DIAGNOSTIC STRATEGIES – COMPARATIVE SPECIFICITY AND SENSITIVITY

In addition to clinical and radiological criteria, several techniques are available for the diagnosis of CADASIL. Full screening for the mutations in exons 2–24 that code for the 34 EGF-like repeats is generally regarded as the 'gold standard' technique and is variously reported to have 95% specificity and 100% sensitivity (Joutel *et al.*, 2001) or 100% specificity and close to 100% sensitivity (Hervé and Chabriat, 2010). Limited genetic testing for mutation hotspots can also be employed, but this strategy can have a false-negative rate of up to 20% (Joutel *et al.*, 2001). For example, testing of exons 3–5 and 8 gives ~62% coverage, but extending screening to include exons 2, 6, 11 and 18 increases the coverage to ~80% (Tikka *et al.*, 2009). Depending on the patient's place of origin, it is also possible that geographic variability in the exons affected may affect the selection of suitable targets. Immunocytochemical labelling for Notch3 in vessel walls (Joutel *et al.*, 2001) is also reported to be highly sensitive (85–95%) and specific (95–100%) (Chabriat *et al.*, 2009). However, immunocytochemistry may not be sensitive in cases of early disease when only small amounts of deposit are present and labelling is reduced (Tikka *et al.*, 2009). The ultrastructural approach centres on the detection of GOM around small arterioles. Muscle and skin biopsies are both suitable for this purpose (Mayer *et al.*, 1999), but skin biopsy is preferred as it is simple to perform and less intrusive. Transmission electron microscopy (TEM) is also 100% specific for CADASIL, but reports of its sensitivity vary from 45% to 100% (see, respectively, Markus *et al.*, 2002; Mayer *et al.*, 1999). An extensive controlled study by Tikka *et al.* (2009) found sensitivity to be well over 90%. TEM is likely to be more sensitive than immunocytochemistry in early disease as TEM allows the detection of small deposits (Tikka *et al.*, 2009). The youngest individual documented with GOM to date was 20 years of age (Brulin *et al.*, 2002).

9.3 DIAGNOSIS BY TEM

For TEM, the recommended strategy is to take a full thickness punch or excision biopsy of skin from the upper arm (Tikka *et al.*, 2009), and it is prudent to avoid sun-damaged areas. Tissue is processed using a standard TEM protocol (primary fixation in glutaraldehyde followed by secondary fixation in osmium tetroxide and embedding in epoxy resin). The tissue should contain deep dermis or upper subcutis containing medium-sized or small arterioles with an outer diameter of 20–40 μm (Tikka *et al.*, 2009). Although GOM may be found adjacent to quite small vessels (~10 – 15 μm), it is best to screen small arterioles first as they have a high incidence of deposits (GOM is not commonly found adjacent to small capillaries and is only rarely found in association with veins). Suitable arterioles can be identified in toluidine blue semithin sections by the presence of an internal elastic membrane which is seen as dark blue dots adjacent to the vessel (Tikka *et al.*, 2009). In our laboratory, to screen efficiently and to include the possibility of sporadic deposits, we screen the block that contains the largest arterioles first. All vessels (of any diameter) are checked, the proviso being that they have associated pericytes or smooth muscle cells. If the first block is negative, we screen an additional two blocks in the same manner up to a minimum of 50 vessels (i.e. a minimum of 50 vessels with associated pericytes or smooth muscle cells in a total of three blocks). No additional tissue is examined once characteristic deposits have been found. Using this protocol, the majority of positive cases have been found in the first block. Only a few positives have been found where screening had progressed to a second block, and no positive cases have been found where screening had progressed to a third block. In most cases, numerous deposits are found. However, in one instance, while the case was found to be positive in the first block, only a single characteristic deposit was found. No additional deposits were located, despite extensive searching in multiple additional blocks.

At the ultrastructural level, characteristic deposits of GOM are seen as extracellular electron-dense granular material located in small indentations of the cytoplasmic membrane of pericytes and VSMC. Typically, they are oval in shape or similar in appearance to a mushroom or jellyfish with a 'tail' extending into the intercellular matrix (Figures 9.1 and 9.2). Typical deposits are also closely apposed to the cell membrane, often with only a narrow electron-lucent zone intervening and no cell lamina (Figure 9.1). Deposits may also be found adjacent to pericytes and VSMC but beyond the cell lamina and within the intercellular matrix (Figure 9.2) (Kanitakis *et al.*, 2002; Tikka *et al.*, 2009). The number of deposits is reported to increase with the patient's age with a peak at ~50 years. Subsequently, and as damage to vessels increases, the number decreases. Deposits in elderly patients may be rare and difficult to locate (Brulin *et al.*, 2002).

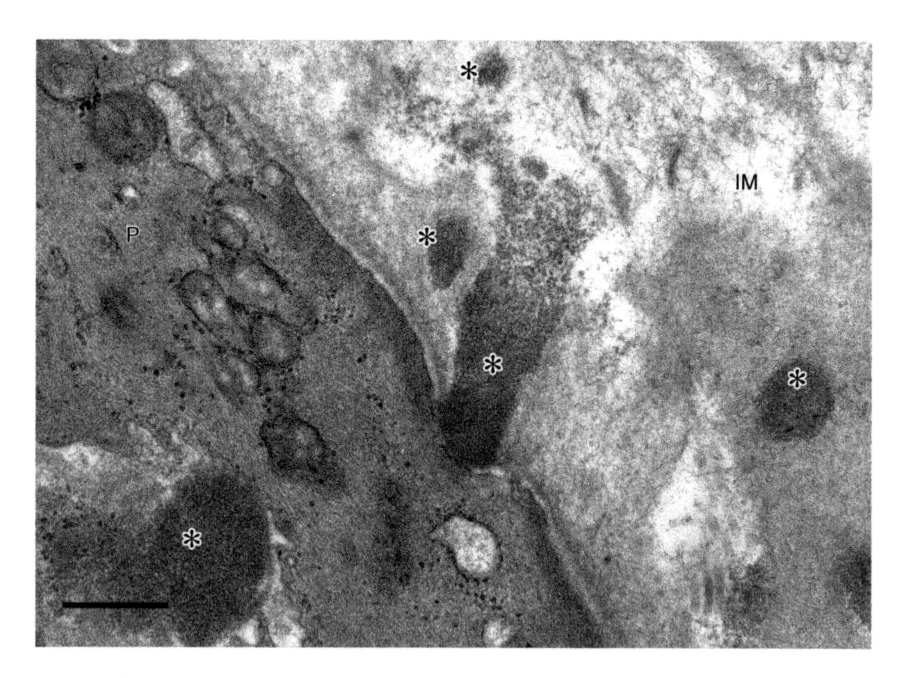

Figure 9.2 Here, deposits of GOM (*) are shown in several locations. Two large typical deposits are seen in close contact with a pericyte (P); the tail of the deposit on the right of the pericyte extends into the intercellular matrix (IM) for some distance. Three other deposits of GOM are sited well within the intercellular matrix on the right and have no obvious contact with a pericyte. Bar = 500 nm.

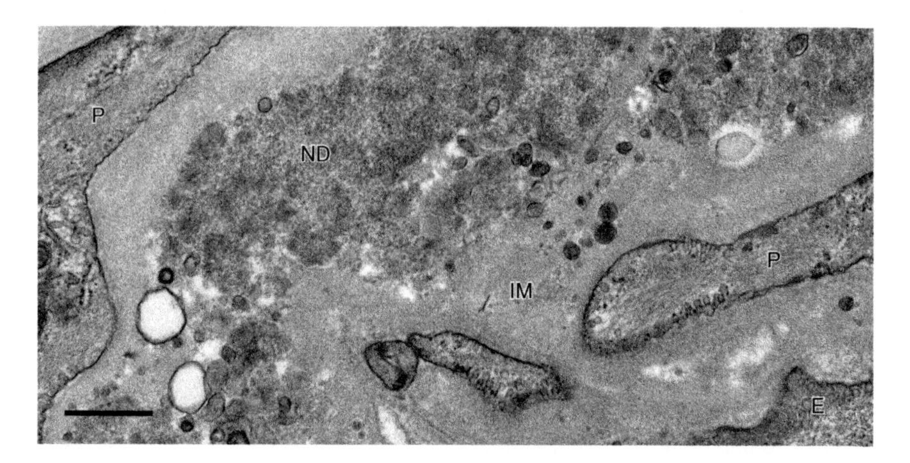

Figure 9.3 A variety of nonspecific deposits (ND) are seen in the intercellular matrix (IM) adjacent to two pericytes (P). E = endothelial cell. The nonspecific deposits include electron-dense granular material, and both dense and pale granules and vesicles. Such nonspecific deposits within the intercellular matrix may be confused with pathognomonic GOM; typical GOM sited in a cell indentation must be located to make a confident diagnosis of CADASIL. Bar = 500 nm.

For a confident diagnosis of CADASIL, it is recommended that characteristic GOM in cell indentations must be found (Figure 9.1). In our laboratory, granular osmiophilic deposits found in the matrix beyond the external lamina (i.e. distant to a pericyte or VSMC and not within a cellular indentation) (Figure 9.2) are regarded as suggestive of CADASIL (rather than diagnostic), and a thorough search for characteristic deposits is made. This strategy is pursued in order to avoid confusion with the wide range of misleading nonspecific material that may be found in the vicinity of blood vessels, including extravascular erythrocytes (which may also be within pericytes), elastin, cell debris and accumulations of granular material of various types and electron density (Figure 9.3) (see also Tikka *et al.*, 2009).

REFERENCES

André, C. (2010) CADASIL: pathogenesis, clinical and radiological findings and treatment. *Arquivos de Neuro-Psiquiatria*, **68** (2), 287–299.

Brulin, P., Godfraind, C., Leteurtre, E. and Ruchoux, M.M. (2002) Morphometric analysis of ultrastructural vascular changes in CADASIL: analysis of 50 skin biopsy specimens and pathogenic implications. *Acta Neuropathologica*, **104**, 241–248.

Chabriat, H., Joutel, A., Dichgans, M. *et al.* (2009) CADASIL. *Lancet Neurology*, **8**, 643–653.

Finsterer, J. (2007) Neuromuscular implications in CADASIL. *Cerebrovascular Diseases*, **24**, 401–404.

Hartley, J., Westmacott, R., Decker, J. *et al.* (2010) Childhood-onset CADASIL: clinical, imaging, and neurocognitive features. *Journal of Child Neurology*, **25** (5), 623–627.

Hervé, D. and Chabriat, H. (2010) CADASIL. *Journal of Geriatric Psychiatry and Neurology*, **23** (4), 269–276.

Ishiko, A., Shimizu, A., Nagata, E. *et al.* (2006) Notch3 ectodomain is a major component of granular osmiophilic material (GOM) in CADASIL. *Acta Neuropathologica*, **112**, 333–339.

Joutel, A., Andreux, F., Gaulis, S. *et al.* (2000) The ectodomain of the Notch3 receptor accumulates within the cerebrovasculature of CADASIL patients. *The Journal of Clinical Investigation*, **105** (5), 597–605.

Joutel, A., Corpechot, C., Ducros, A. *et al.* (1996) Notch3 mutations in CADASIL, a hereditary adult-onset condition causing stroke and dementia. *Nature*, **383** (6602), 707–710.

Joutel, A., Favrole, P., Labauge, P. *et al.* (2001) Skin biopsy immunostaining with a Notch3 monoclonal antibody for CADASIL diagnosis. *The Lancet*, **358**, 2049–2051.

Kalimo, H., Ruchoux, M.M., Viitanen, M. and Kalaria, R.J. (2002) CADASIL: a common form of hereditary arteriopathy causing brain infarcts and dementia. *Brain Pathology*, **12**, 371–384.

Kanitakis, J., Thobois, S., Claudy, A. and Broussolle, E. (2002) CADASIL (cerebral autosomal dominant arteriopathy with subcortical infarcts and leukoencephalopathy): a neurovascular disease diagnosed by ultrastructural examination of the skin. *Journal of Cutaneous Pathology*, **29**, 498–501.

Malandrini, M., Albani, F., Palmeri, S. *et al.* (2002) Asymptomatic cores and paracrystalline mitochondrial inclusions in CADASIL. *Neurology*, **59**, 617–620.

Markus, H.S., Martin, R.J., Simpson, M.A. *et al.* (2002) Diagnostic strategies in CADASIL. *Neurology*, **59**, 1134–1138.

Mayer, M., Straube, A., Bruening, R. *et al.* (1999) Muscle and skin biopsies are a sensitive diagnostic tool in the diagnosis of CADASIL. *Journal of Neurology*, **246**, 526–532.

Tikka, S., Mykkänen, K., Ruchoux, M.M. *et al.* (2009) Congruence between NOTCH3 mutations and GOM in 131 CADASIL patients. *Brain*, **132**, 933–939.

Yamamoto, Y., Craggs, L., Baumann, M. *et al.* (2011) Molecular genetics and pathology of hereditary small vessel diseases of the brain. *Neuropathology and Applied Neurobiology*, **37**, 94–113.

10

Diagnosis of Platelet Disorders by Electron Microscopy

Hilary Christensen[1] **and Walter H.A. Kahr**[1,2]

[1]*Division of Haematology/Oncology, Program in Cell Biology, The Hospital for Sick Children, Toronto, Ontario, Canada*
[2]*Departments of Paediatrics and Biochemistry, University of Toronto, Toronto, Ontario, Canada*

10.1 INTRODUCTION

After erythrocytes, platelets are the next most common cell in human blood, accounting for 150–400 × 10^9 cells per l. Platelets are small anuclear discoid cells measuring 2–4 µm in diameter and 0.5 µm in thickness, with a mean volume of 6–10 fl. Although the primary function of platelets is to establish haemostasis following blood vessel injury, studies suggest that platelets also play roles in development (e.g. closure of ductus arteriosus and lymphatic development), modulating the immune response, fighting infection, healing wounds and regenerating organs (Bertozzi *et al.*, 2010; Echtler *et al.*, 2010; Nurden, 2011; Semple, Italiano and Freedman, 2011).

The human platelet plasma membrane contains important receptors such as the glycoprotein (GP) Ib-IX-V complex and integrin αIIbβ3 (GPIIb-IIIa complex), as well as many others that respond to various platelet-activating molecules to facilitate platelet plug formation and fibrin deposition during blood coagulation. The plasma membrane contains pores that connect to the open canalicular system (OCS), an internal

Diagnostic Electron Microscopy: A Practical Guide to Interpretation and Technique, First Edition. Edited by John W. Stirling, Alan Curry and Brian Eyden.

membrane system that is mobilised when platelets spread on surfaces. The discoid shape of a resting platelet is maintained by a cytoskeleton made up of proteins that include spectrin, actin and tubulin, which interconnect the plasma membrane with the cytoplasm and form dynamic marginal microtubule bands (Patel-Hett *et al.*, 2008, 2011). The dense tubular system (DTS) is a distinct membrane compartment intertwined with the OCS as visualised with electron microscopy (EM) tomography (Van Nispen Tot Pannerden *et al.*, 2010). Platelets contain three major types of organelles that secrete their contents upon activation: (i) α-granules, (ii) dense (δ)-granules and (iii) lysosomes. Other platelet organelles include mitochondria, peroxisomes, occasional multivesicular bodies and Golgi complexes – often prominent in granule-deficient platelets. Since α-granules contain multiple proteins involved in blood clotting (e.g. factor V, von Willebrand factor (VWF) and fibrinogen) and δ-granules contain small molecules involved in platelet activation (e.g. adenosine diphosphate (ADP)), deficiency in either class of granules can cause bleeding.

EM continues to play an important role in the clinical evaluation of platelet disorders (Nurden *et al.*, 2000; Schmitt *et al.*, 2001; White, 2004, 2007b; Cramer and Fontenay, 2006). EM has proven to be a valuable diagnostic tool in disorders characterised by abnormal platelet size, α- or δ-granule deficiencies, cytoplasmic inclusions and membrane defects. In cases of suspected δ-granule deficiencies, a rapid whole-mount technique (see Section 10.2) requiring minimal preparation can provide a quick, reliable platelet dense-granule count. Dense-granule visualisation is possible because of the granules' high concentrations of calcium and polyphosphate, which produce opacity when the granules are visualised with EM. By counting a minimum of 50 well-spread platelets ($\times$15 000–50 000 magnification), the mean number of dense granules per platelet can be readily determined (normally 3–8 per cell – McNicol and Israels, 1999). Transmission electron microscopy (TEM) blocks are usually prepared together with whole-mount grids for simultaneous platelet ultrastructure and δ-granule analysis.

10.2 TEM PREPARATION OF PLATELETS

Since the most important determinant of evaluating platelets by EM is the cell preparation, the methodology is described in detail here. Platelets are readily activated resulting in shape changes and degranulation; therefore, initial handling of unfixed platelets is critical for meaningful analysis. Blood samples must be kept at room temperature before final fixation; and for whole-mount platelet preparation, anticoagulants such as ethylenediaminetetraacetic acid (EDTA) and hirudin should be avoided because they inhibit platelet spreading. Rather, 3.2% sodium citrate or heparin is recommended. Spreading allows the electron beam to penetrate the cytoplasm of each platelet so that clear imaging of the δ-granule contents is possible; if the sample consists mostly of rounded forms, accurate quantification becomes impossible. Platelets are separated from other blood cells by centrifugation at $150 \times g$ ($100 \times g$ for large or giant platelets) for 10 minutes at room temperature. The upper layer, containing the platelet-rich plasma (PRP), is removed carefully using a polypropylene device (not glass), and a 100–200 μl aliquot is set aside for whole-mount grids. The remaining PRP is transferred to a 1.5 ml Eppendorf tube and spun for 20 minutes at $1000 \times g$ to obtain a platelet pellet. The plasma supernatant is then carefully removed before adding a 2.5% glutaraldehyde–phosphate buffered saline (PBS) fixative to the platelet pellet at the bottom of the tube. A fixed platelet pellet remains stable for months and can either be stored at 4 °C until further processing or transferred to a specialised EM facility if the sample was obtained from another hospital. All EM processing including polymerisation is carried out directly inside the Eppendorf tube to avoid cell loss, especially in cases of thrombocytopaenia. An optional method that does not limit the sample to a pure platelet population involves cell separation on a Ficoll–Hypaque gradient. As in the previous method, there is a 5–10-minute preliminary spin at $150 \times g$ to collect an aliquot of PRP from the upper plasma layer for whole mount. The separated blood sample is centrifuged once more at $1000 \times g$ for 20 minutes to bring platelets in the upper layer down into the mononuclear cell layer below, creating a platelet-rich mononuclear cell layer situated just above the gradient. Most of the upper plasma layer is discarded so that the platelet-rich cell layer can be collected and re-suspended in PBS. This method provides additional diagnostic information with regard to inclusions and other changes that may occur within the surrounding cell population, and it also produces a higher cell number that can be processed according to either of the following methods.

The first method involves concentrating the PBS suspension by re-centrifugation 10 minutes at $800 \times g$. Cells are then re-suspended in 1 ml of PBS buffer and aliquotted into one or two BEEM capsules (depending on cell concentration). BEEM capsule trays are spun so that pellets form at the bottom of each BEEM capsule. These cell pellets should be left **undisturbed** and fixed as individual pellets in the BEEM moulds. All steps of routine EM processing can be carried out inside each of the BEEM capsules right up to the final Epon resin without having to spin between each step. This method allows minimal cell loss during processing. Depending on the cell count, it may be necessary to repeat the re-suspension step within each BEEM capsule and further aliquot the suspension into more capsules if the cell pellets exceed 1 mm in size, and to re-centrifuge so that the desired pellet size is obtained. Conversely, if cell pellets are too small and do not cover the base of the BEEM capsule, remove some of the PBS from each capsule, resuspend the incomplete pellets and transfer cells into a single BEEM or conical embedding capsule to concentrate the sample. When cells are in the final resin stage, BEEM capsules can be transferred directly to a 60 °C oven for polymerisation.

If the cell count is high, unfixed cells in PBS or cells fixed in suspension are given a final centrifugation step to create a pellet to which 2.5% glutaraldehyde–PBS is added. However, because the fixed pellet must eventually be scraped from the bottom of the tube to create pellet pieces that are processed and embedded as tissue fragments, some cells will be lost with this method.

Whether fixation is carried out in a tube or inside BEEM capsules, pellet size should never be thicker than 1 mm; otherwise, both glutaraldehyde and osmium fixatives will not penetrate properly.

10.3 WHOLE-MOUNT EM PREPARATION OF PLATELETS

Place 5 or 6 Formvar-coated grids on a sheet of Parafilm. Using a fine-tipped pipette, place a drop of ~25 μl PRP to form a meniscus on top of each Formvar-coated grid. Allow platelets to spread on the Formvar surface for 3 minutes. **Blot each grid thoroughly** by touching the edge of the grid with a piece of filter paper to remove all excess plasma. Excess plasma on the surface of the grid will blanket the platelets under plasma protein debris, contaminating the sample and making it impossible to visualise the δ-granules. Following the blotting step, briefly fix each grid under a drop of 0.5% glutaraldehyde–PBS for 1 minute. Dip-rinse grids with distilled water three times, then blot until they are dry. Store grids on filter paper inside a culture dish together with a specimen label. No further processing is necessary. The polyphosphate–calcium content of electron-dense granules makes them clearly identifiable from their surrounding cytoplasm. To optimise the imaging of smaller sized δ-granules or granules situated at varying focal planes, use the charge-coupled device (CCD) camera on the microscope together with maximum beam current. Results on platelet dense-granule deficiency using whole-mount EM methods show excellent agreement amongst different technicians when performed in experienced centres (Hayward *et al.*, 2009).

Note:

Grids covered with large empty Epon sections (sections without embedded tissue) may be used in place of Formvar for whole-mount supports. Other whole-mount methods describe shorter incubation or spread times of 1–2 minutes **without fixation.**

Quality Control:

1. Establish a range of normal controls. The usual range is 3–8 dense granules per platelet for adult and paediatric samples (neonates may be lower).
2. Avoid electron-dense rounded platelet forms and platelet clusters.
3. Compare counts on the same sample done by different technicians.
4. Compare counts done on the same patient at different times and clinics.

10.4 EM PREPARATION OF BONE MARROW

Platelet studies may also involve a bone marrow biopsy to study associated megakaryocyte morphology. From the first pull (maximum 1 ml), one or two drops of bone-marrow aspirate are quickly added to a vial of 2.5% glutaraldehyde–PBS fixative either before or after slides are prepared. Bone-marrow spicules are then isolated from this solution, transferred to fresh fixative and processed as tissue pieces. Otherwise, bone-marrow blood is separated on a Ficoll–Hypaque gradient at a gentle spin of $150 \times g$ for 15 minutes to capture single megakaryocytes within the mononuclear layer. Ficoll–Hypaque separation is recommended for paediatric samples that often have either smaller or absent spicules. Bone-marrow core biopsies can also be dissected into small fragments ($<1\,mm^3$) and fixed as tissue pieces.

10.5 PRE-EMBED IMMUNOGOLD LABELLING OF VON WILLIBRAND FACTOR IN PLATELETS

The following protocol outlines a method for pre-embed surface labelling of any cell suspension; it is carried out at room temperature **before final fixation and routine TEM processing.** Cells are first separated by centrifugation on a Ficoll–Hypaque gradient followed by a brief 0.1% glutaraldehyde–PBS pre-fixation in suspension for no longer than 10 minutes in order to stabilise the plasma membrane. Subsequently, all rinses and incubations are done in suspension, and cells are pelleted at 200 × *g* between each step. Preliminary rinses in 0.1% glycine–PBS and 2% bovine serum albumin (BSA)–PBS are carried out before cells are incubated in primary antibody (DakoCytomation rabbit antihuman VWF; Code No. A 0082) at a concentration of 1 : 20 to 1 : 50 in 2% BSA–PBS for 60 minutes, followed by three washes in 2% BSA–PBS. Next, cells are incubated in secondary antibody conjugated with 5–10 nm colloidal gold diluted at 1 : 20 in 2% BSA–PBS for 60 minutes. Cells are then given three rinses in PBS, re-suspended in 1 ml of buffer and pelleted inside BEEM capsules or an Eppendorf tube and stored overnight at 4 °C in 2.5% glutaraldehyde–PBS fixative. Next day, cells are post-fixed with 1% osmium tetroxide in distilled water followed by routine TEM processing. Although pre-embed or pre-fixation immunogold labelling is limited to surface markers, it has the advantage of specificity without background, unlike post-embed techniques used for labelling thin sections. Post-embed protocols can also be used for recognising VWF, but these methods involve special embedding techniques and specimen handling.

10.6 ULTRASTRUCTURAL FEATURES OF PLATELETS

It is always best to view samples at lower magnification before focussing on isolated details. Ultrastructural descriptions include observations on platelet size, the presence or absence of granules, lysosomal bodies or fused granules, microtubule arrangement, membrane disruption, cytoplasmic inclusions and prominent dense tubular membrane systems. Although it is important to be aware that some of these features may not in themselves be diagnostically relevant, changes should be noted according to their frequency. However, distinctive features, particularly the absence of granules or the presence of platelet agglutinates or circulating megakaryocytes with megakaryocyte fragments, are diagnostically specific.

10.7 NORMAL PLATELETS

Control samples taken from normal donors not only aid in the technical aspects of laboratory practice but also serve as standard reference points in comparison to clinical samples. Ultrastructural variations are to be expected even within normal populations. For example, there is a certain degree of heterogeneity in platelet size, particularly in adult samples. Some platelets may have occasional large glycogen deposits or lipid inclusions that are not of any diagnostic significance. Platelet activation is another factor that also results in morphological changes. Resting platelets with a diameter of 2–3 µm have a smooth, spherical or disc shape, whereas activated platelets have an irregular cell surface that coincides with the concentration of cytoplasmic organelles in the centre of the platelet due to the rearrangement of the cytoskeleton. α-granules and δ-granules represent the defining morphological features of platelets and megakaryocytes. Not to be confused with mitochondria of similar size and density, membrane-bound α-granules measuring 200–500 nm in diameter number anywhere from 0 to more than 25 per platelet thin section and often contain a dense nucleoid with a 'bull's-eye' appearance (Figure 10.1). VWF can be found in the submembrane zone of the α-granule. Elongated α-granules may branch out and fuse with the OCS or with other α-granules to form larger lysosomal bodies. Rare agranular and organelle-free platelets are of no diagnostic importance and are more often observed in neonate samples. δ-granules or dense bodies are electron-dense membrane-bound granules that often have a surrounding clear halo. Fewer in number than α-granules, they are not necessarily observed in thin section. Based on whole-mount observation, dense granules appear sharply rounded, some having long tail-like extensions (Figure 10.2). A dense-granule count of fewer than 2 per platelet is often found in δ-granule deficiencies and neonate samples. In addition to α- and δ-granules, another unique morphological feature of the platelet is the OCS. Intertwined with the DTS and contiguous with the plasma membrane, the OCS (Figure 10.1) forms a distinct membrane complex which may at times appear somewhat distended as a result of preparation artefact or may contain flocculent-appearing

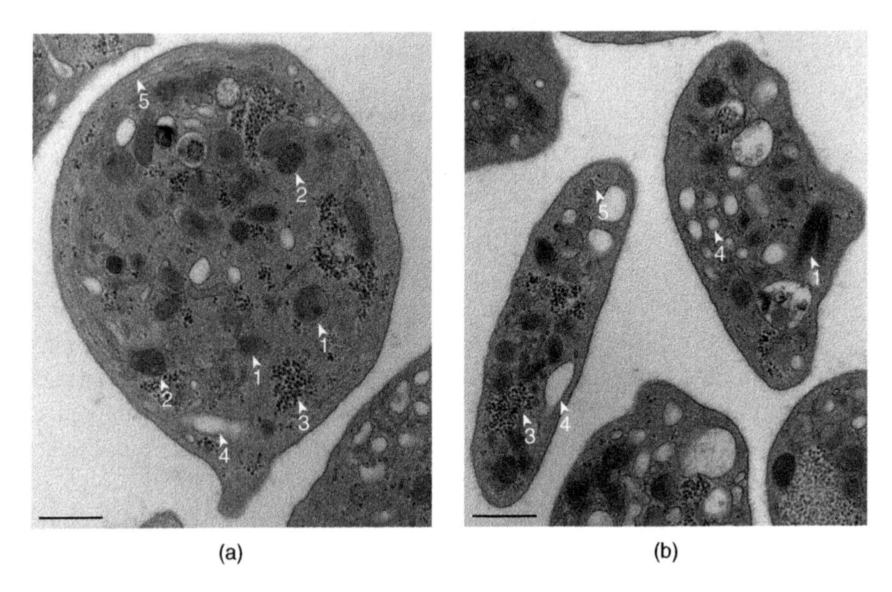

Figure 10.1 Normal platelets. Thin sections are shown for cells cut through the equatorial plane (a) and perpendicular plane (b). Magnification is 30 000×. The black bar represents 500 nm. Platelet structures are indicated by white arrows: (1) α-granules, (2) mitochondria, (3) glycogen stores, (4) open canalicular system and (5) microtubule coil.

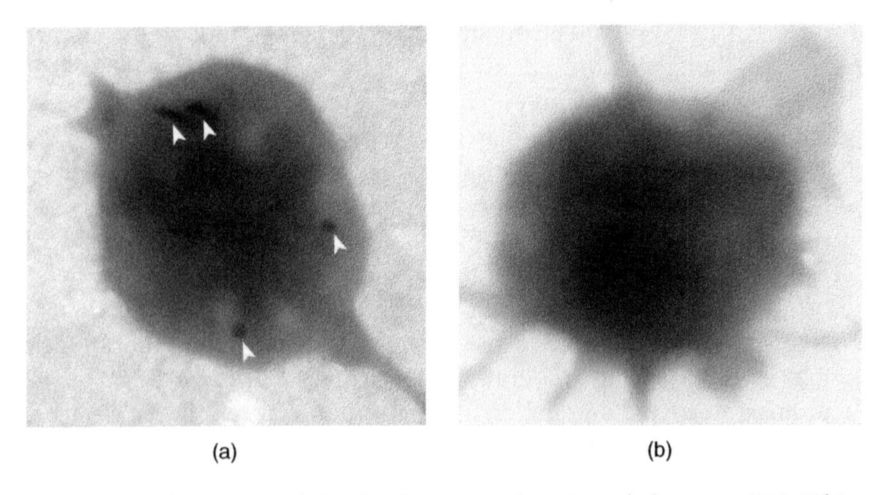

Figure 10.2 Assessment of platelet dense granules using whole-mount EM. White arrows show electron-dense δ-granules seen in normal platelets (a) which are absent in platelets from a patient with Jacobsen syndrome (b). Magnification is 40 000×.

plasma constituents. A platelet sectioned in the equatorial plane will show parallel circumferential microtubules measuring 25 nm in diameter (Figure 10.1b) that are also responsible for maintaining platelet shape. Other cytoplasmic features include glycogen deposits of varying size, free ribosomes with short profiles of smooth endoplasmic reticulum and round or elongated mitochondria that can be distinguished from surrounding α-granules by the presence of a double membrane with internal cristae (Figure 10.1). Lipid inclusions are not unusual but may appear more frequently in abnormal platelets.

10.8 GREY PLATELET SYNDROME

A blood film showing pale-appearing platelets that are large and reduced in number is suggestive of grey platelet syndrome (GPS). The diagnosis is verified by platelet EM, which reveals absent α-granules (Figure 10.3). Instead, the cytoplasm of each platelet contains many empty vacuoles that suggest residual empty granule membranes or incomplete granules unable to store α-granule proteins. Dense granule counts in whole-mount samples are usually in the normal range. Mutations in *NBEAL2* cause GPS (Albers *et al.*, 2011; Gunay-Aygun *et al.*, 2011; Kahr *et al.*, 2011).

10.9 ARTHROGRYPOSIS, RENAL DYSFUNCTION AND CHOLESTASIS SYNDROME

As with GPS, the blood film in arthrogryposis, renal dysfunction and cholestasis (ARC) syndrome shows large pale-appearing platelets, but here the platelet counts are normal. EM confirms the absence of α-granules and the presence of large-appearing platelets (Figure 10.4). Platelets are filled with numerous cytoplasmic vacuoles. Dense granules are often present in the plane of section because of their high numbers. For this reason, accurate dense-granule quantification of ARC whole-mount EM preparations is often difficult, as some have more than 50 per platelet. Mutations in *VPS33B* are associated with absent platelet α-granules (Lo *et al.*, 2005).

10.10 JACOBSEN SYNDROME

Jacobsen syndrome is caused by partial deletion of the long arm of chromosome 11. Platelets from Jacobsen syndrome patients have low

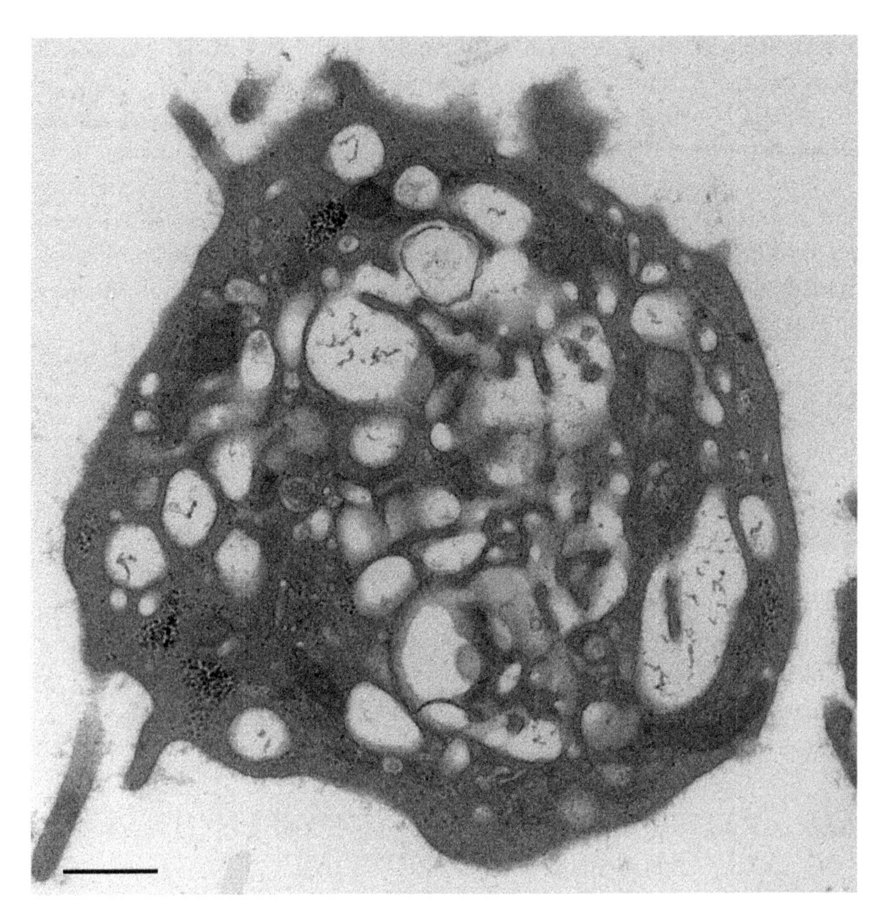

Figure 10.3 Platelet from a patient with grey platelet syndrome (GPS). A large vacuolated cell is shown; GPS platelets lack α-granules but do contain mitochondria (note cristae). Magnification is 30 000×. The black bar represents 500 nm.

dense-granule counts in whole-mount preparations (White, 2007a) as shown in Figure 10.2. Another prominent feature of Jacobsen syndrome is an increase in the number of typically large α-granules that are appearing to either fuse or separate (Figure 10.5). Varying core densities from a single bull's-eye to multiple densities observed in some α-granules possibly represent differences in the distribution of α-granule proteins. Some α-granules measure close to four times the average size (200 nm), while others have the appearance of connected channels. Another α-granule feature is the occasional blebbing of the granule membrane. Membrane

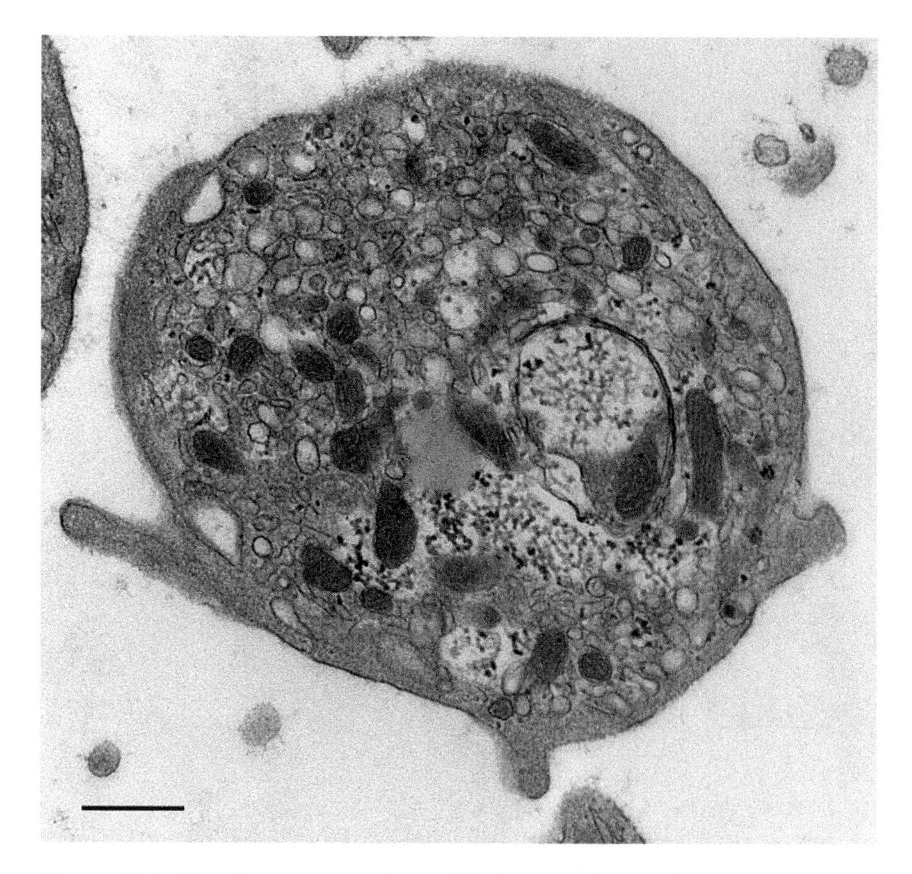

Figure 10.4 Platelet from a patient with ARC syndrome due to mutations in *VPS33B*. A large platelet lacking α-granules but containing multiple mitochondria with cristae is shown. Magnification is 30 000×. The black bar represents 500 nm.

defects appear in some specific areas of the cytoplasm as well as in the breaking apart of the OCS membranes. The OCS often appears as a system of rounded channels giving the cytoplasm a vesicular appearance. Microtubule fragmentation is also evident.

10.11 HERMANSKY–PUDLAK SYNDROME, CHEDIAK–HIGASHI SYNDROME AND OTHER DENSE-GRANULE DEFICIENCIES

Patients presenting with oculocutaneous albinism may have a dense-granule deficiency which is associated with Hermansky–Pudlak and

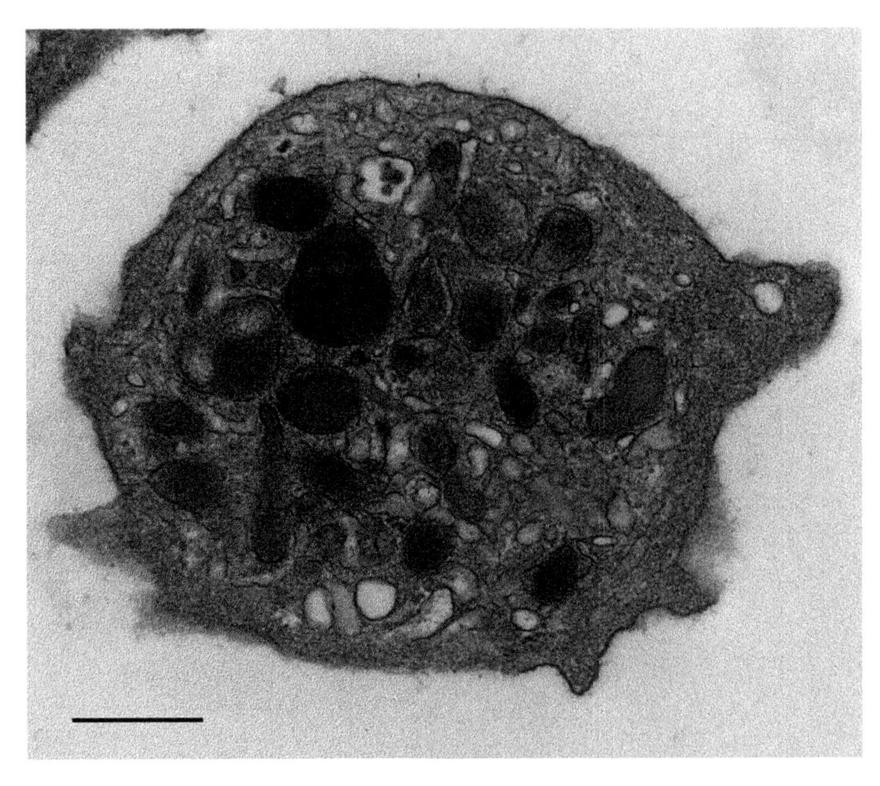

Figure 10.5 Platelet from a patient with Jacobsen syndrome. The cell shown contains α-granules that are abnormally large and appear to fuse or separate. Magnification is 40 000×. The black bar represents 500 nm.

Chediak–Higashi syndromes. The rapid determination of platelet dense-granule deficiency in such patients narrows the diagnostic spectrum since oculocutaneous albinism can be caused by many genetic defects. There are also nonsyndromic isolated cases of dense-granule deficiencies. The whole-mount EM preparations are required to verify the absence of platelet dense granules. The α-granule content and other platelet ultrastructural features are usually normal in these conditions.

10.12 TYPE 2B VON WILLEBRAND DISEASE AND PLATELET-TYPE VON WILLEBRAND DISEASE

Type 2B von Willebrand disease (VWD) and platelet-type von Willebrand disease (PT-VWD) are the result of a gain-of-function interaction between VWF and platelets due to mutations in either VWF (Type 2B

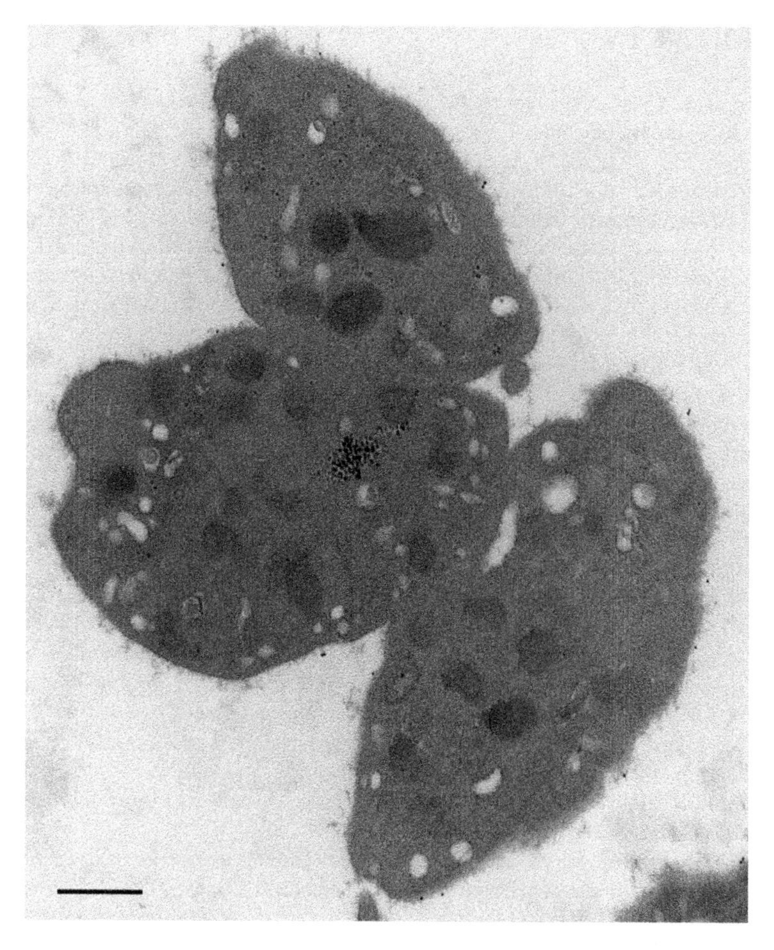

Figure 10.6 Platelets from patient with type 2B VWD. Platelets are agglutinated with each other, and VWF complexes are detected on the platelet surface by VWF immunogold labelling (black dots). Magnification is 25 000×. The black bar represents 500 nm.

VWD) or the cognate receptor on platelets (GPIb-IX-V; PT-VWD). This results in platelet development abnormalities including large or giant platelets (>5 μm) that are decreased in number and are often bound to each other (Nurden *et al.*, 2000). The distinguishing EM features are fused platelet agglutinates (not clots) of varying sizes with protein bridges forming intimate plasma membrane contact points (Figure 10.6; Type 2B VWD). VWF complexes on the platelet surface and sometimes within the OCS are noted by VWF immunogold labelling. In some cases, large α-granules appear within a heterogeneous platelet population.

REFERENCES

Albers, C.A., Cvejic, A., Favier, R. *et al.* (2011) Exome sequencing identifies NBEAL2 as the causative gene for gray platelet syndrome. *Nature Genetics*, **43**, 735–737.

Bertozzi, C.C., Schmaier, A.A., Mericko, P. *et al.* (2010) Platelets regulate lymphatic vascular development through CLEC-2-SLP-76 signaling. *Blood*, **116**, 661–670.

Cramer, E.M. and Fontenay, M. (2006) Platelets: structure related to function, in *Hemostasis and Thrombosis: Basic Principles and Clinical Practice*, 5th edn (eds R.W. Colman, V.J. Marder, A.W. Clowes *et al.*), Lippincott Williams Wilkins, Philadelphia, PA, pp. 463–481.

Echtler, K., Stark, K., Lorenz, M. *et al.* (2010) Platelets contribute to postnatal occlusion of the ductus arteriosus. *Nature Medicine*, **16**, 75–82.

Gunay-Aygun, M., Falik-Zaccai, T.C., Vilboux, T. *et al.* (2011) NBEAL2 is mutated in gray platelet syndrome and is required for biogenesis of platelet alpha-granules. *Nature Genetics*, **43**, 732–734.

Hayward, C.P., Moffat, K.A., Spitzer, E. *et al.* (2009) Results of an external proficiency testing exercise on platelet dense-granule deficiency testing by whole mount electron microscopy. *American Journal of Clinical Pathology*, **131**, 671–675.

Kahr, W.H., Hinckley, J., Li, L. *et al.* (2011) Mutations in NBEAL2, encoding a BEACH protein, cause gray platelet syndrome. *Nature Genetics*, **43**, 738–740.

Lo, B., Li, L., Gissen, P. *et al.* (2005) Requirement of VPS33B, a member of the Sec1/Munc18 protein family, in megakaryocyte and platelet α-granule biogenesis. *Blood*, **106**, 4159–4166.

McNicol, A. and Israels, S.J. (1999) Platelet dense granules: structure, function and implications for haemostasis. *Thrombosis Research*, **95**, 1–18.

Nurden, A.T. (2011) Platelets, inflammation and tissue regeneration. *Thrombosis and Haemostasis*, **105** (Suppl. 1), S13–S33.

Nurden, P., Chretien, F., Poujol, C. *et al.* (2000) Platelet ultrastructural abnormalities in three patients with type 2B von Willebrand disease. *British Journal of Haematology*, **110**, 704–714.

Patel-Hett, S., Richardson, J.L., Schulze, H. *et al.* (2008) Visualization of microtubule growth in living platelets reveals a dynamic marginal band with multiple microtubules. *Blood*, **111**, 4605–4616.

Patel-Hett, S., Wang, H., Begonja, A.J. *et al.* (2011) The spectrin-based membrane skeleton stabilizes mouse megakaryocyte membrane systems and is essential for proplatelet and platelet formation. *Blood*, **118**, 1641–1652.

Schmitt, A., Guichard, J., Masse, J.M. *et al.* (2001) Of mice and men: comparison of the ultrastructure of megakaryocytes and platelets. *Experimental Hematology*, **29**, 1295–1302.

Semple, J.W., Italiano, J.E. and Freedman, J. (2011) Platelets and the immune continuum. *Nature Reviews Immunology*, **11**, 264–274.

Van Nispen Tot Pannerden, H., De Haas, F., Geerts, W. *et al.* (2010) The platelet interior revisited: electron tomography reveals tubular alpha-granule subtypes. *Blood*, **116**, 1147–1156.

White, J.G. (2004) Electron microscopy methods for studying platelet structure and function. *Methods in Molecular Biology*, **272**, 47–63.
White, J.G. (2007a) Platelet storage pool deficiency in Jacobsen syndrome. *Platelets*, **18**, 522–527.
White, J.G. (2007b) Platelet structure, in *Platelets*, 2nd edn (ed. A.D. Michelson), Amsterdam, Elsevier Inc., pp. 45–73.

11

Diagnosis of Congenital Dyserythropoietic Anaemia Types I and II by Transmission Electron Microscopy

Yong-xin Ru

Institute of Haematology & Blood Diseases Hospital, Chinese Academy of Medical Sciences and Peking Union Medical College, Tianjin, China

11.1 INTRODUCTION

The congenital dyserythropoietic anaemias (CDAs) are a group of heterogeneous hereditary diseases characterised by dysplasia of erythroid precursors and ineffective erythropoiesis. The dysplasia includes abnormal morphology and shortened lifespan of nucleated red blood cells in erythroid development and maturation. Ineffective erythropoiesis is marked by the synchronous occurrence of predominantly erythroblastic proliferation and anaemia. Seven types of CDA have been described so far, designated as types I to VII (Crookston *et al.*, 1996; Wickramasinghe, 1997). CDA types I, II and III are the most commonly encountered, and their genetic characteristics have been the most frequently studied (Lind *et al.*, 1995; Iolascon *et al.*, 1996; Tamary *et al.*, 1996). The other types are rare and remain incompletely documented and understood.

Diagnostic Electron Microscopy: A Practical Guide to Interpretation and Technique, First Edition. Edited by John W. Stirling, Alan Curry and Brian Eyden.

In this chapter, the morphological and ultrastructural characteristics of erythroblasts in the various development stages of CDA are described, with a focus on types I and II.

11.2 PREPARATION OF BONE MARROW AND GENERAL OBSERVATION PROTOCOL

Routine transmission electron microscopy (TEM) requires about 10^6 nucleated cells from 3 to 5 ml of anticoagulated bone-marrow aspirate. This is fixed in 2% glutaraldehyde in phosphate buffered saline (pH 7.4) for 1 hour, and processed into epoxy resin according to standard procedures (Pontvert-Delucq, Breton-Gorius and Schmitt, 1993). Floating bone-marrow particles are especially valuable since they often contain complete erythropoietic islands, the examination of which facilitates the diagnosis of CDA and the differentiation between the various anaemias. Also, the observation of bone-marrow slides by light microscopy (LM) is a necessary step carried out before TEM, to exclude such frequent conditions as aplastic anaemia and myelodysplastic syndrome.

11.3 CDA TYPE I

Over 150 cases of CDA type I have been reported – more than 90 cases (mostly with familial incidence) from the Gulf region, more than 40 from Germany and the remaining cases scattered globally (e.g. in Japan and India). More than 10 cases have been reported from China, and one of them was confirmed by DNA sequencing to possess similar mutations to those of the Gulf region cases (Ru *et al.*, 2008). This small Chinese cohort, given the huge population of China, suggests the possibility of under-diagnosis and a role for TEM.

CDA type I patients present with features which include anaemia, splenomegaly, jaundice, occasionally limb deformities and a negative acid haemolytic test. The diagnosis is established from clinical features, morphology and especially ultrastructural examination. Gene sequencing can enhance diagnostic accuracy.

11.3.1 Proerythroblasts and Basophilic Erythroblasts

The enlarged proerythroblasts and basophilic erythroblasts (18–25 μm diameter) in CDA type I have small invaginations of the cell surface membrane and vesicles in the peripheral cytoplasm, and the cell surface also has occasional processes. The distinction between proerythroblasts and basophilic erythroblasts is not always definite, so in CDA type I they are difficult to distinguish by LM because both undergo megalocytic change. Proerythroblasts and basophilic erythroblasts are predominantly euchromatic, apart from a few cells containing a narrow band of heterochromatin on the inner nuclear membrane. Basophilic erythroblasts show denser cytoplasm than proerythroblasts on account of their greater content of haemoglobin, although the appearance is similar with regard to the size of the karyoplasm and nucleolus. Features of nuclear damage (see Section 11.3.2) are often found in these two stages.

11.3.2 Polychromatic and Orthochromatic Erythroblasts

The cytoplasm of polychromatic and orthochromatic erythroblasts in CDA type I is denser than that of their progenitors but the same as that of their normal counterpart. Some erythroblasts show irregular surfaces with protrusions or finger-like processes. Typical nuclear injury – nuclear pore widening, nuclear membrane disruption or dissolution, the so-called Swiss cheese-like nucleus as well as internuclear chromatin bridges – all frequently occur in polychromatic and orthochromatic erythroblasts, and only occasionally in proerythroblasts and basophilic erythroblasts (Figure 11.1).

The Swiss cheese-like or sponge-like nucleus is a distinctive ultrastructural characteristic, composed of condensed chromatin with gross cords or a network-like appearance of heterochromatin interspersed between small electron-lucent dot-like areas within a large nucleus. These clear areas are paler than the cytoplasm. It is presumed that the Swiss cheese-like or sponge-like nucleus may result from uneven heterochromatin shrinkage.

It has been thought that haemoglobin in CDA type I could penetrate the nucleus, leading to early nuclear maturation or condensation when the nuclear pore was widened or disrupted by fundamental pathogenesis (Stohlman, 1970; Heimpel, 1977). In fact, it is usually difficult to distinguish a widened nuclear pore from a focus of nuclear envelope disruption or dissolution. According to my own observation, some nuclear pores appear to be wider than normal, but some illustrate nuclear membrane dissolution. Others show some residual nuclear envelope structure (Figure 11.2c). The erythroblasts are more significantly damaged when cytoplasm comes into contact with the karyoplasm and haemoglobin

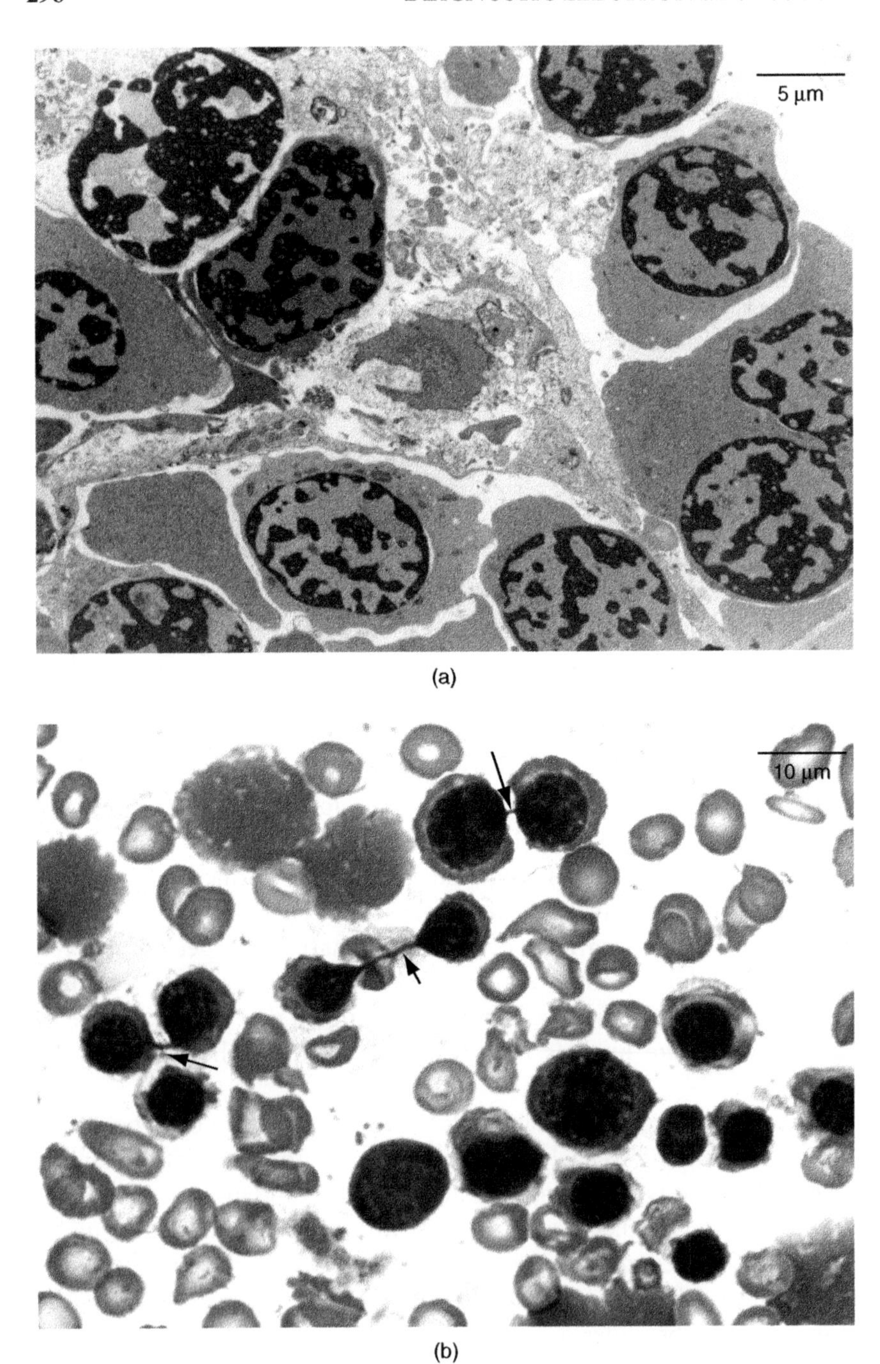

Figure 11.1 (a) Most polychromatic and orthochromatic erythroblasts have Swiss cheese-like or sponge-like nuclei. (b) Three inter-nuclear chromatin bridges are seen (arrows) between orthochromatic erythroblasts.

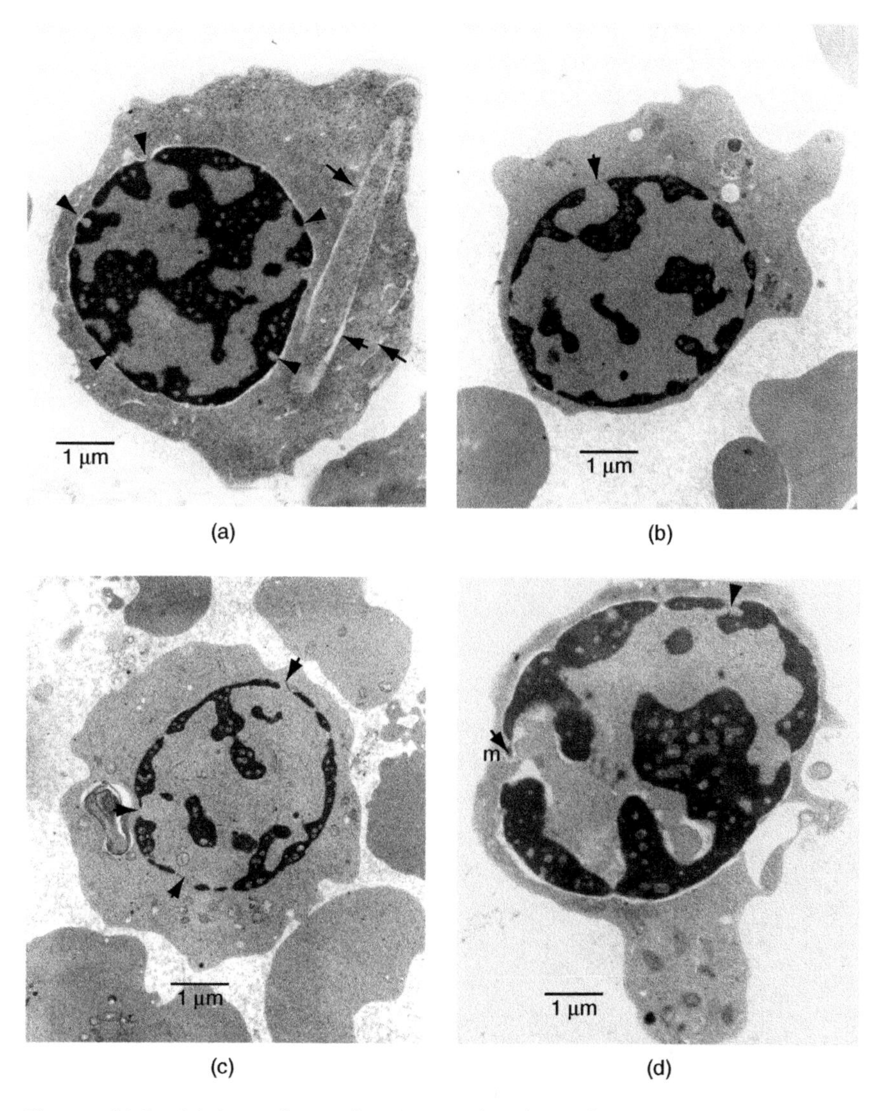

Figure 11.2 (a) A nucleus showing widened nuclear pores (arrowheads). (b) A nucleus with dissolved nuclear membrane and its remnant (arrowheads). (c, d) Focally absent nuclear membrane (arrowheads) and a mitochondria at the entrance.

accesses the nucleus through a widened nuclear pore and damaged nuclear envelope (Figure 11.2a–d).

Other nuclear abnormalities in polychromatic and orthochromatic erythroblasts include a circular, clear cleft around the nucleus, leaving just a little cytoplasm on the nuclear surface (Figure 11.3a), while

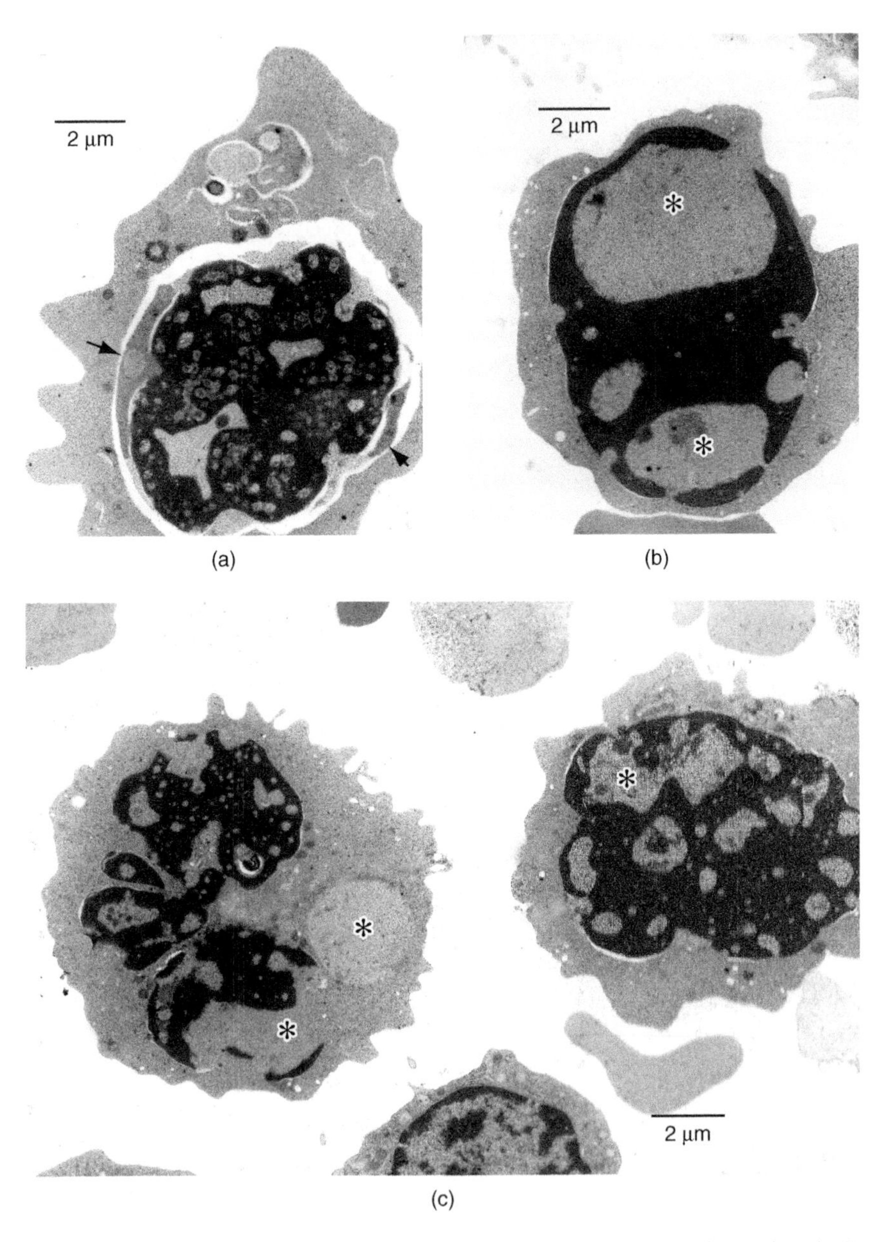

Figure 11.3 (a) A nucleus surrounded by a fissure-like clear circle, with a little cytoplasm remaining on its surface (arrow). (b, c) Two nuclei showing sponge-like changes and the dissolution process (asterisks).

chromatin condensation and karyorrhexis often occur and are associated with nuclear sponge-like changes (Figure 11.3b, c). The inter-nuclear chromatin bridge is a significant marker in the diagnosis of CDA type I, and it is often demonstrable by LM but difficult to encounter in appropriate orientation in TEM (Figure 11.1b). Double nuclei in erythroblasts often occur and occasionally also exhibit sponge-like changes (Figure 11.4a).

Cytoplasmic abnormalities seen by TEM in CDA type I include mitochondria with iron overloading, characterised by the presence of ferritin particles or granules in late-stage orthochromatic erythroblasts (Figure 11.4b). Rough endoplasmic reticulum (rER) sometimes has a degenerated appearance, presenting as an electron-lucent line, indicating the possibility of an overall destruction of the cell membrane system – nuclear, rER and cell surface membranes alike (Figure 11.2a).

The incidence of these abnormalities in polychromatic and orthochromatic erythroblasts varies with the patient. The cytoplasm of orthochromatic erythroblasts is denser than in polychromatic erythroblasts on account of the increasing haemoglobin content, and there are more siderosomes (as seen by LM) in orthochromatic erythroblasts as iron accumulates in mitochondria. Focal necrosis of cytoplasm (e.g. vacuolation) and karyolysis of erythroblasts seldom occur in CDA type I, which marks it as distinct from other anaemias such as haemolytic anaemia and thalassaemia.

11.3.3 Reticulocytes and Erythrocytes

Although most reticulocytes and erythrocytes are similar to their normal counterparts, some of them are macrocytotic. Poikilocytes, elliptocytes and abnormal inclusions such as vacuoles and siderosomes are found.

11.4 CDA TYPE II

CDA-type II, so-called hereditary erythroblastic multinuclearity with a positive acidified serum lysis test (HEMPAS), is the most frequent type of CDA with more than 300 reported cases (Crookston *et al.*, 1969; Wickramasinghe and Wood, 2005). It is an autosomal recessive disorder affecting erythroblast differentiation and proliferation. Diagnostic criteria for CDA type II include a set of principal conditions and one or more secondary conditions. The main problems are hereditary anaemia,

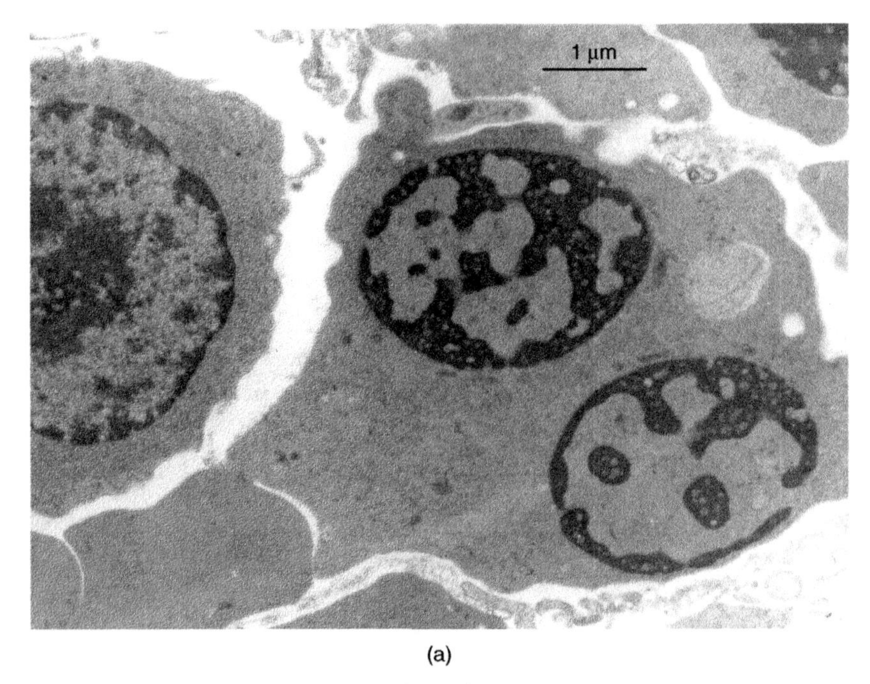

(a)

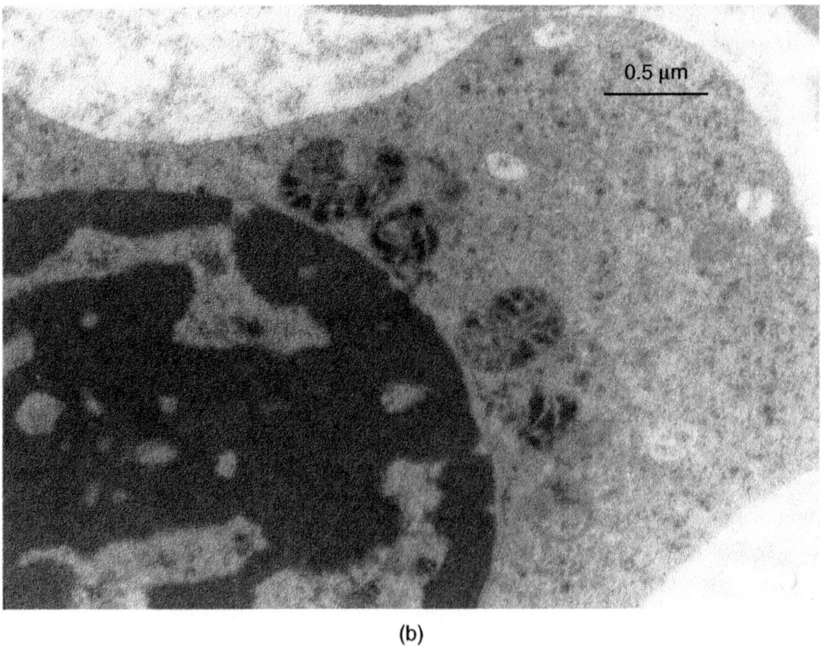

(b)

Figure 11.4 (a) An erythroblast containing a double nucleus with sponge-like changes. (b) An orthochromatic erythroblast containing iron-overloaded mitochondria.

jaundice and ineffective haematopoiesis; 10% of bone-marrow erythroblasts have double nuclei. Secondary conditions include a positive acidified serum lysis test and negative sucrose lysis test, abnormal 3 and 4.5 bands in polyacrylamide gel electrophoresis and erythroblasts with a bilayer membrane structure by TEM (Heimpel *et al.*, 2003) (also called doubling plasma membrane – see Section 11.4.1). Observation of morphological abnormalities of erythroblasts in both peripheral blood and appropriate bone-marrow smears is invaluable for the diagnosis of CDA type II.

11.4.1 Erythroblasts

The erythroid lineage is hyperplastic, and erythroblasts in all stages are increased markedly in bone marrow, the reverse as what is expected in anaemia, indicating ineffective erythropoiesis in CDA type II. Some erythroblasts show irregularities; for example, more than 10% of them are binucleated, and these occur in different developmental stages (Verwilghen *et al.*, 1973). A few polychromatic erythroblasts have multiple nuclei, clover leaf-shaped nuclei and/or karyorrhexis (Figure 11.5).

Most erythroblasts, especially in the basophilic and polychromatic stages, show the 'doubling plasma membrane' phenomenon (Verwilghen *et al.*, 1971; Alloisio *et al.*, 1996) (Figure 11.6a, b). At high magnification, this feature consists of a narrow 40–60 nm clear cisterna running parallel to the cell membrane (Figure 11.7a). This peripheral cisterna appears to be a form of smooth endoplasmic reticulum (sER) (Figure 11.7b), and this is confirmed by immunofluorescence and immunogold electron microscopy (Alloisio *et al.*, 1996).

Some erythroblasts have nuclear abnormalities, including a highly irregular outline, excessive chromatin condensation and micronuclei (Figure 11.8a, b). The erythroblasts sometimes contain long, degenerated cisternae of rER, indicating an overall deterioration of the cell membrane system, including sER and rER (Figure 11.8a). The polychromatic erythroblasts with condensed chromatin illustrate the doubling plasma membrane feature (Figure 11.8b), and proerythroblasts have large cytoplasmic lysosomes and occasionally short cisternae in the peripheral cytoplasm (Figure 11.8c). The number of iron-overloaded mitochondria is increased as a result of the increase in iron deposition, but there are fewer ringed sideroblasts in CDA type II compared with type I. The inter-nuclear chromatin bridges that occur frequently in CDA type I are rarer in type II (Figure 11.9).

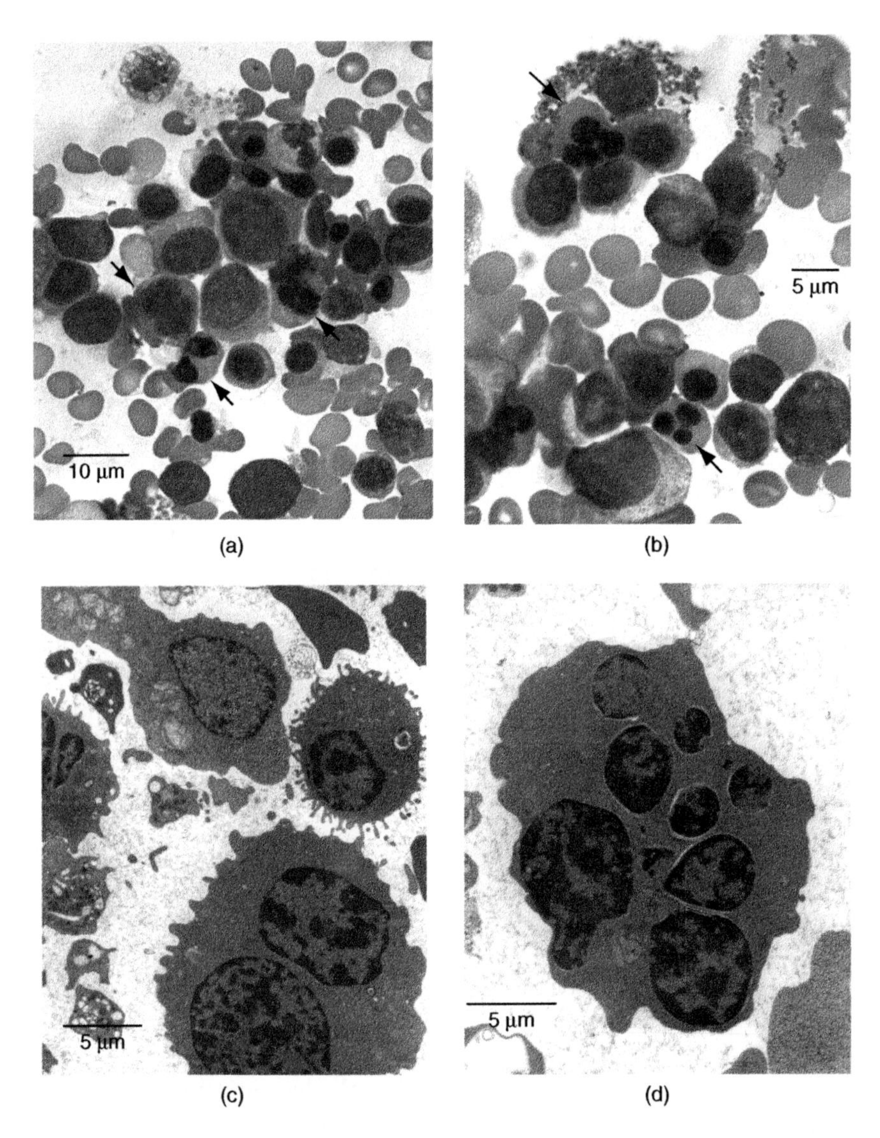

Figure 11.5 (a) Three erythroblasts in different development stages containing double nuclei (arrows). (b) Two polychromatic erythroblasts containing multiple and clover leaf-shaped nuclei. (c) An irregularly shaped basophilic erythroblast (top left) and a binucleated erythroblast (lower right). (d) A polychromatic erythroblast containing multiple nuclei.

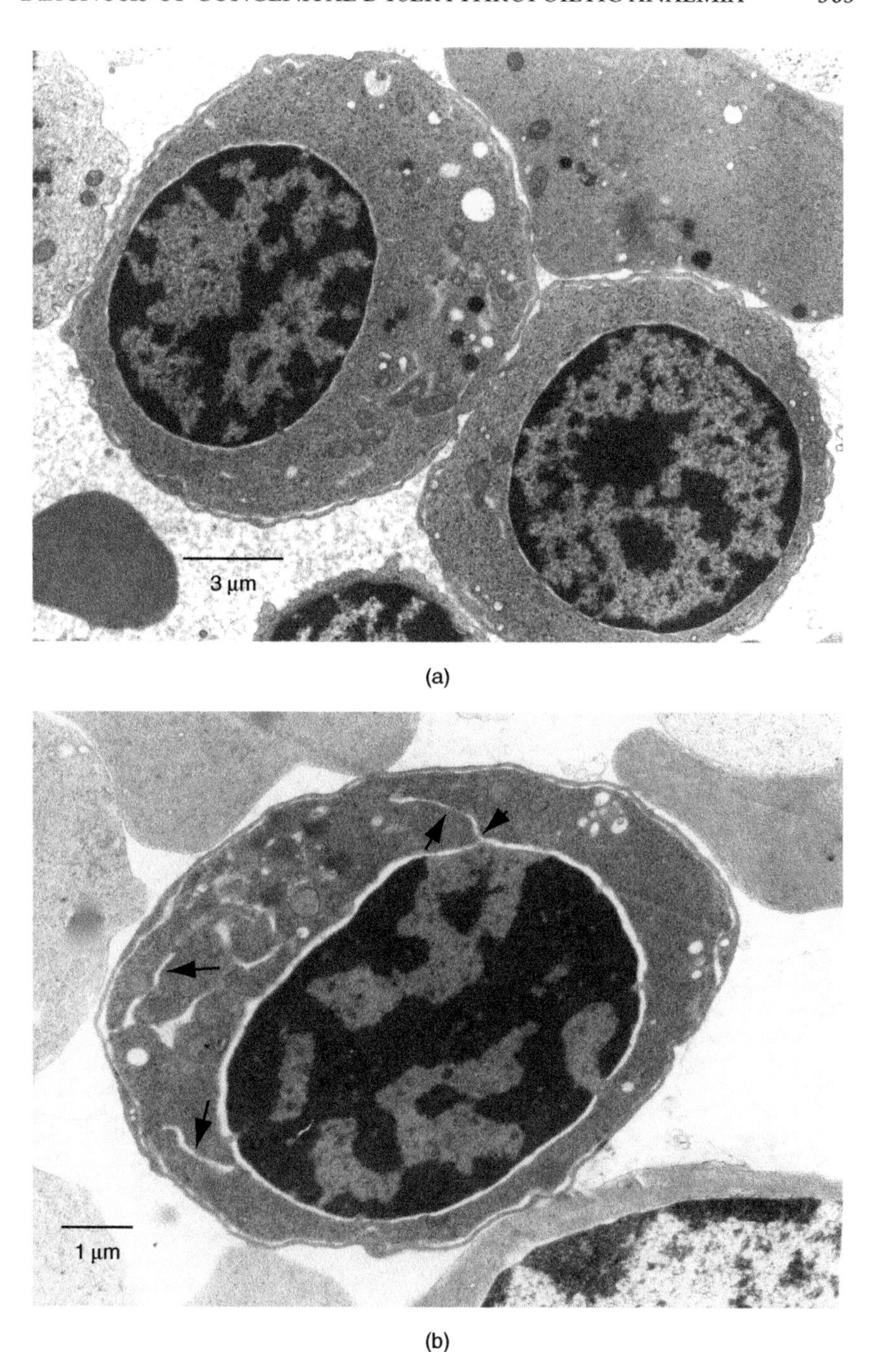

Figure 11.6 (a) Two polychromatophilic erythroblasts showing the doubling plasma membrane phenomenon. (b) A polychromatic erythroblast showing the doubling plasma membrane feature and containing short expanded cisternae (arrows) in the cytoplasm.

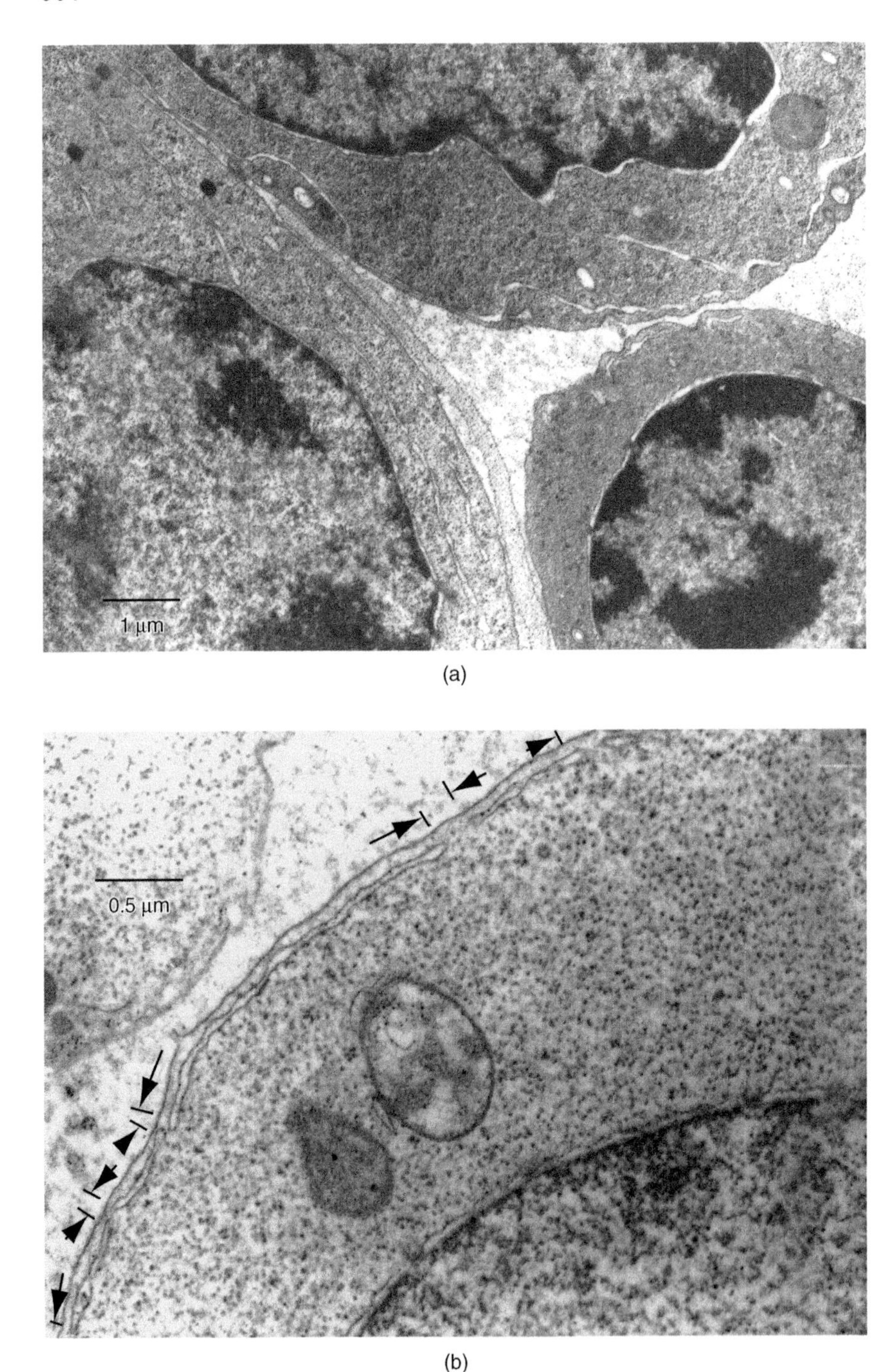

Figure 11.7 (a) Three erythroblasts have thin straps of cytoplasm running parallel to the plasmalemma. (b) A peripheral cisterna of 40–60 nm has the appearance of sER; the arrows indicate two areas of discontinuity.

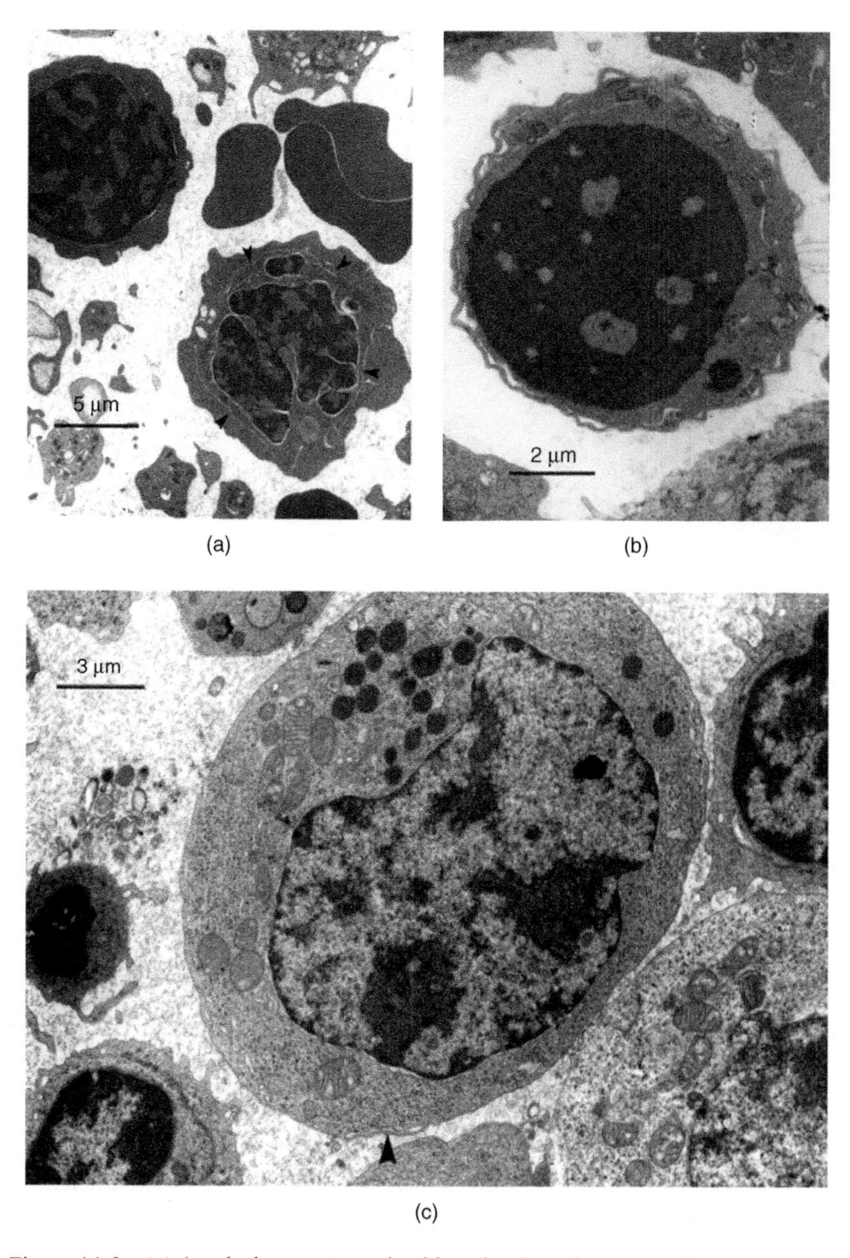

Figure 11.8 (a) A polychromatic erythroblast that has a bizarre nucleus or micronucleus and degenerated rER (arrowheads). (b) A polychromatic erythroblast that has excessive condensation of chromatin and shows the doubling plasma membrane phenomenon. (c) A proerythroblast with large lysosomes and short cisternae in the peripheral cytoplasm (arrows).

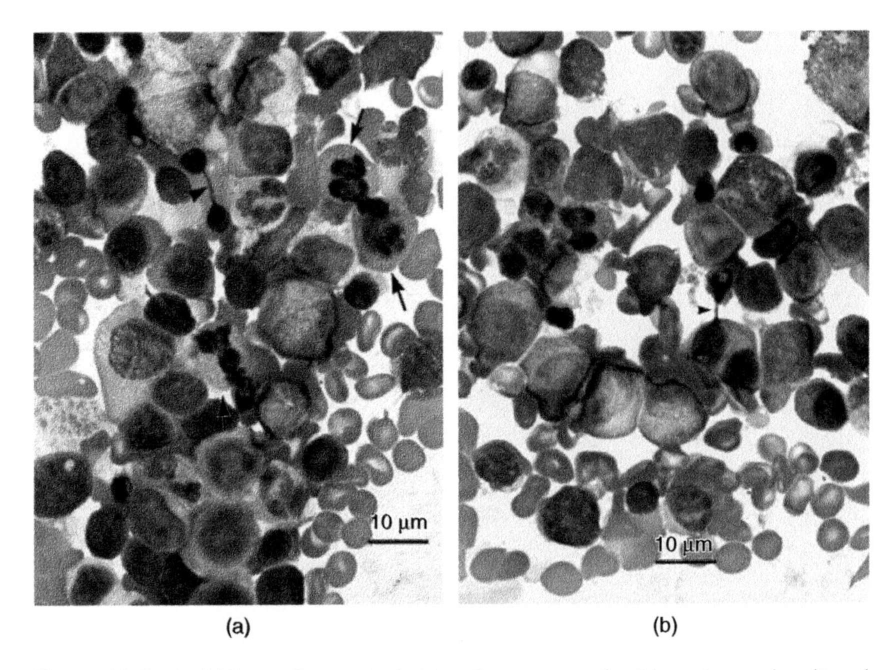

Figure 11.9 (a, b) Inter-chromatin bridges between erythroblasts (arrowhead) and multiple nuclei and/or karyorrhexis (arrows).

11.4.2 Erythrocytes

Reticulocytes are increased inconsistently with the degree of anaemia, and some of them show the doubling plasma membrane feature and iron-overloaded mitochondria. Red cells are usually normocytic, but a few of them show anisocytosis and variable anisochromasia, and appear teardrop-shaped, spherical or stippled, and a few have damaged cell structures suggesting imminent lysis.

11.5 SUMMARY

Diagnosis and typing of CDAs are usually based on clinical presentations, and although DNA sequencing has gained popularity in recent years (Heimpel, Matuschek and Ahmed, 2010), examination by TEM is important in cases showing inconsistency between clinical and laboratory findings, and in rare cases. As always in the clinical environment, TEM can also add to the understanding of these conditions through research.

ACKNOWLEDGEMENTS

I would like to thank Dr Brian Eyden (Department of Histopathology, Christie NHS Foundation Trust, Manchester, United Kingdom) for revising this chapter.

REFERENCES

Alloisio, N., Texier, P., Denoroy, L. *et al.* (1996) The cisternae decorating the red blood cell membrane in congenital dyserythropoietic anemia (type II) originate from the endoplasmic reticulum. *Blood*, **87** (10), 4433–4439.

Crookston, J.H., Godwin, T.F., Wightmann, K.J.R. *et al.* (1996) Congenital dyserythropoietic anemia (abstract), in XIth Congress of the International Society of Hematology, Sydney, p. 18.

Crookston, J.H., Crookston, M.C., Burnie, K.L. *et al.* (1969) Hereditary erythroblastic multinuclearity associated with a positive acidified-serum test: a type of congenital dyserythropoietic anemia. *British Journal of Haematology*, **17** (1), 11–26.

Heimpel, H. (1977) Congenital dyserythropoietic anaemia type I, in *Dyserythropoiesis* (eds S.M. Lewis and R.L. Verwilghen), Academic Press, London, pp. 55–70.

Heimpel, H., Anselstetter, V., Chrobak, L. *et al.* (2003) Congenital dyserythropoietic anemia type II: epidemiology, clinical appearance, and prognosis based on long-term observation. *Blood*, **102** (13), 4576–4581.

Heimpel, H., Matuschek, A. and Ahmed, M. (2010) Frequency of congenital dyserythropoietic anemias in Europe. *European Journal Haematology*, **85** (1), 20–25.

Iolascon, A., D'Agostaro, G., Perrotta, S. *et al.* (1996) Congenital dyserythropoietic anemia type II: molecular basis and clinical aspects. *Haematologica*, **81** (6), 543–559.

Lind, L., Sandstroem, H., Wahlin, A. *et al.* (1995) Localization of the gene for congenital dyserythropoietic anemia type III, CDAN3, to chromosome 15q21–q25. *Human Molecular Genetics*, **4** (1), 109–112.

Pontvert-Delucq, S., Breton-Gorius, J. and Schmitt, C. (1993) Characterization and functional analysis of adult human bone marrow cell subsets in relation to B-lymphoid development. *Blood*, **82** (2), 417–429.

Ru, Y.X., Zhu, X.F., Yan, W.W. *et al.* (2008) Congenital dyserythropoietic anemia in a Chinese family with a mutation of the CDAN1-gene. *Annals of Hematology*, **87** (9), 751–754.

Stohlman, F. (1970) Kinetics of erythropoiesis, in *Regulation of hematopoiesis*, vol. 1 (ed. A.S. Gordon), Appleton-Century-Crofts, New York, pp. 317.

Tamary, H., Shalmon, L., Shalev, H. *et al.* (1996) Localization of the gene for congenital dyserythropoietic anemia type I to chromosome 15q15.1.3 (abstract). *Blood*, **88** (Suppl. 1), 144.

Verwilghen, R.L., Lewis, S.M., Dacie, J.V. *et al.* (1973) HEMPAS: congenital dyserythropoietic anaemia (type II). *Quarterly Journal of Medicine*, **42** (166), 257–278.

Verwilghen, R.L., Tan, P., Wolf-Peeters, C. *et al.* (1971) Cell membrane anomaly impeding cell division. *Experientia*, 27 (12), 1467–1468.
Wickramasinghe, S.N. (1997) Dyserythropoiesis and congenital dyserythropoietic anaemias. *British Journal of Haematology*, **98** (4), 785–797.
Wickramasinghe, S.N. and Wood, W.G. (2005) Advances in the understanding of the congenital dyserythropoietic anaemias. *British Journal of Haematology*, **131** (4), 431–446.

12

Ehlers–Danlos Syndrome

Trinh Hermanns-Lê, Marie-Annick Reginster, Claudine Piérard-Franchimont and Gérald E. Piérard

Department of Dermatopathology, University Hospital of Liège, Liège, Belgium

12.1 INTRODUCTION

Ehlers–Danlos syndrome (EDS) encompasses an heterogeneous group of heritable connective tissue (CT) disorders. It is clinically characterized by variable combinations of skin hyperextensibility and hyperelasticity, joint laxity, and CT fragility. Currently, EDS is classified into six major types (Beighton *et al.*, 1998). The two most frequent conditions are the classic and the hypermobile types. Approximately 50% of patients with a clinical diagnosis of EDS classic type show mutations in the COL5 gene. Sporadic mutations were reported in the hypermobile type. The combination of clinical assessment and family history probably discloses only a minority of EDS patients. Measurements of the mechanical properties of skin and histopathology are required to identify more EDS subjects with subtle skin changes with or without complicated degenerative joint disease. Histopathologic dermal changes (Piérard, Piérard-Franchimont, and Lapière, 1983) with reduction or absence of Factor XIII-a+ type 1 dermal dendrocytes (DD1) are observed in the EDS classic type (Hermanns-Lê and Piérard, 2001) and in the dermatosparaxis type (Piérard *et al.*, 1993).

In EDS, skin ultrastructural changes are more pronounced in the reticular dermis showing alterations in the collagen fibrils (Harvey

Diagnostic Electron Microscopy: A Practical Guide to Interpretation and Technique, First Edition. Edited by John W. Stirling, Alan Curry and Brian Eyden.

and Anton-Lamprecht, 1992; Piérard *et al.*, 1993; Hausser and Anton-Lamprecht, 1994). In addition, other alterations are found in the elastic fibers, as well as in unidentified granulo-filamentous deposits and large stellate hyaluronic acid globules (Harvey and Anton-Lamprecht, 1992; Hermanns-Lê, 2000; Hermanns-Lê and Piérard, 2007a, b). They may vary among tissue sections and microscopic fields. Hence, their identification requires thorough examination of the slides. There are no type-specific ultrastructural alterations in collagen fibrils and elastic fibers, except for the hieroglyphic-shaped fibril cross-sections of the EDS dermatosparaxis type (Piérard *et al.*, 1993). In spite of the heterogeneity in both the structural and molecular abnormalities, the overall architecture and the ultrastructure of the dermal components' changes are of diagnostic relevance, and they may suggest the EDS type.

12.2 COLLAGEN FIBRILS

The number of dermal collagen bundles showing obvious ultrastructural changes is variable among EDS samples. In typical EDS, collagen bundles are composed of fibrils exhibiting variable cross-sectional shape and area with uneven interfibrillar spaces (Figure 12.1). In some cases of EDS hypermobile type, collagen fibrils present homogeneous large diameters. The collagen fibrils may exhibit flower-like cross-sections (Figure 12.2), as well as serrated, irregular contours or hieroglyphic aspects (Figure 12.3). Unusual collagen fibril orientation (Figure 12.4) appears as twisted, hook-like, S-shaped, and ring-shaped structures (Hermanns-Lê and Piérard, 2006).

12.3 ELASTIC FIBERS

The ultrastructure of elastic fibers is commonly altered (Harvey and Anton-Lamprecht, 1992; Hermanns-Lê and Piérard, 2007a, b). They appear fragmented, with frayed contours (Figure 12.5), internal microcavities (Figure 12.6), elastotic changes, or uneven distribution of electron-dense inclusions in the amorphous elastin matrix. The amount in microfibrils at the periphery of the elastic fibers is variable. Discrete spheroid foci are present inside the elastic fibers. They are abutted to the edge of microcavities, or present as homogeneous structures or target-like figures (Figure 12.7). The proportion of altered elastic fibers varies among patients and appears unrelated to age and photo exposure.

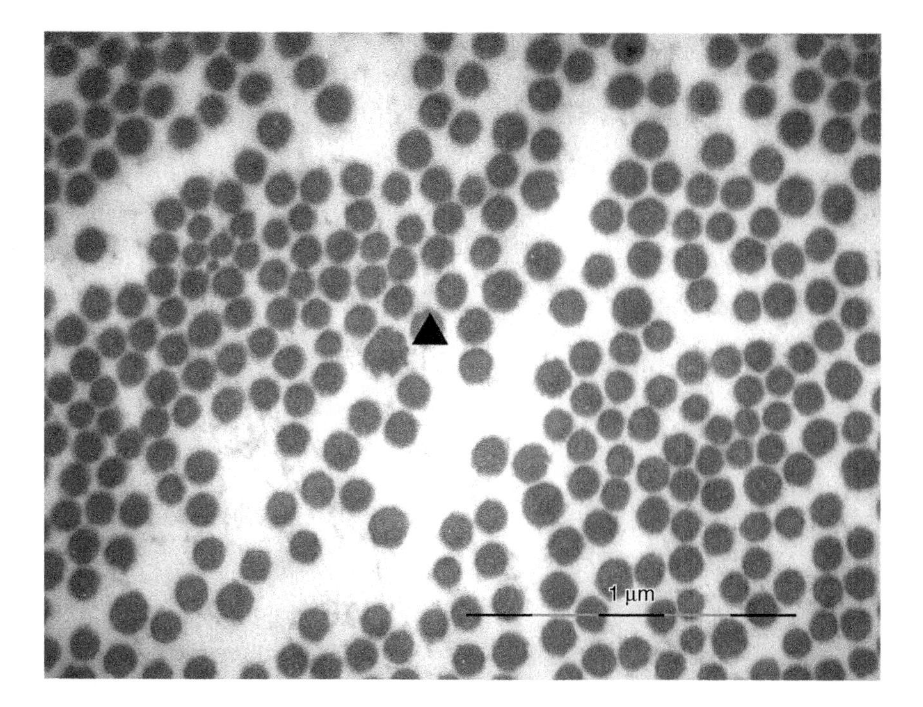

Figure 12.1 Variable diameter collagen fibrils and flower-like fibril (►) with uneven interfibrillar spaces (EDS hypermobile type).

12.4 NONFIBROUS STROMA AND GRANULO-FILAMENTOUS DEPOSITS

Granulo-filamentous deposits are found in variable amounts inside the interfibrillar area of the collagen bundles (Figure 12.8). In addition, large stellate globules, presumably of hyaluronic acid (Figure 12.9), and focal accumulations of granulo-filamentous deposits are commonly disclosed in the interstitial matrix. The hyaluronic acid globules are generally single, but they may form chains or networks.

12.5 CONNECTIVE TISSUE DISORDERS

12.5.1 Ehlers–Danlos Syndrome

The EDS classic type shows major alterations in the collagen scaffolding. In the gravis variant, disorganized bundles are composed of collagen fibrils showing prominent variations in shape and diameter

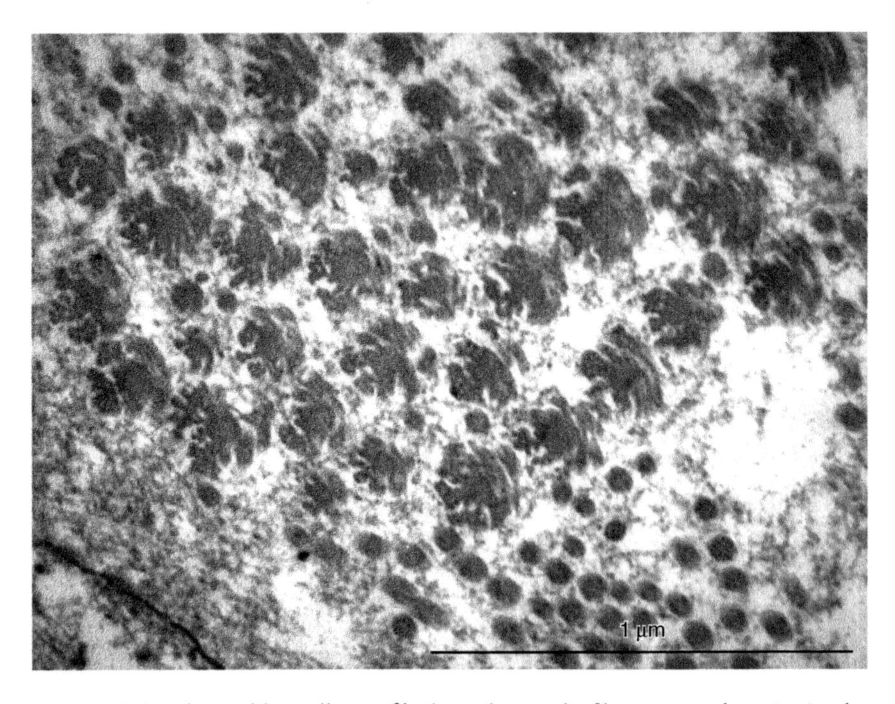

Figure 12.2 Flower-like collagen fibrils and granulo-filamentous deposits in the interfibrillar area (EDS classic (mitis) type).

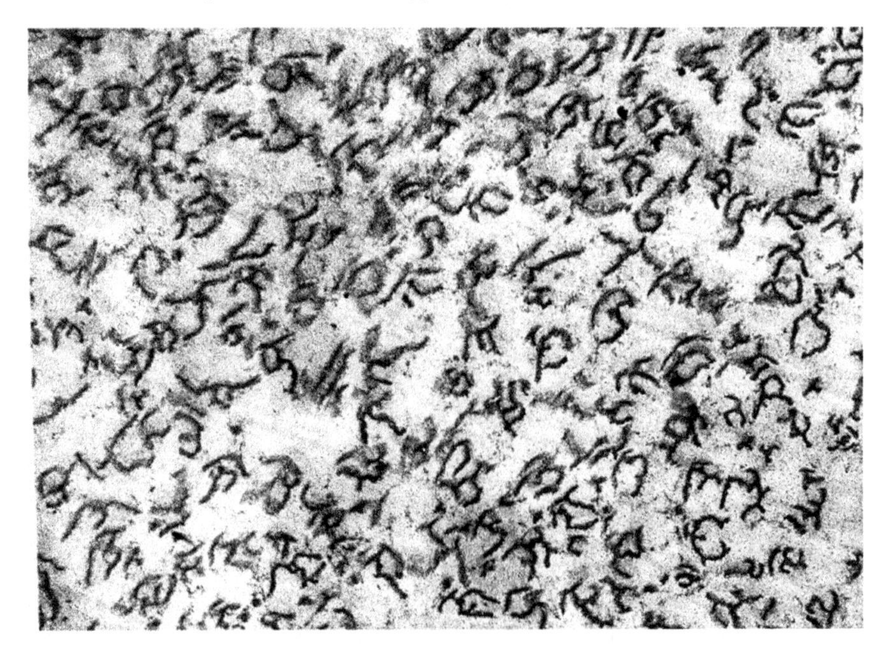

Figure 12.3 Hieroglyphic collagen fibrils (EDS dermatosparaxis type).

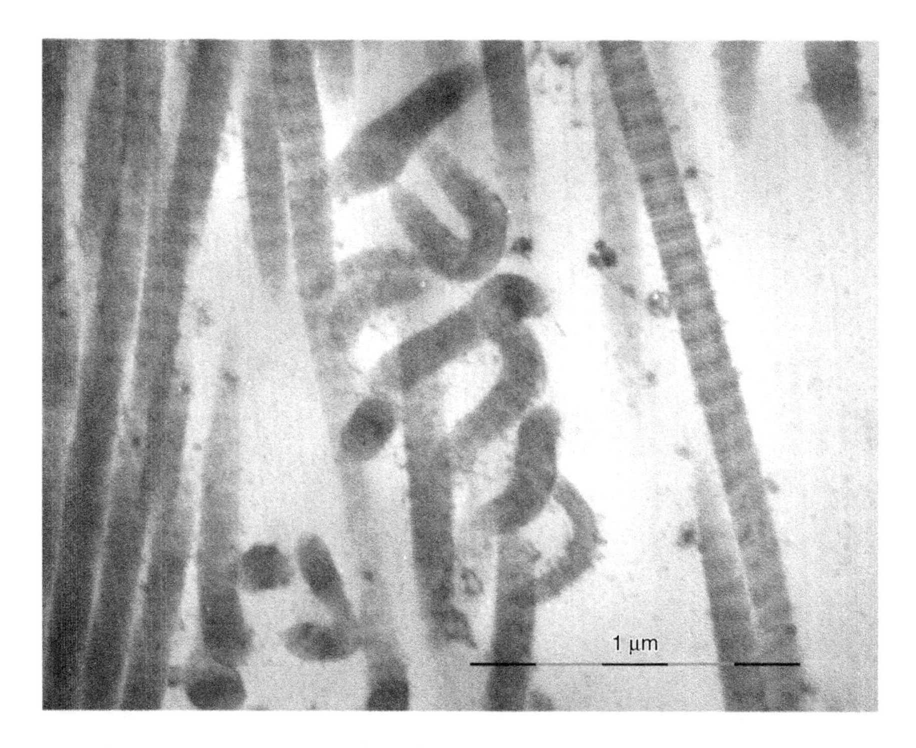

Figure 12.4 Unusual collagen fibril orientation (EDS hypermobile type).

(Figure 12.10). The cauliflower-shaped cross-section areas of the fibrils are about fivefold larger than the regular rounded collagen fibrils (Piérard *et al.*, 1988). In the mitis variant, elastic fibers are altered in size, distribution, and organization (Piérard, Piérard-Franchimont, and Lapière, 1988). Granulo-filamentous deposits are abundant.

In the EDS hypermobile type (Harvey and Anton-Lamprecht, 1992; Hausser and Anton-Lamprecht, 1994; Kobayasi, 2004), collagen bundles are composed of fibrils with either variable diameter or uniform but larger diameter. Some fibril outlines are serrated, and flower-like collagen fibrils are fewer and smaller (Figure 12.1) than in the classic type (Figure 12.2). Some collagen fibrils are whirled inside the bundles (Figure 12.11). Variable combinations of altered aspects of elastic fibers (Figures 12.5–12.7) may be present (Hermanns-Lê and Piérard, 2007a, b). Variable amounts of granulo-filamentous material are often present within the collagen bundles and dispersed the interstitial matrix enriched in large stellate globules of hyaluronic acid (Hermanns-Lê and Piérard, 2007a, b).

In EDS related to tenascin-X deficiency (Zweers *et al.*, 2004, 2005), collagen fibrils are small and normal in shape. Elastic fibers show

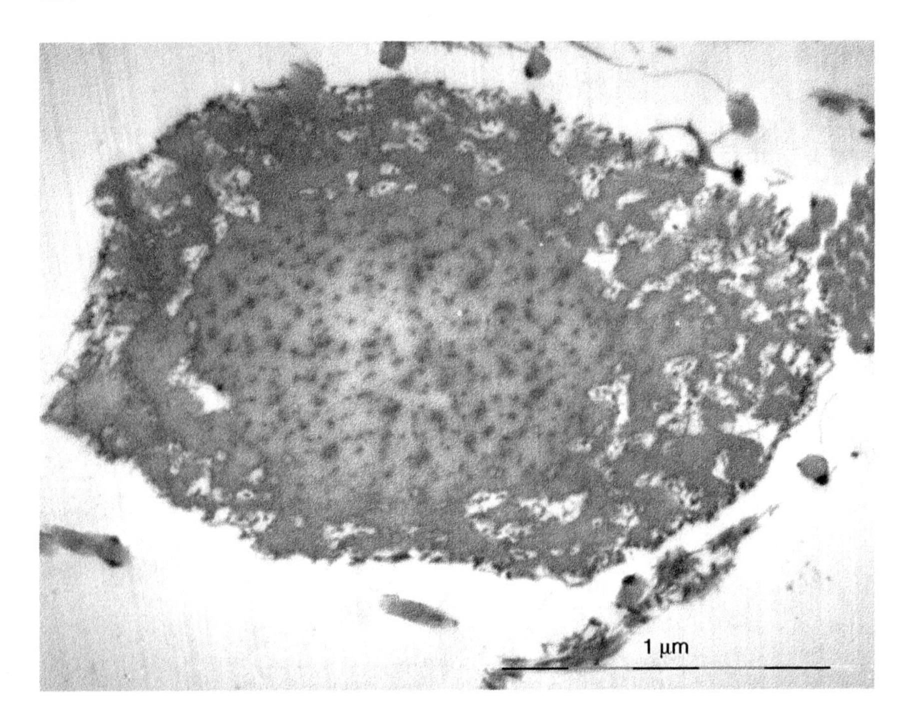

Figure 12.5 Frayed contour elastic fiber.

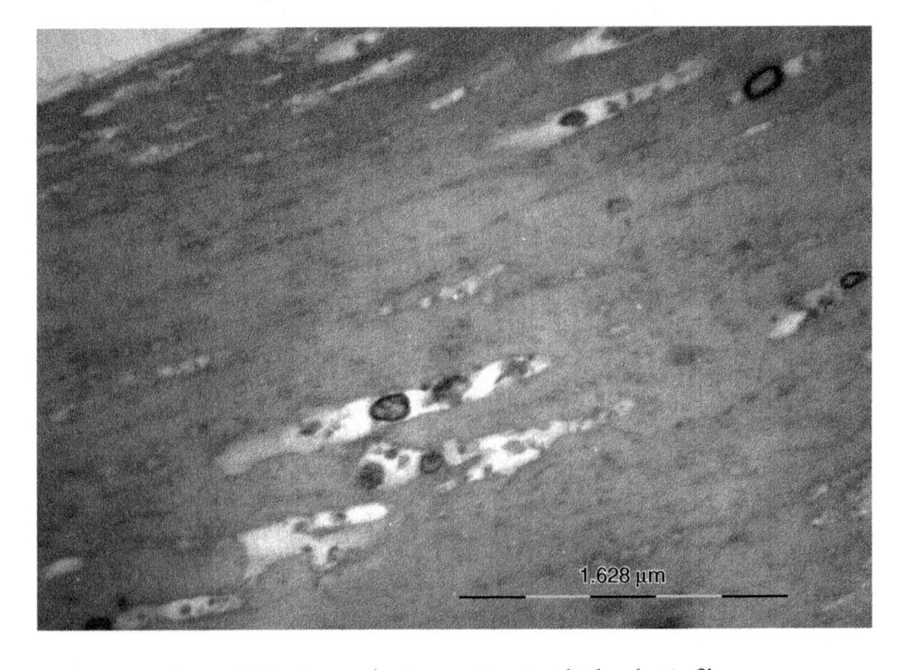

Figure 12.6 Internal microcavities inside the elastic fiber.

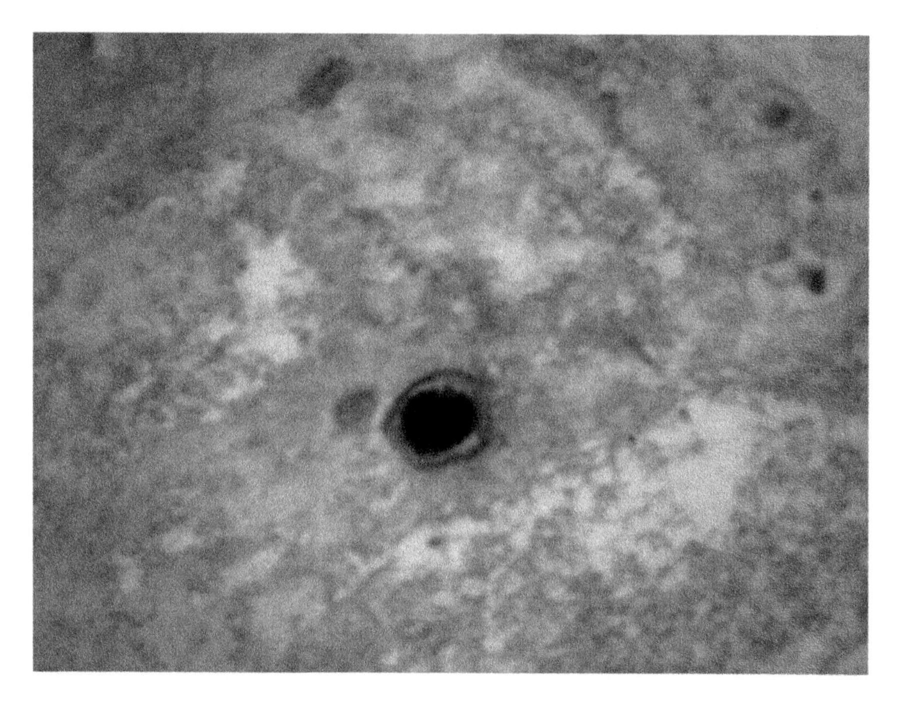

Figure 12.7 Speroid foci inside the elastic fiber.

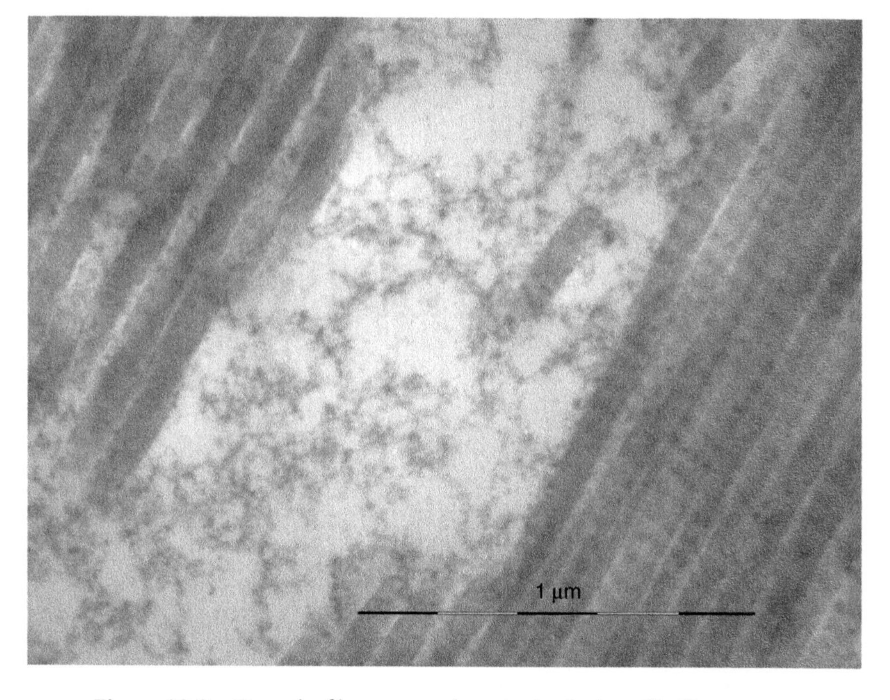

Figure 12.8 Granulo-filamentous deposits in the interfibrillar spaces.

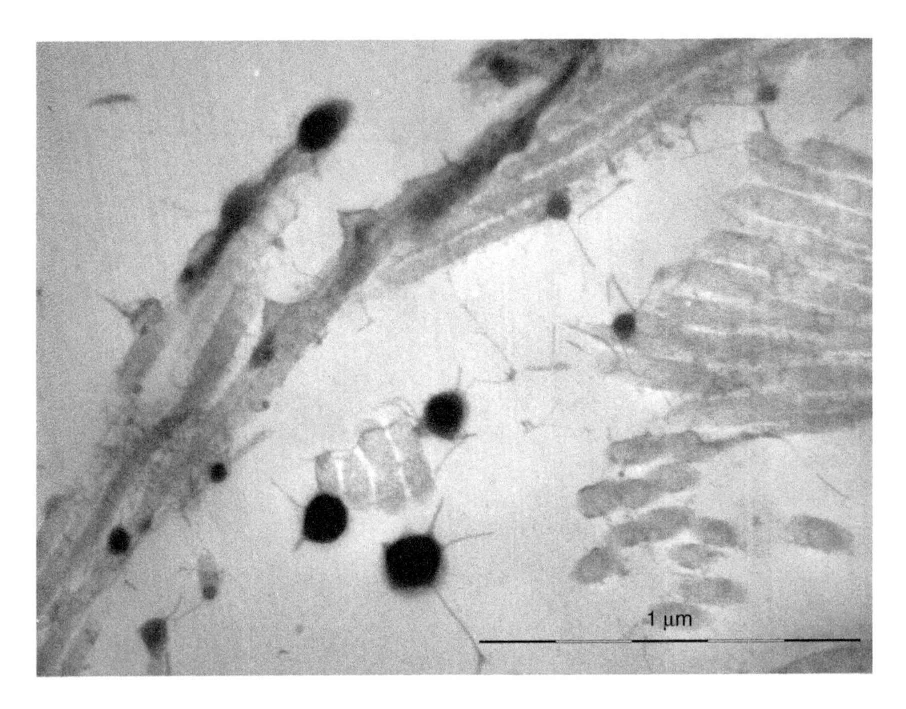

Figure 12.9 Large stellate globules of hyaluronic acid in the interstitial matrix.

fragmentation with irregular distribution of the osmiophilic components inside the amorphous elastin matrix.

In the EDS vascular type, the dermis is thinned to about one-third. The collagen bundles and the other dermal components are loosely arranged. Both the collagen bundles and their fibrils are thin (Harvey and Anton-Lamprecht, 1992, 1994; Smith *et al.*, 1997). Composite, notched, or hieroglyphic fibrils are absent (Smith *et al.*, 1997). The elastic fibers appear branched or fragmented and seem increased in amount. Fibroblasts exhibit dilated endoplasmic reticulum filled with a granular material (Figure 12.12).

In the EDS kyphoscoliosis type, the collagen bundles are thin and poorly organized and contain flower-like fibrils. Excess in granulo-filamentous deposits is found inside both the collagen bundles and the interstitial matrix (Figure 12.13).

The collagen fibrils in the EDS arthrochalasis type (VIIA and VIIB) (Giunta *et al.*, 2008) are more loosely and randomly organized, and show highly irregular contours. In the type VIIB, the collagen fibrils exhibit far less abnormal contour but more variable diameters than those of the EDS type VIIA. The EDS dermatosparaxis type is characterized by

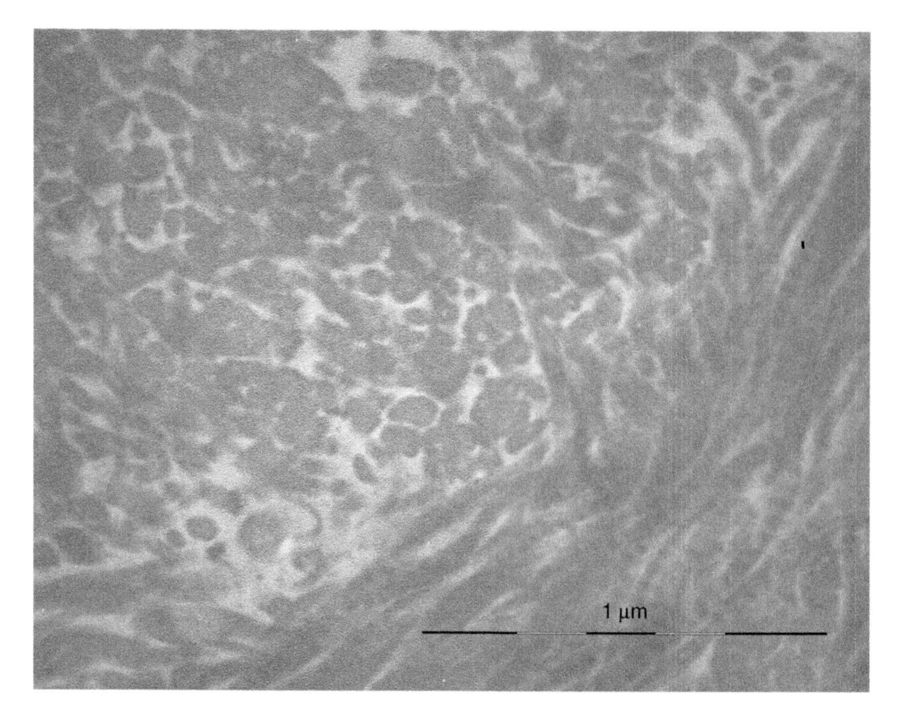

Figure 12.10 Disorganized collagen bundles with abnormal fibrils (EDS classic (gravis) type).

the presence of unique hieroglyphic collagen fibrils (Piérard *et al.*, 1993) (Figure 12.3).

12.5.2 Spontaneous Cervical Artery Dissection

Spontaneous cervical artery dissection (SCAD) syndrome is clinically characterized by stroke in young and middle-aged patients (Ulbricht *et al.*, 2004; Brandt, Morcher, and Hausser, 2005; Martin *et al.*, 2006). In most patients, morphologic alterations of the dermis are reminiscent of those found in the EDS hypermobile type. Flower-like collagen fibrils are associated with fragmented and microcalcified elastic fibers. Other SCAD cases exhibit aspects similar to the EDS vascular type with a thinned dermis and small-diameter collagen fibrils loosely packed in bundles. Still other cases show only elastic fiber abnormalities.

In some families with vascular dissection (Flagothier *et al.*, 2007), the ultrastructural changes range from those commonly found in the EDS classic (mitis) type to those in the EDS vascular type. In some cases of arterial rupture (Hermanns-Lê and Piérard, 2007a, b) with type

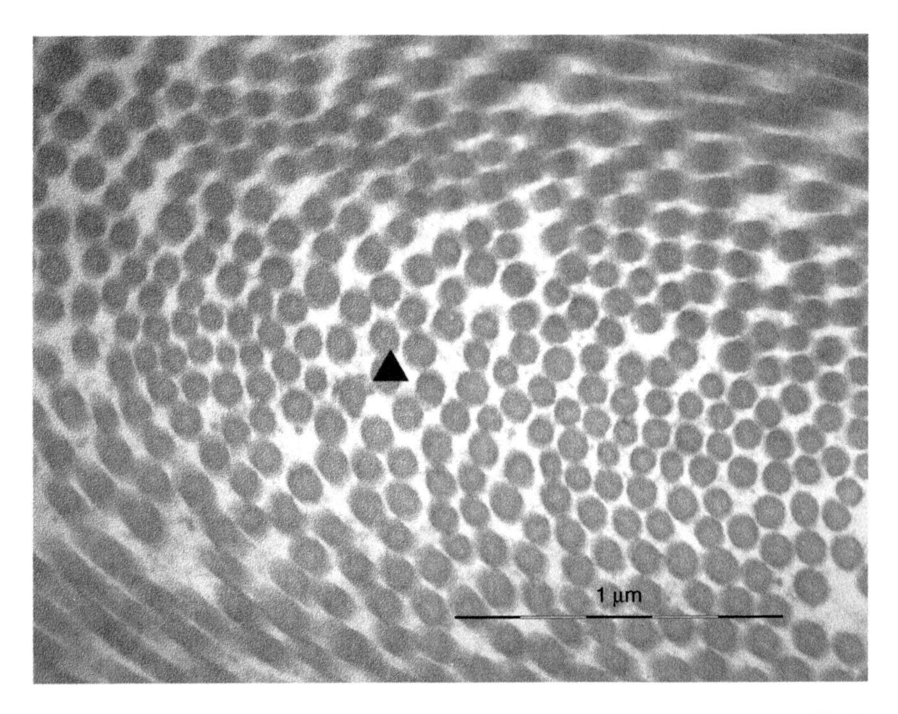

Figure 12.11 Collagen fibrils whirling in the bundles with small flower-like fibrils (►) (EDS hypermobile type).

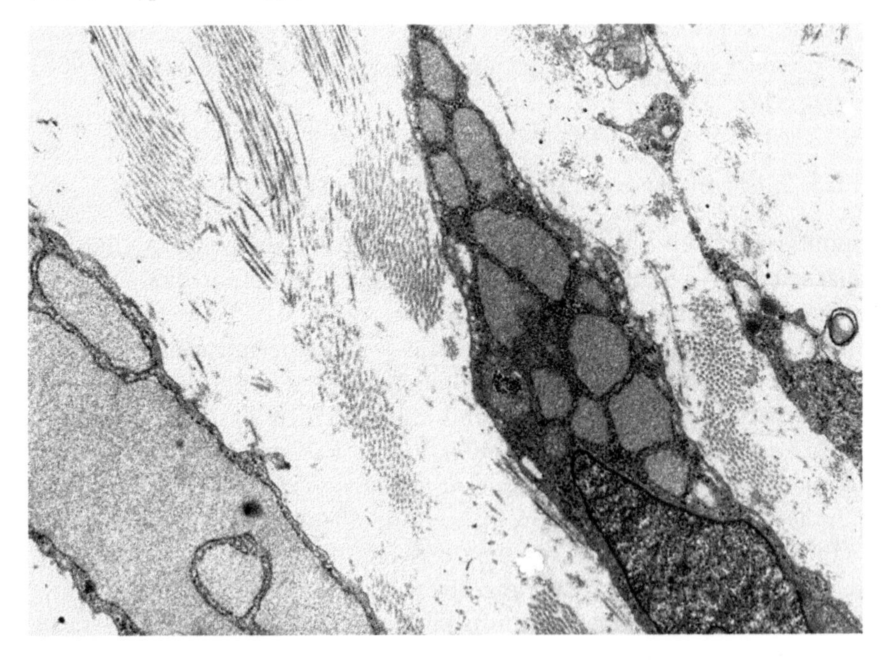

Figure 12.12 Fibroblasts with granular material in dilated reticulum endoplasmic and small collagen bundles (EDS vascular type).

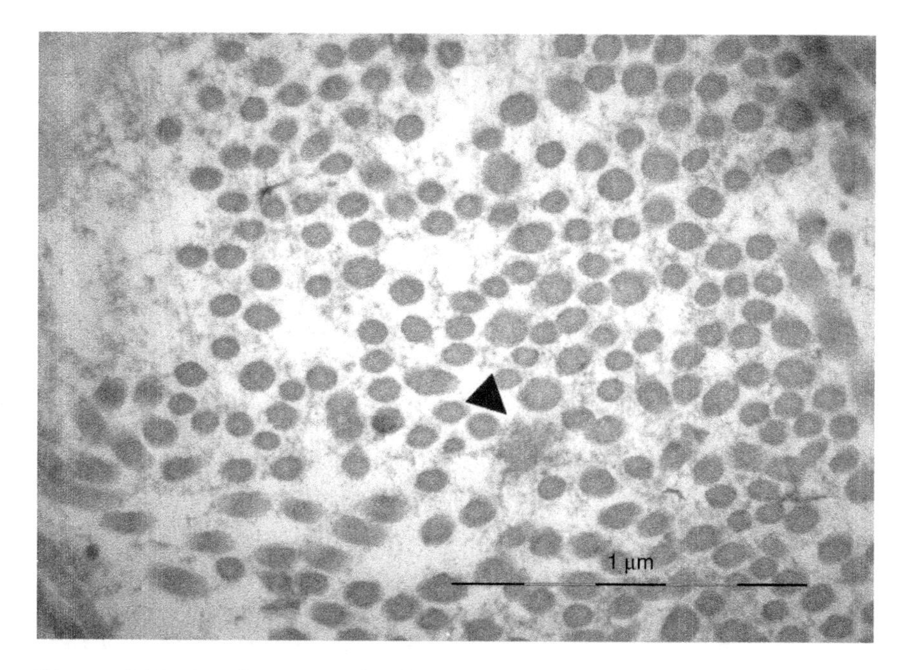

Figure 12.13 Poorly organized collagen bundles with flower-like fibrils (▶) and granulo-filamentous deposits (EDS kyphoscoliosis type).

I collagen mutation, diverse diameter and flower-like collagen fibrils, large hyaluronic acid globules, and annular microcalcified elastic fibers are observed.

12.5.3 Recurrent Preterm Premature Rupture of Fetal Membrane Syndrome

In the recurrent preterm premature rupture of fetal membrane (PPROM) syndrome, the skin may present EDS-like dermal abnormalities (Hermanns-Lê, Piérard, and Quatresooz, 2005) without any clinical signs. The CT alterations resemble those found in the EDS classic, hypermobile, and, more rarely, vascular types. The ultrastructural EDS vascular type changes, when present, are restricted to some areas, and the dermal thickness is normal.

REFERENCES

Beighton, P., De Paepe, A., Steinmann, B. *et al.* (1998) Ehlers-Danlos syndromes: revised nosology, Villefranche 1997. *American Journal of Human Genetic*, 77, 31–37.

Brandt, T., Morcher, M., and Hausser, I. (2005) Association of cervical artery dissection with connective tissue abnormalities in skin and arteries. *Frontiers of Neurology and Neuroscience*, 20, 16–29.

Flagothier, C., Goffin, V., Hermanns-Lê, T. *et al.* (2007) A four-generation Ehlers-Danlos syndrome with vascular dissections. Skin ultrastructure and biomechanical properties. *Journal of Medical Engineering and Technology*, **31**, 175–180.

Giunta, C., Chambaz, C., Pedemonte, M. *et al.* (2008) The arthrochalasia type of Ehlers-Danlos syndrome (EDSVIIA and VIIB): the diagnostic value of collagen fibril ultrastructure. *American Journal of MedicalGenetics, Part A*, **146A**, 1341–1346.

Harvey, J.M. and Anton-Lamprecht, I. (1992) Stromal aberrations, in *Diagnostic Ultrastructure of Non-Neoplastic Diseases* (eds J.M. Papadimitriou, D.W. Henderson, and D.V. Spagnolo), Churchill Livingstone Medical Division of Longman Group, UK Ltd, Edinburgh, pp. 389–401.

Hausser, I. and Anton-Lamprecht, I. (1994) Differential ultrastructural aberrations of collagen fibrils in Ehlers-Danlos syndrome types I-IV as a means of diagnostic and classification. *Human Genetics*, **3**, 394–407.

Hermanns-Lê, T. (2000) How I explore ... some cutaneous disorders using ultrastructural examination of the skin. *Revue Medicale de Liège*, **55**, 954–956.

Hermanns-Lê, T. and Piérard, G.E. (2001) FactorXIII a positive dendrocyte rarefaction in Ehlers-Danlos syndrome, classic type. *American Journal of Dermatopathology*, **23**, 427–430.

Hermanns-Lê, T. and Piérard, G.E. (2006) The collagen fibril arabesques in connective tissue disorders. *American Journal of Clinical Dermatology*, **7**, 323–326.

Hermanns-Lê, T. and Piérard, G.E. (2007a) Multifaceted dermal ultrastructural clues for Ehlers-Danlos syndrome with arterial rupture and type I collagen R-to-C substitution. *American Journal of Dermatopathology*, **29**, 449–451.

Hermanns-Lê, T. and Piérard, G.E. (2007b) Ultrastructural alterations of elastic fibers and other dermal components in Ehlers-Danlos syndrome of the hypermobile type. *Americna Journal of Dermatopathology*, **29**, 370–373.

Hermanns-Lê, T., Piérard, G.E., and Quatresooz, P. (2005) Ehlers-Danlos-like dermal abnormalities in women with recurrent preterm premature rupture of fetal membranes. *American Journal of Dermatopathology*, **27**, 407–410.

Kobayasi, T. (2004) Abnormality of dermal collagen fibrils in Ehlers-Danlos syndrome. Anticipation of the abnormality for the inherited hypermobile disorders. *European Journal of Dermatology*, **14**, 221–229.

Martin, J.J., Hausser, I., Lyrer, P. *et al.* (2006) Familial cervical artery dissections: clinical, morphologic, and genetic studies. *Stroke*, **37**, 2924–2929.

Piérard, G.E., Hermanns-Lê, T., Arrese Estrada, J. *et al.* (1993) Structure of the dermis in type VIIc Ehlers-Danlos syndrome. *American Journal of Dermatopathology*, **15**, 127–132.

Piérard, G.E., Lê, T., Piérard-Franchimont, C., and Lapière, C.M. (1988) Morphometric study of cauliflower collagen fibrils in Ehlers-Danlos syndrome type I. *Collagen Related Research*, **8**, 453–457.

Piérard, G.E., Piérard-Franchimont, C., and Lapière, C.M. (1983) Histopathology aid at the diagnosis of the Ehlers-Danlos syndrome, gravis and mitis types. *International Journal of Dermatology*, **22**, 300–304.

Smith, L.T., Schwarze, U., Goldstein, J., and Byers, P.H. (1997) Mutations in the COL3A1 gene result in the Ehlers-Danlos Syndrome type IV and alterations in the size and distribution of the major collagen fibrils in the dermis. *Journal of Investigative Dermatology*, **108**, 241–247.

Ulbricht, D., Diederich, N.J., Hermanns-Lê, T. *et al.* (2004) Cervical artery dissection: an atypical presentation with Ehlers-Danlos-like collagen pathology? *Neurology*, **63**, 1708–1710.

Zweers, M.C., Dean, W.B., Van Kuppevelt, T.H. *et al.* (2005) Elastic fibers abnormalities in hypermobility type Ehlers-Danlos syndrome patients with tenascin-X mutations. *Clinical Genetics*, **67**, 330–334.

Zweers, M.C., van Vlijmen-Willems, I.M., van Kuppevelt, T.H. *et al.* (2004) Deficiency of tenascin-X causes abnormalities in dermal elastic fiber morphology. *Journal of Investigative Dermatology*, **122**, 885–891.

13

Electron Microscopy in Occupational and Environmental Lung Disease

Victor L. Roggli

Department of Pathology, Duke University Medical Center, Durham, North Carolina, United States

13.1 INTRODUCTION

The respiratory system is exposed to between 10 000 and 20 000 l of air per day. This air is contaminated with various amounts of particulate material, including organic and inorganic dusts, fibers, metals, and other pollutants. A variety of mechanisms have evolved to deal with these contaminants, and the consequences of these exposures depend upon the nature of the dust, its biopersistence in the lung, the efficiency of host clearance mechanisms, and the susceptibility of the individual. For example, considerable amounts of carbonaceous material may accumulate in lung tissue with relatively little effect on respiratory function. On the other hand, exposure to relatively low amounts of beryllium in the susceptible host can result in serious medical consequences and impairment of respiratory function.

Diagnostic Electron Microscopy: A Practical Guide to Interpretation and Technique, First Edition. Edited by John W. Stirling, Alan Curry and Brian Eyden.

A number of respiratory diseases may occur as a consequence of the accumulation of inorganic particulate material within the lungs and adjacent tissues, and many of these have well-defined characteristics (Sporn and Roggli, 2008; Butnor and Roggli, 2010). The identification, and in some instances quantification, of this material in lung samples may provide useful information regarding the diagnosis and prevention of these diseases. Several techniques have been developed to aid in the identification of such particulates within the lung. One of the most powerful and useful approaches involves the use of an electron microscope (EM) equipped with an energy-dispersive spectrometer (Ingram *et al.*, 1999). It is the purpose of this chapter to provide a survey of this methodology and its application to the diagnosis of diseases caused by occupational and environmental exposures to inorganic particulates. A summary of diseases for which analytical electron microscopy (also EM) is often useful is provided in Table 13.1.

13.2 ASBESTOS

A variety of diseases may be related to asbestos exposure, including asbestosis, benign asbestos-related pleural diseases, malignant mesothelioma, and carcinoma of the lung (Roggli, Oury, and Sporn, 2004). Analysis of the mineral fiber content of lung tissue by means of analytical EM has provided useful information pertaining to the concentrations, dimensions, and types of fibers that are associated with these various diseases and with various occupational and non-occupational exposures. For example, analysis of Canadian chrysotile miners and millers with mesothelioma in the 1970s provided surprising information that these patients had primarily tremolite (now recognized as a contaminant of Canadian chrysotile) in their lung samples (McDonald, 2010). Similarly, although insulators were thought initially to have relatively light and intermittent exposures to chrysotile asbestos, analysis of lung samples have shown heavy burdens of commercial amphibole fibers (mostly amosite) with little if any correlation of disease with chrysotile or tremolite (Churg and Vedal, 1994; Roggli, Oury, and Sporn, 2004).

13.2.1 Preparatory Techniques

The preparation of samples for fiber analysis involves several steps. These include sample selection, digestion procedure, and recovery of the residue on a suitable substrate (Roggli, Oury, and Sporn, 2004). Lung

Table 13.1 Use of analytical electron microscopy in diagnosis of dust-related diseases

Disease category	Particle types	Analytical methodology
Asbestos-related diseases *Asbestosis* *Mesothelioma* *Asbestos-related lung cancer*	Asbestos fibers: amosite, crocidolite, chrysotile, tremolite, actinolite, and anthophyllite	Lung fiber burden Analytical SEM and TEM
Hypersensitivity pneumonitis Sarcoidosis	Calcium oxalate, calcium carbonate, and calcium phosphate	Analytical SEM and BEI
Silicosis	Silica (SiO_2)	Analytical SEM and BEI
Silicate pneumoconiosis		Analytical SEM and BEI
Talcosis	Talc (Mg and Si)	
Kaolin worker's pneumoconiosis	Kaolin (Al and Si)	
Mica or feldspar pneumoconiosis	Mica and feldspar (K, Al, and Si)	
Mixed dust pneumoconiosis	Silica, various silicates, and metals	
Metal-induced diseases		Analytical SEM and BEI
Siderosis	Iron (Fe)	
Aluminosis	Aluminum (Al)	
Hard metal lung disease	Tungsten (W), tantalum (Ta), tin (Sn), ± cobalt (Co)	
Berylliosis	Beryllium (Be)	
Rare-earth pneumoconiosis	Cerium (Ce), samarium (Sa), neodymium (Nd), and lanthanum (La)	Analytical SEM and BEI

Note: BEI = backscattered electron imaging; SEM = scanning electron microscopy; and TEM = transmission electron microscopy.

parenchyma is the type of tissue typically analyzed, and this should be free from tumor, congestion, and consolidation as much as possible. Formalin fixed tissue is preferred, but fresh tissue or paraffin blocks work just as well. Because tissue is lost during processing in paraffin, a correction factor must be applied for such samples. Either wet chemical digestion, using sodium hypochlorite solution or potassium hydroxide, or low-temperature plasma ashing may be used to extract particulates from the organic matrix in which they are embedded. The particulate residue may then be collected on the surface of a polycarbonate or acetate filter of suitable pore size, which may then be mounted on a substrate for examination by microscopy. Other samples such as lymph nodes or pleura may also be examined. The advantage of lung tissue is that it concentrates dusts from ambient air, and there is more information available regarding background values for lung than for other tissues. Tumor tissue is unsuitable because of the dilutional effect of neoplastic growth.

13.2.2 Analytical Methodology

Once the fibers have been collected on the filter surface, they may be examined by various forms of microscopy to identify the amounts and types of fibers present. The author uses both light microscopy (to quantify asbestos bodies) and scanning electron microscopy (SEM) (to quantify coated and uncoated fibers), counting fibers that are 5 μm or greater in length (Figure 13.1a). An energy-dispersive spectrometer is then used to analyze individual fibers and to classify them as chrysotile, amosite, crocidolite, tremolite, actinolite, anthophyllite, or various non-asbestos mineral fibers (Figure 13.1b) (Roggli, Oury, and Sporn, 2004). Other investigators prefer to use analytical transmission electron microscopy to count and identify fiber types. Because there are so many variables in the processing and analysis of tissues for determination of fiber burden, results between different laboratories may not be comparable. This was well shown in the international interlaboratory counting trial reported by Gylseth *et al.* (1985). The results of this study strongly indicate that each laboratory performing such analyses should establish its own reference range using a suitably selected control group analyzed by the same methodology.

13.2.3 Asbestos-Related Diseases

Analysis of tissue mineral fiber content has proved informative concerning the pathology and pathogenesis of the various diseases caused by asbestos, and it is complementary to the findings in epidemiological studies. For example, we have shown that analytical SEM can accurately distinguish asbestosis from other fibrotic interstitial lung disorders with which it may be confused (Schneider, Sporn, and Roggli,

2010). Whereas asbestosis typically occurs at very high lung asbestos burdens, mesothelioma may occur as a consequence of considerably less exposure. Both mesothelioma and asbestosis in the United States are predominately associated with elevated levels of commercial amphibole fibers (Roggli *et al.*, 2002; Schneider, Sporn, and Roggli, 2010). Increased lung cancer risk is generally associated with fiber levels similar to those that cause asbestosis (Roggli, Oury, and Sporn, 2004). Interestingly, although chrysotile was by far the predominant fiber type used in the United States commercially, most asbestos bodies occur on commercial amphibole (amosite or crocidolite) cores. This observation

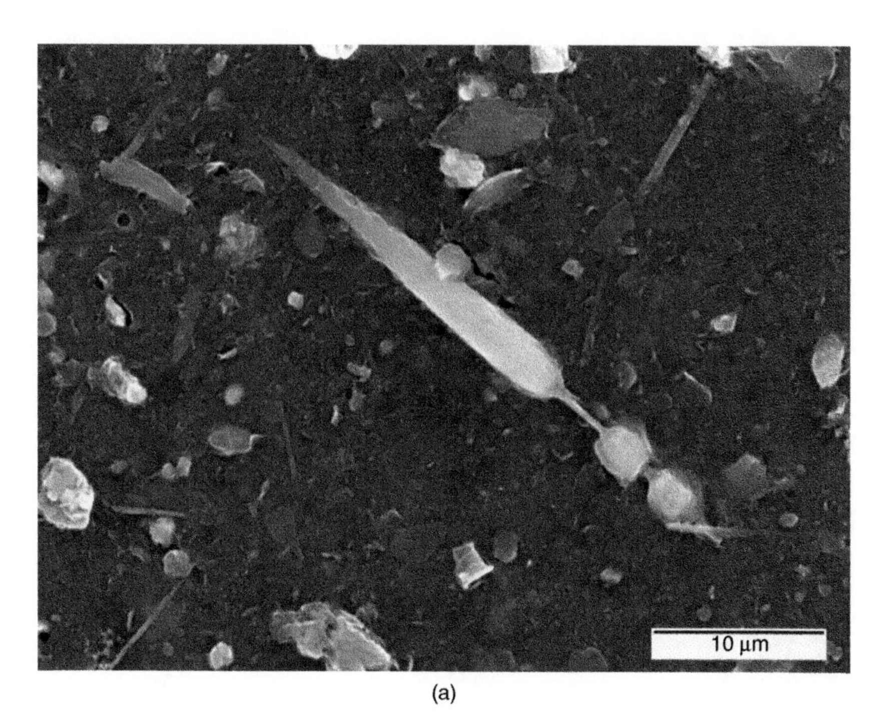

Figure 13.1 (a) Scanning electron micrograph showing numerous fibers on the filter surface, prepared from digestion of a lung tissue sample. This patient had asbestosis as a consequence of environmental exposures to tremolite and actinolite asbestos in the southern Anatolian region of Turkey. (b) Energy-dispersive spectrum from one of the fibers from (a), showing peaks for silicon, magnesium, iron, and calcium, which are typical for actinolite. (c) Backscattered electron image of giant cell in patient with cholesterol interstitial granulomas. Note the asteroid body in the cytoplasm of the giant cell (negative polarity BEI). (d) Backscattered electron image of nodule in lung of individual with silicosis, showing numerous particulates (negative polarity BEI). (e) Energy-dispersive spectrum from particle in (e), showing peak for silicon only, consistent with α-quartz

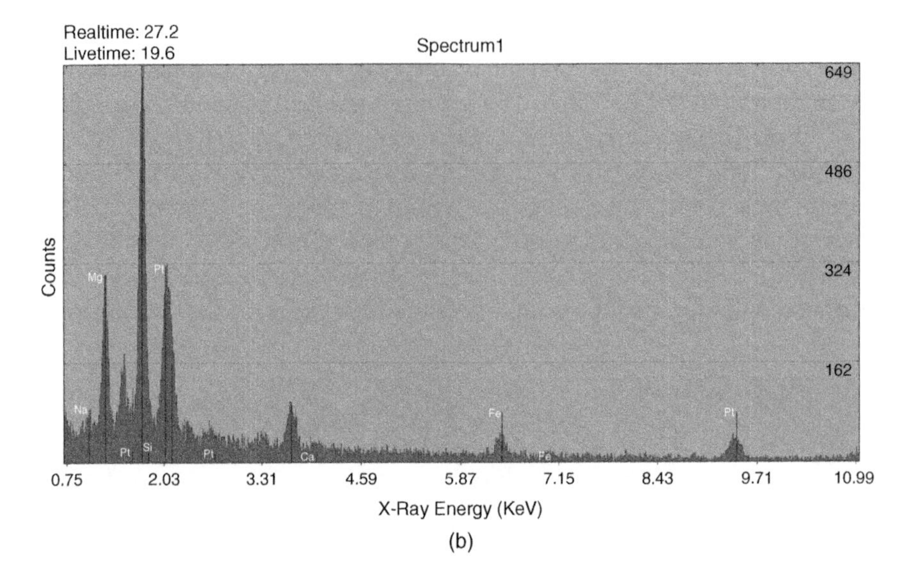

(b)

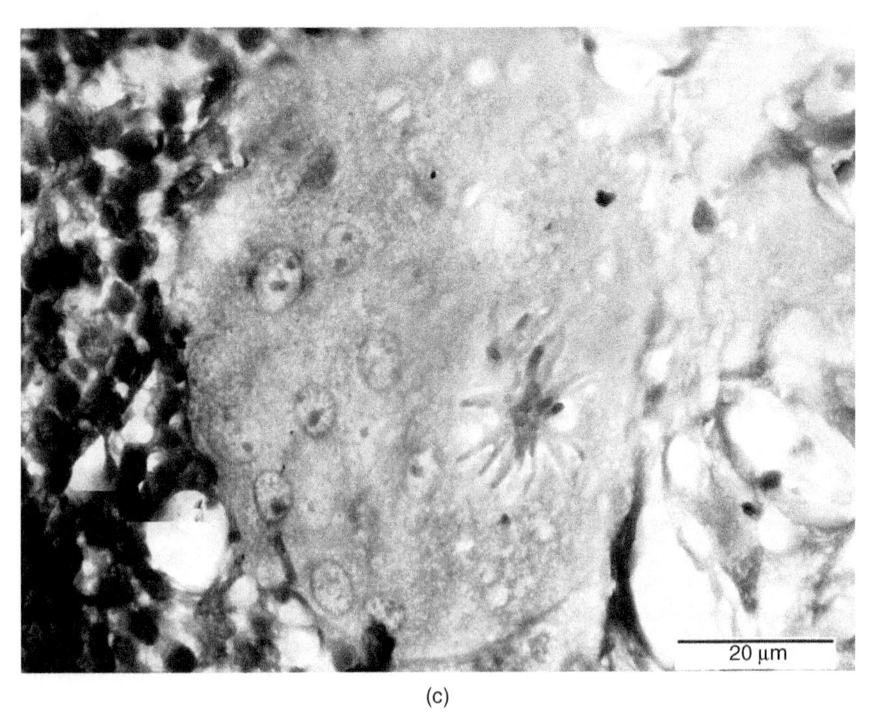

(c)

Figure 13.1 (*continued*)

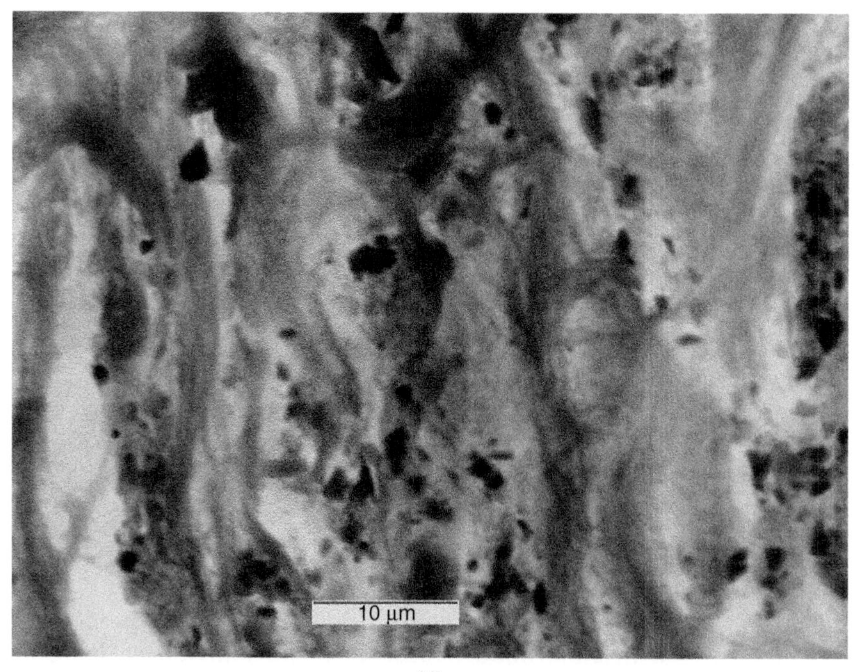

(d)

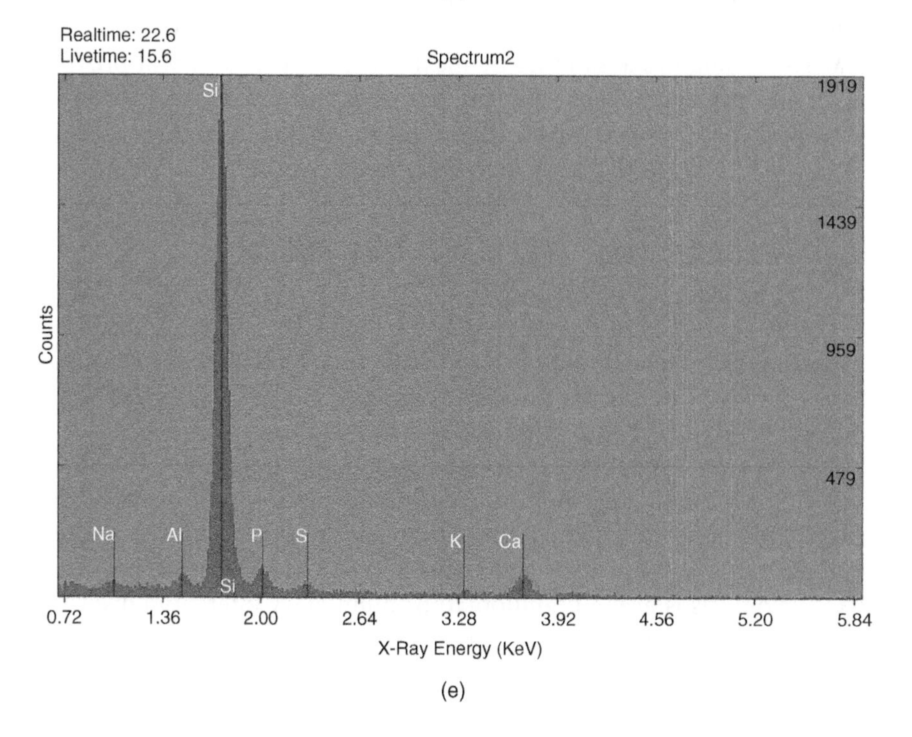

(e)

Figure 13.1 *(continued)*

is likely related to the biopersistence of commercial amphiboles and the tendency for chrysotile to break down in tissues into fibers or fibrils too small to form asbestos bodies (Roggli, Oury, and Sporn, 2004).

13.2.4 Exposure Categories

Early studies of asbestos-related diseases focused on workers in the manufacturing sector and on asbestos miners and millers. Subsequent studies identified "end users" of asbestos products as being at risk for disease, with asbestos insulators being prime examples. As more and more occupational exposures have been identified, analytical EM has provided useful insight into the extent of these exposures and the types of fibers most often found in lung samples of exposed individuals (Roggli *et al.*, 2002; Roggli, Oury, and Sporn, 2004). These include studies of household contacts of individuals who were exposed to asbestos in the workplace, in which the laundering of the worker's clothes resulted in para-occupational or domestic exposures to asbestos and, in some cases, asbestos-related disease (Roggli *et al.*, 2002). Current investigations have focused on neighborhood exposures (from individuals living in the vicinity of manufacturing plants) or environmental exposures (from naturally occurring mineral fibers). In both of these circumstances, analysis of lung tissue samples by means of analytical EM will provide useful information to help prevent future exposures. In addition, this technology assists courts in their attempt to resolve disputes regarding alleged exposures and disease in individual cases.

13.3 HYPERSENSITIVITY PNEUMONITIS AND SARCOIDOSIS

Hypersensitivity pneumonitis is a granulomatous pulmonary interstitial disorder caused by the inhalation of organic antigens. Common inciting agents include thermophilic actinomycetes in moldy hay (farmer's lung), avian proteins (bird fancier's or pigeon breeder's lung), mycobacteria (hot tub lung), and mold spores (humidifier lung). The pathologic reaction consists of airway-centered chronic inflammation associated with scattered multinucleate giant cells and patchy organizing pneumonia. Sarcoidosis is also a granulomatous pulmonary interstitial disorder of unknown etiology. Aberrant response to environmental allergens is suspected. The pathologic reaction consists of well-formed nonnecrotizing granulomas distributed along lymphatic pathways within the lung and with prominent involvement of regional lymph nodes. In either condition, the giant cells may contain asteroid bodies (Figure 13.1c) (Barrios, 2008; Farver, 2008).

Since these conditions are not caused by inorganic particulates, analytical EM is not useful in their diagnosis (other than to exclude metal-induced diseases: see Section 13.6). However, in about 10% of sarcoidosis cases and a smaller percentage of hypersensitivity pneumonitis cases, the giant cells are associated with prominent birefringent particulates. These take the form of either large platy crystals or smaller needle-like particles, and may or may not be associated with laminated basophilic structures (Schaumann bodies). The large platy crystals are calcium oxalate monohydrate, and the small needle-like particles are calcium carbonates. Schaumann bodies consist primarily of calcium phosphate (Ingram *et al.*, 1999). These particles are of endogenous origin but may be confused with foreign particulate matter. Consequently, these cases may be referred for analysis to rule out aspiration or some form of pneumoconiosis.

13.3.1 Preparatory Techniques and Analytical Methodology

Analysis of particulate material in lung tissue samples may be readily accomplished by means of SEM equipped with backscattered-electron imaging (Figure 13.1d) and energy-dispersive spectrometry (Figure 13.1b,e) (Ingram *et al.*, 1999). Paraffin sections are prepared for routine histology with a serial section mounted unstained on a carbon disc or stub. Correlation of histologic sections with the secondary electron image facilitates identification of areas of interest (such as granulomas or giant cells), and particulate material can then be identified in the SEM with back-scattered electron imaging. Spectra obtained from individual particulates provide guidance for the identification of these particulates (Ingram *et al.*, 1999). Calcium oxalate plates and calcium carbonate needle-like particles yield peaks for calcium in patients with sarcoidosis or hypersensitivity pneumonitis. Schaumann bodies yield peaks for calcium and phosphorus.

13.4 SILICOSIS

Lung disease caused by the inhalation of crystalline silica has been recognized since ancient times. The predominant form is α-quartz with tridymite, cristobalite, coesite, and stishovite constituting less common polymorphs (i.e., crystalline variants) (Butnor and Roggli, 2010; Sporn and Roggli, 2008). Due to the abundance of silica in the earth's crust, exposure occurs as a consequence of all sorts of mining operations (including coal mining). Exposures also occur from secondary usage, such as sandblasting, a practice which has been banned in many countries. Fine particles with a diameter of 1–5 μm and with freshly cleaved surfaces (as produced by sandblasting) are particularly toxic. The histologic hallmark of silicosis is the silicotic nodule, rounded well-circumscribed collagenous structures which are typically most numerous in the upper lung zones. Examination by polarizing microscopy invariably demonstrates tiny birefringent particulate matter within the nodules. As the disease progresses, the nodules (typically a few millimeters in size) coalesce to produce irregular masses referred to as conglomerate silicosis. With particularly heavy exposures, the disease may progress quite rapidly (accelerated silicosis) or present as an alveolar filling process referred to as silicoproteinosis (acute silicosis) (Sporn and Roggli, 2008).

In the typical case, the morphologic features are characteristic and analytical EM is not required. However, we have seen cases with atypical presentations or unusual features in which SEM provided useful and diagnostic information. In one case, biopsies of mediastinal lymph nodes and liver demonstrated fibrotic nodules that were interpreted as healed sarcoidosis. Since the patient had a history of working in the "dusty trades" for some 30 years and radiographic findings suggested progressive massive fibrosis (conglomerate silicosis), consultation for further analysis was requested. Alveolar macrophages obtained by broncho-alveolar lavage were placed on Thermonox® coverslips and examined with backscattered-electron imaging, demonstrating backscatter-positive particles within the cytoplasm of many cells. In addition, an aliquot of bronchoalveolar lavage fluid was spun down to produce a pellet, which was then digested in sodium hypochlorite solution and the residue collected on a polycarbonate filter (see Section 13.2.1). Numerous silicon particles (silica) within the respirable size range were identified by energy-dispersive spectrometry. It should be noted that silicotic nodules have been described in the liver and spleen of heavily exposed individuals (Sporn and Roggli, 2008).

In another example, a patient presented with numerous calcified nodules within the lungs and calcified hilar and mediastinal lymph nodes. The patient had a significant exposure history to silica, but the radiographic interpretation favored histoplasmosis (a common fungal infection in the Ohio River Valley in the United States, where the patient lived).

Examination of lung sections by SEM demonstrated the presence of discrete nodules, and back-scattered electron imaging showed numerous backscatter-positive particulates (Figure 13.1d). Although some calcium phosphate particles were encountered, the vast majority of the particles had peaks for silicon only (Figure 13.1e), consistent with α-quartz and confirming a diagnosis of silicosis. It should be noted that calcification and even ossification may occur within silicotic nodules (Sporn and Roggli, 2008; Butnor and Roggli, 2010).

13.4.1 Preparatory Techniques and Analytical Methodology

The preparatory techniques for examination of a silicosis case by analytical SEM are the same as those described in the section on hypersensitivity pneumonitis and sarcoidosis. As noted, silica particles are birefringent and often angulated. Examination of bronchoalveolar lavage fluid may also provide useful information, and whole cells may be mounted on cover slips and examined in the SEM. Alternatively, an aliquot of fluid may be centrifuged to produce a pellet, which may be embedded in paraffin and sectioned or alternatively digested in sodium hypochlorite solution and examined for its particulate content. We typically examine 50–100 particles and report results as percentages of each particulate type.

13.5 SILICATE PNEUMOCONIOSIS

Lung diseases have been reported in patients with exposure to various silicates (Table 13.1). These exposures typically take place during mining and milling but in some cases may also occur among end users. Some of the more common silicates are reviewed throughout the remainder of this section.

13.5.1 Talc Pneumoconiosis

Talc is a magnesium silicate that has widespread usage in our society because of its lubricating properties. Talc occurs in platy, granular, and fibrous forms. In some deposits, the talc is closely associated with other minerals such as silica or asbestos (anthophyllite, tremolite, or actinolite), which serve as contaminants. Other deposits contain relatively pure talc. Long-term exposure to high levels of talc may cause interstitial fibrosis (talc pneumoconiosis) (Sporn and Roggli, 2008;

Butnor and Roggli, 2010). In some cases, a granulomatous reaction has been reported. We have seen such an example in a patient who used talc for personal hygiene purposes (Ingram *et al.*, 1999). Talc is also used for pleurodesis in patients with malignant pleural effusions. In addition, talc is used as a filler in some tablets. When these tablets are crushed, suspended in an aqueous medium, and injected intravenously, they can produce intravascular lesions resulting in pulmonary hypertension (intravenous talcosis). In some cases, progressive massive fibrosis has been reported. Examination of sections of lung from patients with talcosis by polarizing microscopy shows numerous platy or needle-like brightly birefringent particulates (Sporn and Roggli, 2008; Butnor and Roggli, 2010). Analysis of serial sections by backscattered electron imaging and energy-dispersive spectrometry shows platy particulates that are backscatter positive and composed of magnesium and silicon (Ingram *et al.*, 1999).

13.5.2 Kaolin Worker's Pneumoconiosis

Kaolinite is an aluminum silicate that is a major component of some clay minerals. In some deposits, kaolinite is closely associated with silica. Individuals involved with the mining and processing of kaolinite may develop kaolin worker's pneumoconiosis. This consists of deposits of fine brown dust within the cytoplasm of macrophages, primarily in a perivascular and peribronchiolar distribution (Sporn and Roggli, 2008; Butnor and Roggli, 2010). In some cases, silicotic nodules or progressive massive fibrosis may occur, and this is related to contaminating silica. Examination of sections of lung from patients with kaolin worker's pneumoconiosis by polarizing microscopy shows numerous birefringent particulates, and analysis of serial sections by analytical SEM shows backscatter-positive particles composed of aluminum and silicon (Ingram *et al.*, 1999). If there is a significant component of silica, silicon-only particles may be identified as well.

13.5.3 Mica and Feldspar Pneumoconiosis

Mica is a potassium aluminum silicate which is used primarily in thermal and electrical insulation. Feldspars are common aluminum silicates that are used in the ceramics industry. Exposure to these minerals can in some cases lead to pneumoconiosis. Significant amounts of silica may also be found in dusts containing these minerals. Histologic sections of lung from individuals with heavy exposures to mica or feldspars show dust macules with numerous birefringent particles observed with polarizing microscopy (Sporn and Roggli, 2008; Butnor and Roggli, 2010). Silicotic nodules or massive fibrosis signals the presence of a significant silica component. Analysis by SEM shows numerous backscatter-positive particulates composed of potassium, aluminum, and silicon. If there is a significant component of silica, silicon-only particles may be identified as well.

13.5.4 Mixed Dust Pneumoconiosis

Exposure to silicates typically results in the formation of dust macules, which consist of dust-laden macrophages within the interstitium in a peribronchiolar and perivascular distribution. Exposure to silica results in the formation of silicotic nodules. If an individual is exposed to dusts that contain a mixture of silica and silicates, both lesions may be noted. If silicotic nodules predominate, the process is referred to as silicosis. If dust macules predominate, the term "mixed dust pneumoconiosis" is preferred (Honma *et al.*, 2004). Examination with polarizing microscopy demonstrates a mixture of brightly birefringent needle-like particles and tiny, weakly birefringent rounded particles. Analysis by SEM shows a mixture of silica (silicon-only particles) and silicates (silicon associated with various cations such as magnesium, potassium, aluminum, calcium, or iron). Metallic particles are also commonly observed (see Section 13.6).

13.5.5 Preparatory Techniques and Analytical Methodology

The preparatory techniques and analytical methodology for cases with silicate pneumoconiosis are the same as those described in this chapter for silica and silicosis.

13.6 METAL-INDUCED DISEASES

Exposure to a variety of metals may occur in the workplace. The lung reaction to these metals ranges from accumulation of metallic dust with little if any response to granulomatous disease to diffuse pulmonary fibrosis. These disorders are the subject of the remainder of this section (see also Table 13.1).

13.6.1 Siderosis

The accumulation of iron oxides within the lung is one of the more common dust-related diseases encountered by the author. Most of the cases result from exposure to welding fumes, and the condition is referred to as welder's pneumoconiosis (Sporn and Roggli, 2008; Butnor and Roggli, 2010). The characteristic finding is that of peribronchiolar and perivascular deposits of iron oxides, which consist of dark refractile particles surrounded by a golden-brown halo of variable thickness. There is usually little or no reaction to the dust, and the finding of significant amounts of fibrosis suggests concomitant exposure to silica or asbestos. The latter most often occurs among shipyard welders, and the former among hematite (iron sesquioxide) miners. The histologic appearance is sufficiently characteristic that analysis of the pigment by analytical SEM is seldom warranted. Such analysis when performed demonstrates spherical backscatter-positive particles that yield a peak for iron only (Sporn and Roggli, 2008; Butnor and Roggli, 2010). Among hematite miners with silicotic nodules or progressive massive fibrosis, significant numbers of silica particles are identified as well.

13.6.2 Aluminosis

Aluminum is an abundant component of the earth's crust, and exposure seldom causes significant lung disease. Individuals working in aluminum-processing plants (e.g., pot room workers), aluminum (bauxite) miners, or aluminum arc welders may have substantial workplace exposures to aluminum. Both fibrosis and granulomatous reactions have been reported in some individuals, and an intraalveolar accumulation of proteinaceous material resembling that seen in silicoproteinosis has also been observed (Sporn and Roggli, 2008; Butnor and Roggli, 2010). In histologic sections, the dust has a gray-brown refractile appearance and tends to localize around bronchioles and blood vessels. Analytical SEM demonstrates spherical backscatter-positive particles that yield a peak for aluminum only.

13.6.3 Hard Metal Lung Disease

Tungsten carbide is used in the manufacture of alloys, armaments, ceramics, cutting tools, and drilling equipment. Cobalt is used as a binder, comprising 5–25% of the final product by weight. Exposure occurs most commonly in the manufacturing process or during polishing of cutting tools (such as circular saw blades) or drilling equipment. An aggressive diffuse interstitial lung disease known as giant cell interstitial pneumonia has been described in a small proportion (<1%) of exposed individuals, and has been ascribed to an allergic response to cobalt. The disease is characterized by diffuse interstitial inflammation and fibrosis in association with numerous intra-alveolar and alveolar septal multinucleated giant cells. These giant cells may also be identified in bronchoalveolar lavage fluid (Sporn and Roggli, 2008; Butnor and Roggli, 2010). Analysis of lung tissue samples by analytical SEM typically demonstrates numerous backscatter-positive particles yielding peaks for tungsten. Tantalum and titanium particles may be identified as well (Ingram *et al.*, 1999). Cobalt may or may not be observed since it is water soluble and may be removed *in situ* from the lung tissue or by fixation in formalin solution. Giant cell interstitial pneumonia has also been described in diamond polishers who used a cobalt-containing rouge.

13.6.4 Berylliosis

Beryllium is a light element that has been used in the manufacture of structural elements, guidance systems, optical devices, rocket motor parts, and heat shields. Exposures to beryllium can result in granulomatous disease that closely resembles sarcoidosis. The distinction between berylliosis and sarcoidosis is usually made on clinical grounds including exposure history and *in vitro* blast transformation assays performed on blood or bronchoalveolar lavage lymphocytes. Because beryllium is lighter than carbon, it is difficult to detect by means of analytical SEM. Other methodologies that have been used to analyze beryllium in tissues include chemical analysis, laser microprobe mass analyzer, and secondary ion mass spectrometry (Ingram *et al.*, 1999). Butnor *et al.* (2003) reported a single case in which beryllium was detected by analytical SEM of a lung tissue sample. More studies are needed to see if there is a role for routine application of analytical SEM in the diagnosis of berylliosis.

13.6.5 Preparatory Techniques and Analytical Methodology

The preparatory techniques and analytical methodology for cases with metal-induced lung disease are the same as those described in this chapter for silica and silicosis.

13.7 RARE-EARTH PNEUMOCONIOSIS

Rare earths (predominately cerium oxide) are used in carbon arc lamps or as a rouge for polishing lenses and glass tubes. A few cases of pneumoconiosis have been reported in individuals involved in occupations using these materials or employed in an extraction plant (Sporn and Roggli, 2008; Butnor and Roggli, 2010). Both diffuse pulmonary fibrosis and a granulomatous reaction have been reported. The pattern of fibrosis is nondescript and can be confused with various forms of idiopathic pulmonary fibrosis, so knowledge of the patient's exposure history and a high index of suspicion are required. Analytical SEM demonstrates backscatter-positive particulates yielding peaks for cerium and occasionally other rare earths (lanthanum, neodymium, or samarium) (McDonald *et al.*, 1995).

13.8 MISCELLANEOUS DISORDERS

Coal worker's pneumoconiosis, also known as black lung, is a malady of coal miners caused by the inhalation of coal dust, a complex mixture of carbonaceous, silicate, and silica particles. The morphologic features of this disorder are so well characterized that analytical EM is not necessary for diagnostic purposes (Sporn and Roggli, 2008; Butnor and Roggli, 2010). In some individuals with silicone breast implants, rupture of the capsule has resulted in the migration of silicone to regional lymph nodes and even into the lungs (Roggli, McDonald, and Shelburne, 1994). Analytical SEM may be of assistance in the identification of this material, which is weakly backscatter-positive and yields a peak for silicon. Recent studies have identified a peculiar lung disease associated with the inhalation of indium and/or indium tin oxide dust (Cummings *et al.*, 2010). This material is used in the manufacture of liquid crystal displays which are becoming more widespread. The pathologic reaction shows various stages of alveolar proteinosis, cholesterol granulomas, and interstitial fibrosis. Analytical SEM showed numerous backscatter-positive particulates yielding peaks for indium with or without tin (Cummings *et al.*, 2010). Finally, it should be noted that a variety of fillers other than talc (discussed in this chapter) are used in medications intended for oral consumption, including microcrystalline cellulose, crospovidone, and corn starch. When such tablets are crushed and suspended in an aqueous medium for intravenous injection, a foreign body granulomatous vasculitis may result which may culminate in pulmonary hypertension and cor pulmonale (Tomashefski and Felo, 2004). These materials can usually be identified by light microscopic techniques, and analytical SEM has a limited role in the differential diagnosis, since these materials are organic. However, it may be useful to exclude other inorganic foreign material (such as talc).

REFERENCES

Barrios, R. (2008) Hypersensitivity pneumonitis (extrinsic allergic alveolitis), in *Dail and Hammar's Pulmonary Pathology*, 3rd edn, Chapter 17 (eds J. Tomashefski, P. Cagle, C. Farver and A. Fraire), Springer-Verlag, New York, pp. 650–667.

Butnor, K.J. and Roggli, V.L. (2010) Pneumoconioses, in *Practical Pulmonary Pathology*, 2nd edn, Chapter 9 (eds K. Leslie and M. Wick), Elsevier, New York, pp. 311–337.

Butnor, K.J., Sporn, T.A., Ingram, P. *et al.* (2003) Beryllium detection in human lung tissue using electron probe x-ray microanalysis. *Modern Pathology*, **16**, 1171–1177.

Churg, A. and Vedal, S. (1994) Fiber burden and patterns of asbestos-related disease in workers with heavy mixed amosite and chrysotile exposure. *American Journal of Respiratory and Critical Care Medicine*, **150**, 663–669.

Cummings, K.J., Donat, W.E., Ettensohn, D.B. *et al.* (2010) Pulmonary alveolar proteinosis in workers at an indium processing facility. *American Journal of Respiratory and Critical Care Medicine*, **181**, 458–464.

Farver, C.F. (2008) Sarcoidosis, in *Dail and Hammar's Pulmonary Pathology*, 3rd edn, Chapter 18 (eds J. Tomashefski, P. Cagle, C. Farver and A. Fraire), Springer-Verlag, New York, pp. 668–694.

Gylseth, B., Churg, A., Davis, J.M.G. *et al.* (1985) Analysis of asbestos fibers and asbestos bodies in human lung tissue samples: an international interlaboratory trial. *Scandinavian Journal of Work, Environment & Health*, **11**, 107–110.

Honma, K., Abraham, J.L., Chiyotani, K. *et al.* (2004) Proposed criteria for mixed dust pneumoconiosis: definition, descriptions, and guidelines for pathologic diagnosis and clinical correlation. *Human Pathology*, **35**, 1515–1523.

Ingram, P., Shelburne, J., Roggli, V. and Le Furgey, A. (eds) (1999) *Biomedical Applications of Microprobe Analysis*, Academic Press, San Diego.

McDonald, J.C. (2010) Epidemiology of malignant mesothelioma – An outline. *The Annals of Occupational Hygiene*, **54**, 851–857.

McDonald, J.W., Ghio, A.J., Sheehan, C.E. *et al.* (1995) Rare earth (cerium oxide) pneumoconiosis: analytical scanning electron microscopy and literature review. *Modern Pathology*, **8**, 859–865.

Roggli, V.L., McDonald, J.W. and Shelburne, J.D. (1994) The detection of silicone within tissues. *Archives of Pathology & Laboratory Medicine*, **118**, 963–964.

Roggli, V.L., Oury, T.D. and Sporn, T.A. (eds) (2004) *Pathology of Asbestos-Associated Diseases*, 2nd edn, Springer, New York.

Roggli, V.L., Sharma, A., Butnor, K.J. *et al.* (2002) Malignant mesothelioma and occupational exposure to asbestos: a clinicopathological correlation of 1445 cases. *Ultrastructural Pathology*, **26**, 55–65.

Schneider, F., Sporn, T.A. and Roggli, V.L. (2010) Asbestos fiber content of lungs with diffuse interstitial fibrosis: an analytical scanning electron microscopic analysis of 249 cases. *Archives of Pathology & Laboratory Medicine*, **134**, 457–461.

Sporn, T.A. and Roggli, V.L. (2008) Pneumoconioses, mineral and vegetable, in *Dail and Hammar's Pulmonary Pathology*, 3rd edn, Chapter 26 (eds J. Tomashefski, P. Cagle, C. Farver and A. Fraire), Springer-Verlag, New York, pp. 911–949.

Tomashefski, J.F. and Felo, J.A. (2004) The pulmonary pathology of illicit drug and substance abuse. *Current Diagnostic Pathology*, **10**, 423–426.

14

General Tissue Preparation Methods

John W. Stirling

Centre for Ultrastructural Pathology, IMVS – SA Pathology, Adelaide, Australia

14.1 INTRODUCTION

Overall, the fixation and processing methods used for diagnostic transmission electron microscopy (TEM) mirror those used for the majority of animal tissues: primary fixation in glutaraldehyde followed by secondary fixation in osmium tetroxide, then dehydration and embedding in epoxy resin. There are innumerable fixation and processing strategies for biomedical tissues, and almost every individual has his or her own preferred method. The reality is that there is no single 'superior' technique – only methods that work for specific individuals, or specific tissues, at a particular time and place: what gives the best results may change with circumstances. The methods and guidelines described here should be adequate for the majority of diagnostic specimens prepared singly or by batch processing. Note that techniques may need minor modification when using a tissue processor, and a small number of tissues (e.g. nerve) may require specialised treatment to achieve a high-quality result.

14.1.1 Specimens Suitable for Diagnostic TEM

Fresh solid tissue from biopsies or surgical procedures, fluids (e.g. blood) and suspensions of particulate material (e.g. nasal brush biopsies) may

Diagnostic Electron Microscopy: A Practical Guide to Interpretation and Technique, First Edition. Edited by John W. Stirling, Alan Curry and Brian Eyden.

all be used for diagnostic TEM. Tissue can also be salvaged from formalin-fixed material or reprocessed from a wax block or histology slide. Formalin-fixed specimens may be quite well preserved, but many will show a significant loss of fine structural detail. Tissue reprocessed from a wax block or histology slide will invariably show significant damage and may be difficult to interpret (especially material from histology slides). While useful diagnostic detail may still be visible in such material, it is important to remember that the tissue may not be directly comparable to optimally preserved tissue, and the results must therefore be treated with caution. Specifically, reprocessed tissue is not suitable for making critical measurements of the glomerular basement membrane (GBM) (Nasr *et al.*, 2007). Frozen tissue is generally too badly damaged at the cellular level to be useful for diagnostic purposes, although some gross morphological structure may still be interpretable.

14.2 TISSUE COLLECTION AND DISSECTION

Specimen collection, preparation and dissection must be carried out in a way that facilitates tissue screening and the rapid location of diagnostic features. On collection, tissue must be immediately placed in fixative: this is critical for preserving ultrastructural detail. When this is not possible (e.g. while a renal biopsy is being divided for light microscopy (LM), TEM and immunofluorescence (IF)), the material may be placed in cold sterile saline or transport gel for a short period prior to fixation (no more than ~15–30 minutes). Primary (and secondary) fixation may be carried out from ~4 °C to 20 °C (room temperature); in respect to diagnostic applications, this temperature range produces a negligible difference in the quality of ultrastructural preservation. Tissue may be dissected immediately; however, soft tissues can be hardened by a short period of initial glutaraldehyde fixation (~15 minutes). The final size of tissue blocks is also important because TEM fixatives do not penetrate tissue quickly to any great depth. For example, standard texts such as Hayat (1970) suggest that osmium tetroxide may penetrate effectively only to ~0.25 mm. On this basis, Hayat (1970) suggests that while aldehydes may penetrate larger specimens, blocks no larger than 0.5 mm^3 are required to achieve full uniform fixation with osmium tetroxide. While small blocks are recommended, for practical purposes (tissue sampling, orientation and structural integrity) pieces of tissue larger than this are often required. Tissues vary in their requirements, but 1 mm cubes or thin slices ~1–2 mm^2 long and 0.5–1 mm thick will often give acceptable results. Muscle and nerve should be left in pieces large enough to preserve tissue integrity, and skin biopsies for the diagnosis of cerebral autosomal dominant arteriopathy with subcortical infarcts and leukoencephalopathy (CADASIL) can be left in large thin sheets to facilitate the location and selection of blood vessels.

14.2.1 Tissue Cut-Up

14.2.1.1 Renal Biopsies

Renal tissue is usually divided for LM, IF and TEM following a set protocol. Here we provide only general guidelines as local practice varies considerably.

On collection (or before dissection), check that there is an adequate amount of cortex with glomeruli present and that when the tissue is divided there are enough glomeruli for LM (if in doubt, check the allocation of tissue with the reporting pathologist). Typically, a core of native kidney consisting entirely of cortex can be divided approximately as follows:

TEM	IF	LM	IF	TEM

The total overall allocations are generally ~50% for LM, 30% for IF and 20% for TEM. The amount for LM may be increased to ~75% for transplant wedge biopsies with the remainder divided equally for TEM and IF. The tissue should be kept in cold sterile saline or transport gel while it is transported and dissected and should never be allowed to dry out. Place the dissected portions of tissue into transport medium (for IF) or fixative (for LM and TEM) as appropriate. **The IF tissue must be collected first so that it is not contaminated by fixative.** For TEM, do not cut the cores too small or thin as the glomeruli may fall out of the periphery of the tissue. For example, cores up to ~3–4 mm in length can be left whole, but longer cores can be cut transversely into ~3 mm pieces. This division means that the entire core can be evaluated in a single semithin section.

14.2.1.2 Skin Biopsies: CADASIL

For the diagnosis of CADASIL, a 3–4 mm punch biopsy of skin is normally adequate for TEM. On collection, the tissue may be hardened by a brief period of fixation (~15 minutes), then the full depth of the biopsy is cut into vertical slices ~0.5 mm thick. This is best done from the side to avoid crushing the tissue (Figure 14.1).

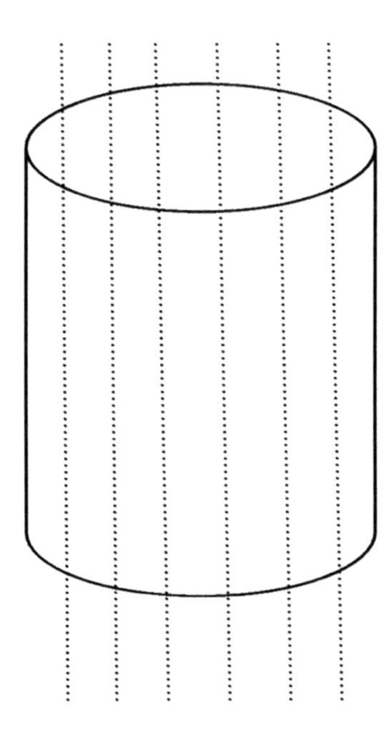

Figure 14.1 Punch biopsy of skin for CADASIL. Cut the tissue from the side into full-depth slices ~0.5 mm thick.

To facilitate tissue screening, the full-depth slices are left whole and embedded flat: this allows the entire tissue block to be evaluated from a single semithin section (see Chapter 9). When transport in glutaraldehyde is not possible, skin biopsies taken at a distant site can be fixed whole for ~15–18 hours, then transferred into buffer before they are sent to the receiving laboratory. Biopsies fixed in formaldehyde or formalin are not suitable for CADASIL evaluation.

14.2.1.3 Small or Delicate Specimens, Fluids and Suspensions

To prevent loss, material extracted from fluids and suspensions can be embedded in a medium such as gelatine or bovine serum albumin (BSA) to form a 'tissue block' for ease of handling. Similarly, fragile sections on slides and small or delicate specimens can be embedded in BSA or wrapped in lint-free paper.

14.2.1.4 Cornea

Corneal specimens are best processed so that the full depth of the cornea (from the epithelial surface to Descemet's membrane) can be evaluated

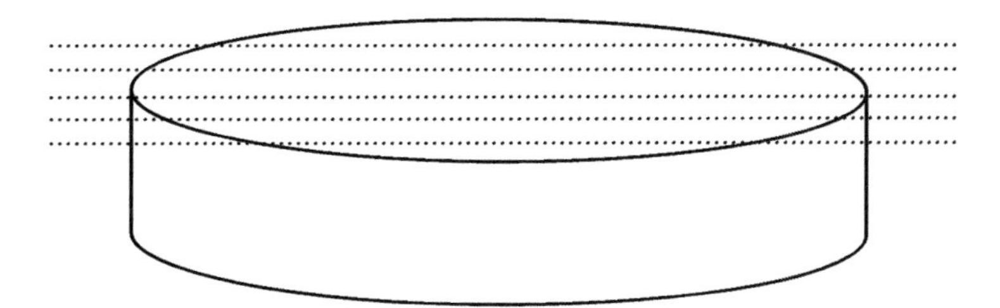

Figure 14.2 Cornea. Cut the tissue from above (the epithelial surface) into vertical full-depth slices ~1 mm wide.

in a single section. To achieve this, cut the cornea (or corneal segment) into long, thin, vertical 'full-depth' slices ~1 mm wide (like the slices of a cake; Figure 14.2), and embed flat.

14.2.1.5 Liver

Some individuals find that liver fixes and processes poorly; this presumably reflects differences in practice. To ensure adequate fixation and processing, the size of tissue samples may need to be minimised. For example, needle biopsies can be divided longitudinally to allow full penetration of reagents (particularly osmium tetroxide).

14.2.1.6 Muscle and Nerve

See Chapter 3 on muscle and Chapter 4 on nerve. Cardiac muscle does not need to be held under tension but is otherwise treated in the same way as striated muscle.

14.2.1.7 Solid Tissue and Lymph Nodes

Solid tissue such as tumours and lymph nodes should be dissected and sampled to include material from different areas (so that all possible variations in the tissue are represented). Where different areas are clearly visible in the tissue, each of these should be sampled.

14.3 TISSUE PROCESSING

14.3.1 Fixatives and Fixation

Historically, several fixatives have been used for diagnostic TEM, either alone or as mixtures (principally aldehydes). Several fixative additives have also been used to improve the fixation of specific cell components (Glauert, 1975; Hayat, 1981; Bullock, 1984). Physico-chemical fixation combining microwave techniques with glutaraldehyde and osmium tetroxide fixation is increasingly being used for diagnostic TEM in conjunction with rapid processing (Hayat, 2000).

In some instances, fixatives and buffers are tailored to specific tissues (e.g. nerve). For general purposes, fixation in glutaraldehyde and osmium tetroxide (double fixation) is adequate for most specimens. For an overview of fixation, see Grizzle, Fredenburgh and Myers (2008); and for an in-depth discussion of fixatives and fixation as applied to routine TEM, see Glauert (1975) and Hayat (1970, 1981); the text by Hayat (2000) is comprehensive and includes information on the storage and purity of glutaraldehyde, fixation reactions and fixation techniques in general. Microwave techniques are described in Chapter 15 and by Hayat (2000).

During processing, it is critical that tissues are always held in a volume of fluid that is adequate for the purpose intended; this is particularly important when using a rotator or tissue processor and when changing fluids during hand processing. Specimens must always be completely immersed in fixative, buffer, dehydrant, transition fluid or resin and must remain 'enveloped' in reagent during fluid exchange steps (particularly when using volatile solvents such as epoxy propane). For full fixation to occur, the tissue must also be immersed in an adequate volume of fixative: in general, this should be at least 10 times the volume of the specimen(s). For transport purposes, specimen containers should be completely filled with fluid.

General indicators of poor fixation and osmotic stress are swollen mitochondria and endoplasmic reticulum (both of these organelles are potent indicators of post-mortem change) followed by an increasing level of damage to other organelles. In renal biopsies, delayed fixation is initially signalled by damage to the foot processes which lift from the GBM and show signs of osmotic stress: a significant delay will result in more widespread damage and cell extraction (loss of cell contents). When tissues are poorly fixed, cells will appear swollen and pale, their contents will be obviously damaged or extracted and cell membranes will be damaged or lost. Poor or uneven fixative penetration is characterised by damage in the poorly infiltrated area. Furthermore, there may be pale or poorly stained areas visible in the semithin sections by LM, or the tissue may show post-mortem changes at the ultrastructural level. Poor osmium tetroxide penetration is characterised at the ultrastructural level by pale nuclei and lipid extraction.

14.3.2 Primary Fixation: Glutaraldehyde

Glutaraldehyde may be used at a range of concentrations in phosphate or cacodylate buffer (pH 7.2–7.4): a 2.5% solution is recommended as this is easy to prepare from the commercial stock solutions generally available. Glutaraldehyde is not osmotically active and does not contribute to the osmolarity of the solution (Glauert, 1975). The length of fixation (time) can be varied according to specimen type. Small blocks and cores of standard tissues (e.g. renal biopsies) can be fixed at room temperature for a minimum of 1.5–2.0 hours. Large tissue blocks and dense tissues may require longer. Fluids and tissue suspensions may be fixed for 30–60 minutes if no large pieces of material are present. The majority of specimens do not appear to be significantly damaged by long periods (e.g. 2 weeks) in glutaraldehyde.

14.3.3 Secondary Fixation (Post-fixation): Osmium Tetroxide

Osmium tetroxide may be prepared at a range of concentrations in water (the compound destroys the osmotic activity of membranes) or in phosphate or cacodylate buffer, pH 7.2–7.4 (Glauert, 1975). (Note that osmium tetroxide is also soluble in nonpolar solvents.) A solution of 1% osmium tetroxide in water is recommended as aqueous solutions are stable and can be stored at 4 °C for long periods; solutions in buffer are unstable and will deteriorate (within ~24 hours) unless frozen. The length of fixation in osmium tetroxide can be varied, but 60 minutes at room temperature is adequate for most specimens. After osmium tetroxide fixation, the usual practice is to immediately process the tissue into resin (while storage for a short period in buffer is permissible, retaining specimens in osmium tetroxide or storing them for any length of time after osmium tetroxide fixation is not generally recommended). Potassium ferricyanide may be added to osmium tetroxide fixatives to improve the fixation of glycogen and other organelles (De Bruijn, 1973; Hayat, 2000).

14.3.4 Fixative Vehicles and Wash Buffers

Buffers are used as fixative vehicles and washing fluids; they are usually made with an osmolarity similar to that of normal tissue (300–330 mOsm) and a pH in the range of 7.2–7.4. The most common buffers are 0.1 M phosphate and 0.1 M sodium cacodylate. Phosphate is recommended for general use as it is the most 'physiological', easy to prepare and nontoxic, and in comparison to cacodylate it is inexpensive. The drawbacks of phosphate buffer are that it can support the growth of microorganisms and, with some specimens, may cause a fine precipitate within the tissue (especially nerve). Cacodylate is toxic (it contains arsenic) and does not support the growth of microorganisms (Glauert, 1975).

For the majority of specimens, the same buffer that is used in the primary fixative may be used for subsequent washing steps. Various authorities (e.g. Glauert, 1975) suggest the addition of sucrose to fixatives and wash buffers for osmotically sensitive specimens, and this is standard practice in some laboratories, especially for cacodylate buffer. However, other practitioners have found that sucrose can be omitted from cacodylate buffer in diagnostic applications with no deleterious effects.

14.3.5 En Bloc Staining with Uranyl Acetate

The application of uranyl acetate *en bloc* has both staining and fixative effects, and its use can improve the visibility of fine structural detail. However, the stain may change the appearance of glycogen, and prolonged treatment may cause the extraction of some components. Furthermore, the compound reacts strongly with cacodylate and phosphate buffers, so specimens must be washed appropriately before application (Lewis and Knight, 1977; Hayat, 2000), usually with a 0.1 M solution of sodium acetate (B. Eyden, personal communication). *En bloc* staining with uranyl acetate can be carried out either before or after osmium tetroxide fixation, and it can be applied in either aqueous or alcoholic solution (Hayat, 1993). Alcoholic solutions can be incorporated into the dehydration series, and this may be the most convenient strategy considering the possibility of buffer interactions. Hayat (1993) recommends two alternative methods: (i) 2% uranyl acetate in acetone or ethanol (as appropriate) for ~15 minutes as the first step of the dehydration series or (ii) a 0.5% aqueous solution at pH 3.9 applied for 10 minutes immediately before dehydration. Some practitioners have used a 0.5% aqueous solution successfully for long periods, including overnight (B. Eyden, personal communication).

14.3.6 Dehydrant and Transition Fluids

After secondary fixation, tissues are dehydrated by passing them through an increasing concentration (graded series) of dehydrant followed by infiltration using epoxy resin mixtures. Resin mixtures may be made using the dehydrant or a 'transition fluid' such as propylene oxide (1–2, epoxy propanol or 'epoxy propane'). Note that the choice of dehydrant, transition fluid and resin may have an impact on the dimensions of tissue structures such as the GBM and can affect the outcome of morphometric studies (Edwards *et al.*, 2009).

Ethanol and acetone are the two most popular dehydrants. Ethanol is preferred over acetone as it is less hygroscopic and is compatible with *en bloc* staining. The use of methanol as a dehydrant is not widely described in standard texts; however, it gives similar results to ethanol and is also compatible with *en bloc* staining. Ethanol and methanol are usually used in conjunction with propylene oxide, the latter as the transition fluid for resin infiltration (Hayat, 1970; Glauert, 1975).

Acetone can be used as both the dehydrant and transition fluid, or in conjunction with propylene oxide. Acetone is slightly more miscible with epoxy resins than ethanol, but its hygroscopic properties may result in incomplete dehydration. Traces of acetone remaining in the tissue may also interfere with polymerisation: this can result in a soft block, or soft areas within the block (Glauert, 1975).

Propylene oxide is popular as a transition fluid because it penetrates tissue very easily, thus removing traces of dehydrant and improving resin infiltration. Furthermore, propylene oxide is a reactive diluent and has no effect on block quality if small amounts are left in the tissue (Glauert, 1975).

14.3.7 Resin Infiltration and Embedding Media

When dehydration is complete, infiltration is carried out using a graded resin series: typically 50%, 75% then 100% resin. Resin mixtures are made using the transition fluid (propylene oxide) or dehydrant as appropriate. Gentle agitation is highly recommended for resin infiltration, and if available, vacuum embedding is recommended for the final resin steps.

14.3.7.1 Epoxy Resin Embedding Media

Epoxy resins are the most popular embedding medium for standard biomedical applications. Epoxy resins are all multi-element resins that may include a number of components (resin, hardener, plasticiser, flexibiliser and accelerator) in the final mixture. The ratio of these components is varied to change the hardness of the block and its sectioning qualities. The three common resins, Araldite, Epon and Spurr's, are all available commercially as kits with instructions for producing blocks of varying hardness (soft, medium and hard). Medium-hardness resin should be suitable for most diagnostic applications. Resin qualities can be adjusted more specifically by adjusting the anhydride–epoxy ratio: such calculations are complex and not recommended for the general user (Glauert, 1975). Spurr's, the least viscous resin, originally contained vinyl cyclohexene dioxide (ERL 4200) which was found to be highly toxic; this component has now been replaced with the safer ERL 4221 (Ellis, 2006). In all cases, the complete resin mix (including accelerator) is recommended for infiltration and final embedding. Accelerator is required to initiate polymerisation which is completed using heat at ~70 °C.

14.3.7.2 Alternative Embedding Media

There are several embedding media that may be used as alternatives to epoxy resin for specialist procedures such as low-temperature processing, histochemistry and immunogold labelling (Acetarin, Carlemalm and Villiger, 1986; Carlemalm, Garavito and Villiger, 1982; Newman and Hobot, 1987, 1993). Many of these resins can be photo-polymerised. Photo-polymerisation is not generally compatible with pigmented or osmicated material because the colouration interferes with light penetration so that the resin in the centre of the tissue may not be completely polymerised. In such cases, chemical polymerisation must be used (Acetarin, Carlemalm and Villiger, 1986). In addition, oxygen interferes with the polymerisation of a number of these resins, so to ensure polymerisation, sealable embedding capsules are required.

14.3.7.2.1 Low-Acid Glycol Methacrylate

Low-acid glycol methacrylate (LA-GMA) is useful for immunogold labelling studies because, in many cases, it has a minimal effect on epitope immunoreactivity. LA-GMA may be used in conjunction with 'inert dehydration' using ethanediol to process unfixed tissue for immunogold labelling (Stirling, 1992). Even with conventional techniques, the use of LA-GMA can result in extremely high label densities, in some cases outperforming frozen sections (Bendayan, Nanci and Kan, 1987; Stirling, 1992, 1995).

14.3.7.2.2 LR White™ and LR Gold™ (the London Resin Company)

LR White is a low-viscosity (8 cps), single-component acrylic resin suitable for a wide range of applications, particularly immunogold labelling. A feature of LR White is that it is compatible with ~12% by volume of water which allows partial tissue dehydration. The polymerised resin is hydrophilic (Newman, 1987; Newman and Hobot, 1987). Polymerisation is effected by blue light (365 nm) at low temperature in combination with an accelerator, or by heating in an oven. A method of polymerising LR White in 7 minutes for immunocytochemistry is described by Hillmer, Joachim and Robinson (1991). The resin can be used for both light and electron microscopy. LR White is incompatible with acetone, traces of which interfere with polymerisation.

LR Gold™ is an acrylic resin designed for processing fresh unfixed tissue at low temperature, which makes it ideal for histochemistry and immunogold labelling. The unfixed tissue is dehydrated through a graded series of methanol and polyvinyl pyrollidone (PVP), then embedded in pure resin monomer at −25 °C. The PVP is added to protect the unfixed tissue from osmotic changes. Low-temperature polymerisation is carried out using blue light in conjunction with benzil as the initiator. LR Gold polymerisation at room temperature is initiated using benzoyl peroxide; LR White accelerator can be used to increase the curing speed. Note that polymerisation is exothermic, so blocks must be cooled. Polymerised blocks can be stored at room temperature, but cold storage may help preserve enzyme activity. LR Gold is oxygen sensitive.

14.3.7.2.3 To the Lowicryl® Series

The Lowicryl resins are low-viscosity acrylics designed for low-temperature embedding (Carlemalm, Garavito and Villiger, 1982; Acetarin, Carlemalm and Villiger, 1986). K4M (useable at −35 °C) and K11M (useable at −60 °C) are hydrophilic and can be polymerised with up to 5% by weight of water. K4M and K11M are particularly useful for immunolabelling studies as they retain epitope immunoreactivity in combination with low background. HM20 (useable to −70 °C) and HM23 (useable to −80 °C) are hydrophobic and have a low density, properties that make them useful for producing high-contrast images of unstained material with minimal denaturation. All the Lowicryl resins can be polymerised in the cold or at room temperature (or higher) using blue light (360 nm) and an initiator. Blocks polymerised at room temperature may be ready for sectioning in only a few hours. Lowicryl is oxygen sensitive.

14.3.7.2.4 The Nanoplast® Series

The Nanoplast resins FB101 and AME01 are melamine–formaldehyde (a 'technical aminoplast') that has been formulated to give hard 'glass-like' blocks for extremely thin sectioning and a level of ultrastructural resolution that cannot be achieved with other resins (Bacchuber and Frösch, 1982; Frösch and Westphal, 1985). Unfixed material embedded at low temperature in AME01 shows extremely good ultrastructural detail (Frösch and Westphal, 1986). Nanoplast can be used for applications that require extremely thin sections (down to ~10 nm), including conventional high-resolution TEM and electron phase-contrast studies of unstained material. Nanoplast is not suitable for immunolabelling studies.

14.3.8 Tissue Embedding

After epoxy resin impregnation, samples are embedded in fresh resin in embedding capsules or flat silicone moulds. Tissue should be orientated so it can be sectioned from the appropriate direction. Polymerisation is achieved using heat. Commercial embedding capsules of various shapes are available in two materials: polyethylene (soft: resistant to 75 °C) and polypropylene (hard: resistant 100 °C). The tops of large polyethylene specimen tubes can also be used for flat embedding: this can be useful when specimen orientation is problematic. A layer of resin is pre-polymerised in the lid, the specimen is placed on this layer of resin and the lid is then filled with fresh resin to a suitable depth.

All specimen moulds and capsules must contain a label with the relevant specimen identifiers. A strip of paper placed in the resin around the perimeter of the mould is adequate: the labels can be laser-printed, or the details written in pencil (ink may leach into the embedding medium).

14.4 TISSUE SECTIONING

14.4.1 Ultramicrotomy

Ultramicrotomy is usually a two-step process: (i) semithin sections are cut and stained to survey the tissue, and (ii) the specific area of interest is ultrathin sectioned for diagnostic screening. For a detailed discussion of all aspects of sectioning, see Reid and Beesley (1991).

14.4.1.1 Semithin 'Survey' Sectioning

To cut survey sections, the block is minimally trimmed with a clean degreased razor blade to form a regular, shallow and flat-topped pyramid. The face to be sectioned should have a square or trapezoidal surface and should include as much of the tissue as possible. The overall dimensions should generally be no more than $\sim 2.0 \times 3.0$ mm in size. The upper and lower edges of the block face must be straight and parallel if ribbons of sections are required. In respect to height, the pyramid should be as low as possible to maximise block stability and to retain the maximum amount of tissue in case additional levels are required: a height of ~ 1 mm is usually adequate.

Semithin survey sections are cut on a glass knife at $\sim 0.5-1.0$ µm: this can be done dry, or the sections can be cut onto water (a drop held at the knife edge or a water-filled trough). Thick sections should never be cut on a diamond knife unless the knife has been specifically designed for this purpose. Initial block 'facing' and subsequent coarse sectioning (trimming) can be done on the right-hand side of the knife edge; better quality sections for viewing can be cut from the left-hand edge. (The right-hand edge of a glass knife usually has numerous small imperfections or chips – these decrease towards the left so that the extreme left-hand edge is usually 'chip-free' and may be used to cut either semithin or ultrathin sections.) Sections are collected using clean fine forceps (or a wire loop or damp fine brush), transferred to a glass slide and stained for LM (e.g. toluidine blue).

14.4.1.2 Survey Sections: Choice of Tissue

The sections should be reviewed (preferably with the reporting pathologist) to ensure that the feature (or tissue) of interest is present and that the material corresponds to the LM findings. Where only a small amount of useable tissue is present, it is best to proceed with ultrathin sectioning rather than risk trimming deeper and losing the diagnostic tissue.

14.4.1.3 Ultrathin Sectioning

Prior to ultrathin sectioning, excess tissue and resin are removed from the edges of the block to leave a cutting face of $\sim 1\,mm^2$; the block should also be shaped and orientated to facilitate ultrathin sectioning. For example, tissue and embedding medium do not generally compress equally during cutting, and wherever possible, tissue-free resin should be excluded from the block face. Alternatively, if straight ribbons of sections are required, areas of plain resin can be located at the top or bottom edge of the block face so that the sections expand without sideways distortion. If the tissue contains a hard inclusion that will damage the section, the inclusion should be removed or, alternatively, positioned at the edge of the block so that the area of damage is limited.

Occasionally, block faces larger than 1 mm^2 may be required for some purposes. For example, it is useful to cut the full depth of skin biopsies for the diagnosis of CADASIL; this allows arterioles to be located easily and reduces the number of blocks that must be sampled. For screening, sections can be observed across multiple grids, or two sections can be positioned on the grid so that the bottom of one and the top of the other can be viewed.

Ultrathin sections are cut on a diamond (or glass) knife. During sectioning, section thickness is monitored using the interference colour of the sections and the specimen feed is adjusted as required: grey is <60 nm, silver is 60–90 nm and gold is 90–150 nm (Hayat, 1970; Reid and Beesley, 1991). It is common practice to cut sections that are thicker than required, that is, ~100–130 nm: after flattening, these will reduce to a useable thickness. Section thickness is determined by the level of contrast and resolution required: contrast increases with section thickness, while resolution decreases (Frösch and Westphal, 1985; Reid and Beesley, 1991). High-resolution studies require sections of ~60–70 nm (or thinner), but for diagnostics and immunogold labelling, more robust sections with a thickness of ~80–90 nm are preferred.

14.4.1.4 Specimen Grids

Sections are collected on bare (uncoated) 3 mm grids. Grids are available with different composition, mesh size and geometry, and the choice of style will depend on the size of the diagnostic features of interest, section size and stability and application (e.g. diagnostics or immunolabelling). For example, relatively large (150) mesh grids may be preferred for standard applications (e.g. renal tissues), but a smaller mesh (400) may be more appropriate for small unstable or fragile sections and immunolabelling. Copper, copper–rhodium and copper–palladium grids are all acceptable, but nickel grids, commonly used for immunolabelling, are not recommended for general use as they can become charged and cause image astigmatism. Sections may be collected on either side of the grid according to personal preference. Note that on one side, the grid bars are domed with a rough surface that reputedly gives excellent section adhesion (the dull side on plain copper grids). On the other side, the grid bars are flat and smooth with sharp edges; these edges can cause the section to tear (Summers and Rusanowski, 1973).

14.4.1.5 Section Flattening and Collection

Conventionally, sections are flattened before they are collected in order to relieve sectioning compression which may be up to 35% (Studer and Gnaegi, 2000). Flattening is achieved using chloroform, ether or amyl–acetate vapour applied with a sharp wooden stick (preferred) or cotton bud; heat is also effective and may be applied using a 'heat pen' (available commercially). Some practitioners do not flatten their sections in this way; they rely on the heating effect of the electron beam which also has a modest flattening or 'stretching' effect. Whichever strategy is used, it is important to maintain a consistent protocol if the sections are to be used for measuring purposes as the degree of flattening will determine the dimensions of the section's contents. After flattening, sections are collected from below onto a grid held with a pair of clean fine forceps. Sections can be manipulated while they are in the knife trough by using a fine hair or eyelash that has been mounted on a wooden stick with dental wax. The hair should be cleaned before use to avoid contaminating the trough fluid.

14.4.1.6 Knife Troughs and Trough Fluid

Disposable troughs are available for glass knives, and a simple trough can also be made from metal tape (available commercially). Troughs can be fixed and sealed using molten dental wax or clear nail varnish.

For epoxy and acrylic resins, clean distilled or de-ionised water is normally recommended as the trough fluid. During sectioning the water level should be monitored to ensure that the knife edge is wet at all times. A variety of additives and alternative trough fluids have been recommended for reducing surface tension and for special applications (such as preventing the extraction of tissue components) (Reid and Beesley, 1991). A 10–20% solution of ethanol or acetone has been advocated for methacrylate resins: these reagents lower surface tension and help to flatten the sections (they have less effect on epoxy resins). Historically, the use of solvents was not recommended for use in conjunction with diamond knives because of the danger of damaging the knife cement. Some diamond knives now have solvent-resistant cement – check with the knife manufacturer.

14.4.2 Sectioning Technique and Ultramicrotome Setup

For a full discussion of sectioning and sectioning techniques, see Reid and Beesley (1991).

14.4.2.1 Choice of Knife and Knife Angle

Diamond knives in a variety of designs are available for both general and specialist use (e.g. cryo-ultramicrotomy and cutting large resin sections for LM). The knife angle (i.e. the angle between the front and back surfaces of the knife) is the major factor determining knife application: 35 ° is recommended for soft biological blocks and extremely thin sectioning (~10 nm), 45 ° is recommended for general use (including diagnostics) and 55 ° is recommended for hard samples. Use of an oscillating knife reduces compression (Studer and Gnaegi, 2000).

14.4.2.2 Clearance Angle

The clearance angle is the angle between the front face of the knife and the block face when the block is set up for sectioning. In general, compression increases with the clearance angle and knife angle combined (Reid and Beesley, 1991). A recommended clearance angle is usually supplied by the knife manufacturer for each knife (generally 5–6 °). Manufacturers generally advise that clearance angles of 2–10 ° are safe but that angles greater than 10 ° may cause damage to the knife edge.

14.4.2.3 Cutting Speed

The cutting speed is the rate at which the specimen passes the knife during the cutting stroke and is usually expressed in millimetres per second (mm/s). The optimum cutting speed will depend on the hardness of the block: relatively hard specimens require a slow speed, and softer specimens a higher speed. Standard diagnostic material in epoxy resin generally cuts well at ~2 mm/s. The optimum cutting speed for a specific tissue should produce ribbons of smooth sections with minimal compression.

14.4.3 Common Sectioning Problems and Artefacts

14.4.3.1 The Knife Edge Fails to Wet

The knife edge fails to wet either partly or fully when the trough is filled with fluid.

- Check the knife edge for contamination, and clean it according to the manufacturer's instructions
- Over-fill the trough with fluid (convex meniscus), and leave it for 5–10 minutes. Reduce the fluid to the required level – draw the fluid across the knife edge using a clean hair mounted on a wooden stick (if required).
- Cut a section dry so that it 'bunches' on the edge of the knife, then over-fill the trough with water. Slowly lower the water to the required height, and start regular sectioning.

- Make the surfaces close to the diamond hydrophilic. Apply a thin layer of mild surfactant such as Photo-Flo 200^{TM} (Eastman Kodak, Rochester, NY) to the inside of the trough adjacent to the diamond (not on the diamond). Allow the detergent to dry thoroughly before refilling the trough and commencing sectioning.
- Use a 10–20% solution of ethanol or acetone as the trough fluid (see Section 14.4.1.6 above).

14.4.3.2 Irregular Sectioning

Sections are produced erratically or with alternative cutting strokes, or thick and thin sections are cut alternatively. Check that the specimen is locked in place and that the fine feed is working and not at the end of its working range. Slowly increase the specimen advance until sections cut with every stroke. In some cases, the cutting speed may be inappropriate – adjust as required.

14.4.3.3 The Block Face Becomes Wet

Epoxy resin blocks may attract water because the fluid level in the knife trough is too high or because of electrostatic charging, particularly if humidity is low. Dry the block and the back of the knife using a sliver of clean filter paper (avoid dust), then adjust the level of the trough fluid. Reduce static with an antistatic gun. In some cases, increasing the cutting speed may be helpful.

14.4.3.4 Sections Pull over the Back of the Knife

Sections may be pulled over the back of the knife, either wholly or partially; at the same time, the block face may become wet and the section may be pulled onto the block face on the return stroke. The underlying cause of this problem is complex and may include a soft block, electrostatic charging, low humidity, the level of fluid in the knife trough and slow cutting speed. In the first instance, dry the back of the knife and the block (as discussed in Section 14.4.3.3), then adjust the fluid level in the knife trough. Subsequently, slowly increase the cutting speed and decrease static with an antistatic gun.

14.4.3.5 Poor Ribbon Formation

The principal reason that sections do not form ribbons is because the upper and lower edges of the block face are not straight and parallel with each other or with the knife edge. The latter is particularly important: as each new section starts, the whole edge of the block face must be in contact with the previous section (and the knife edge) for ribbons to form effectively. The upper and lower edges of the block must also be smooth, or the sections may not fully adhere to each other. Poor ribbon formation may also occur because the trough fluid is too high.

14.4.3.6 Chatter

Chatter has the appearance of alternating thick and thin bands (for a venetian-blind effect) across the entire width of the section (parallel to the knife edge). Firstly, check for vibrations in the ultramicrotome and in the general environment. Secondly, ensure that:

- The specimen block is held firmly in the specimen holder (chuck).
- The specimen holder is locked on the cutting arm.
- The knife is securely locked.
- The diamond knife is firm in its mounting.

Finally, check that the ultramicrotome has not developed a mechanical fault. Chatter may also be caused by:

- A blunt knife.
- Incorrect cutting speed, clearance angle or knife angle (which may relate to the hardness of the block) – adjust as required.
- A soft or unevenly polymerised block.
- A large block face. Retrim the block, and reduce the size of the block face.

14.4.3.7 Wrinkles and Folds

Wrinkles and folds are generally the result of compression or localised differences in block hardness (due to uneven polymerisation or the presence of discrete structures such as lipid droplets). Sections should be flattened before collection.

14.4.3.8 Compression

Compression occurs during cutting and results in sections that are shorter than the original block face. Small wrinkles and folds may also be the result of compression. Compression is caused by:

Table 14.1 Section contamination: causes and remedies.

Contamination	Cause	Remedy
Oily deposits	Dirty sectioning equipment	Check and clean all equipment as appropriate.
		Use clean blades for block trimming.
	Knife trough contaminated during sectioning or section flattening	Check and clean equipment; avoid contaminating the knife trough.
	Dirty grids or forceps	Clean grids and forceps.
	Staining: poor washing technique or contaminated wash water	Improve washing technique. Clean the wash container.
	Dirty staining or washing equipment	Replace wash water. Check and clean washing equipment.
Amorphous small particulates	General environment	Remedy as appropriate – avoid air currents, etc.
Crystalline particulates		Store grids in a clean container.
	Poor staining technique: inadequate washing and/or contaminated washing water	Improve washing technique during grid staining. Clean wash water container. Replace wash water.
		Check and clean washing equipment.
	Dirty grids	Clean grids.
	Poor technique when collecting and drying sections	Use fresh dust-free filter paper for drying grids. Avoid touching the sections with the filter paper.
Amorphous or small flat particulates Small round or crystalline particulates	Lead stain artefact due to contaminated stain, poor staining technique or inadequate washing	Check stain: discard old or cloudy solutions. Filter or centrifuge stain before use. Reduce exposure to CO_2 during staining. Improve washing technique during grid staining.
Pale amorphous or 'network- like' plates of deposit	Interaction between uranium and lead stains	Remove residual uranyl acetate after staining sections: improve washing technique during grid staining.

(continued overleaf)

Table 14.1 (*continued*)

Contamination	Cause	Remedy
Needle-like crystalline particulates	Uranyl acetate stain artefact due to poor staining technique or inadequate washing	Check stain: discard old or cloudy solutions. Filter or centrifuge stain before use. Improve washing technique during grid staining. Ensure staining is carried out in the dark to prevent photosensitive reactions.
Bacteria	Poor staining technique	Improve washing technique during grid staining.
	Contaminated wash water and, rarely, contaminated stain	Clean wash water container. Replace wash water and/or stain. Check and clean washing equipment.
	Dirty washing equipment General environment	Check and clean as appropriate.

- A block that is soft or unevenly polymerised.
- A sectioning angle that is too large (adjust as required).
- A cutting speed that is too fast (adjust as required).
- A blunt knife.

14.4.3.9 Uneven Thickness within the Section

Sections may be of uneven thickness after they are cut. The uneven areas may be irregular or in bands.

- Irregular patches of uneven thickness are often due to the block face being of uneven hardness (uneven polymerisation or non-uniform tissue composition).
- Irregular thick and thin bands that run parallel to the knife edge (similar to chatter) are caused by vibrations. Check and remedy as required (see Section 14.4.3.6).

14.4.3.10 Score Marks

Score marks are unevenly spaced lines (scratches) that go from the top to the bottom of the section in the direction of cutting (perpendicular to the knife edge). Score marks are normally caused by:

- Imperfections along the knife edge, principally damage to the knife edge (small chips) or dirt. Inspect the knife edge and move to an undamaged area, or clean the knife.
- Hard inclusions within the specimen – in which case the marks may appear only above the inclusion. Inspect the block face to identify the problem, and if possible, retrim the block face to remove the inclusion or re-orientate it to limit the area of section damage.

14.4.3.11 Section Contamination: Unstained Sections

Sectioning should be carried out in a clean, dust-free environment. Also see Table 14.1.

Section contamination can be caused by:

- A dirty blade during initial block trimming. Clean the blade before use.
- A dirty knife trough or knife. Check before use, and clean as required.
- Contaminants in the sectioning trough fluid (or introduced during section flattening). Check the fluid and replace it; clean the boat.
- Dirty grids and forceps. Check equipment before use. Wash and degrease grids before use.
- Careless drying and blotting during section collection. Use fresh dust-free filter paper for drying grids. Avoid touching the sections with the filter paper.

14.4.3.12 Sections That Contain Holes or Tear Easily

Sections are easily damaged by over-vigorous washing and rough handling during staining procedures; they may also be prone to tearing adjacent to grid bars when the sections are collected on the side of the grid where the bars have flat surfaces. In the electron beam, holes may form in poorly infiltrated areas, or where the resin has not fully bonded with specific tissue structures or components (e.g. microsporidial spores). Holes may also form in sections of tissue reprocessed from wax if the wax has not been fully removed.

14.4.3.13 Remedying Problems Due to Soft or Unevenly Polymerised Blocks

The sectioning qualities of soft or unevenly polymerised blocks may be improved by re-trimming and/or re-orientating the tissue to remove (or minimise the impact of) the problem area. Soft blocks may be hardened by re-incubating them for ~30–60 minutes at 90 °C, or at 60–70 °C overnight.

It is important to remember that soft or unevenly polymerised blocks may be the result of poor processing technique and/or problems with the embedding resin. In such cases, the processing technique should be reviewed to ensure that dehydration and resin infiltration are being carried out correctly. The resin formulation and thoroughness of the mixing procedure should also be checked.

14.4.3.14 Minimising General Sectioning Problems

To minimise sectioning problems:

- The specimen should be held firmly in the specimen block holder and on the ultramicrotome cutting arm. Check that all locking devices are tight.
- Prior to cutting, the block should be aligned so that: (i) the upper and lower edges of the block are parallel to the knife edge; (ii) the block face is flat to the knife edge, both horizontally and vertically (i.e. as the block is moved up and down, it stays the same distance away from the knife edge in all directions) and (iii) the clearance angle is set as recommended for the knife in use (see 14.4.2.2).
- The specimen height should be adjusted so that the sections are cut in the middle of the cutting stroke (when the specimen arm is locked, the specimen should also be just above the knife edge).
- The knife should be clean (front and back).
- The cutting speed should be set appropriately for the knife and block (hardness).
- The knife edge should be wet, and the surface of the trough fluid should slope down from the knife edge (a concave meniscus). If required, fluid can be drawn across the knife edge using a hair mounted on a wooden stick (as for moving sections on the trough fluid).
- Adjacent to the knife edge, ensure the trough fluid is well lit (reflective) so that section thickness can be monitored easily.

14.4.4 Section Staining

Sections are usually stained with uranyl acetate followed by lead citrate. Staining is done at room temperature or by using heat (~40 – 50 °C). Sections of tissue previously stained *en bloc* with uranyl acetate may need to be stained with lead only. Note that sections that have been previously viewed in the TEM may not stain subsequently. Stains and staining for TEM are described in depth by Hayat (1975, 1993) and Lewis and Knight (1977).

Uranyl acetate can be used in aqueous or alcoholic solution. A 5% or saturated aqueous solution (which is light sensitive and must be stored in the dark) is adequate; however, a 5% to ~7% (saturated) solution in 50% ethanol (or methanol) stains more rapidly (Lewis and Knight, 1977). Methanolic solutions are reported to stain faster than ethanolic solutions (Hayat, 1993). Uranyl acetate is toxic and radioactive, and it must be handled with care. Oolong tea extract is said to produce a similar staining effect to uranyl acetate and has been promoted as a nontoxic alternative (Sato *et al.*, 2008).

Lead citrate is applied after the sections have been stained with uranyl acetate. The common formulation is that of Reynolds (1963), and a simple formulation using solid lead citrate has been described by Venable and Coggeshall (1965). A lead stain containing a combination of lead salts (nitrate, acetate and citrate) was introduced by Sato (1968). Sato's method was later improved by Hanaichi *et al.* (1986), who described a formulation said to be stable for approximately one year. Lead citrate may be employed *en bloc* as described by Hayat (1975). Lead stains are toxic and must be handled with care.

14.4.5 Section Contamination and Staining Artefacts

The section surface is easily contaminated during sectioning, staining and general handling by a variety of particulates, oily films and microorganisms. Uranyl acetate and lead citrate may also interact to form deposits (Kuo, Husca and Lucas, 1981). The main sources of contamination are usually sectioning and staining, but the general environment and block trimming may contribute. To prevent contamination, all materials and equipment should be kept clean: razor blades, glass and diamond knives, forceps, grids and staining equipment. During staining, grids should be washed with clean distilled or de-ionised water after each step to remove residual staining solution and prevent stain interactions. To remove particulates, stains may be centrifuged or filtered with a disposable 0.2 μm filter prior to use. When staining is complete, the grids should be stored in a clean dry container.

When contamination occurs, the cause or source must be systematically identified and eliminated (Table 14.1). The first step is to examine an unstained grid: this will determine whether the contamination has occurred before or during staining. Careful observation should also show if the offending material is in, or on, the section.

In addition to surface deposition, precipitates may also form within the tissue due to a variety of complex interactions involving the tissue, fixative, buffer, dehydrant and *en bloc* stain (Hendriks and Eestermans, 1982; Louw *et al.*, 1990). Phosphate buffer in particular can cause fine granular deposits, including when it used as a vehicle for osmium tetroxide (Gil and Weibel, 1968; Glauert, 1975).

14.4.5.1 Removing Particulate Contamination

Stain deposits on the section surface may be removed by careful rinsing or treatment with dilute oxalic acid, acetic acid or aqueous or ethanolic uranyl acetate solution. Treatment with acetic acid can also be used to de-stain sections. Treatments are specific to the type of deposit present and are fully described by Kuo, Husca and Lucas (1981). Precipitates within the tissue caused by glutaraldehyde–osmium tetroxide fixation may be removed using oxidising agents such as periodic acid and hydrogen peroxide (Ellis and Anthony, 1979). Note that these treatments may have a deleterious effect on copper grids.

Processing Schedules

The following schedules are given as general guidelines for tissue processing: methods may need to be modified to suit specific applications. Processing can be done in one day (unfixed tissue to resin embedding) followed by overnight resin polymerisation to give a block ready for sectioning.

Standard Fixation and Processing: Surgical Tissues and Biopsies

This method is an example of a protocol that can be used for hand processing most diagnostic specimens. When material is processed in a tissue processor, the incubation times may need to be extended to reduce the effects of fluid 'carry-over'.

DO NOT ALLOW THE TISSUE TO DRY OUT AT ANY STAGE.

Use a fume cupboard when handling hazardous chemicals.

Fixation

Incubate at room temperature, and use a rotator where the fluid volume is adequate (see notes). Water is distilled or de-ionised in all cases.

1. Primary fixation: fix for 1.5–2.0 hours at room temperature in 2.5% glutaraldehyde in 0.1 M phosphate buffer, pH 7.4. Small specimens may be adequately fixed in 30–60 minutes.

2. Buffer wash: wash in 0.1 M phosphate buffer, pH 7.4; two changes, each of 10 minutes. Use a rotator.
3. Secondary fixation: fix for 60 minutes at room temperature in 1% aqueous osmium tetroxide.
4. Wash in water: two changes, each of 10 minutes. Use a rotator.

Dehydration

Incubate at room temperature, and use a rotator if available.

5. 70% ethanol: 10 minutes (optional alternative *en bloc* stain: a 2% solution of uranyl acetate in 70% ethanol: 15 minutes).
6. 90% ethanol: 10 minutes.
7. 95% ethanol: 10 minutes.
8. 100% ethanol: 15 minutes.
9. 100% dry ethanol: two changes, each of 20 minutes.

Transition Fluid and Infiltration

Incubate at room temperature, and use a rotator if available. Full resin mixture (including accelerator) is used for all steps.

10. Epoxy propane: two changes each of 15 minutes.
11. 50:50 epoxy propane–resin mixture: 60 minutes.
12. 25 : 75 epoxy propane–resin mixture: 60 minutes.
13. 100% resin: 60 minutes (use vacuum if available).

Embedding and Polymerisation

14. Using fresh resin (i.e. not previously used for infiltration), embed the tissue in appropriate embedding capsules or moulds. All specimens should be individually labelled.
15. Polymerise for 12–24 hours at 60–70 °C depending on the resin mixture (see the supplier's guidelines).

Notes:

- A small volume (~10 times the specimen volume) of osmium tetroxide fixative can be used to minimise cost; however, this may not be compatible with the use of a rotator.
- Tissue may be held overnight in 100% resin if required.
- Alcohol is dried using molecular sieve.

Alternative Fixative Vehicles and Buffers

Cacodylate buffer (0.1 M, pH 7.2–7.4) may be used in place of phosphate buffer as the fixative vehicle (for glutaraldehyde and osmium tetroxide) and for the intermediate washing step between glutaraldehyde and osmium tetroxide fixation (see Section 14.3.4).

Osmium tetroxide may be made up in water, phosphate or cacodylate buffer (Glauert, 1975). Osmium tetroxide in buffer is unstable and must be stored frozen. A solution of 1% aqueous osmium tetroxide stored at 4 °C in a clean, tightly sealed glass container will keep indefinitely (~12 months or more) providing the container and the solution remain clean. After osmium tetroxide fixation, specimens may be washed with water (osmium tetroxide destroys osmotic activity).

Rapid Processing

Rapid 'same-day' processing may be used for urgent specimens. Tissue block size may need to be reduced to ensure adequate reagent penetration. For a 4-hour protocol and brief methodological review, see Baic and Baic (1984).

Incubate at room temperature, and use a rotator where the fluid volume is adequate. The total time required to give a polymerised block is ~5 hours.

1. Primary fixation: fix for 30 minutes at room temperature in 2.5% glutaraldehyde in 0.1 M phosphate buffer, pH 7.4. The tissue must be cut into thin slices (~1 mm).
2. Buffer wash: wash in 0.1 M phosphate buffer, pH 7.4; two changes, each of 2 minutes. Use a rotator.
3. Secondary fixation: fix for 30 minutes at room temperature in 1% osmium tetroxide in water.
4. Rinse in water, then give three changes, each of 2 minutes. Use a rotator.
5. 70% ethanol: 15 minutes.
6. 70% ethanol: 15 minutes (or use a saturated solution of uranyl acetate in 70% ethanol if *en bloc* staining is required).
7. 100% dry ethanol: two changes, each of 30 minutes.
8. Epoxy propane: 15 minutes.
9. 50 : 50 epoxy propane–resin mixture: 15 minutes.
10. 100% resin: 30 minutes (use vacuum if available).
11. Embed (use heat-resistant capsules) and polymerise for ~60 minutes at ~90 – 100 °C (this may depend on the specific resin used; see the supplier's guidelines).

Fluid Samples and Suspensions

Specimens concentrated from fluids and suspensions can be embedded in BSA after glutaraldehyde fixation to facilitate specimen handling.

1. Centrifuge the material to form a loose pellet.
2. Primary fixation: remove excess fluid, and resuspend the specimen in standard 2.5% glutaraldehyde fixative in buffer (pH 7.4). Incubate at

room temperature for ~30–60 minutes. If fixative is added to a fluid, the final fixative concentration should not be less than 50% (~1.25% glutaraldehyde).

3. Centrifuge the specimen to form a loose pellet.
4. Buffer wash: remove the fixative, and wash the specimen by resuspending it in buffer. Hold in buffer for 10–15 minutes. A single wash is usually adequate, but give additional washes if there is a large amount of material. If the specimen is in a centrifuge tube, transfer it to an Eppendorf tube or soft (polyethylene) embedding capsule that can be cut open to release the final pellet. Depending on specimen volume, divide the specimen as required.
5. Centrifuge the specimen to form a loose pellet.
6. Embed in BSA: remove as much buffer from the tube as possible, then introduce ~0.5 ml of 15% aqueous BSA. Resuspend the specimen, and infiltrate for ~1 hour or more.
7. Centrifuge the specimen to form a loose pellet.
8. Remove excess BSA, leaving ~3 ml depth of BSA (enough to form a small pellet), then gently introduce ~0.5 ml of standard 2.5% glutaraldehyde fixative so that it forms a layer above the BSA. If the pellet resuspends, centrifuge to reform a loose pellet (this will not affect the solidification process). Allow the specimen to solidify for ~2 hours or more (overnight if required).
9. When the BSA has solidified, cut open the tube and divide the material into small portions. Partially solidified pellets may be further treated with 2.5% glutaraldehyde until they have hardened.
10. Buffer wash: give two to four washes in buffer, each of 5 minutes. After this step, ciliary biopsies may be mordanted with tannic acid (see 'Cilia and Sperm', below); specimens can also be held and stored briefly at this point (until further processing).
11. Secondary fixation: remove excess fluid, and fix the specimen in 1% osmium tetroxide in water (or buffer). Incubate at room temperature for ~30–60 minutes.
12. Wash: remove the excess osmium tetroxide by swilling the tissue in water (or buffer). Wash the tissue for a total of 10 minutes using ~4–6 changes of water. Use a rotator.
13. Dehydrate and embed using a standard solid-tissue processing schedule. The blocks may swell during dehydration; this should not have a major effect on tissue integrity.

Notes:

- Particulate contaminants, especially glass fragments, can cause knife damage during sectioning. Wherever possible, use plastic-ware for all steps and ensure that containers and equipment are dust and particle free. Wash all containers before use, particularly centrifuge tubes.

- Centrifugation at 3000 rpm for 10 minutes will sediment most specimens (including ciliary specimens). Fine particulates may need to be spun at higher speeds.
- Large pieces of material in fluid samples may be separated and processed as normal tissue. Check that the material is actually solid (e.g. not mucous) and will not disintegrate during processing.
- Ensure tissue pellets are not too hard: reagents will not penetrate densely packed material.
- After centrifugation steps, check for tissue loss: particulates may stick to the sides of the centrifuge tube.
- Depending on the initial volume of the specimen (divide as required), use an Eppendorf tube, embedding capsule or centrifuge tube to hold it. For convenience, Eppendorf tubes and BEEM®capsules may be held in the neck of a centrifuge tube.

Cilia and Sperm

Cilia and sperm may be mordanted with tannic acid to improve the visibility of the components of the axoneme. The tannic acid is applied between steps 10 and 11 of the instructions given at the beginning of this section, as follows:

1. After step 10, remove the wash buffer and incubate the material for ~15 minutes in 0.1% tannic acid in a compatible buffer (pH 7.4).
2. Wash: remove the tannic acid and give two to four washes in buffer, each of 5 minutes.
3. Proceed to osmium tetroxide fixation (step 11).

Blood and Platelets

Whole blood (with anticoagulant) is centrifuged to form a 'buffy coat' which can be fixed and processed in a similar manner to solid tissue. To reduce platelet activation, blood should be fixed as quickly as possible, and citrate anticoagulant is preferred over ethylenediaminetetraacetic acid (EDTA). Small volumes of blood can be processed in microcentrifuge tubes, and Eppendorf tubes may be used for slightly larger volumes; however, the buffy coat will be quite thin (and less easy to handle) owing to the greater width of the tube.

Processing Large Samples (~10 ml)

1. Place the blood in a 10 ml centrifuge tube, and centrifuge as required to form a buffy coat (~600 g for 10 minutes; do not over-compact).

2. Remove the plasma from above the buffy coat, leaving ~1 mm of plasma to avoid accidentally removing the platelet layer. Fix the buffy coat *in situ* using standard glutaraldehyde fixative: fix for 1–2 hours or until the layer forms a 'solid' disc (up to 24–48 hours).
3. Remove the buffy coat from the tube (as a solid disk), cut it into ~1 mm cubes or columns, then post-fix and process similarly to solid tissue. Columns of tissue are flat-embedded so that the whole profile of the buffy coat is visible in a single section. If a discrete solid disc does not form, the material can be processed as fluid or suspension (steps given at the beginning of the 'Fluid Samples and Suspensions' section).

Processing Small Samples (~1 ml)

(Bart Wagner, Sheffield Teaching Hospitals, Sheffield, UK)

Small volumes of blood (~1 ml or less) can be processed using polyethylene microcentrifuge tubes (polyethylene tubes are easy to cut and are preferred over higher density polypropylene).

1. Using a pipette, transfer the blood to a 0.4 ml polyethylene microcentrifuge tube (avoid air bubbles: a plastic pipette may cause problems, so a glass pipette is recommended).
2. Centrifuge as required to form a buffy coat (~600 g for 10 minutes; do not over-compact).
3. Remove the plasma from above the buffy coat, leaving ~1 mm of plasma to avoid accidentally removing the platelet layer.
4. Gently pipette standard glutaraldehyde fixative into the tube, and fix for 48 hours at 4 °C (or until the layer forms a solid disc).
5. Remove the buffy coat by cutting the microcentrifuge tube above and below the material – push the layer out from the direction of the red blood cells (this may be done with a 1 ml plastic pipette).
6. Wash to remove the fixative, then cut the buffy coat into vertical 'columns', post-fix and process as a standard tissue sample. Flat-embed the columns so that the entire profile of the buffy coat can be viewed in a single section. If a discrete solid disc does not form, the material can be processed as fluid or suspension (steps given at the beginning of this section).

Reprocessing Tissue from Wax Blocks

Material can be retrieved from a wax block and processed into resin using a variation of the standard TEM protocol (Widéhn and Kindblom, 1988). Before proceeding, it is important to check that the tissue of interest is still present in the block. It is also important to check whether the tissue is solid: if required, suspensions can be dewaxed, rehydrated and embedded in BSA to facilitate handling.

1. Select the tissue: correlate the block face with the area of interest in the histology slide, and cut out the material or extract a 'core' of tissue with a wide-bore needle (take care not to split or damage the block). If possible, cut the tissue into smaller pieces (~1 mm slices) so it will de-wax adequately.
2. De-wax: incubate in xylene for ~60 minutes at room temperature using a minimum of three changes. Use a rotator. Inspect the tissue to ensure it is fully de-waxed, and give additional changes of xylene if required. Large blocks can be treated overnight. Dissect large pieces of material into thin slices (~1 mm) as soon as the tissue is soft enough to be cut easily (without splitting).
3. Remove the xylene from the tissue, and briefly rinse in fresh xylene.
4. Place in 50 : 50 propylene oxide–xylene for 10 minutes.
5. Place in 100% propylene oxide: two changes, each of 5 minutes.
6. Place in 50 : 50 propylene oxide–resin for 30 minutes.
7. Place in 100% resin at 40 °C for 30 minutes.
8. Embed and polymerise.

Notes:

- The original method (Widéhn and Kindblom, 1988) utilised Agar 100 epoxy resin so that, depending on the resin used, the incubation temperature for step 7 may need to be modified. Originally, final polymerisation was carried out at 100 °C for 60 minutes.
- Pale tissues that are difficult to locate can be stained after step 3 for 10–30 minutes in 0.01% toluidine blue in 100% ethanol. Subsequently, the material is taken to 100% propylene oxide (step 5).

Additional Fixation with Osmium Tetroxide

Widéhn and Kindblom (1988) found that fixing reprocessed tissue with osmium tetroxide made no difference to the appearance of the material after double staining with uranyl acetate and lead citrate. However, if the tissue does require additional fixation with osmium tetroxide, it can be done in xylene or water. For aqueous fixation, rehydrate the tissue using a graded series of alcohols, then treat with water-based fixatives. After fixation, the material must be taken back to 100% alcohol and transitioned to resin through propylene oxide–resin mixtures for embedding. Alternatively, fix the tissue with 1% osmium tetroxide in xylene during the de-waxing step as in the original method (de-wax using 1% osmium tetroxide in xylene for 10 minutes at 40 °C) (van den Bergh Weerman and Dingemans, 1984) or after the tissue has been de-waxed when the treatment can be carried out at room temperature. Resin infiltration is through 50 : 50 propylene oxide–xylene as given at the beginning of this section. Whichever method is used, osmication at room temperature is probably best, considering the health and safety concerns associated with heating osmium tetroxide solutions in the laboratory environment.

Reprocessing Tissue from Histology Sections: The 'Pop-Off' Technique

Several techniques have been described for reprocessing sections on glass slides, a method which is useful not only for reviewing tissue structure or searching for viruses, but also for viewing immunocytochemistry preparations and material processed using special stains (Yaoita, Gullino and Katz, 1976; Bretschneider, Burns and Morrison, 1981; di Sant Ágnese and De Mesy-Jensen, 1984). Sections may be stained or unstained, although unstained material can be difficult to locate during processing. A drawback of the pop-off technique is that the final resin block can be extremely difficult to separate from the glass slide, so, whenever possible, it is better to cut fresh sections and collect them on Thermanox® (polyethylene terephthalate) cover slips. Unlike glass, these separate from the resin block easily. This method can also be used for cell cultures grown on Thermanox® coverslips. Thermanox® is resistant to many chemicals including those used for TEM (ethanol, methanol, propylene oxide, acetone and xylene). Aclar® film (a flexible thermoplastic, chlorotrifluoroethylene) can be used similarly.

Histology Slides (Bretschneider, Burns and Morrison, 1981)

1. Remove the coverslip. Soak the slide in xylene until the cover slip can be removed easily. Do not use force as the section may be damaged.
2. Wash in fresh xylene to remove traces of mountant: use one to two changes, each of 5 minutes.
3. 50 : 50 propylene oxide and xylene: 5–10 minutes.
4. Propylene oxide: 5–10 minutes.
5. 75 : 25 propylene oxide and resin: 5–10 minutes.
6. 50 : 50 propylene oxide and resin: 5–10 minutes.
7. 25 : 75 propylene oxide and resin: 5–10 minutes.
8. 100% resin: 5–10 minutes.
9. Clean excess resin from the slide, taking care not to damage the section.
10. Completely fill (to overflowing) an embedding capsule with resin (or use a pre-made block: see notes following these steps). Place the slide upon the full capsule, covering the area of interest in the section. Invert the slide, and press the capsule down onto the section. Ensure no air is trapped in the capsule, and do not allow the capsule to slip across the slide and damage the section. The capsule should be labelled appropriately.
11. Polymerise the resin block: incubate the slide and capsule (time and temperature as required). The slide must be kept completely flat to stop

the block drifting during the polymerisation process. Alternatively, the embedding capsule can be placed in a block holder and the slide placed downwards (inverted) so that it sits on the capsule as described by Bretschneider, Burns and Morrison (1981); additional support can be placed on the holder to keep the slide level.

12. Separate the block from the slide: when polymerisation is complete, place the slide on a hotplate at 100 °C for ~15 seconds. Remove the slide from the hot plate, and rock the capsule until it separates. **Take care:** the slide may snap or shatter. Alternatively, freeze the slide with liquid nitrogen and try to prise the block free.

Notes:

- Check the block face using a dissecting microscope: glass shards must be removed before sectioning to avoid knife damage. It is advisable to use a glass or old diamond knife for cutting reprocessed sections.
- For embedding, pre-made capsules containing polymerised resin can be used instead of capsules with fresh resin. Fill a capsule completely with resin to give a level surface, and polymerise the resin. To embed, place some fresh resin on the area of interest, followed by the pre-made capsule. Ensure all air is excluded and that there is enough liquid resin to allow the tissue to become incorporated into the surface of the pre-made block after final polymerisation.

New (Freshly Cut) Wax Sections

1. Ensure the tissue required remains in the wax block (recut a survey section if necessary).
2. Collect 4–8 μm wax sections on one end of a Thermanox® coverslip that has been coated in a suitable section adhesive (see note following these steps). The sections should be pressed flat but not heated (the coverslip may be damaged). Mark (scratch) the location of the area of interest on the back of the coverslip, or stain the section (see 'Staining Sections to Improve Tissue Visibility').
3. Dewax the section: incubate in xylene using three changes, each of 5 minutes. Use a rotator. If required, trim the coverslip so that the specimen is fully immersed in the xylene. Check that the section is dewaxed; incubate further if required.
4. Infiltrate with resin: infiltrate and embed in the same way as for an existing histology slide (as discussed in the 'Histology Slides' section; Bretschneider, Burns and Morrison, 1981).
5. Embed: place the coverslip in a Petri dish on a filter paper with the section upwards. Invert a resin-filled capsule (or pre-made resin block; see 'Histology Slides'), and quickly and firmly place it over the section

on the coverslip. Do not allow the capsule to slip as the section may be damaged.

6. Polymerise the block. After polymerisation, the cover slip should simply peel off.

Note:

- Prior to use, cover slips can be coated with a section adhesive such as 3-aminopropyltriethoxysilane (APES) which will help to prevent section loss during processing (Maddox and Jenkins, 1987).

Additional Fixation

In general, histology sections should not need additional fixation as components such as lipid will have been extracted during preparation for LM (see 'Reprocessing Tissue from Wax Blocks').

Staining Sections to Improve Tissue Visibility

When processing material using the pop-off technique, the tissue may need to be stained to identify specific areas for embedding; this can be done prior to resin infiltration by taking the sections to water and staining with toluidine blue. Alternatively, sections can be taken to ethanol immediately after they have been dewaxed and stained in 0.01% toluidine blue in ethanol (Widéhn and Kindblom, 1988). Subsequently, an intermediate 50 : 50 propylene oxide–ethanol step can be used (instead of 50 : 50 propylene oxide–xylene) prior to resin infiltration.

Loose or Friable Sections

Loose or friable sections may be covered with BSA to prevent loss. Note that BSA gives a fine granular background at the ultrastructural level.

1. Apply a drop of 15% aqueous BSA solution to the section: apply just enough to give a thin film on the section only (±30 µl on a section 3 mm × 3 mm). Infiltrate for 5 minutes.
2. Gently add an equal volume of 2.5% glutaraldehyde solution.
3. Allow the BSA to solidify for 30–60 minutes.
4. Check that the BSA is solid, then trim off the excess.
5. Wash off excess fluid with water, then give three changes of water, each of 3 minutes.
6. Dehydrate, process into resin, embed and polymerise.

Formulae and Reagents

Use a fume cupboard when handling hazardous chemicals.

Buffers

0.1 M phosphate buffer pH 7.4 (after Sörensen; see Glauert, 1975)

Phosphate buffer for use as a fixative vehicle or wash buffer during fixation. For a full range of phosphate buffer formulations, see Glauert (1975).

Stock Solutions

Solution A: 0.2 M dibasic sodium phosphate

	$Na_2HPO_4{\cdot}2H_2O$:	35.61 g
Or	$Na_2HPO_4{\cdot}7H_2O$:	53.65 g
Or	$Na_2HPO_4{\cdot}12H_2O$:	71.64 g
	Distilled (or de-ionised) water:	1 000 ml

Solution B: 0.2 M monobasic sodium phosphate

	$NaH_2PO_4{\cdot}H_2O$:	27.60 g
Or	$NaH_2PO_4{\cdot}2H_2O$:	31.21 g
	Distilled (or de-ionised) water:	1 000 ml

Method

Prepare stock solutions A and B.

To make standard 0.1 M phosphate buffer (pH 7.4), mix 40.5 ml of solution A with 9.5 ml of solution B, and make up to 100 ml with distilled or de-ionised water.

The pH should be checked and adjusted using hydrochloric acid or sodium hydroxide as required.

0.1 M sodium cacodylate pH 7.2–7.3 (Glauert, 1975)

Cacodylate buffer for use as a fixative vehicle.

Dissolve 21.4 g of sodium cacodylate ($Na(CH_3)_2AsO_2 \cdot 3H_2O$) in distilled or de-ionised water to make 1000 ml. Adjust the pH as required with hydrochloric acid.

Fixatives

Primary Fixative: 2.5% Glutaraldehyde

Standard primary fixative for general use. Handle with care in a fume cupboard.

To 25 ml of 25% glutaraldehyde (electron microscopy (EM) grade), add 225 ml of *either* 0.1 M phosphate buffer *or* 0.1 M sodium cacodylate buffer (no sucrose). Store at 4 °C in a tightly sealed bottle: the solution should not be used if it becomes discoloured.

Secondary Fixative: 1% Osmium Tetroxide

Standard secondary fixative for general use. Handle with care in a fume cupboard.

Dissolve 1.0 g of osmium tetroxide (OsO_4) in 100 ml of distilled or deionised water (the solution can also be made using 0.1 M phosphate buffer *or* 0.1 M sodium cacodylate buffer with or without sucrose). Alternatively, for greater convenience and safety, aqueous solutions of osmium tetroxide are available from commercial suppliers – dilute as required.

Before making the fixative from crystalline solid, remove the label from the ampoule and clean the ampoule thoroughly (remove all glue and grease – handle with disposable gloves). Score the ampoule with a clean diamond pencil, then place it in a clean strong glass bottle with a top that will seal tightly (a glass-stoppered bottle is *not* adequate). Break the ampoule inside the bottle with a clean glass rod. Osmium tetroxide dissolves slowly: ultrasonicate to dissolve within a few minutes. Uncontaminated aqueous osmium tetroxide solutions will last for at least a year if prepared in clean glassware and stored at 4 °C in a tightly sealed bottle. Buffered osmium tetroxide solutions must either be made fresh or frozen immediately for storage. Frozen buffered fixative will last for a period of months; solution that has turned brown or black should not be used.

Osmium tetroxide is highly volatile, and solutions decrease in concentration if not stored in a tightly sealed container (Hayat, 1970). Bottles of fixative should be kept in a secure, tightly sealed, second (outer) container to stop vapours escaping; a layer of milk powder (covered in paper) can be placed in the bottom of the outer container to neutralise vapour that does escape.

Osmium Tetroxide and Potassium Ferricyanide

Potassium ferricyanide (and ferrocyanide) may be added to osmium tetroxide fixatives to improve the fixation of glycogen, phospholipids (and membranes in general), glycoproteins, elastin, myelin and other structures (De Bruijn, 1973; Goldfischer *et al.*, 1981; Hayat, 1981). The method was first described as a method for the fixation of phospholipids by Elbers, Ververgaert and Demel (1965), and several formulations and techniques have been published subsequently (Glauert, 1975). For a full discussion of the chemistry involved in both ferricyanide and ferrocyanide fixatives, see Hayat (2000) who warns that the method is 'capricious' because of uneven tissue penetration. A typical formulation is described by De Bruijn (1973): 0.05 M potassium ferricyanide $K_3Fe(CN)_6$ in 1% osmium tetroxide in 0.1 M sodium cacodylate buffer pH 7.0 (this was used for aldehyde-fixed tissues that were post-fixed for 24 hours at 0–4 °C). More practically, Hayat (2000) outlines several methods utilising fixation for 1–2 hours with 0.8% potassium ferricyanide (or ferrocyanide) in 1% osmium tetroxide. In this author's laboratory, secondary fixation for 60 minutes with 1.66% potassium ferricyanide in 1% aqueous osmium tetroxide has been routinely used to improve glycogen fixation in a wide range of diagnostic specimens.

Resins

Medium-hardness epoxy resin: either of the formulations given in these tables is suitable for embedding diagnostic specimens using conventional polymerisation in an oven. Formulations may need to be modified for microwave polymerisation.

Araldite–Procure Resin Mixture

Araldite 502	7.0 g	(6 ml)
Procure 812	11.0 g	(10 ml)
DDSA	22.0 g	(22 ml)
DMP-30 (accelerator)	0.56 ml	
Total volume: 38 ml		

Spurr's Resin Mixture

NSA	12.0 g	(12 ml)
ERL 4221	9.0 g	(8 ml)
DER 732	2.0 g	(2 ml)
DMAE (accelerator)	0.3 ml	
Total volume: 22 ml		

Method

Estimate the total amount of resin required for embedding and infiltration. Weigh the resin components sequentially to a lidded disposable polypropylene beaker. Ensure the weights are as accurate as possible, and tare the balance between components if required. Alternatively, add the components sequentially to a measuring cylinder ensuring the volumes are measured as accurately as possible. Finally, measure and add the appropriate accelerator using a disposable syringe, then stir or shake (ensure the container is fully sealed) until the resin is completely mixed (inadequate mixing may result in uneven polymerisation). Allow the mixture to stand until bubbles have dispersed, and check to ensure that the resin is completely mixed. Resin that contains bubbles can be used for solvent–resin mixtures providing that the resin is added to the solvent – this allows the bubbles to disperse so that the volume of resin can be measured accurately. Ensure that resin infiltration mixtures are also made accurately and fully mixed. Clean reusable equipment after use with alcohol or acetone. Polymerise excess resin and contaminated items such as disposable beakers and pipettes before disposal.

Stains

Uranyl Acetate

Standard stain for general use. Uranyl acetate may be used in aqueous solution if preferred; however, staining times will need to be modified.

Make a saturated (~7%) solution of uranyl acetate in 50% aqueous ethanol or methanol. Add 0.7 g of uranyl acetate ($UO_2(CH_3COO)_2$) to a 10 ml centrifuge tube, and make up to 10 ml with the 50% alcohol solution of choice. Stopper the tube, and shake well. Ultrasonicate to ensure the compound is fully dissolved (optional).

Lead Citrate

Standard stain for general use. Formulations are given by Lewis and Knight (1977), as shown here.

Lead citrate: Reynolds (1963)

Lead nitrate $Pb(NO_3)_2$:	1.33 g
Tri-sodium citrate $Na_3(C_6H_5O_7).2H_2O$:	1.76 g
1 M sodium hydroxide NaOH (preferably carbonate free and freshly prepared):	8.0 ml

Dissolve the lead nitrate and the citrate separately, each in ~10–15 ml of distilled or de-ionised water. Mix the solutions together in a 50 ml volumetric flask, and shake vigorously for 1 minute. Stand for 30 minutes with occasional shaking. A cloudy precipitate will appear. After 30 minutes, add the sodium hydroxide and shake to give a colourless solution. Add distilled or de-ionised water to make up to a final volume of 50 ml. Decant the solution into clean 10 ml plastic centrifuge tubes, then wash the volumetric flask immediately (use weak sodium hydroxide if required). Store tubes of stain (firmly stoppered) at 4 °C to prolong shelf life. Stain 'in use' may be stored, firmly stoppered, at room temperature.

1 M Sodium Hydroxide

Dissolve 0.4 g of dry sodium hydroxide (molecular weight 40) in distilled or de-ionised water to give a final volume of 10 ml. To reduce the amount of carbon dioxide in the water, it can be boiled immediately before use (stopper and allow to cool).

Lead Citrate (Venable and Coggeshall, 1965)

Standard stain for general use. Alternative formulation using lead citrate.

Add 0.01–0.04 g of lead citrate to 10 ml of distilled or de-ionised water, and shake well (the water may be boiled before use to reduce dissolved carbon dioxide). Add 0.1 ml of 10 M sodium hydroxide solution (carbonate free), stopper and shake until the solution is clear.

General TEM Staining Procedure

1. Place two clean sheets of Parafilm (or a sheet of dental wax) in a clean glass Petri dish. Wash the Parafilm with warm water, then distilled or de-ionised water. Put drops of water in the dish adjacent to the film to create a humid chamber (the surface of the Parafilm must be dry). Some practitioners place pellets of sodium hydroxide in the Petri dish to absorb carbon dioxide which, theoretically, can react with the lead stain to form insoluble carbonates (seen as a white precipitate). However, other individuals find that such precautions are not essential.
2. Place separate drops (~50–100 µl) of uranyl acetate and lead stain on the sheets of Parafilm. Ensure the drops of stain cannot mix. If required, the stains may be centrifuged before use, or dispensed using a disposable filter, to prevent particulate contamination.
3. Stain the sections (as given in the next set of steps) by floating the grids with the sections down and fully in contact with the stain solution. The lid should be kept on the Petri dish during staining to prevent evaporation and to minimise the exposure of the lead stain to carbon dioxide. Standard texts suggest that photosensitive uranyl acetate solutions should be covered to exclude light; some practitioners have found that this is not essential.
4. Wash with distilled or de-ionised water by dipping the grids in a series of beakers of water. Use separate beakers for initial washes after the uranium and lead stains (to prevent precipitates), then an additional ~3 beakers subsequently. Alternatively, drip water onto the grids from a wash bottle (with care!).

Stain the sections as follows:

1. Stain for 3 minutes with 7% uranyl acetate in 50% ethanol (the uranyl acetate *must* be applied first).
2. Wash with water.
3. Remove excess water with a piece of filter paper.
4. Stain for 1.5 minutes with lead citrate (minimise exposure to carbon dioxide to prevent the formation of precipitates, and do not breathe on the sections). Do not use the stain if there is a white precipitate present.
5. Wash with water.
6. Remove excess water with a piece of filter paper.
7. Place the stained grids on filter paper in a clean Petri dish. Ensure the dish is labelled clearly.

Note:

- The staining times given in the preceding steps are a guide only and may be modified to suit.

REFERENCES

Acetarin, J-D., Carlemalm, E. and Villiger, W. (1986) Developments of new Lowicryl ®resins for embedding biological specimens at even lower temperatures. *Journal of Microscopy*, **143**, 81–88.

Bacchuber, K. and Frösch, D. (1982) Melamine resins, a new class of water-soluble embedding media for electron microscopy. *Journal of Microscopy*, **130** (1), 1–9.

Baic, D. and Baic, B. (1984) A fast method for processing biopsy material for electron microscopy. *Ultrastructural Pathology*, **6**, 347–349.

Bendayan, M., Nanci, A. and Kan, F.W.K. (1987) Effect of tissue processing on colloidal gold cytochemistry. *Journal of Histochemistry and Cytochemistry*, **35**, 983–986.

van den Bergh Weerman, M.A. and Dingemans, K.P. (1984) Rapid deparaffinization for electron microscopy. *Ultrastructural Pathology*, 7, 55–57.

Bretschneider, A., Burns, W. and Morrison, A. (1981) 'Pop-off' technic. The ultrastructure of paraffin-embedded sections. *American Journal of Clinical Pathology*, **76**, 450–453.

Bullock, G.R. (1984) The current status of fixation for electron microscopy: a review. *Journal of Microscopy*, **133** (1), 1–15.

Carlemalm, E., Garavito, R.M. and Villiger, W. (1982) Resin development for electron microscopy and an analysis of embedding at low temperature. *Journal of Microscopy*, **126** (2), 123–143.

De Bruijn, W.C. (1973) Glycogen, its chemistry and morphologic appearance in the electron microscope I. A modified OsO_4 fixative which selectively contrasts glycogen. *Journal of Ultrastructure Research*, **42**, 29–50.

Edwards, K., Griffiths, D., Morgan, J. *et al.* (2009) Can the choice of intermediate solvent or resin affect glomerular basement membrane thickness? *Nephrology Dialysis Transplantation*, **24**, 400–403.

Elbers, P.F., Ververgaert, P.H.J.T. and Demel, R. (1965) Tricomplex fixation of phospholipids. *Journal of Cell Biology*, **24**, 23–30.

Ellis, E.A. (2006) Solutions to the problem of substitution of ERL 4221 for vinyl cyclohexene dioxide in Spurr low viscosity embedding formulations. *Microscopy Today*, **14** (4), 32–33.

Ellis, E.A. and Anthony, D.W. (1979) A method for removing precipitate from ultrathin sections resulting from glutaraldehyde-osmium tetroxide fixation. *Stain Technology*, **54** (5), 282–285.

Frösch, D. and Westphal, C. (1985) Choosing the appropriate section thickness in the melamine embedding technique. *Journal of Microscopy*, **137** (2), 177–183.

Frösch, D. and Westphal, C. (1986) Improved ultrastructure and histochemistry in unfixed cyanobacteria embedded at 191 K in Nanoplast® AME 01. Proceedings of the XIth International Congress on Electron Microscopy, Kyoto, Japan, pp. 2171–2172.

Gil, J. and Weibel, E.R. (1968) The role of buffers in lung fixation with glutaraldehyde and osmium tetroxide. *Journal of Ultrastructure Research*, **25** (5), 331–348.

Glauert, A.M. (ed.) (1975) *Fixation, Dehydration and Embedding of Biological Specimens*, vol. 3 of *Practical Methods in Electron Microscopy* (ed. A.M. Glauert), North Holland Publishing Company, Amsterdam.

Goldfischer, S., Kress, Y., Coltoff-Schiller, B. and Berman, J. (1981) Primary fixation in osmium potassium ferrocyanide: the staining of glycogen, glycoproteins, elastin, an intranuclear reticular structure and intercisternal trabeculae. *Journal of Histochemistry and Cytochemistry*, **29**, 1105–1111.

Grizzle, W.E., Fredenburgh, J.L. and Myers, R.B. (2008) Fixation of tissues, in *Theory and Practice of Histological Techniques*, Chapter 4, 6th edn (eds J.D. Bancroft and M. Gamble), Churchill Livingstone, Edinburgh, pp. 53–74.

Hanaichi, T., Sato, T., Iwamoto, T. *et al.* (1986) A stable lead by modification of Sato's method. *Journal of Electron Microscopy*, **35** (3), 304–306.

Hayat, M.A. (1970) *Principles and Techniques of Electron Microscopy: Biological Applications*, vol. 1, Van Nostrand Reinhold Company, New York.

Hayat, M.A. (1975) *Positive Staining for Electron Microscopy*, Van Nostrand Reinhold Company, New York.

Hayat, M.A. (1981) *Fixation for Electron Microscopy*, Academic Press, New York.

Hayat, M.A. (1993) *Stains and Cytochemical Methods*, Plenum Press, New York.

Hayat, M.A. (2000) *Principles and Techniques of Electron Microscopy: Biological Applications*, 4th edn, Cambridge University Press, Cambridge.

Hendriks, H.R. and Eestermans, I.L. (1982) Electron dense granules and the role of buffers: artefacts from fixation with glutaraldehyde and osmium tetroxide. *Journal of Microscopy*, **126** (2), 161–168.

Hillmer, S., Joachim, S. and Robinson, D.G. (1991) Rapid polymerisation of LR-White for immunocytochemistry. *Histochemistry*, **95**, 315–318.

Kuo, J., Husca, G.L. and Lucas, L.N.D. (1981) Forming and removing stain precipitates on ultrathin sections. *Stain Technology*, **56** (3), 199–204.

Lewis, P.R. and Knight, D.P. (1977) *Staining Methods for Sectioned Material*, vol. 5 of *Practical Methods for Electron Microscopy* (ed. A.M. Glauert), North-Holland Publishing Company, Amsterdam.

Louw, J., Williams, K., Harper, I.S. and Walfe-Coote, S.A. (1990) Electron dense artefactual deposits in tissue sections: the role of ethanol, uranyl acetate and phosphate buffer. *Stain Technology*, **65** (5), 243–250.

Maddox, P.H. and Jenkins, D. (1987) 3-Aminopropyltriethoxysilane (APES): a new advance in section adhesion. *Journal of Clinical Pathology*, **40**, 1256–1260.

Nasr, S.H., Markowitz, G.S., Valeri, A.M. *et al.* (2007) Thin basement membrane nephropathy cannot be diagnosed reliably in deparaffinized, formalin-fixed tissue. *Nephrology Dialysis Transplantation*, **22**, 1228–1232.

Newman, G.R. (1987) Use and abuse of LR white. *Histochemical Journal*, **19**, 118–120.

Newman, G.R. and Hobot, J.A. (1987) Modern acrylics for post-embedding immunostaining techniques. *Journal of Histochemistry and Cytochemistry*, **35**, 971–981.

Newman, G.R. and Hobot, J.A. (1993) *Resin Microscopy and On-section Immunocytochemistry*, Springer-Verlag, Berlin.

Reid, N. and Beesley, J.E. (1991) *Sectioning and Cryosectioning for Electron Microscopy*, vol. 13 of *Practical Methods in Electron Microscopy* (ed. A.M. Glauert), Elsevier, Amsterdam.

Reynolds, E.S. (1963) The use of lead citrate at high pH as an electron opaque stain based on metal chelation. *Journal of Cell Biology*, **17**, 208–212.

di Sant Ágnese, P.A. and De Mesy-Jensen, K.L. (1984) Diagnostic electron microscopy on reembedded ("popped off") areas of large Spurr epoxy sections. *Ultrastructural Pathology*, **6**, 247–253.

Sato, T. (1968) A modified method for lead staining of thin sections. *Journal of Electron Microscopy*, **17** (2), 158–159.

Sato, S., Adachi, A., Sasaki, Y. and Ghazizadeh, M. (2008) Oolong tea extract as a substitute for uranyl acetate in staining of ultrathin sections. *Journal of Microscopy*, **229** (1), 17–20.

Stirling, J.W. (1992) Unfixed tissue for electron immunocytochemistry: a simple preparation method for colloidal gold localization of sensitive epitopes using ethanediol dehydration. *Histochemical Journal*, **24**, 190–206.

Stirling, J.W. (1995) Immunogold labelling: resin sections, in *Laboratory Histopathology: A Complete Reference*, vol. 2, Section 9, Part 3 (eds A.E. Woods and R.C. Ellis), Churchill Livingstone, New York, pp. 9.3-1–9.3-21.

Studer, D. and Gnaegi, H. (2000) Minimal compression of ultrathin sections with use of an oscillating diamond knife. *Journal of Microscopy*, **197**, 94–100.

Summers, R.G. and Rusanowski, P.C. (1973) A scanning electron microscopic evaluation of section and film mounting for transmission electron microscopy. *Stain Technology*, **48** (6), 337–342.

Venable, J.H. and Coggeshall, R. (1965) A simplified lead citrate stain for use in electron microscopy. *Journal of Cell Biology*, **25**, 407–408.

Widéhn, S. and Kindblom, L-G. (1988) A rapid and simple method for electron microscopy of paraffin-embedded tissue. *Ultrastructural Pathology*, **12**, 131–136.

Yaoita, H., Gullino, M. and Katz, S.I. (1976) Herpes gestationis. Ultrastructure and ultrastructural localisation of in vivo–bound complement. *Journal of Investigative Dermatology*, **66**, 383–388.

15

Ultrastructural Pathology Today – Paradigm Change and the Impact of Microwave Technology and Telemicroscopy

Josef A. Schroeder

Zentrales EM-Labor, Institut für Pathologie, Klinikum der Universität Regensburg, Regensburg, Germany

15.1 DIAGNOSTIC ELECTRON MICROSCOPY AND PARADIGM SHIFT IN PATHOLOGY

The basis of pathological tissue diagnosis is the morphological analysis of biopsied cells and matrix; examination by transmission electron microscopy (TEM) extends this analysis to the ultrastructural level, providing information not discerned by other methods, for example on the basis of the antigens expressed by a neoplasm (Eyden, 2002). Additional stains and techniques, such as immunohistochemistry, flow cytometry, cytogenetics and molecular techniques (gene rearrangement analysis and fluorescence *in situ* hybridisation (FISH) and polymerase

Diagnostic Electron Microscopy: A Practical Guide to Interpretation and Technique, First Edition. Edited by John W. Stirling, Alan Curry and Brian Eyden.

chain reaction (PCR) analysis), provide additional information to refine the understanding of a disease and strengthen the diagnosis.

In today's era of reliable biomarkers, genomics and high-throughput molecular technologies, a paradigm shift is occurring in pathology towards predictive individual patient pathology (Saidi, Cordon-Cardo and Costa, 2007; Dietel and Schafer, 2008). The last 30 years have shown, for example, that cancer patients benefit from a targeted cancer therapy based on recognition of a molecular biomarker profile of the tumour in the context of the individual personal condition, age and organ function. The development of agents (e.g. small molecule drugs and monoclonal antibodies) able to specifically block the growth pathways and spread of cancer cells, interfere with tumour angiogenesis, promote their apoptosis or deliver toxic molecules to neoplastic cells has the potential to reduce the side effects of an anti-tumour therapy and increase the life quality of affected patients (NIH, 2012). Recently, *in silico* models of diseases (computer-software simulated) are creating a novel holistic scientific approach called systems pathology which is necessary to meet the goals of modern predictive pathology (Kitano, 2002; Grabe, 2008; Faratian *et al.*, 2009; Morange, 2009).

Unquestionably, the number of diagnostic tools has increased, and consequently, in recent decades, there has been a marked decline in the significance of 'traditional' TEM in pathology (Mierau, 1999). This has not been helped by the intrinsic limitations of TEM methodology – for example sampling error, the need for adequate tissue fixation, expensive microscopes, long turnaround times (TATs), and the requirement of a high level of staff skills and interpretational expertise (Papadimitriou, Henderson and Spagnolo, 1992; Erlandson, 1994; Ghadially, 1997; Dickersin, 2000; Eyden, 2007; Cheville, 2009). Apart from the development of high-throughput molecular technologies suitable for efficient screening of thousands of samples and the paradigm shift in pathology mentioned in this section, the basis for evidence-related medicine still remains the morphological tissue section diagnosis (Rosai, 2007). In this context, the high resolution of TEM can provide unique morphological data at the ultrastructural level which can be significant for personalised patient care management (e.g. early amyloid deposit detection, ciliopathy and renal thin membrane syndrome) (Eyden, 1999; Tucker, 2000; Wagner and Curry, 2002; Erlandson, 2003; Turbat-Herrera, D'Agostino and Herrera, 2004; Woods and Stirling, 2008).

However, another aspect is posed by emerging infectious agents that are now capable of rapid intercontinental dissemination on account of globalised trade and travel, animal vectors or bioterrorist attack

(Madeley, 2003; Morens, Folkers and Fauci, 2004). The rapid and very simple negative-staining sample preparation method (approximately 30 minutes) without the necessity of specific reagents, and the 'open view' of a TEM examination (capable of revealing unexpected findings), exemplify the impact of the unrivalled resolution and speed of some aspects of TEM technology (Hazelton and Gelderblom, 2003; Miller, 2003; Bannert, Schroeder and Hyatt, 2009; Stieger *et al.*, 2009; Robert-Koch-Institute, 2012).

Based on the example of our centralised electron microscopy (EM) unit, in this chapter we share our experience with measures addressed to overcome the main technological limitations of traditional diagnostic TEM and present examples of rapid turnaround techniques of samples processed in other laboratories.

15.2 STANDARDISED AND AUTOMATED CONVENTIONAL TISSUE PROCESSING

The key issue in our laboratory was to automate and standardise the routine sample processing with the LYNX (Leica) tissue processor to prepare resin blocks with embedded tissue for ultrathin section examination in the TEM (Figure 15.1). This device is based on a carousel with 20 vials containing the appropriate reagents for fixation, dehydration and resin infiltration of tissue samples for subsequent manual resin embedding (in Epon) in flat moulds (which allows sample pre-orientation for the preferred section plane). The vials are arranged in an order following the processing protocol steps given in Table 15.1. Tissue samples are placed in small round disposable plastic baskets (we prefer the four-shared, with one biopsy or sample in one division); usually, four or five of them are arranged on a rod which is mounted on a robotic arm of the processor to provide continuous agitation (Figure 15.2). This assembly is automatically immersed in the reagents step by step according to the protocol being used and facilitates the batch sample processing by precise microcomputer control of time, temperature and agitation.

The automated batch sample processing reduces the amount of reagent for each step (12–20 ml), minimises health risks for the staff (the device must be operated in a ventilated environment to work, and it eliminates exposure to osmium tetroxide and resin component vapours), runs overnight to minimise labour and also complies with the guidelines of good laboratory practice defined for laboratory accreditation (our laboratory is accredited according to DIN EN ISO/IEC 17 020) (Ong

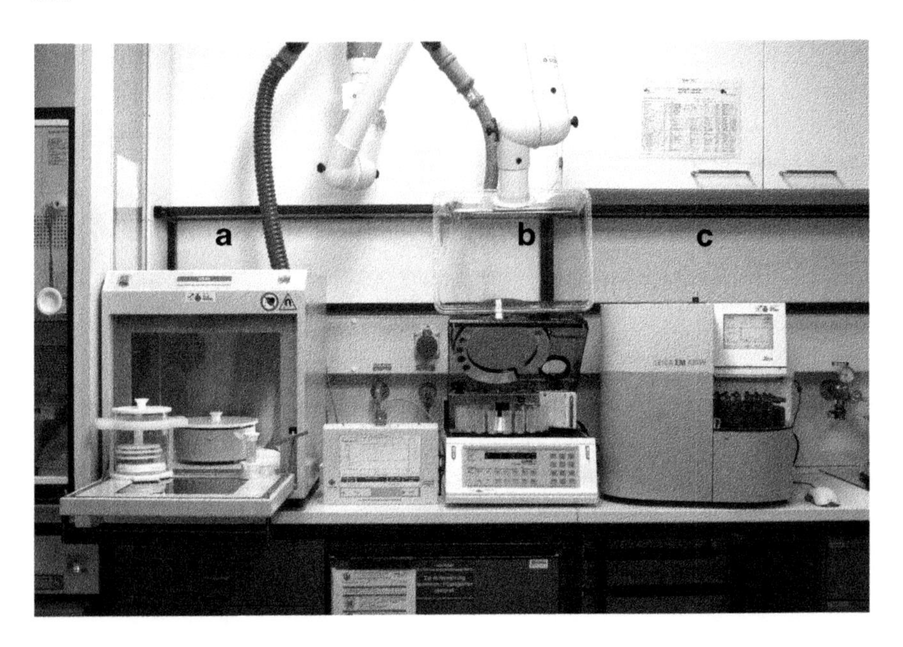

Figure 15.1 Tissue processors on the workbench: (a) REM/Milestone MW tissue processor – note, on the left, the glass chamber with BEEM holder for under-water polymerisation; in the middle of the MW cavity, the round chamber for dry flat mould polymerisation and, at right (arrow), the vial with the sample immersed in the appropriate solution for processing, and the panel with a touchscreen for device control and process monitoring. (b) LYNX/Leica tissue processor for automated conventional processing. Note under the raised cover the carousel bearing vials filled with the solutions according to the steps defined in the protocol. (c) AMW/Leica fully automated processor for MW-assisted and conventional tissue processing.

and Looi, 2001; Hotaling, 2006; Rauch and Nichols, 2007; Stirling and Curry, 2007; Rocken and Manke, 2010).

Additionally, in our opinion, only automated batch sample processing can provide the uniformity of material processing in translational research experiments with differently treated objects for comparative microscopic morphology assessment (e.g. in basic research and in animal or cell culture trials) (Reipert *et al.*, 2008).

In a period of over 18 years of routine diagnostic and basic research material embedding (mostly twice a week), we experienced only three equipment malfunctions of the automated process, and these were without sample damage or loss. The laboratory TAT of this routine conventional EM sample processing, including a 48-hour heat resin polymerisation at 60 °C, is 3 or 5 days (including the weekend and holidays).

Table 15.1 Protocol outline for conventional and MW-assisted tissue processing

Step and medium		LYNX–EPON		AMW–EPON		KOS–LR-White		KOS–Spurr's-Kit	
		Time	Temperature	Time	Temperature	Time	Temperature	Time	Temperature
1	Karnovsky-Fixative	4 h[a]	RT	4 h[a]	RT	4 h[a]	4 °C	4 h[a]	4 °C
2	Buffer	30 min	RT	5 min	35 °C	15 min	4 °C	3 min	RT
3	Buffer	30 min	RT	5 min	35 °C	15 min	4 °C	3 min	RT
	Buffer			5 min	35 °C	15 min	4 °C	3 min	RT
4	1% OsO_4 in buffer	120 min	RT	60 min	35 °C	14 min	37°C	15 min	50 °C
5	Buffer	10 min	RT	5 min	37°C	12 min	RT	3 min	RT
	Buffer					12 min	RT	3 min	RT
	Buffer					12 min	RT	3 min	RT
	Buffer					12 min	RT		
6	Aqua bidest	10 min	RT	5 min	37°C	–	–	–	–
7	Aqua bidest	10 min	RT	5 min	37°C	–	–	–	–
8	Aqua bidest	10 min	RT	–	–	–	–	–	–
9	50% ethanol	15 min	RT	5 min	37°C	–	–	–	–
10	70% ethanol	15 min	RT	5 min	37°C	10 min	37°C	5 min	40 °C
	2% uranyl acetate in 70% ethanol					15 min	37°C	–	–
	70% ethanol					10 min	37°C	–	–
	80% ethanol					10 min	37°C	–	–

(*continued overleaf*)

Table 15.1 (*continued*)

Step and medium		LYNX–EPON		AMW–EPON		KOS–LR-White		KOS–Spurr's-Kit	
		Time	Tempe-rature	Time	Tempe-rature	Time	Tempe-rature	Time	Tempe-rature
11	90% ethanol	15 min	RT	–	–	–	–	–	–
12	95% ethanol	15 min	RT	5 min	37°C	–	–	–	–
	96% ethanol					10 min	37°C	–	–
	96% ethanol					10 min	37°C	–	–
13	100% ethanol	15 min	RT	5 min	37°C	–	–	10 min	40°C
14	100% ethanol	15 min	RT	–	–	–	–	10 min	40°C
	100% acetone			5 min	37°C	–	–	–	–
15	100% ethanol	15 min	RT	–	–	–	–	10 min	40°C
	100% acetone			5 min	37°C	–	–	–	–
16	Propylene oxide	30 min	RT	–	–	–	–	5 min	30°C
17	Propylene oxide	30 min	RT	–	–	–	–	–	
18	Resin:propylene oxide (1+1)	3 h	RT	–	–	–	–	20 min	30°C
	Resin:acetone (1+1)			15 min	40°C	–	–	–	–
	Resin:acetone (3+1)			20 min	40°C	–	–	–	–

	Resin:ethanol (1+2)			–	–	20 min	40 °C	–	–
	Resin:ethanol (2+1)			–	–	20 min	40 °C	–	–
19	Resin (pure)	15 h	RT	20 min	50 °C	30 min	40 °C	60 min	40 °C
	Resin (pure)			20 min	50 °C	–	–	–	–
20	Polymerisation	2 d (48 h)[b]	60 °C	5 min	63 °C	17 min	60 °C	105 min	90–97 °C[f]
				5 min	75 °C	15 min	70 °C		–
				15 min	83 °C	28 min	80 °C		–
				105 min	83 °C				–
		Total[c] 3 d		Total[c] 5 h 25 min		Total[c] 5 h 2 min		Total[c] 4 h 18 min	–

[a]Fixation outside the processor for 4 hours to 3 days depending on the specimen and other factors. Urgent samples (size $\leq$ 1 mm^3): MW-assisted Karnovsky fixation for 20–60 minutes at 35 °C. Karnovsky fixative formulation (Regensburg Central EM-Lab): 2% PFA + 2.5% GA in 0.1 M cacodylate buffer, pH = 7.3 (aliquots of 6 ml in glass vials and stored at −20 °C). RT = room temperature.
[b]Outside LYNX in a bench incubator.
[c]Without fixation time.
[d]Kindly donated by Dra Rocio Guevara de Bonis and Dra Marta E. Couce, Servei d'Anatomia Patologica, Hospital Universitari Son Espases, Palma, Baleares.
[e]Kindly donated by MLT Jacqueline Pittman, Advanced Bioimaging Centre, Pathology and Laboratory Medicine, Mount Sinai Hospital, Toronto, Canada.
[f]Six-step phase of temperature from 90 – 97 °C.

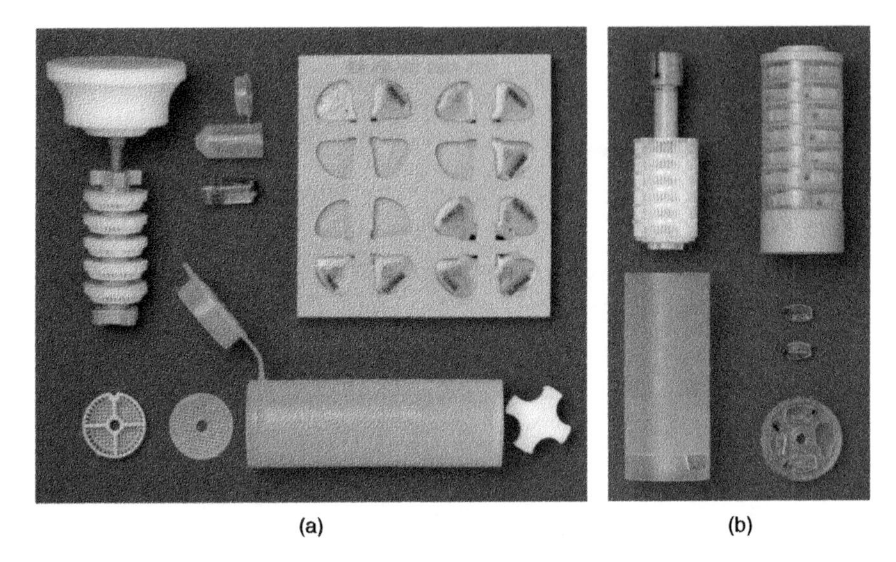

Figure 15.2 Equipment for MW-assisted sample processing: (a) Assembly for disposable baskets on a rod for the REM and LYNX processor. The magnetic stirrer placed at the bottom of the vial ensures uniform temperature distribution in the irradiated solution. Note the round BEEM capsule and the flat Chien embedding moulds used for under-water and dry resin polymerisation, respectively. (b) AMW equipment: left, the sample basket stem assembly, and right, the dedicated silicon moulds rod assembly for polymerisation.

15.3 MICROWAVE-ASSISTED SAMPLE PREPARATION

In the clinical context, there are a number of urgent diagnostic cases for which long TATs are unacceptable, where we have reduced the TAT by applying microwave (MW) technology (Giberson *et al.*, 2003; Schroeder *et al.*, 2006).

This technology has been successfully applied in fields as diverse as domestic cooking, organic chemistry and nanoparticle synthesis (De la Hoz, Diaz-Ortiz and Moreno, 2005), while MW ovens have also been in use in histopathology laboratories since the early 1970s to speed up a range of processes – tissue fixation, decalcification, antigen retrieval, section staining, immunolabelling and *in situ* hybridisation (Shi, Gu and Taylor, 2000; Giberson and Demaree, 2001; Nordhausen and Barr, 2001; Kok and Boon, 2003; Morel and Raccurt, 2003).

The effect of MW irradiation on polar molecules is now well understood (Kok and Boon, 2003). It is attributed to dielectric heating – also called the thermal effect – causing a temperature rise in the whole sample ('internal heating', in contrast to conventional heating which starts at the specimen surface). This in turn significantly accelerates the diffusion of reagents into the tissue and chemical reactions, for example the fixative interaction with cell components. The existence of an additional 'nonthermal' direct MW energy effect, which may be particularly effective in biological hydrated material (and hence relevant for tissue fixation), is still a matter of controversy (Marani and Feirabend, 1993; Galvez, Giberson and Cardiff, 2004; Wendt *et al.*, 2004).

We report the results of MW-assisted rapid tissue sample processing in a semi-automatic (Milestone) and a fully automatic (Leica) MW tissue processor for routine use. This technology cuts the usual TAT of 3–5 days down to approximately 5 hours, enabling 'same-day' TEM diagnosis in urgent clinical cases or infectious agents (e.g. *Bacillus anthracis* (which causes anthrax) or SARS) if a verification of the negative-staining results is mandatory from thin sections. The robust embedding protocol based on the EMbed-812 epon resin (Science Services, Electron Microscopy Sciences (EMS)) is given in Table 15.1. For very urgent samples, it is possible to shorten the total MW-assisted processing time to approximately 3 hours 20 minutes. Because of such significantly reduced material extraction from the specimen, cell structures appear relatively dark in the sections (Schroeder *et al.*, 2006).

The composition of the resin mixture is crucial for reproducible and high-quality embedding, sectioning (with a diamond knife) and staining properties (with toluidine blue or basic fuchsin in semithin sections, and aqueous uranyl acetate and lead citrate for ultrathin sections). We have achieved consistently good results with the Epon resin mix EMbed-812 (Table 15.2) making light of soft- or hard-tissue components (e.g. in soft- and hard-tissue tumours, cell culture or blood pellets, skin, nondecalcified bone, insects and biomaterials) in both conventional and MW-assisted processing (Franzen *et al.*, 2005; Schroeder *et al.*, 2006).

This resin formulation was also found to be successful for tissue embedding intended for elemental analysis by the electron energy loss spectroscopy (EELS) and electron spectroscopic imaging (ESI) methods, which need very thin stable sections (approximately 40 nm) because the EM is operated at very high electron-beam brightness (Schroeder

Table 15.2 Resin formulation for MW-assisted tissue embedding

EPON resin components (Regensburg mix)[a]		Supplier[b]	Spurr resin with Quetol (Toronto mix)		Supplier[c]
EMbed-812	90 g	EMS# 14 900	NSA	84 g	EMS# 19 050
DDSA	60 g	EMS# 13 710	ERL 4 221	29 g	EMS# 15 004
NMA	40 g	EMS# 19 000	DER	15 g	EMS# 13 000
DMP30	3 g	EMS# 13 600	QUETOL-651	18 g	CANEMCO # 049-5
			BDMA	3 g	EMS# 11 400

[a]The final amount is approximately 170 ml. After mixing thoroughly (at least 90 minutes using a magnetic stirrer), the **resin mixture needs to be degassed three times** (desiccator and water pump) and can then be aliquoted in 5 or 10 ml volumes in disposable dry plastic syringes. The loaded syringe must be tightly secured by a plastic cork. Store at −20 °C (the frozen resin can be kept for over 6 months). To use, warm up to room temperature for 1–2 hours, taking care to ensure that the outside of the syringe is thoroughly dry and **free from condensation** (to avoid contact of any moisture with resin).

[b]EMS = Electron Microscopy Sciences (Science Services, Munich/Germany).

[c]EMS = Electron Microscopy Sciences (Cedarlane Labs, Burlington, Ontario/Canada); CANEMCO Inc, Lakefield, Quebec/Canada.

et al., 2008). Examples of MW-assisted processed tissues showing excellent ultrastructural preservation are presented in Figures 15.3 and 15.4.

In Milestone's Rapid Electron Microscope MW device (REM), the vial containing the basket assembly (Figure 15.2a) with the samples is placed in a specially designed carrier which locates the vial in a defined position in the MW cavity of the device. A noncontact infrared temperature sensor measures the reagent temperature in the vial, which is the critical parameter to monitor the magnetron wattage power output (maximum, 700 W). This is controlled via an electronic circuit feedback loop during the continuous MW irradiation of the sample. The slope of the temperature rise and stabilisation and the time for each processing step can be defined and displayed on a dedicated touch-screen monitor. **Each change to the next processing step in the REM device must be done manually by the user.**

The processed samples can be embedded in resin in round BEEM capsules and polymerised when immersed in water (40 at once, MW irradiation exposure = 105 minutes). For over 4 years, we have preferred oriented sample embedding in silicone-rubber flat moulds (so-called Chien moulds; Science Services EMS) and use this in a dedicated round

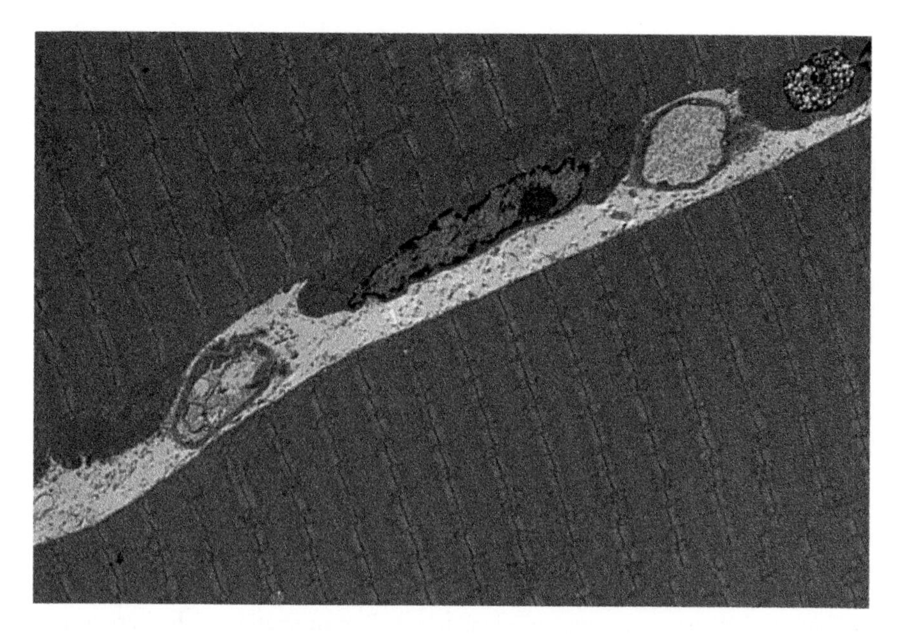

Figure 15.3 Control skeletal muscle, mouse (original magnification × 1250). AMW/Leica, Epon-resin embedding. Complete MW-assisted tissue processing (including 20 minutes Karnovsky fixation and 105 minutes polymerisation).

polymerisation chamber appropriately positioned in the REM device (16 specimen blocks at once, MW irradiation exposure = 105 minutes) (Figure 15.1).

The merging of the proven LYNX carousel concept with a smart MW device led to the Leica Automatic Microwave tissue processor (AMW) shown in Figure 15.1. The next step change in the process is carried out automatically by a robotic reagent system in this bench-sized device, which has the great advantage of being labour saving. The smart-designed monomode MW chamber provides homogeneous MW distribution at the sample location without the hot and cold spots known to occur with other devices. Thus, water loads are not required, and virtually 100% of the MW radiation energy (restricted to 30 W) is absorbed by the processing reagents and immersed specimens (Figure 15.2b). Additionally, the user can choose a pulse or continuous irradiation mode to maximise the benefits of the MW-assisted processing.

In both devices, the complete MW-assisted process is controlled by a microprocessor and dedicated software for individual programming

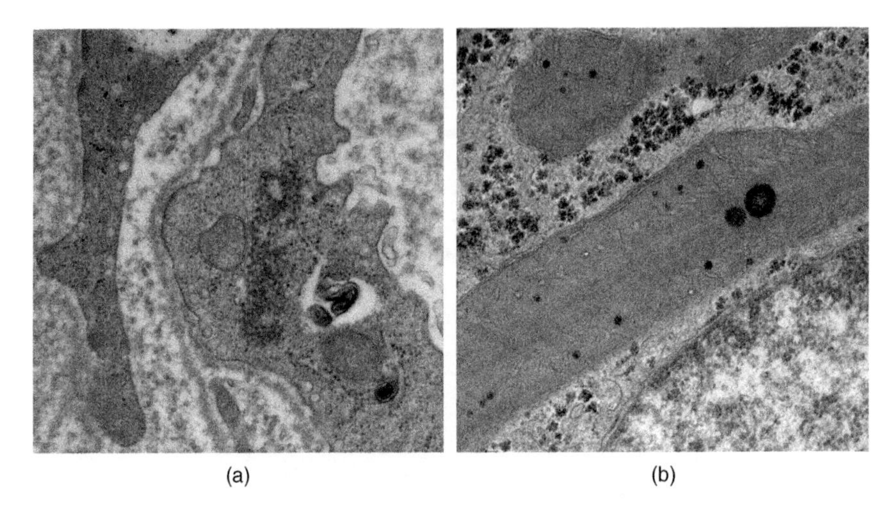

Figure 15.4 (a) Skeletal muscle (human), dermatomyositis–note tubuloreticular inclusions in the epithelial cell (original magnification × 16 000). AMW/Leica, Epon-resin embedding. Four hours of Karnovsky fixation at RT, and subsequently MW-assisted tissue processing (osmium tetroxide post-fixation, dehydration and polymerisation). (b) Liver (human), mitochondriopathy–note the paracrystalline inclusions in the mitochondria and excellent membrane and glycogen visualisation (original magnification ×20 000). Epon-resin embedding. Conventional LYNX processing, and rapid MW-assisted REM/Milestone polymerisation.

of the MW irradiation conditions for each step. Both operate with the same sample baskets and similar vials as used in the EM tissue processor (LYNX) for routine conventional tissue embedding at room temperature. We noticed that this shared equipment has a great advantage for EM and laboratory workflow because one can combine MW-assisted steps (fixation = 20–60 minutes or shorter, resin block polymerisation = 105 minutes) with conventional overnight processing, generating more flexibility in handling urgent clinical samples.

A number of laboratories using MW-assisted sample processing for routine diagnostic and research material recently reported similar significant TAT reductions, excellent results and improved embedding quality of challenging material (Webster, 2007). Examples of animal and plant samples embedded in Epon resin from other laboratories are illustrated in Figures 15.5–15.7 (McDonald, 2002; Zechmann and Zellnig, 2009; LEICA, 2012).

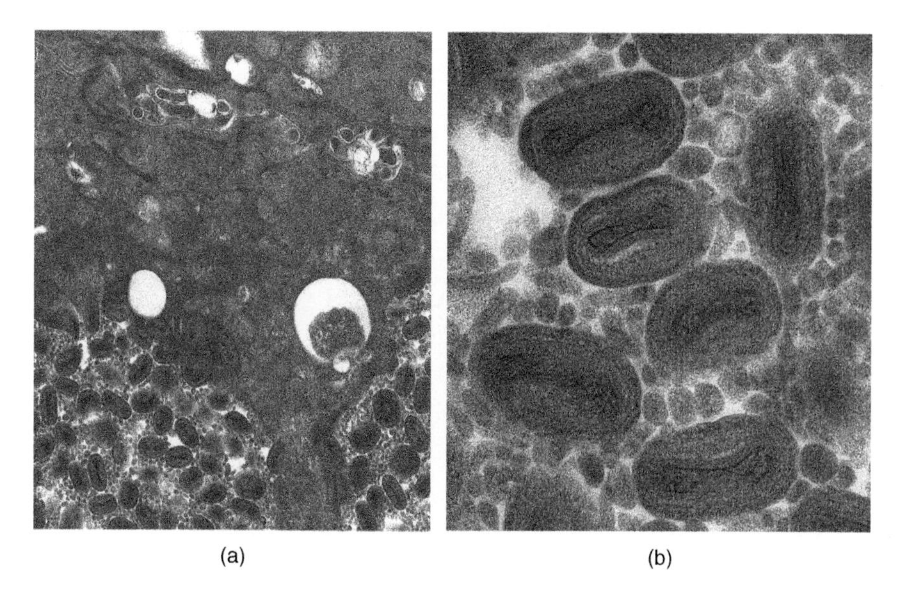

Figure 15.5 (a) Fowlpox-infected skin (cocks comb), stratum transitivum with mature virions inclusion. Note the good preservation of the epidermal layers and of the infectious agent (original magnification ×1600). AMW/Leica and Epon-resin embedding. Overnight Karnovsky fixation and subsequent MW-assisted AMW/Leica tissue processing. (b) Mature fowlpox virus particles; same specimen as (a) and AMW processing (original magnification ×40 000). Courtesy of Dr Susanne Richter, Austrian Agency Food & Health Safety, Vienna.

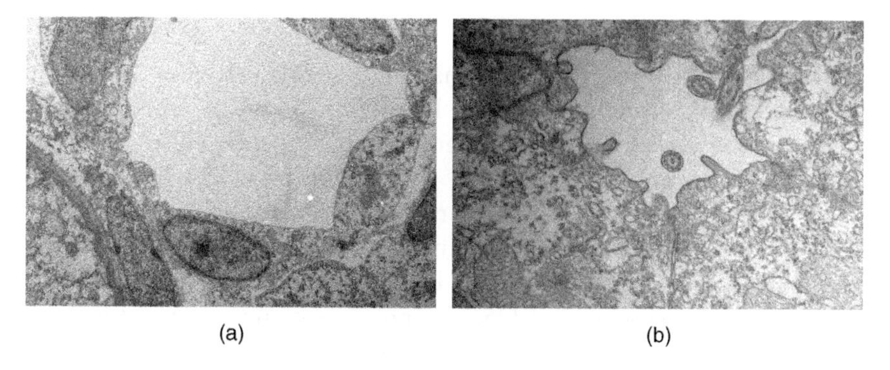

Figure 15.6 Zebra fish embryo. AMW/Leica, experimental non-automated step-by-step Epon-resin processing; protocol adopted from Kent McDonald (2002). (a) Endothelial cells of the dorsal aorta (original magnification × 8800) and (b) kidney tubule: note excellent preservation of the primary cilium (original magnification × 31 000). Courtesy of Dr Yannick Schwab, IGBMC, France.

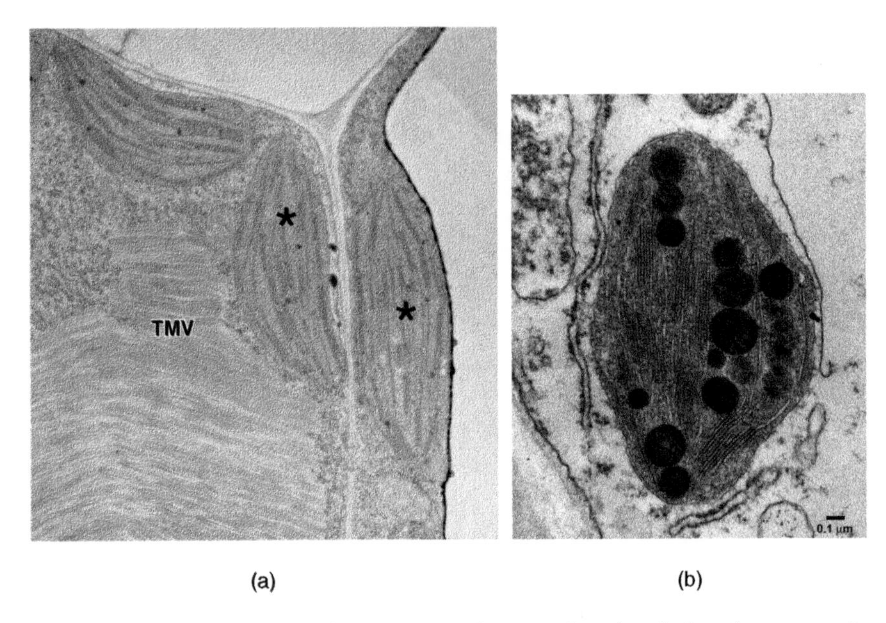

Figure 15.7 (a) Plant material: *Nicotiana tabacum* infected with the tobacco mosaic virus (TMV); TMV = virus agglomerate, asterisk = chloroplasts (original magnification × 6600). AMW/Leica, Epon-resin embedding. Courtesy of Dr Bernd Zechmann, Institute of Plant Sciences, University Graz, Austria. (b) Plant material: a chloroplast from the leaf of *Ginkgo biloba L.* (original magnification × 50 000). AMW/Leica, Epon-resin embedding. Courtesy of Chantal Cazevieille, INSERM and Jean-Marc Brillouet, INRA, France.

In addition to Epon, most resins traditionally used for TEM can be modified for MW techniques. LR White is useful especially for researchers interested in immunogold labelling (Rangell and Keller, 2000; Leria, Marco and Medina, 2004; Munoz *et al.*, 2004). This medium was also reported for the rapid detection of infectious agents (Laue, Niederwöhrmeier and Bannert, 2007). Examples of routine diagnostic applications from other laboratories processing renal biopsies and other tissue with KOS, the novel semi-automatic Milestone MW processor, using LR White or Spurr's resin are presented in Figure 15.8 (see Table 15.1 for an outline of the respective processing protocols, and Table 15.2 for the Spurr resin with Quetol formulation). Obviously, the standard formulation of the less viscous Spurr's resin needs the addition of Quetol (Toronto mix) or a plasticiser such as dibutyl phthalate (DBP) for satisfactory polymerisation (John W. Stirling, personal communication).

15.4 CYBERSPACE FOR TELEPATHOLOGY VIA THE INTERNET

Just as in light microscope (LM) histopathology, consultation among experts is essential for complex TEM cases, and by telepathology controversial findings and interpretations can be resolved in the context of an original specimen being examined directly rather than being interpreted on the basis of pre-selected images (mostly sent as email attachments) (Weinstein *et al.*, 2001; Kumar and Dunn, 2009).

For semithin section evaluation and the selection of the appropriate area for ultratrimming of the block, we apply – sometimes in consultation with the external material sender – a commercial, web-browser-based, LM remote system (Digital Sight DS-L1, Nikon) (Figure 15.9). The remote participant enters the valid device IP address into the URL field in his or her desktop browser and can immediately see the same image live (Figure 15.10) as the observer at the microscope and monitor in our laboratory. Using the microscope pointer, the microscope operator can delineate the relevant tissue area and discuss the details by a parallel phone connection with the pathologist in charge.

For the TEM examination, we established a dynamic remote EM diagnostic system for 'second-opinion' consultation (Schroeder and Voelkl, 1998; Schroeder, Voelkl and Hofstaedter, 2001) based on the LEO912AB transmission electron microscope (Zeiss) equipped with side- and bottom-mounted 1k × 1k-pixel charge-coupled device (CCD) cameras (TRS) (Figure 15.11). The system is controlled by the 'iTEM' software (OSIS) from a server PC and linked via the Internet to locations throughout Europe (Figure 15.12).

This server–client arrangement of the TEM–telepathology system enables the remote participant to perform stage navigation and live searching for the area of interest at low magnification; select adequate magnification (from 18× to 4 00 000×); adjust focus, beam brightness and exposure time and store images at full resolution on the local and remote computer hard drive (Schroeder, 2009). Included overlay features, such as direct size measurement and discussion tools (pointing arrows, drawing marks and annotations) used in the active remote session hooked up by a parallel phone connection, ensure a real telepresence feeling at both collaborating locations (Figure 15.13).

The use of such novel remote active telemicroscopy systems for diagnostic purposes often poses a psychological challenge to some colleagues (Kumar and Dunn, 2009), although we noted that the 'Internet-imprinted' generation have no such reservations to practising

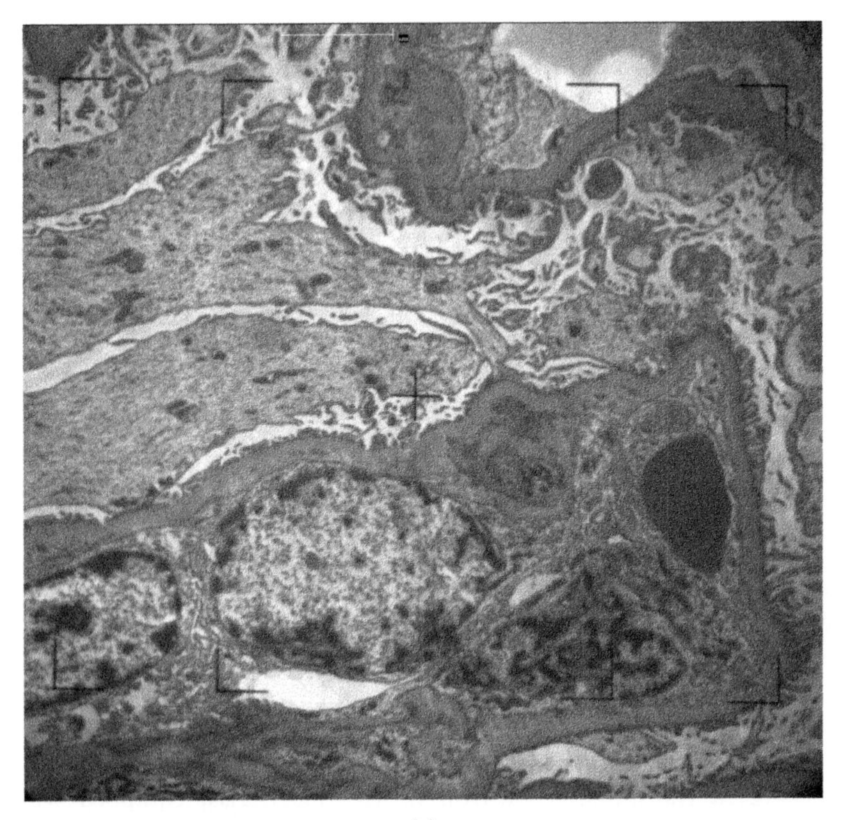

(a)

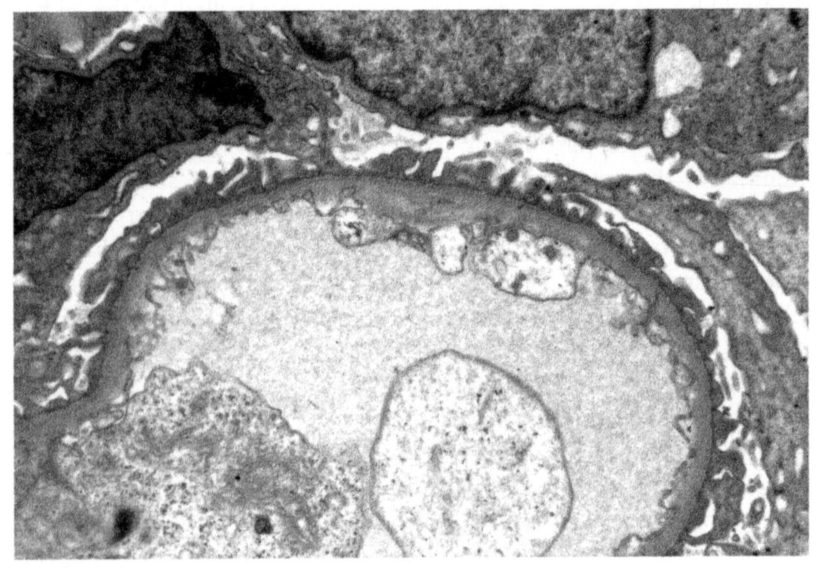

(b)

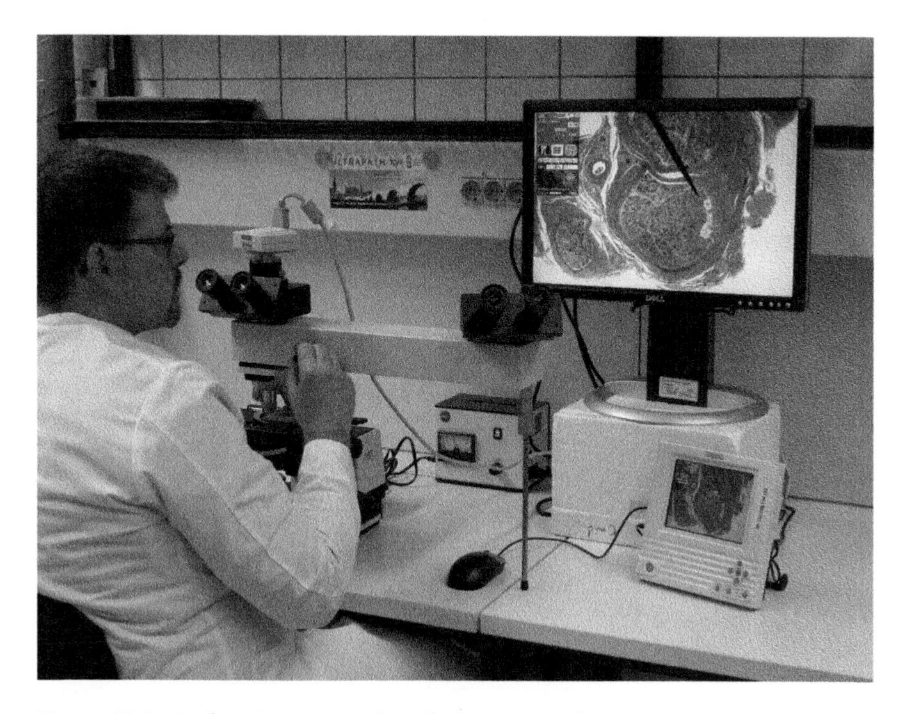

Figure 15.9 Light microscopy telepathology set-up for semithin section consultation. The system is based on the Nikon Digital Sight DS-L1 device and is used for area selection for subsequent tissue block trimming for cutting ultrathin sections.

telepathology at the LM or TEM levels. The term telepresence was coined in 1998 by Nestor Zaluzec (Draper, Kaber and Usher, 1998; Zaluzec, 1998), whose multi-user Telepresence Microscopy Collaboratory at the Argonne National Laboratory, Illinois, United States, drives a number of remote-controlled microscopes for collaborative purposes

←

Figure 15.8 (a) Renal biopsy, glomerulus (human). Note capillary loop with endocapillary cells, podocytes with well-preserved foot processes and inconspicuous glomerular basement membrane (original magnification × 2200). KOS/Milestone, LR-White resin embedding using protocol in Table 15.1. Courtesy of Dr Rocio Guevara de Bonis, Dr Marta E. Couce, and Dr Carles Saus, Hospital Universitari Son Espases, Baleares. (b) Renal biopsy, glomerulum (human). Detail of well-preserved components of the capillary loop displaying effacement of podocyte foot processes (original magnification × 9700). KOS/Milestone, Spurr/Quentol resin embedding as in Table 15.1. Courtesy of MLT Jacqueline Pittman, Advanced Bioimaging Centre, Mount Sinai Hospital, Canada.

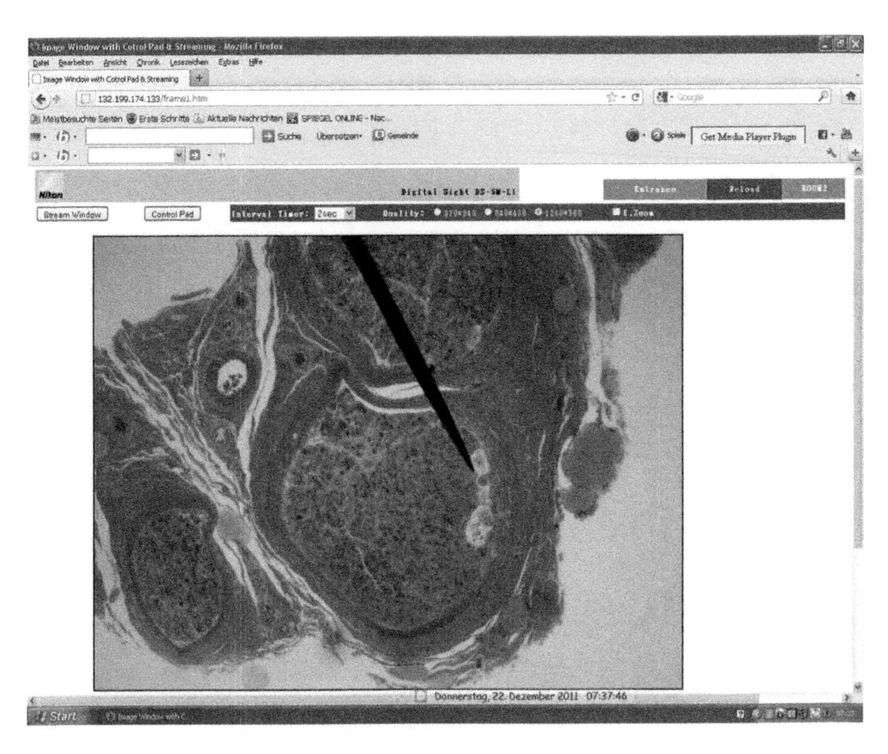

Figure 15.10 Screenshot of the monitor of the remote participant pathologist showing the transmitted image (peripheral nerve) of the semithin section under discussion (in parallel phone connection).

(TMP, 2012). In addition to the mostly technically complex ‘active telemicroscopy’ systems, some ‘passive telepathology’ solutions have been reported, such as the ‘Diagnostic Imaging Network System’ established at the Australian Animal Health Laboratory by Alex Hyatt, which enables an interaction in diagnostic cases between the users without high bandwidth requirements for the connection or special software (Bannert, Schroeder and Hyatt, 2009).

15.5 CONCLUSIONS AND FUTURE PROSPECTS

TEM, with its 1000 times higher resolving power compared with classic LM, is still used as an ancillary tool, quality control method or gold standard to complement, support or confirm the results of specific histopathological diagnoses (Eyden, 2005).

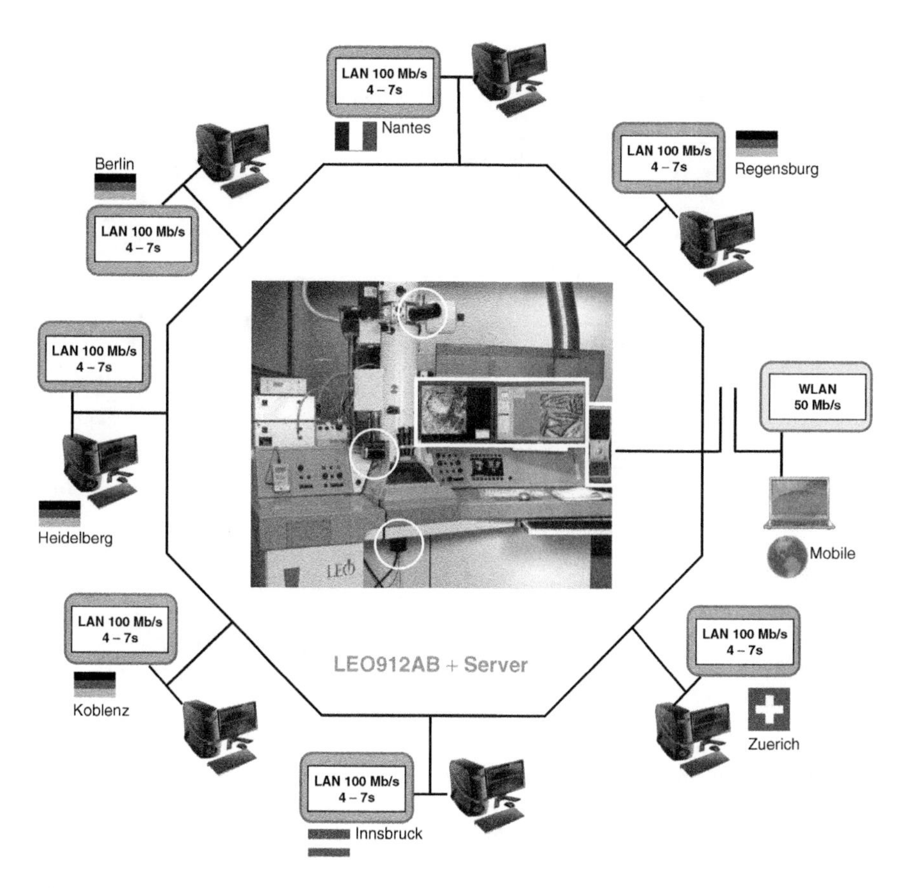

Figure 15.11 Electron microscopy active telepathology network at the University of Regensburg based on the server–client architecture. The LEO912 EFTEM is retrofitted with a motorised thermal drift-minimised objective aperture, and a 1k × 1k pixel CCD USB-connected camera is bottom- and side-entry mounted on the EM column (white circle). The system is controlled by a Windows server running the 'iTEM' software and handling the communication via standard LAN or WLAN links to the Internet. The image transmission performance of the system in binned 'live mode' is 2–4 frames per s; in the 'frozen snapshot mode' (uncompressed high-resolution 16-bit images for storage), the transfer needs 3–7 seconds (dependent on the daytime bandwidth).

MW technology can significantly reduce the sample TAT from days to hours, providing excellent ultrastructural preservation of animal, human and plant material. Most resins used routinely in TEM are compatible with MW-assisted processing or need only small modifications of standard formulations. Rapid MW-assisted tissue processing combined with digital image acquisition make 'same-day' TEM diagnosis a reality, which can be crucial in urgent clinical cases (Nordhausen and

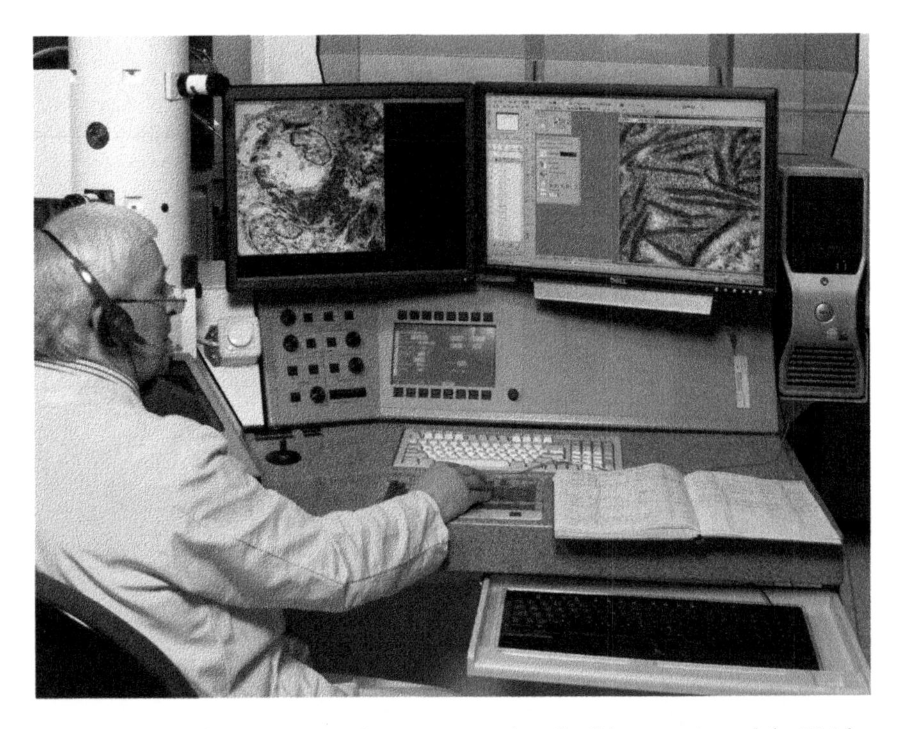

Figure 15.12 The operator at the 'server site' has flexible operation of the EM by the software button controls and/or the microscope panel. The software supports a two-monitor image display; note the left screen showing the live image (deposits in a glomerular capillary); the large right-hand screen displays the control panels for the EM and camera, and the saved image (higher resolution of the deposits) or the used database. The sense of telepresence in a telepathological consultation can be markedly increased by using a parallel standard phone connection–note the headset for freehand EM operation.

Barr, 2001; Giberson *et al.*, 2003; Leong and Leong, 2005; Schroeder *et al.*, 2006).

In a relatively short period of time, the digital era has witnessed a number of new applications in medical care (e.g. telediagnosis, teletherapy and telemonitoring) and in LM pathology, such as the 'virtual slide' and 'virtual microscopy' technique (Gu and Ogilvie, 2005; Kayser, Molnar and Weinstein, 2006). The knowledge coming from cyberspace and novel technologies (e.g. Stimulated Emission Depletion (STED) microscopy, resolution = 16–80 nm) (Hell, 2007)) has considerable impact on the practice of medicine and science, and, in tandem with holistic concepts of systems biology and targeted therapy, is inducing a paradigm change in

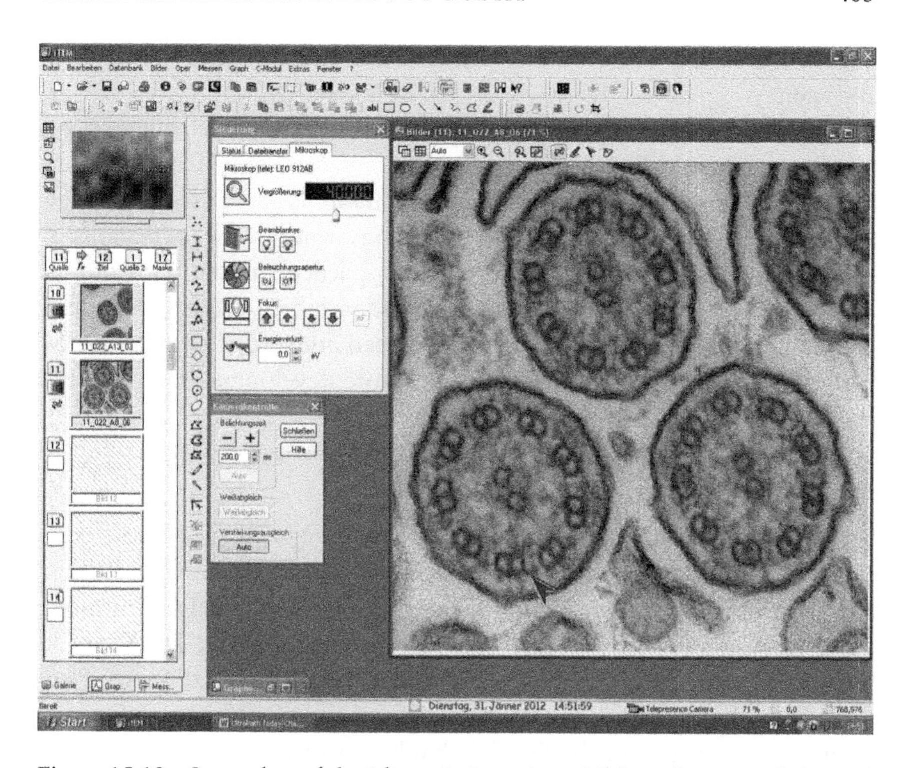

Figure 15.13 Screenshot of the 'client site' monitor visible to the remotely located expert. Note the transmitted image during a ciliopathy examination and two panels for the EM and CCD camera control for the EM operation by the expert. Synchronised software object measurement and discussion tools like pointing (arrow) and annotation functions significantly enhance the consultation performance.

many disciplines, including pathology, as mentioned in the Section 15.1 (Rashbass, 2000; Wells and Sowter, 2000). It is wise to realise this new potential, and practitioners are encouraged to implement these new ideas and solutions (Riley *et al.*, 2004; Leong and Leong, 2005). In this context, one should note that it is mandatory for the survival of a diagnostic TEM laboratory that it is in compliance with accreditation guidelines and maintains a high-quality (Stirling and Curry, 2007) and timely diagnostic service.

'Passive' and 'active' ultrastructural telepathology bridge space and time, can provide access to unique instruments, and are novel tools for instant live second-opinion retrieval and to share interesting findings worldwide (Sawai, Uzuki and Watanabe, 2000; Takaoka *et al.*, 2000; Telescience, 2012). Rapid advances in Internet technology and bandwidth enable a high level of telepresence collaboration in diagnostic ultrastructure, translational research and distance teaching, saving time and money.

Rapid MW-assisted sample processing and interactive telemicroscopy solutions, using the Internet, will boost the potential of EM as one of the tools in the *ensemble* of modern diagnostic methods being used to achieve the goals of emerging predictive pathology and future personalised healthcare strategies. Networking and experience sharing with others at workshops and conferences (www.ultrapath.org) are of additional proven help to keep the practitioner informed and at the cutting edge of pathological diagnostic technology and future developments.

ACKNOWLEDGEMENTS

Many thanks to Dr Chantal Cazevieille, Montpellier, France; Dra Rocio Guevara de Bonis, Palma, Baleares; Dra Marta E. Couce, Palma, Baleares; Dr Carles Saus, Palma, Baleares; MLT Jacqueline Pittman, Toronto, Canada; Dr Susanne Richter, Vienna, Austria; Dr Yannick Schwab, Strasbourg, France and Dr Bernd Zechmann, Graz, Austria, for providing images for Figures 15.5–15.8, and the LR-White and Spurr's resin processing protocols. Special thanks go to my co-worker Heiko Siegmund, Regensburg, for excellent technical support as well as to Dr Brian Eyden, Manchester, England, for linguistic help.

REFERENCES

Bannert, N., Schroeder, J.A. and Hyatt, A. (2009) The role of passive and active microscopy in diagnostic electron microscopy. *AM&M Newsletter*, (104), 20–21.

Cheville, N.F. (2009) *Ultrastructural Pathology: The Comparative Cellular Basis of Disease*, Wiley-Blackwell, Oxford.

De la Hoz, A., Diaz-Ortiz, A. and Moreno, A. (2005) Microwaves in organic synthesis. Thermal and non-thermal microwave effects. *Chemical Society Reviews*, **34** (2), 164–178.

Dickersin, G.R. (2000) *Diagnostic Electron Microscopy: A Text/Atlas*, Springer, New York.

Dietel, M. and Schafer, R. (2008) Systems pathology – or how to solve the complex problem of predictive pathology. *Virchows Archiv*, **453** (4), 309–312.

Draper, J.V., Kaber, D.B. and Usher, J.M. (1998) Telepresence. *Human Factors*, **40** (3), 354–375.

Erlandson, R.A. (1994) *Diagnostic Transmission Electron Microcsopy of Tumors: With Clinicopathological, Immunohistochemical, and Cytogenetic Correlations*, Raven Press, New York.

Erlandson, R.A. (2003) Role of electron microscopy in modern diagnostic surgical pathology, in *Modern Surgical Pathology*, vol. 1 (eds W. Cote and S. Weiss), Saunders, Philadelphia, PA, pp. 81–89.

Eyden, B. (1999) Electron microscopy in tumour diagnosis: continuing to complement other diagnostic techniques. *Histopathology*, **35** (2), 102–108.

Eyden, B. (2002) Electron microscopy in the diagnosis of tumours. *Current Diagnostic Pathology*, **8**, 216–224.

Eyden, B. (2005) Electron microscopy in pathology, in *The Science of Laboratory Diagnosis* (eds J. Crocker and D. Burnett), John Wiley & Sons, Ltd, Chichester, pp. 43–60.

Eyden, B. (2007) *The Myofibroblast: A Study of Normal, Reactive and Neoplastic Tissues with an Emphasis on Ultrastructure*, Nuova Immagine Editrice, Siena.

Faratian, D., Clyde, R.G., Crawford, J.W. and Harrison, D.J. (2009) Systems pathology – taking molecular pathology into a new dimension. *Nature Reviews Clinical Oncology*, **6** (8), 455–464.

Franzen, C., Fischer, S., Schroeder, J. *et al.* (2005) Morphological and molecular investigations of *Tubulinosema ratisbonensis* gen. nov., sp. nov. (Microsporidia: Tubulinosematidae fam. nov.), a parasite infecting a laboratory colony of *Drosophila melanogaster* (Diptera: Drosophilidae). *The Journal of Eukaryotic Microbiology*, **52** (2), 141–152.

Galvez, J.J., Giberson, R.T. and Cardiff, R.D. (2004) Microwave mechanisms – the energy/heat dichotomy. *Microscopy Today*, **12** (2), 18–23.

Ghadially, F.N. (1997) *Ultrastructural Pathology of the Cell and Matrix*, 4th edn, Butterworth-Heinemann, Boston.

Giberson, R.T. and Demaree, R.S. (2001) *Microwave Techniques and Protocols*, Humana Press, Totowa, NJ.

Giberson, R.T., Austin, R.L., Charlesworth, J. *et al.* (2003) Microwave and digital imaging technology reduce turnaround times for diagnostic electron microscopy. *Ultrastructural Pathology*, **27** (3), 187–196.

Grabe, N. (2008) Virtual microscopy in systems pathology. *Pathologe*, **29** (Suppl. 2), 259–263.

Gu, J. and Ogilvie, R.W. (2005) *Virtual Microscopy and Virtual Slides in Teaching, Diagnosis, and Research*, Taylor & Francis, Boca Raton, FL.

Hazelton, P.R. and Gelderblom, H.R. (2003) Electron microscopy for rapid diagnosis of infectious agents in emergent situations. *Emerging Infectious Diseases*, **9** (3), 294–303.

Hell, S.W. (2007) Far-field optical nanoscopy. *Science*, **316** (5828), 1153–1158.

Hotaling, M. (2006) Managing hazardous waste in the laboratory. *Clinical Leadership & Management Review*, **20** (5), E5.

Kayser, K., Molnar, B. and Weinstein, R.S. (2006) *Virtual Microscopy Fundamentals, Applications, Perspectives of Electronic Tissue-based Diagnosis*, Veterinärspiegel Verlag GmbH, Berlin.

Kitano, H. (2002) Computational systems biology. *Nature*, **420** (6912), 206–210.

Kok, L.P. and Boon, M.E. (2003) *Microwaves for the Art of Microscopy*, Coulomb Press Leyden, Leiden.

Kumar, S. and Dunn, B.E. (2009) *Telepathology*, Springer, New York.

Laue, M., Niederwöhrmeier, B. and Bannert, N. (2007) Rapid diagnostic thin section electron microscopy of bacterial endospores. *Journal of Microbiological Methods*, **70** (1), 45–54.

LEICA (2012) AMW Embedding Protocols, http://www.leica-microsystems.com/products/electron-microscope-sample-preparation/biological-specimens/room-temperature-techniques/tissue-processing/details/product/leica-em-amw/downloads/ (accessed July 2012).

Leong, A.S. and Leong, F.J. (2005) Strategies for laboratory cost containment and for pathologist shortage: centralised pathology laboratories with microwave-stimulated histoprocessing and telepathology. *Pathology*, **37** (1), 5–9.

Leria, F., Marco, R. and Medina, F.J. (2004) Structural and antigenic preservation of plant samples by microwave-enhanced fixation, using dedicated hardware, minimizing heat-related effects. *Microscopy Research and Technique*, **65** (1–2), 86–100.

Madeley, C.R. (2003) Diagnosing smallpox in possible bioterrorist attack. *Lancet*, **361** (9352), 97–98.

Marani, E. and Feirabend, H.K. (1993) A non-thermal microwave effect does not exist. *European Journal of Morphology*, **31**, 141–144.

McDonald, K. (2002) Microwave-assisted embedding of tissue culture cell monolayers. *Microscopy and Microanalysis*, **8** (Suppl. 2), 144–145.

Mierau, G.W. (1999) Electron microscopy for tumour diagnosis: is it redundant? *Histopathology*, **35** (2), 99–101.

Miller, S.E. (2003) Bioterrorism and electron microscopic differentiation of poxviruses from herpesviruses: dos and don'ts. *Ultrastructural Pathology*, **27** (3), 133–140.

Morange, M. (2009) A new revolution? The place of systems biology and synthetic biology in the history of biology. *EMBO Reports*, **10** (Suppl. 1), S50–S53.

Morel, G. and Raccurt, M. (2003) *PCR/RT-PCR in Situ Light and Electron Microscopy*, CRC Press, Boca Raton, FL.

Morens, D.M., Folkers, G.K. and Fauci, A.S. (2004) The challenge of emerging and re-emerging infectious diseases. *Nature*, **430** (6996), 242–249.

Munoz, T.E., Giberson, R.T., Demaree, R. and Day, J.R. (2004) Microwave-assisted immunostaining: a new approach yields fast and consistend results. *Journal of Neuroscience Methods*, **137** (2), 133–139.

NIH (2012) Targeted Therapy, http://www.cancer.gov/cancertopics/factsheet/Therapy/targeted (accessed July 2012).

Nordhausen, R.W. and Barr, B.C. (2001) Specimen preparation for thin-section electron microscopy utilizing microwave-assisted rapid processing in a veterinary diagnostic laboratory, in *Microwave Techniques and Protocols* (eds R.T. Giberson and R.S. Demaree), Humana Press, Totowa, NJ, pp. 49–66.

Ong, B.B. and Looi, L.M. (2001) Medico-legal aspects of histopathology practice. *The Malaysian Journal of Pathology*, **23** (1), 1–7.

Papadimitriou, J.M., Henderson, D.W. and Spagnolo, D.V. (1992) *Diagnostic Ultrastructure of Non-Neoplastic Diseases*, Churchill Livingstone, Edinburgh.

Rangell, L.K. and Keller, G.A. (2000) Application of microwave technology to the processing and immunolabeling of plastic-embedded and cryosections. *The Journal of Histochemistry and Cytochemistry*, **48** (8), 1153–1159.

Rashbass, J. (2000) The impact of information technology on histopathology. *Histopathology*, **36** (1), 1–7.

Rauch, C.A. and Nichols, J.H. (2007) Laboratory accreditation and inspection. *Clinics in Laboratory Medicine*, **27** (4), 845–858, vii.

Reipert, S., Kotisch, H., Wysoudil, B. and Wiche, G. (2008) Rapid microwave fixation of cell monolayers preserves microtubule-associated cell structures. *The Journal of Histochemistry and Cytochemistry*, **56** (7), 697–709.

Riley, R.S., Ben-Ezra, J.M., Massey, D. *et al.* (2004) Digital photography: a primer for pathologists. *Journal of Clinical Laboratory Analysis*, **18** (2), 91–128.

Robert-Koch-Institute (2012) Consultant Lab for Diagnostic EM in Infectious Diseases, http://www.rki.de/cln_234/nn_197444/sid_B67DAE4861E6C59BB8704DE081F5E2AB/DE/Content/Institut/OrgEinheiten/ZBS/ZBS4/zbs4__org.html?__nnn=true (accessed July 2012).

Rocken, C. and Manke, H. (2010) Accreditation in pathology. Systematic presentation and documentation of activities in pathology. *Pathologe*, **31** (4), 268–278.

Rosai, J. (2007) Why microscopy will remain a cornerstone of surgical pathology. *Laboratory Investigation*, **87** (5), 403–408.

Saidi, O., Cordon-Cardo, C. and Costa, J. (2007) Technology insight: will systems pathology replace the pathologist? *Nature Clinical Practice Urology*, **4** (1), 39–45.

Sawai, T., Uzuki, M. and Watanabe, M. (2000) Telepathology at presence and in the future. *Rinsho Byori*, **48** (5), 458–462.

Schroeder, J. (2009) Ultrastructural telepathology: remote EM diagnostic via internet, in *Telepathology* (eds K. Kumar and B.E. Dunn), Springer, Berlin, pp. 179–204.

Schroeder, J.A., Gelderblom, H.R., Hauroeder, B. *et al.* (2006) Microwave-assisted tissue processing for same-day EM-diagnosis of potential bioterrorism and clinical samples. *Micron*, **37** (6), 577–590.

Schroeder, J.A. and Voelkl, E. (1998) Electron microscopy examination of pathological samples in Oak Ridge/USA by remote control via Internet from Regensburg/Germany: a live ultrastructure-telepathology presentation. G7SP4 Conference: The Impact of Telemedicine on Health Care Management, Regensburg/Germany.

Schroeder, J.A., Voelkl, E. and Hofstaedter, F. (2001) Ultrastructural telepathology – remote EM-diagnostic via Internet. *Ultrastructural Pathology*, **25** (4), 301–307.

Schroeder, J.A., Weingart, C., Coras, B. *et al.* (2008) Ultrastructural evidence of dermal gadolinium deposits in a patient with nephrogenic systemic fibrosis and end-stage renal disease. *Clinical Journal of the American Society of Nephrology*, **3** (4), 968–975.

Shi, S.R., Gu, J. and Taylor, C.R. (2000) *Antigen Retrieval Techniques: Immunohistochemistry and Molecular Morphology*, Eaton Publications, Natick, MA.

Stieger, K., Schroeder, J., Provost, N. *et al.* (2009) Detection of intact rAAV particles up to 6 years after successful gene transfer in the retina of dogs and primates. *Molecular Therapy*, **17** (3), 516–523.

Stirling, J.W. and Curry, A. (2007) Quality standards for diagnostic electron microscopy. *Ultrastructural Pathology*, **31** (5), 365–367.

Takaoka, A., Yoshida, K., Mori, H. *et al.* (2000) International telemicroscopy with a 3 MV ultrahigh voltage electron microscope. *Ultramicroscopy*, **83** (1–2), 93–101.

Telescience (2012) Grid Supercomputing, http://www.sdsc.edu/pub/envision/v16.2/telescience.html (accessed July 2012).

TMP (2012) Telepresence Microscopy Collaboratory, Argonne NL, http://tpm.amc.anl.gov/TPMSelect.html (accessed July 2012).

Tucker, J.A. (2000) The continuing value of electron microscopy in surgical pathology. *Ultrastructural Pathology*, **24** (6), 383–389.

Turbat-Herrera, E.A., D'Agostino, H. and Herrera, G.A. (2004) The use of electron microscopy to refine diagnoses in the daily practice of cytopathology. *Ultrastructural Pathology*, **28** (2), 55–66.

Wagner, B.E. and Curry, A. (2002) The usefulness of electron microscopy in diagnosing and investigating infectious diseases. *Current Diagnostic Pathology*, **8**, 232–240.

Webster, P. (2007) Microwave-assisted processing and embedding for transmission electron microscopy, in *Methods in Molecular Biology*, vol. 369 (ed. J. Kuo), Humana Press, Clifton, NJ, pp. 47–65.

Weinstein, R.S., Descour, M.R., Liang, C. *et al.* (2001) Telepathology overview: from concept to implementation. *Human Pathology*, **32** (12), 1283–1299.

Wells, C.A. and Sowter, C. (2000) Telepathology: a diagnostic tool for the millennium? *The Journal of Pathology*, **191** (1), 1–7.

Wendt, K.D., Jensen, C.A., Tindall, R. and Katz, M.L. (2004) Comparison of conventional and microwave-assisted processing of mouse retinas for transmission electron microscopy. *Journal of Microscopy*, **214** (Pt 1), 80–88.

Woods, A.E. and Stirling, J.W. (2008) Electron microscopy, in *Theory and Practice of Histological Techniques* (eds J.D. Bancroft and M. Gamble), Churchill Livingstone, Edinburgh, pp. 601–640.

Zaluzec, N.J. (1998) Tele-presence microscopy: a progress report. *Microscopy and Microanalysis*, **4** (Suppl. 2), 18–19.

Zechmann, B. and Zellnig, G. (2009) Rapid diagnosis of plant virus diseases by transmission electron microscopy. *Journal of Virological Methods*, **162** (1-2), 163–169.

16

Electron Microscopy Methods in Virology

Alan Curry
Health Protection Agency, Clinical Sciences Building, Manchester Royal Infirmary, Manchester, United Kingdom

16.1 BIOLOGICAL SAFETY PRECAUTIONS

All samples processed for virus identification should be handled with the utmost precautions for health and safety, on account of their potential for infection. Some laboratories now require that stool samples be 'inactivated' prior to preparation for electron microscopy (EM), as they could contain pathogenic bacteria in addition to viruses. Inactivation normally requires fixation in buffered neutral formaldehyde (Gelderblom, 2006) or 2–4% paraformaldehyde in buffer (Laue, 2010). However, this procedure may sometimes adversely affect virus morphology, so use caution with interpretation. In some centres, skin lesions are still processed in safety cabinets without inactivation, and this can expedite results and enhance confidence in the results obtained (in this area, EM is a rapid diagnostic technique). However, if samples are from suspected bioterrorism-related events, they should be handled only where the highest containment facilities are available and should be inactivated prior to EM examination (i.e. examination for such agents is restricted to specialised centres).

Diagnostic Electron Microscopy: A Practical Guide to Interpretation and Technique, First Edition. Edited by John W. Stirling, Alan Curry and Brian Eyden.

16.2 COLLECTION OF SPECIMENS

If one is investigating the possible viral causes of diarrhoea and/or vomiting, specimens need to be collected as soon as possible after symptoms start. In cases of viral gastroenteritis, viruses are usually most abundant early in the course of enteric symptoms, and this is particularly important if one is trying to detect noroviruses. It is therefore important that faecal samples are obtained as quickly as possible, so that there is a reasonable chance of detecting viruses by EM. With urine samples, the urine should be an 'early morning' sample taken soon after waking to avoid the dilution effects on samples taken later in the day.

16.3 PREPARATION OF FAECES, VOMITUS OR URINE SAMPLES

Viruses contained in faeces, vomitus or urine samples need to be concentrated and partially purified for visualisation under a TEM. A suspension (~10%) of faecal material needs to be available for further processing. Samples should initially be clarified by low-speed centrifugation to sediment large debris and most bacteria, followed by high-speed centrifugation (at about 65 000 g) to concentrate any contained virus particles. Following ultracentrifugation, the supernatant is removed and the pellet re-suspended in a small volume of buffer before being adsorbed onto the coated EM grid.

If an ultracentrifuge is not available, the ammonium sulphate precipitation method can be used (Caul, Ashley and Eggleston, 1978). This virus precipitation method may be a gentler method of virus concentration compared with ultracentrifugation and is reported to be of equal sensitivity in concentrating faecal viruses (Caul, Ashley and Eggleston, 1978).

16.4 VIRUSES IN SKIN LESIONS

Normally, vesicle fluid or a piece of skin lesion is provided for preparation and examination. If dried onto a microscope slide, vesicle fluid should be re-suspended in buffer and a drop placed on a small sheet of parafilm. A coated grid should then be floated on this drop to allow any contained viruses to adhere. After washing, grids should be stained with a negative stain, such as phosphotungstic acid (PTA).

Tissue samples of skin lesions are prepared by placing them in a drop of buffer and macerating the piece to release any viruses from cells and produce a suspension. As with the method described in the previous paragraph, a coated grid should then be floated on the suspension, and this should be washed and stained after a suitable adsorption period.

Samples containing few virus particles or very dilute preparations may require direct centrifugation using an Airfuge (Beckman) of any contained virus particles onto the grid surface. This centrifuge can be fitted with a fixed-angle rotor and adaptors that allow the coated grid surface to remain parallel to the axis of rotation (Gelderblom, 2006; Laue and Bannert, 2010).

16.5 REAGENTS AND METHODS

16.5.1 Negative Stains

Negative staining of virus suspensions is the method of choice for diagnostic EM of enteric viruses. However, the choice of negative stain has important implications, as it has been demonstrated that, for example, some rotaviruses (Group B rotaviruses) show morphological degradation after negative staining with PTA (Suzuki *et al.*, 1987), and alternative stains should be used (e.g. uranyl acetate). The pH of PTA can affect the morphology of rotaviruses, as a study by Nakata *et al.* (1987) showed normal morphology at a pH of 4.5 but degraded virus particles at higher pH values. Methylamine tungstate is a negative stain that may also preserve fragile viruses, such as noroviruses, and thus may enhance detection by EM (Patrick Costigan, personal communication). See Harris (1997) for a comprehensive overview of negative stains and negative staining.

16.5.1.1 Examples of Negative Stains

1. **PTA.** 3% aqueous, pH 6.3–6.5 (adjust with NaOH). A variable pH range can be used with PTA from about pH 4.5 to pH 8.
2. **Uranyl acetate.** 1 or 2% aqueous, pH 4.5. Uranyl acetate should not be used with phosphate buffers (Hayat, 1970), and it is slightly radioactive (safety precautions are needed in handling and preparation).
3. **Ammonium molybdate.** 2% aqueous, pH 5.3–8 (adjust pH with NaOH or NH_4OH).
4. **Methylamine tungstate.** 2% aqueous, pH 6.8.
5. **Sodium silicotungstate.** 2% aqueous, pH 5–8 (adjust pH with NaOH).

16.6 COATED GRIDS

Grids should have a small mesh size (e.g. 400 mesh) to give adequate support and stability to the thin but electron-transparent support film. A number of plastic support films are available, including Formvar and Pioloform, and the grid-coating process should be performed in advance of preparing the virus specimen samples for examination. Solutions of either coating plastic can be made up in ethylene dichloride. If necessary, plastic support films can be stabilised by depositing a layer of carbon reinforcement on the surface, but this can produce a hydrophobic surface to which viruses will not easily adhere. Hydrophilic surfaces can be produced by pre-treatment of the grid with bovine serum albumin, bacitracin or Alcian blue (Lang, Nermut and Williams, 1981; Laue and Bannert, 2010), or the grids can be exposed to a glow discharge.

For high-resolution EM of viruses, a carbon-coated grid can be used. Such grids tend to be fragile and need careful handling, but the low granularity of carbon can enhance visualisation of virus morphology.

16.7 IMPORTANT ELEMENTS IN THE NEGATIVE STAINING PROCEDURE

This procedure can be modified depending on the requirements or initial results obtained:

1. Produce a concentrated and partially purified virus suspension.
2. Place a drop of virus suspension on a strip of parafilm or a clean glass microscope slide.
3. Float an inverted coated grid on the surface of the virus suspension, and allow a period for viruses to adhere to the grid coating (times will vary, but 5–10 minutes is reasonable).
4. Remove grid, and wash on a drop of distilled water.
5. Remove excess water with filter paper, but do not let the grid dry out completely.
6. Repeat the above washing cycle.
7. Float the grid on a drop of negative stain. Again, staining time is variable, but it should be from about 20 seconds to about 2 minutes.
8. Remove excess stain using filter paper, and allow a brief drying period.
9. Load the grid into a microscope, and examine it at an appropriate instrument magnification.
10. It is good practice to photograph the first virus particle encountered, then to look for better stained or better preserved particles.

16.8 TEM EXAMINATION

Specimen grid examination requires both knowledge and experience and is fundamentally dependent on the observational skills of the electron microscopist. With faecal viruses, a screening magnification of 40 000–60 000 is recommended as this allows location of viruses on the grid surface and recognition of the virus morphology. Generally, the viruses found in skin lesions are larger than those found in diarrhoea samples, and, therefore, a lower screening magnification can be used (about 40 000×).

16.9 IMMUNOELECTRON MICROSCOPY

There are several methods available for detecting viruses by using specific antibodies (Doane and Anderson, 1987).

16.9.1 Immune Clumping

One of the simplest methods involves incubating a virus suspension with a virus-specific antibody, then looking for virus clumps (immune clumping) after negative staining (Milne and Luisoni, 1977). This method can obscure morphology through excess antibody coating. Optimising the concentration of antibody used is important. Excess antibody results in small clumps of virus heavily coated with antibody, producing a 'fuzz' over the particles, while optimal concentrations of antibody result in large 'rafts' of virus particles where morphology may be more easily seen. Immune clumping can be used with many viruses, but it is particularly useful in detecting those that lack characteristic morphological features (such as featureless enteroviruses) or viruses that have many serotypes, such as adenoviruses (Wood and Bailey, 1987). However, the use of control reagents (such as a pre-immune serum from the animal used to raise the specific antibody) is essential. In addition, viruses lying close together by chance ('pseudoclumps') must be differentiated from true immune virus clumping.

16.9.2 Solid-Phase Immunoelectron Microscopy

Solid-phase immunoelectron microscopy (SPIEM) is an alternative immunological method where the specimen support grid is incubated with a specific antibody before being floated on the virus suspension. Virus particles adhere to the antibody and, after negative staining, are visible as individual particles adhering to the grid support film (Gerna *et al.*, 1984). The attraction of this method is that the virus particles are not coated with antibody and therefore retain their morphology.

16.9.3 Immunogold Labelling

Viruses can also be specifically labelled using immunogold particles if an antigen-specific antibody is available (Hopley and Doane, 1985; Hawkins, Rehm, and Zhu, 1992). This procedure produces a specific electron-dense label and is relatively simple when labelling of whole viruses is attempted on coated grids.

16.9.4 Particle Measurement

Particle size, morphology and clinical source can all aid virus identification under the electron microscope (also EM). All EMs have marked magnification steps, but these are not entirely accurate because of the phenomenon of lens hysteresis. Modern EMs have more accurate magnification values than older instruments, but the marked magnification is accurate only to a value of ±5%. In addition, most microscopes are now fitted with digital imaging cameras, with images being displayed on a monitor. Again, this can make sizing potentially inaccurate unless suitable calibration specimens are used for the inbuilt measurement feature provided with most digital imaging computers. A line grating specimen is useful for calibration up to about 20 000× magnification, but for higher magnifications (which are more useful in the sizing of virus particles), a fixed beef catalase calibration specimen is more appropriate.

16.10 THIN SECTIONING OF VIRUS-INFECTED CELLS OR TISSUES

Tissues can normally be fixed and processed using conventional histological EM methodology (see Chapter 14). Infected cell cultures can be more problematic to handle. Cells may have to be released from the surfaces on which they are growing by either trypsinisation before fixation or scraping from the culture vessel walls after fixation. Once released and in suspension, the cells can be centrifuged in, for instance, an Eppendorf tube to produce a pellet for the remainder of the resin embedding preparation process. If the resultant pellet is small, post-fixation in osmium tetroxide can make the pellet visible, thus allowing easier preparation.

Immunogold labelling of viruses in thin sections is a more elaborate procedure than that required for labelling unfixed viruses on support films (Hawkins, Rehm and Zhu, 1992; Stirling, 1995). It fundamentally requires 'light fixation' (e.g. paraformaldehyde) to preserve antigenicity as well as preserve cell organelles and contained assembling viruses. If antigenicity is poorly preserved, some antigen retrieval or antigen unmasking methods may have to be applied (Stirling, 2000). In addition, the embedding resin used must normally be hydrophilic (e.g. LR White or Lowicryl) rather than hydrophobic (e.g. Araldite).

16.11 VIROLOGY QUALITY ASSURANCE (QA) PROCEDURES

16.11.1 External QA

The Robert Koch Institute in Berlin, Germany, periodically distributes formalin-fixed virus preparations and collates the results of those who participate. A variety of viruses are prepared for each (approximately yearly) distribution, and these can be challenging to identify. This external quality assurance (EQA) scheme is highly recommended.

16.11.2 Internal QA

Stored, known positive samples should be periodically reprocessed and re-examined as an internal QA procedure. However, some viruses, such as noroviruses, are relatively fragile and do not survive freezing (unlike rotaviruses). Norovirus-positive faecal samples should be stored at +4° C under liquid paraffin to inhibit fungal growth.

If more detail is required of the virus preparation methods or reagents and stains used, the references Doane and Anderson (1987), Madeley and Field (1988), Harris (1997) and Laue (2010) should be consulted.

ACKNOWLEDGEMENTS

Sincere thanks are given to the following people for information for and constructive comments on this chapter: Dr Michael Laue (Robert

Koch Institute, Berlin), Dr E. Owen Caul (Bristol, United Kingdom) and Mr. Patrick Costigan (Dublin, Ireland).

REFERENCES

Caul, E.O., Ashley, C.R. and Eggleston, S. (1978) An improved method for the routine identification of faecal viruses using ammonium sulphate precipitation. *FEMS Microbiology Letters*, **4**, 1–7.

Doane, F.W. and Anderson, N. (1987) *Electron Microscopy in Diagnostic Virology. A Practical Guide and Atlas*, Cambridge University Press, Cambridge.

Gelderblom, H.R. (2006) Virus enrichment using the Airfuge for rapid diagnostic EM in infectious diseases. *Rotor*, **4**, 4–5.

Gerna, G., Passarani, N., Battaglia, M. and Percivalle, E. (1984) Rapid serotyping of human rotavirus strains by solid-phase immune electron microscopy. *Journal of Clinical Microbiology*, **19**, 273–278.

Harris, J.R. (1997) *Negative Staining and Cryoelectron Microscopy. Royal Microscopical Society Microscopy Handbooks*, vol. 35, BIOS Scientific Publishers Ltd, London.

Hawkins, H.K., Rehm, L.S. and Zhu, J.Y. (1992) Colloidal gold labelling of sections and cell surfaces. *Ultrastructural Pathology*, **16**, 61–70.

Hayat, M.A. (1970) *Principles and Techniques of Electron Microscopy*, Biological Applications, vol. 1, Van Nostrand Reinhold Company, New York.

Hopley, J.F.A. and Doane, F.W. (1985) Development of a sensitive protein A-gold immunoelectron microscopy method for detecting viral antigens in fluid specimens. *Journal of Virological Methods*, **12** (1–2), 135–147.

Lang, R.D.A., Nermut, M.V. and Williams, L.D. (1981) Ultrastructure of sheep erythrocyte plasma membranes and cytoskeletons bound to solid supports. *Journal of Cell Science*, **49**, 383–399.

Laue, M. (2010) Electron microscopy of viruses, in *Methods in Cell Biology*, vol. 96, Chapter 1 (ed. T. Muller-Reichert), Academic Press, New York, pp. 1–20.

Laue, M. and Bannert, N. (2010) Detection limit of negative staining electron microscopy for the diagnosis of bioterrorism-related micro-organisms. *Journal of Applied Microbiology*, **109**, 1159–1168.

Madeley, C.R. and Field, A.M. (1988) *Virus Morphology*, 2nd edn, Churchill Livingstone, Edinburgh.

Milne, R.G. and Luisoni, E. (1977) Rapid immune electron microscopy of virus preparations, in *Methods in Virology*, vol. 6 (eds K. Maramorosch and H. Koprowski), Academic Press, New York, pp. 265–281.

Nakata, S., Petrie, B.L., Calomeni, E.P. and Estes, M.K. (1987) Electron microscopy procedure influences detection of rotaviruses. *Journal of Clinical Microbiology*, **25**, 1902–1906.

Stirling, J.W. (1995) Immunogold labelling for electron microscopy. Basic techniques for resin sections, in *Laboratory Histopathology – A Complete Reference*, Chapter 9.3 (eds T. Woods and R. Ellis), Churchill Livingstone, Edinburgh, pp. 9.3-1-9.3-21.

Stirling, J.W. (2000) Antigen retrieval and unmasking for immunoelectron microscopy, in *Antigen Retrieval Techniques: Immunocytochemistry and Molecular*

Morphology, Chapter 5 (eds S.R. Shi, J. Gu, and C.R. Taylor), Eaton Publishing, Natick, MA, pp. 93–113.

Suzuki, H., Chen, G.M., Hung, T. *et al.* (1987) Effects of two negative staining methods on the Chinese atypical rotavirus. *Archives of Virology*, **94**, 305–308.

Wood, D.J. and Bailey, A.S. (1987) Detection of adenovirus types 40 and 41 in stool specimens by immune electron microscopy. *Journal of Medical Virology*, **21**, 191–199.

17

Digital Imaging for Diagnostic Transmission Electron Microscopy

Gary Paul Edwards
Chelford Barn, Stowmarket, Suffolk, United Kingdom

17.1 INTRODUCTION

Selecting the right camera is more than just choosing a camera with the maximum number of pixels. Many factors, such as mounting position, chip technology, speed, resolution, software and support, need to be taken into consideration.

There is no ideal camera suitable for all applications. Some microscopists need the highest possible sensitivity for low-dose work, while others need a camera with the fastest frame rate to observe dynamic processes. These two requirements are almost mutually exclusive, and what may be the ideal camera for one application may turn out to be inadequate for another application.

17.2 CAMERA HISTORY

Since Ernst Ruska built the first electron microscope in the 1930s (Encyclopædia Britannica, 2012), we endured a near 50-year period where images could be viewed only on a fluorescent viewing screen coated

Diagnostic Electron Microscopy: A Practical Guide to Interpretation and Technique, First Edition. Edited by John W. Stirling, Alan Curry and Brian Eyden.

with phosphor, or photographically recorded with a direct-exposure camera to film or photographic plate. Steven Sasson, an electrical engineer working for Kodak, invented the first digital still camera using a Fairchild charge-coupled device (CCD) chip in 1975 (Pavlus, n.d.). But it was a further 10 years before this technology was advanced enough for use in transmission electron microscopy (TEM). In the mid-1980s, Gatan revolutionised TEM imaging with the introduction of the first TV rate camera for surveying and imaging specimens electronically. The quality of these cameras was adequate for surveying but not suitable for printing.

In the early 1990s, the first commercial slow-scan CCD cameras became available from both Gatan and Advanced Microscopy Techniques giving 1k × 1k (1 megapixel (Mp)) digital images which could be printed and used in place of film. Since this time, pixel count has increased from 1 to 64 Mp (Tietz Video and Image Processing Systems (TVIPS) launched a 64 Mp 8k × 8k camera in 2006). In this chapter, the technology available will be explained, and a broad understanding given of TEM cameras and their application in the diagnostic field.

17.3 THE PIXEL DILEMMA

Most people purchasing a camera for TEM use believe that more pixels will mean more resolution; however, this is not necessarily true. When purchasing a TEM, the choice would not be based solely on maximum specified magnification because ultimate resolution is not linked to magnification. If the detail required is not resolved with the objective lens, subsequent magnification by the projector lenses will simply make the image larger without improving resolution. Similar effects can occur in a TEM CCD camera system, which will affect ultimate pixel resolution.

All TEM CCD camera systems have an inherent resolution loss caused by losses in the collection system. The primary electrons that form an image on film or photographic plates cannot be detected directly by a CCD camera. The electrons must first be converted to photons (light) visible to the CCD sensor using a thin layer of phosphor called a scintillator. As electrons strike the scintillator and generate light, they deviate slightly from their original path and scatter, causing the resolution of the photon image to be less than that of the electron image. The design of the scintillator is therefore critical to the design of a CCD camera system; if the scintillator is too thick, there will be inadequate resolution, but if it is too thin, there will be inadequate electron-to-photon conversion or insufficient brightness.

17.4 CAMERA POSITIONING

Most TEMs have two positions for mounting cameras (Figure 17.1). These positions are generally known as side mount or bottom mount. Side-mount cameras are installed on the side of the column at the 35 mm port where, historically, a 35 mm camera would have been mounted. The port is located above the viewing chamber and provides access to the beam as it leaves the final lens. This position provides cameras with a large field of view (often larger than the negative), which is a side-mount camera's biggest advantage, although there are compromises.

With side-mount cameras, the physical size of the image projected onto the phosphor screen is small, which can lead to limited pixel size compared with bottom-mount cameras; this in turn can limit ultimate resolution.

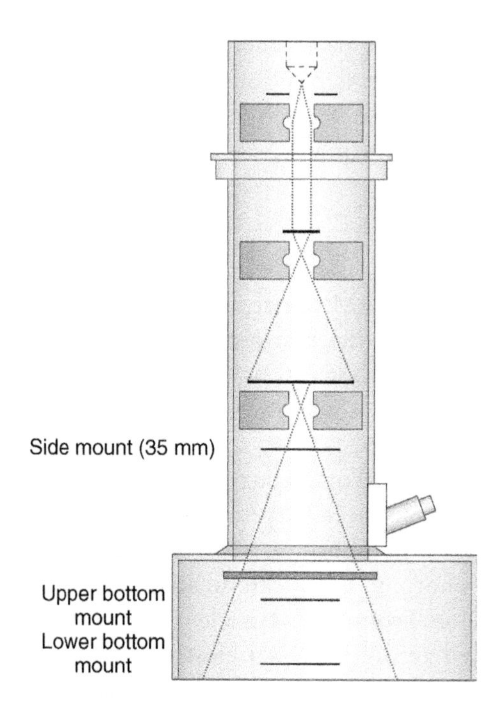

Figure 17.1 Column and lens (TEM CCD camera mounting positions).

Side-mounted cameras require the scintillator to move in and out of the beam path, so a moving mechanism is required. All TEM cameras use stored reference images and background subtraction algorithms to correct for uneven brightness and contrast caused by uneven illumination of the beam, artefacts or contamination on the phosphor. To achieve the highest quality final image, these reference images must not shift with respect to the actual image by more than 1 pixel. With side-mount cameras moving over relatively long distances and with pixel size in the order of a few μm, this is not a trivial task, and special care needs to be taken to ensure the phosphor screen repositions accurately.

Because side-mount cameras often see a field of view larger than the original film area, they are prone to distortion at the edges, especially at low magnifications.

Bottom-mount cameras are essentially complementary to side-mount cameras. They are installed below the viewing chamber, so they do not interfere with normal TEM operation. To expose the camera, the viewing screen needs to be lifted up; on computer-controlled TEMs, this can be controlled automatically from the camera software.

Because of their position, bottom-mount cameras are able to see a physically larger image, which allows a higher ultimate resolution, but field of view will normally be smaller than with a side-mount camera. Some bottom-mount cameras position the scintillator high up inside the film chamber, which allows a much larger field of view compared with traditional bottom-mount cameras.

17.5 RESOLUTION

Ultimate resolution of a TEM camera is defined by a number of factors. Clearly, the resolution of the microscope is important, as is the pixel resolution of the camera, but just as important is the thickness of the phosphor, the size of the CCD pixels and the way the scintillator is coupled to the CCD.

Striking a balance with all of the factors affecting resolution is essential. To understand this, it is necessary to consider what happens when a single electron strikes the phosphor. The electron excites the phosphor, causing it to emit photons, and it can do so many times as it is scattered within the phosphor. From this, it is clear that even if the microscope itself had unlimited resolution, two electrons hitting the phosphor in close proximity may not be individually distinguishable if their paths cross while scattering inside the phosphor. The volume that a single

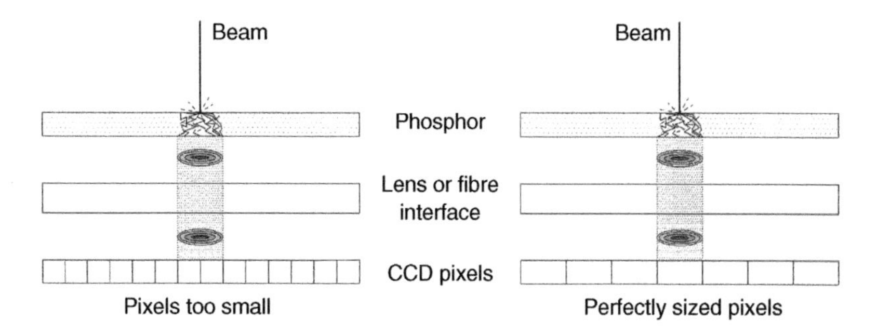

Figure 17.2 CCD pixels (importance of pixel size in the design of a TEM CCD camera).

electron takes up while scattering inside the phosphor will start to define the ultimate resolution. The volume itself will depend on various parameters, of which the energy of the electron is the most important. In general, the excitation volume ranges from several μm in diameter to several tens of μm. This provides a fairly accurate definition of the CCD pixel size that should be used for imaging the phosphor. It should be in the order of the width of the scattering volume. Larger pixels will lead to loss of information, while smaller pixels will lead to duplicated information. In most cases, the pixel size of the CCD chip used can be adjusted to the desired size by use of the optical elements which would normally be either optical lenses or fibre-optics (Figure 17.2).

With pixel size determined by the physics of the collection system, the task of the camera designer is to match the pixel size of the CCD chip to the desired effective pixel size. This can be achieved with an optical lens or fibre-optics. Lenses have the advantage that they can be designed with more flexibility, but signal attenuation and distortions are concerns. Fibre-optic systems historically provide the best signal-to-noise ratio, but they can be difficult to manufacture, thus making them expensive.

17.6 FIBRE COUPLED OR LENS COUPLED?

Lens-coupled systems are, as mentioned in Section 17.5, more flexible to design and ultimately less expensive to produce. Standard off-the-shelf lenses used in low-cost TEM CCD camera systems are likely to be fairly insensitive, be prone to distortions and have full-field focussing problems, that is, it is difficult to focus all four corners and the centre of the image at the same time. High-performance lens-coupled systems, however, use custom-designed finite-conjugate lenses with very low distortion figures and excellent full-field focussing properties.

Lenses do have a tendency to show more vignetting (i.e. the images become darker towards the edges), although with a TEM, the beam itself is more often a source of intensity variation. Background correction will negate the effect of this.

Fibre-coupled systems have to achieve a magnification or de-magnification with a fibre-optic bundle, and the fibre needs to be 'tapered'. In this process, the fibre bundle is heated and stretched. Stretching thins the fibres so the cross-section can be reduced. This is a complex process with high failure rates, so these tapered fibres are expensive to produce.

As optical fibres arrange in a hexagonal pattern and pixels are in a square pattern, artefacts caused by the fibres are visible on the images. These artefacts take the shape of dark lines, arranged in a hexagonal format (the so-called chicken wire effect). This is a fixed artefact that can be eliminated through appropriate correction images.

Most TEM camera systems have Peltier-cooled CCDs; fibre-coupled systems tend to be cooled to a lower temperature than lens-coupled systems; because of this, they require water cooling, whereas lens-coupled cameras do not.

17.7 SENSITIVITY, NOISE AND DYNAMIC RANGE

Each electron creates many photons within the phosphor while it is being scattered. The scattering process reduces the resolution of the resultant photon image, as discussed in this chapter. A thicker phosphor will create more scattering and more photons. While this will give a brighter resultant image, it will also reduce resolution. A thin phosphor can be used to increase resolution, but it will produce fewer photons. Scintillator efficiency needs to be optimised for acceleration voltage by carefully choosing the correct phosphor type and thickness.

Each CCD pixel has a maximum amount of charge it can hold. This is the so-called full well capacity and is expressed as the number of electrons the pixel can hold. Depending on the size of the pixel, this value typically ranges from several tens of thousands to several hundreds of thousands. Over-filling the pixel (saturation) must be avoided because it diminishes the quantitative ability of the CCD and produces image smearing on account of a phenomenon known as blooming (Photometrics, 2012).

Scientific-grade CCD sensors incorporate features to prevent blooming effects; this is normally stated as 'anti-blooming architecture'.

Noise is the random part of the image signal that is not specifically related to the sample. A significant source of noise is the so-called dark current signal in a CCD chip, which is created when a photon strikes the silicon in a pixel and creates an 'electron hole pair'. One of the two (depending on the type of silicon) is then retained until the pixel is read out, while the other is discarded. The hotter the CCD chip, the more carriers are created, which are then registered as signal. This effect worsens with increasing temperature of the CCD. Most TEM cameras use cooling of the CCD to reduce the effect of this noise.

So-called electron-beam shot noise will also affect imaging, especially when working in low-dose mode with very weak beams. This noise can be reduced by either longer exposure times or higher beam currents, both of which increase the electron dose on the sample. It is therefore important to reduce instrument noise as much as possible while at the same time increasing the sensitivity of the camera.

'Read-out' noise is another source of noise. To read a CCD chip, the charge carriers have to be shifted from one pixel to the next until they reach a read-out gate. At higher speeds, this transfer becomes increasingly difficult to control, and charge carriers remain behind and are eventually measured in the wrong pixels.

Each pixel on the CCD chip will have a slightly different characteristic. Some may be slightly more sensitive, some are slightly less sensitive, and with so many pixels there are normally some which will be fully on or off (fixed black or white). Once again, this is a fixed structure, and reference images can eliminate these differences in pixel characteristics.

As the scintillator phosphor layer is normally sintered and has a grain structure, some areas may be less efficient at producing photons than others. This leads to a visible grain structure. As the structure is fixed, it can be eliminated (together with fibre artefacts) through reference images.

The dynamic range of a camera will be determined by all of the factors discussed here, as well as the resolution of the analogue-to-digital converter in the camera electronics. Most scientific-grade camera systems will have a dynamic range of between 12 and 16 bits, which relates to between 4096 and 65 536 grey levels. However, for most TEM imaging 8 bits (255 grey levels) will be more than adequate.

17.8 CCD CHIP TYPE (FULL FRAME OR INTERLINE)

CCD chips, although performing the same task, can achieve this task in various ways. One major difference is the way they output the image data. There are principally two different techniques, and it is important to understand them and their implications for imaging.

The original CCD technology is known as frame transfer. This technology is straightforward – all pixels of the CCD chip are exposed to the signal for a pre-defined exposure time. The pixels are then cut off from further exposure by using a shutter or blanking the beam; the image is then read into memory. During this read-out time, no new signal can be acquired. This technology has the advantage that nearly 100% of the chip surface can be used for image acquisition, allowing the highest possible sensitivity. Two major drawbacks are that a shutter or beam blanker is required to shield the pixels during the read-out period and that this typically results in fairly low read-out times. These two factors make full-frame chips less desirable for cameras with live imaging capabilities. The potentially higher sensitivity, however, makes this the preferred technology for techniques like cryomicroscopy or low-dose microscopy.

Progressive scan interline transfer CCD chips trade some sensitivity for higher read-out speeds. In these CCDs, each line of pixels has an extra hidden line which is covered and does not receive a signal. Once the exposed line has accumulated sufficient signal, the information is quickly transferred to the hidden line. The hidden line is then read out during the next exposure cycle. Because of the speed with which the transfer can be processed internally, there is no shuttering or blanking required, that is, higher read-out speeds can be achieved. The drawback is that interline chips trade active area for read-out speed; the hidden lines take up space on the CCD which is not sensitive to any signal.

17.9 BINNING AND FRAME RATE

As indicated in Sections 17.7 and 17.8, the read-out speed is an important criterion that needs to be evaluated to find the optimum camera. However, resolution and noise also play important roles in the selection of camera operating parameters.

For live imaging, all TEM CCD cameras employ a technique known as binning. Binning is just as it sounds – during each frame, a number of pixels are discarded or 'binned'. This allows the CCD to run at much higher speeds than would otherwise be possible. For example, a 2048×2048 CCD may have a full resolution frame rate of only 2.5 frames per second, but if every other pixel is binned (2×2 binning), the visual resolution will drop to 1024×1024 pixels and the speed could increase by four times to 10 frames per second. Most computer monitors do not have a vertical resolution of much more than 1000 pixels, so trying to display more than this will only result in lost speed without any extra visual resolution.

In an ideal world, we would display an image with a resolution that matches the pixel resolution of the monitor, which is noise-free with a frame rate of at least 25 frames per second. In the real world, however, there have to be compromises. Most TEM operators would sacrifice speed for less noise and more resolution because the quality of the image is of paramount importance. In cryo or low-dose applications, exposure times have to be very long to acquire images with such a small amount of signal, so the frame rate has to be extremely slow. For live surveying and focussing (with a progressive scan CCD), it is generally accepted that 8–10 frames per second would be the slowest suitable frame rate, while for a full-frame camera, this would be 12–15 frames per second.

17.10 SOFTWARE

With TEM cameras being available from many different suppliers, not only does the camera itself need careful consideration but also software needs to be carefully evaluated. Clearly, the software requirements for a TEM camera in a university research laboratory are different from those in a routine diagnostic laboratory; in clinical TEM, image analysis, montaging and image stitching may be seen as more of a luxury than as day-to-day requirements. Of primary importance is a software package which is easy to learn and use, and intuitively provides all the features that are needed on a day-to-day basis, without the need for staff to spend time on technical manuals or attend training courses.

Most TEM cameras will give the operator software options for adjusting binning, exposure time and any additional filtering; thus, the camera parameters can be optimised for particular applications. Ideally, there will be default settings so that minimal adjustment is needed on a day-to-day basis when looking at similar samples.

It is clear that many factors affecting the performance of a camera can be controlled through software. Reference images providing shading correction are a point in case. Typically, cameras need several sets of reference images for different operating conditions and the correct images must be applied, often at high speeds. This requires sophisticated software algorithms. Good software will shield the user from behind-the-scenes complexity and provide superior images.

17.11 CHOOSING THE RIGHT CAMERA

It is clear from the explanations given in this chapter that an ideal TEM camera system cannot be assembled without skill. High levels of knowledge of both technology and physics are required during the design process, and many parameters have to be evaluated against each other.

Many of the technical problems will be managed by the camera suppliers. Let them know which types of samples are being examined, and what results are to be expected. Do not be unrealistic – accept that different applications will require different cameras and that for varied applications, compromises will need to be made when choosing a single camera. Choices such as phosphor type and thickness are best left to the camera supplier, but the purchaser can make some choices.

Of the three camera positions shown in Figure 17.1, either side mount or upper-bottom mount would generally be considered acceptable for diagnostic TEM. Side-mount cameras will give the largest field of view, but they can be affected by distortion, although on modern TEMs this is normally limited to low magnifications. Upper-bottom-mount cameras will see a smaller field of view but give better performance at higher magnifications and provide a neater installation with all camera mechanics being under the console. This mounting position would also suit mixed-use applications where some higher magnification (>150 000) work will be carried out.

There will be almost no detectable performance differences between quality lens-coupled and fibre-coupled cameras when used for diagnostic TEM. For cryo-applications and low-dose work, a fibre-coupled camera would be recommended, but for general microscopy, a good-quality lens-coupled system would be just as effective.

Pixel resolution is important. While a 1024 × 1024 (1 Mp) camera will give a fairly good on-screen image, it will not allow any degree of zooming once the final image is acquired. However, if cost is of paramount importance, a 1 Mp camera will clearly be satisfactory, and in fact, there are many hundreds of diagnostic laboratories using such cameras. Most TEM operators would prefer at least a 2048 × 2048

(4 Mp) camera, and these cameras offer a good price-to-performance ratio with speeds up to 30 frames per second (×4 binned). Clearly, technology is constantly improving, and 8 and 16 Mp cameras are now being offered at competitive prices. CCD pixel size will eventually provide the limit on resolution for TEM cameras, although, because most new commercial large-format CCDs have pixels of only a few square μm, this will be too small to be utilised usefully in TEM camera systems.

Software is of primary importance when choosing a camera. The operator may be sitting in front of a computer running this software for many hours a day. Therefore, it is appropriate to make sure the software chosen provides all the features required in a convenient and accessible format and at an acceptable resolution and speed.

Choosing a camera is not easy. To help make this decision, discuss options with colleagues, identify other users who have already purchased a camera and take note of their experience. Service and support from your camera supplier are also important, especially if the laboratory lacks a parallel conventional film processing facility to resort to in the event of digital camera failure.

REFERENCES

Encyclopædia Britannica (2012) Ernst Ruska, *Encyclopædia Britannica Online*, http://www.britannica.com/EBchecked/topic/513086/Ernst-Ruska (accessed July 2012).

Photometrics (2012) Full Well Capacity, http://www.photometrics.com/resources/learningzone/fwellcapacity.php (accessed July 2012).

Pavlus, J. (N.d.) Steven Sasson Invented the World's First Digital Camera, video, http://www.fastcodesign.com/1663611/how-steve-sasson-invented-the-digital-camera-video (accessed July 2012).

18

Uncertainty of Measurement

Pierre Filion

Electron Microscopy Section, Division of Anatomical Pathology, PathWest Laboratory Medicine, QE II Medical Centre, Nedlands, Australia

18.1 INTRODUCTION

The measurement of a morphological structure in a sample provides a quantitative estimate of its 'real' size in the population; repeating the measurements in the same sample can lead to a slightly different value of this estimate. This leaves a doubt about the accuracy of the measurement, where accuracy is a qualitative assessment of the agreement between a measured value in the sample and its 'true' value. This doubt can be quantified by collating the variation due to individual factors used in the measurement process to calculate a single uncertainty of measurement (UM) (Bell, 2011). UM is due to many factors: the biological variation between similar structures in a sample can be quantified statistically by repeated measurements; rulers and other measuring tools are not perfect, and imply a limit of precision to be considered, and errors due to the experimental protocol, equipment used and method of analysis should be controlled and quantified. UM collates these variations in a single quantitative estimate of the spread of measurements around the mean value; it is then used to determine a confidence level or confidence interval for a desired degree of certainty (Bell, 2011). Synonyms for UM include 'measurement of uncertainty' and 'measured uncertainty'.

Diagnostic Electron Microscopy: A Practical Guide to Interpretation and Technique, First Edition. Edited by John W. Stirling, Alan Curry and Brian Eyden.

18.2 PURPOSE

18.2.1 Diagnostic Value

The quantitative measurement of structural features may contribute significantly to the diagnostic exercise: for example, measuring the diameter of fibrils deposited in the glomeruli to identify a case of renal amyloidosis or measuring the thickness of the glomerular basement membrane (GBM) in a suspected renal laminopathy. If these measurements are included in the pathology report, the associated uncertainty should be calculated and added to the report to qualify the diagnostic decision by providing a quantitative basis for the accuracy of a measurement.

18.2.2 Internal Quality Control

The calculation of UM for all quantitative measurements is an important component of the diagnostic exercise. It requires a thorough understanding of the correct calibration and use of all equipment, optimal training of staff and vigilance in applying analytical standards and quality procedures in the laboratory. In this respect, both internal and external quality assurance programs in a diagnostic laboratory benefit from a thorough understanding of the scope and limitations of quantitative measurements. In addition to its significance for the reliability of a measurement, the calculation of UM also identifies the most significant source of variation associated with a measurement; this can serve as an indicator for the performance of laboratory equipment procedures and staff.

18.2.3 External Quality Control and Accreditation

The International Organisation for Standardisation (ISO) standards for the accreditation of medical testing laboratories (ISO15189, and more generally ISO/IEC 17025:2005) require that quantitative measurements performed in the laboratory be reported with an estimate of UM described in the ISO *Guide to Uncertainty of Measurement* (ISO): this guide was derived through international consultation and forms the foundation of UM. These standards are also described in a publication by the International Organisation of Legal Metrology (OIML, 2010).

National and international organisations governing the accreditation and quality of testing laboratories recommend similar standards.

Although most are based on quantitative chemical or physical testing, the standards can be adapted to testing in a pathology laboratory. The following are a sample of recommended standards:

- In Australia, documents describing UM and the relevant standards of ISO 15 189 are both jointly (AACB, 2005) and individually issued by the National Association of Testing Authorities (NATA) (Cook, 2002; NATA, 2012), the Australian Institute of Medical Scientists (AIMS), the Royal College of Pathologists of Australasia (RCPA) (RCPA, 2009) and the National Pathology Accreditation Advisory Council (NPAAC).
- In Canada, standards are available through the Canadian Association for Environmental Analytical Laboratories (CAEAL) (CALA, 2005; Gravel, 2011).
- In the United Kingdom, the Clinical Pathology Accreditation (CPA UK) and the United Kingdom Accreditation Society (UKAS) offer publications concerning UM (UKAS, 2007).
- In the United States, the National Institute of Standards and Technology (NIST) standards are oriented to physical testing (NIST, 2004).

18.3 FACTORS THAT INFLUENCE QUANTITATIVE MEASUREMENTS

18.3.1 Sources of Variation

The sources of variation can be intrinsic to the method of preparation of a biopsy specimen: this can be controlled, if not easily quantified, by the use of a standard and reliable method of preparation of the specimen and its examination with both the transmission electron microscope (TEM) and the scanning electron microscope. The regular calibration and optimal performance of the analytical equipment, especially of the electron microscope (EM), are essential components of the measurement process, which in turn monitors the performance of the equipment. An understanding of the methods used to measure structures and calculate their UM, combined with the periodic review of known measurements, allow the performance of staff to be monitored and provide information on any deviation from the expected result (e.g. a controlled reference value). In this respect, the measurements must be compared with known reference values obtained with the same analytical protocol (intra-laboratory reference).

The quantitative measurement of a structure in an image (micrograph) can be influenced by the following three factors:

- Primary changes in the intrinsic dimensions of the structure
- Variation due to the analytical method used in imaging and measurement
- Variation due to a bias in selecting and measuring the structure.

18.3.2 Alteration of the Intrinsic Dimension of the Structure

18.3.2.1 Shrinkage Due to the Chemical Reagents Used to Prepare the Specimen

The chemical preparation of biological samples may alter their cellular and subcellular dimensions to an irregular degree. Tissues, cells and their organelles are variably affected by fixation and dehydration during specimen preparation, depending on their molecular composition and water content. For example, fibrils and microtubules remain mostly unchanged from their natural state, while nuclei, mitochondria and the whole cell may vary by up to 15% in their linear dimension (Crang and Klomparens, 1988). The factors discussed here contribute to this shrinkage to a degree not easily measured in the general practice of a diagnostic pathology laboratory. They can be controlled by adherence to a set, documented protocol for the uniform preparation of specimens.

The factors due to preparation include:

- **Fixation:** time elapsed before fixation, type and concentration of the fixative, osmolarity of buffer, duration of fixation, rinses and use of additional fixatives (post-fixation)
- **Specimen:** absolute size (thickness, area) of the specimen, aspect ratio (thin core vs. bulky sample), surface-to-volume ratio available for the exchange of chemicals and speed of penetration of the chemicals
- **Dehydration:** type of chemical dehydrating agents and their temperature; it has been suggested, however, that the concentration, number of steps and timing of dehydration may have limited influence on final dimensions in cells (Crang and Klomparens, 1988).
- **Resin:** type and concentration of solvents, duration of infiltration, composition of the resin, duration and temperature of infiltration of the resin, polymerisation and hardness of the resin. A resin of low molecular weight and low viscosity is thought to cause fewer dimensional changes to the cells and their constituent organelles (Crang and Klomparens, 1988).

18.3.2.2 Expansion Due to Sectioning of the Specimen Block

Sectioning of resin-embedded tissues results in further changes in their intrinsic dimensions, principally compression during sectioning and expansion of the 'free' section thereafter. These changes are also difficult to quantify at the cellular and subcellular levels, but they can be controlled by adhering to a consistent technical preparation.

The factors due to sectioning include:

- The knife used in sectioning: type (glass or diamond), knife angle, sharpness and defects
- Speed of sectioning
- The resin block: dimension of the specimen being cut, shape of the face, type and hardness of the resin, age of the block and conditions of storage (humidity)
- The ultramicrotome settings used for sectioning: angle of attack, sectioning speed, section thickness selected, degree of compression in the section and temperature of the water bath receiving the section
- Ambient conditions during sectioning: temperature, humidity and exposure of the block to changing environment (e.g. vacuum and temperature)
- The handling of the section after sectioning: spreading with chloroform, other solvent or heat; degree of folding when deposited on the TEM grid and type of grid (support for the section)
- Staining the grids: duration, incubation temperature, rinses and drying method (hot plate or air dried)
- Storage of the section: conditions, duration and exposure to vacuum or humidity.

18.3.2.3 Expansion Due to Examination with the TEM

The TEM settings used during examination and the exposure of the section to the TEM's electron beam during an extended period of time expand the resin sections and will alter the dimensions of structures embedded therein. As this influence on ultrastructural dimensions is not easily quantified, the settings for examination and imaging with the TEM should be documented and strictly followed; deviation from the documented procedure may influence the measurements of structures and invalidate their comparison with previously prepared specimens. In this respect, the periodic measurement of a well-defined, standard structure and its comparison with a known reference value may help to correct a systematic deviation from the expected value.

The factors due to TEM viewing include:

- The quality of the vacuum in the TEM
- The duration of exposure (during examination of structures) to the beam
- The number of times the sections are examined (repeated exposure to beam)
- The general settings of the TEM: accelerating voltage, beam current, beam saturation, intensity and focusing of the beam, filters and apertures, strength of condenser and objective lenses.

More factors are considered in the ISO 29 301 guide to calibration for microbeam analysis (ISO, 2010).

18.3.3 Variation Due to the Analytical Equipment and Method

18.3.3.1 Magnification and Calibration of the TEM

The TEM is the most critical component of the analytical protocol: its optimal performance must be ensured before attempting a quantitative measurement. The magnification (or range of magnifications) used in measurement must be verified, and an estimate of the variation encountered should be included in the calculation of UM. Where images are recorded to film, the scale bar (or other measuring device) embedded in the image serves to verify the precise magnification on the final image (film or printed micrograph). In the absence of a scale bar, an estimate of the magnification can be obtained by the photographic record of a known calibrating grid. Digital cameras can be calibrated directly using the same grids (at low magnifications) or a crystal lattice of known periodicity (at higher magnifications).

Verification of the magnification and calibration of a digital camera are essential components of the quality assurance program in the EM laboratory. It should be a regular, well-documented and comprehensive exercise that covers all steps of magnification likely to be used for a subsequent measurement. This verification and calibration serves a number of purposes:

- Monitoring the optimal performance of the TEM, and identifying any deviation requiring prompt attention

- Providing readily available data for the calibration to be used at any time for quantitative measurements
- A valuable quality procedure documenting the TEM's calibration to satisfy both internal and external audits (i.e. accreditation).

18.3.3.2 Illumination and Contrast in the TEM

An optimal illumination of the structure is necessary for a precise measurement of its defining boundaries. Both the brightness and contrast should be sufficient for the structure to be clearly identified and precisely defined.

18.3.3.3 Calibration of Measuring Equipment

All equipment used for the purpose of the measurement should be fully calibrated and maintained in optimal working condition. This includes all measuring tools such as the photographic enlarger, calibrated ocular lens and ruler. An electronic spread sheet (e.g. Microsoft Excel), created to maintain a record of this calibration, can also serve as a reminder for a regular schedule of calibration of any equipment subject to change in its performance over time (e.g. the height of a photographic enlarger).

18.3.3.4 Calibration of Images Recorded on Film and Printed Micrographs

The verification of the calibration on images recorded on film can be carried out by the photographic recording of a calibration grid. As the image may not be symmetrically recorded on film, the template must then be measured in both directions (perpendicular axes), and the mean length averaged; the deviation from that mean is an estimate of the imprecision of the measurement and is included in the UM. If printed micrographs are prepared from film-recorded images, the setting (height) of the photographic enlarger used to print must be calibrated regularly to ensure a uniform enlargement factor from film to print. This can be simply done using an image (on film) of a precisely calibrated ruler that is measured on both film and print. The variation in measuring this enlargement factor is then incorporated in the calculated UM.

18.3.3.5 Calibration of Images on Laser Printed Images

Printing a digital image using a laser printer distorts the original image by compressing or enlarging the original image to fit the printed page at the size and especially the aspect ratio (width-to-height ratio) selected. Care must be taken to select a printed format (ratio) that matches that of the image captured on film (if scanned) or digital camera. This effect can be verified by recording an image of a calibration grid and comparing its enlargement and preserved aspect ratio on the printed image to that of the original image. The measurement on each image of a scale bar, if present, also helps to verify the calibration of the printed image, if only in one dimension. Most flat-bed scanners used to digitise an image recorded on film produce a distortion of linear dimensions along the periphery of the field (edge effect), again corrected by measuring a pre-calibrated image and also by restricting measurements to the more centrally positioned structures; ISO 29 301 (ISO, 2010) suggests a possible correction for the edge effect. Ultimately, a sufficiently high dot-per-inch (dpi) setting should be selected for scanning to allow the clean definition of the boundaries of the structures to measure.

18.3.4 Variation Due to Selection Bias

18.3.4.1 Limited Size of the TEM Specimen

Given the often limited amount of material available to measure in a routine EM specimen, care must be taken to avoid a sampling bias. The small size of a biopsied specimen examined with the TEM limits the number and range of structures measured; this should be mentioned as a qualifier of the measurement reported.

18.3.4.2 Approaches to Avoid a Selection Bias

A selection bias is often embedded in the request for a measurement. Thus, it is not best practice to measure only the thinnest lamina in order to document a case of suspected glomerular laminopathy: although indeed yielding a low thickness, it may not represent adequately the range of basement membrane thickness present in the specimen. The range of diameters of deposited fibrils in the glomerulus generally follows a normal distribution, and the selection of the thickest (or smallest) fibril does not reflect the range of fibril diameters present, a point of diagnostic interest.

To avoid such bias, one of several sampling methods should be investigated, and the most appropriate adopted for each type of analysis:

- **Universal sampling:** every structure (capillary loop) is sampled, no matter how many are encountered. This has the disadvantage of increasing the work carried out to a needless extent.
- **Systematic sampling:** a rule establishes which limited portion of the grid square or visual fields is sampled, and which structures therein are measured: for example, measure all fibrils within a defined quadrant or area of the visual field. This method is robust if systematically applied, but remains subject to a residual selection bias ('cheating').
- **Random sampling:** the structure targeted (e.g. portions of a capillary loop) is selected according to a pre-determined rule, for example by placing a lined template traced on a transparent acetate sheet over the printed micrograph or computer display monitor, and measuring only those structures intercepted by the test lines. Various templates such as those used in stereological analysis can be prepared to measure different structures. This method eliminates selection bias while retaining a wide representation of structures to measure.

18.3.5 Measurement Using a Digital Camera

The advent of the digital camera to capture and display images introduces a new source of bias in the measurement of structure. The regular calibration of the digital camera at appropriate TEM settings and at all magnifications used in measurements is essential to ensure a reliable analysis. The sharpness of the digital image displayed depends on the resolution of the digital camera (charge-coupled device (CCD)). This may severely limit the precise localisation of the boundaries of structures such as fibrils with a diameter measured in nanometres.

Equally important is the resolution of the computer display screen (monitor): a display capable of the best resolution possible is essential, as is a very precise computer-driven cursor on the image display. The precise location of the cursor and its shape (e.g. arrow) influence the measurement of finer structures such as slender fibrils at both the starting and end points of the measurement. In this regard, it may be difficult to know when to end a measurement on the image displayed as some cursors partially obscure the image they overlay, a factor of significance in measuring small structures. Finally, the measurement of a diameter requires the precise traversing of the structure across its perpendicular axis, a minute movement that requires some practice (and honesty) to obtain reliably. As a useful aid to this problem, some cursors change to a set of parallel lines that can be aligned parallel to a long structure (fibril or membrane) to establish a true perpendicular axis of measurement.

Finally, the 'live' measurement on a digital image provides an immediate feedback of the measurement as it evolves (i.e. lengthens) on the monitor, which may tempt the user to end a measurement at the point of greatest expectation (when the expected value is displayed). This bias is minimised with practice and experience with measurement on a digital image.

18.4 HOW TO CALCULATE THE UM

18.4.1 Steps Required to Analyse and Calculate the UM

The calculation of UM is approached in five successive steps:

1. Define precisely define what quantity is to be measured.
2. Determine all steps to be taken in the measurement.
3. From these, identify all possible influences on the measurement, their type and how to quantify their influence.
4. Quantify in turn each source of variation, documenting these for future reference (e.g. magnification of the microscope).
5. Combine the quantified variations in a single numerical value to report.

18.4.2 Type of Error and Distribution of Measurements

Two types of error are commonly associated with quantitative measurements of ultrastructural features: each is associated with a different type of evaluation that depends on the method used for the measurements and reflects the distribution of values measured. If they are expressed in the same units, both types of error can be combined in a single compounded calculation as estimates of a single deviation interval around the mean.

18.4.2.1 Type A Error – Repeated Measurements

The evaluation of a type A error concerns the description of the innate variation obtained in repeated measurements in a population or sample, where the measurements assume a normal distribution. This type of error can be compensated by obtaining repeated measurements from the sample to assess the spread of values, calculated as a standard deviation (SD), from which the standard error of the mean (SEM) or estimated standard deviation of the mean (ESDM) is calculated by correcting the SD for the number of measurements performed.

The following statistical parameters calculated from the measurements are described in most general guidelines on UM (Cook, 2002; Bell, 2011) and in basic texts on statistics:

- The arithmetic mean: calculated as the sum of all measurements divided by the number of measurements

$$\overline{x} = \sum_{1}^{n} \frac{x_i}{n}$$

- The standard deviation: calculated as the square root of the sum of the squared deviations between each measurement and the mean (discussed in this section), divided by the degree of freedom (1 less than the number of measurements)

$$s = \sqrt{\frac{\sum_{1}^{n} (x_i - \overline{x})^2}{n - 1}}$$

- Hence the SEM, also known as the ESDM, is calculated by dividing the SD (given in the last list item) by the square root of the number of measurements.

$$ESDM = \frac{s}{\sqrt{n}}$$

- This SEM or ESDM is used for the standard uncertainty u(x) for the measurements.

18.4.2.2 Type B Error – Precision of a Measuring Scale

The type B error concerns measurements assuming a rectangular distribution, where any value between the upper and lower boundaries is equally likely to occur. A type B error is most commonly associated with pre-calibrated equipment with a set precision (e.g. a ruler). This type of error is not corrected by repeated measurements but is calculated from a single set value: for example the half-width, or half the smallest unit of measurement on a ruler. The standard uncertainty u(x) for this type of error is calculated as half the smallest unit of measurement divided by the square root of 3.

$$u(x_i) = \frac{a}{\sqrt{3}}$$

18.4.2.3 Calibrated Equipment Used for Measurements

Pre-calibrated equipment used for precision measurement provides an estimate of their precision as a percentage of the measurement, along with coverage factor k (e.g. $k = 2$ assuming a normal distribution). The precision expressed as a percentage can be transformed to one expressed in units of the measurement by multiplying the percentage and the mean measurement. Dividing this scaled precision, or estimate of the deviation around the mean, by the coverage factor of $k = 2$ (for a normal distribution) gives the equivalent to a single SD around the mean, used as an estimate of the uncertainty u(x).

18.4.3 Calculating the UM

- **Step 1: Individual standard uncertainty:** For each component or source or error, where applicable, calculate the standard uncertainty specific to that component of the total uncertainty:
  - by repeated measurements to yield a SEM, if the measurements are normally distributed (type A), or
  - by deriving the single value from the precision of the scale or calibration if they follow a rectangular distribution (type B).

  The standard uncertainty calculated here for a set of measurements or source of variation is commonly assigned the label U(x), where x stands for an individual set of measurements characterising a single source of variation.
- **Step 2: Combined standard uncertainty:** When expressed using the same units of measurement, the individual standard uncertainties are now combined to calculate a single, combined standard uncertainty U_c (or $U_c(y)$). This combined uncertainty is calculated by taking the square root of the sum of all squared standard uncertainties calculated previously.

$$u_c(y) = \sqrt{\sum_{1}^{n} [c_i u(x_i)]^c}$$

- **Step 3: Expanded uncertainty:** Finally, with the assumption that the original measurements are normally distributed and that a student's *t*-test can be applied to test the significance of the deviation, the combined uncertainty is multiplied by a coverage factor to obtain the final, expanded UM. This coverage factor *k* is generally set at 2, rounded from the student's *t*-table value of 1.96 for a 95% level of certainty. Other values from the student's *t*-table can be used as coverage factors for a different level of certainty.

$$U_{95} = ku_c(y)$$

18.4.4 Precision of Measurement and Biological Significance

The measuring software used for on-screen measurements with a digital camera gives a precision (typically down to 0.01 nm) greatly exceeding that of the theoretical resolution of the TEM (roughly 0.2 nm) or that of the imaging potential of a resin-embedded standard section (typically 2 nm). The statistical calculations performed on the measurements can use the full precision of values obtained for the latter to avoid an error of calculation due to rounding; however, the final values reported may be rounded to the nearest significant number, roughly to 1 nm precision.

Reported values should be rounded to reflect the scale of the structure measured. As the diameter of cellular fibrils differs by a few nanometres, their measurement may be reported with that precision. The thickness of a GBM of approximately 300 nm does not require such precision, and the reported values could then be rounded to the nearest 10 nm. Reporting quantitative measurements with minute precision (e.g. at the 0.1 nm scale) does not add significantly to the biological or medical information it provides.

18.4.5 The Electronic Spread Sheet as an Aid to Calculating UM

A useful adjunct to the calculation of UM is the creation and upkeep of an electronic calculator tool, such as a computerised spread sheet. This can be used to store all data pertaining to the periodic verification of magnifications used in analysis, calibrate ancillary equipment (e.g. scales and rulers) and enter new data pertaining to a specific measurement. Pre-programmed equations relating these various values to derive the combined standard uncertainty can greatly enhance the performance of the laboratory by automating calculations and facilitating the task of staff at all levels. It also provides a useful document for the quality assurance program by collating disparate information in one location for easy reference by laboratory staff and auditors alike.

18.4.6 Reporting the UM

The use of a standard template for reporting quantitative measurements and UM offers to the laboratory (scientific) staff a document that is simple to prepare and update in daily practice; and to the reader, it offers one that is easy to read. The simple presentation of the quantitative data is especially important to clinical consultants less acquainted with the specialist jargon of EM and the interpretation of statistical analysis.

The report of quantitative measurements can simply state the arithmetic mean of measurements and either the uncertainty (±UM) or the level of confidence (e.g. a 95% confidence interval is generally acceptable in medical laboratory practice) calculated from the UM. A brief description of the method used to derive the UM, or a cited reference, may be useful. The list of all sources of variation and their relative contribution to the combined uncertainty is useful only in the laboratory as an internal quality assurance document. However, it can be used to monitor the preparation of specimens and performance of equipment and staff engaged in the analysis.

As an example, the suggested text reported could read as follows: 'The diameter of fibrils was 10 ± 2 nm. The reported uncertainty was calculated using a combined standard uncertainty multiplied by a coverage factor of 2 to provide a level of confidence of 95%'.

18.5 WORKED EXAMPLES

18.5.1 Diameter of Fibrils in a Glomerular Deposit

The diameter of fibrils present in the renal glomerular mesangium was measured electronically using a calibrated digital camera of 11 MP resolution calibrated at a verified TEM magnification of 80 000×. Fibrils were randomly selected within two mesangial masses, measured in two separate visual fields also selected randomly.

The following sources of variation were identified and measured:

- Calibration of the digital camera: The digital camera was recently calibrated using a diffraction grating pattern grid. Repeated measurements (n = 9) of the scale bar on captured images gave a mean of 499.0 nm and a SD of 0.65 nm. Assuming a type A error, the standard uncertainty of calibration is the SEM = 0.22 nm.

- Measurement on display screen: To compensate for the uncertain position of the cursor used for measuring on the computer display screen, a single fibril was measured repeatedly (n = 4) on two occasions; the pooled results gave a mean of results to give a mean of 19.9 nm and a SD of 0.7 nm. Assuming a type A error, the standard UM onscreen is the SEM = 0.25 nm.
- Biological variation: A total of 50 fibrils were measured (25 on each of two fields); the mean diameter was 10.6 nm with a SD of 3.8 nm. Assuming a type A error, the standard uncertainty for fibril diameter is the SEM = 0.54 nm.
- The combined uncertainty is calculated as the square root of the sum of squared individual uncertainties, combining the three standard uncertainties above: $\sqrt{(0.22^2 + 0.25^2 + 0.54^2)} = 0.634$ nm.
- The expanded combined uncertainty for 95% confidence is twice the combined uncertainty calculated above: $2 \times 0.634 = 1.3$ nm (rounded).

The values reported were a mean diameter of 10.6 nm with a UM of ±1.3 nm, giving a range of values of 9–12 nm for a 95% confidence interval.

18.5.2 Thickness of the Glomerular Basement Membrane

The thickness of the GBM was measured on film-recorded negative images. The image magnification (as stated by the TEM setting) was 19 000×, and the magnification had been verified recently in a yearly exercise. The thickness of the GBM was measured from the cell membrane of the capillary endothelium to the cell membrane of the epithelial cell foot process. Measurements were made directly on film using a calibrated ocular lens with a precision of 0.1 mm. The GBM to be measured was selected by overlaying the film with a line template and measuring the thickness of the membrane where it intersected the test line, in an axis perpendicular to the GBM at that point.

- Calibration of the TEM: The TEM was recently calibrated using a diffraction grating pattern grid of 463 nm bar length. Ten measurements of the grid bar gave a mean of 463.2 nm with a SD of 2.65 nm. Assuming a type A error, the standard uncertainty for the TEM calibration is the SEM = 0.83 nm.

- Calibration of the measuring graticule: The graticule in the ocular lens used for measurement has a stated precision of 0.1 mm. The half-width is 0.05 mm, giving an uncertainty on the film (transformed for the magnification) of 3.2 nm. Assuming a type B error, the standard uncertainty for the calibration of the graticule is given by $3.2/\surd 3 = 1.85$ nm.
- Biological variation: Fifty measurements in total were made on 10 capillary loops to give a mean thickness of 244.3 nm with a SD of 78.5 nm. Assuming a type A error, the standard uncertainty for the biological variation of GBM thickness is the SEM = 11.1 nm.
- The combined standard uncertainty is calculated as in the previous example: $\surd(0.83^2 + 1.85^2 + 11.1^2) = 11.3$ nm.
- The expanded combined uncertainty for 95% confidence is twice the combined uncertainty given above: $2 \times 11.3 = 22.6$ nm (rounded to 23 nm).

The values reported were a mean thickness of 245 nm with a UM of 23 nm, giving a range of values from 220 to 270 nm (all values rounded) for a 95% confidence interval.

18.6 CONCLUSION

The measurement of uncertainty should be added to any quantitative measurement performed as part of a diagnostic exercise. The mean value of the measurements quoted in the diagnostic report is thereby qualified by an estimate of the variability of the measurements made. The UM can be simply derived by the application of a few basic procedures in the EM laboratory:

- A standard, documented method for the preparation of samples
- The periodic verification of the magnification of the EM as an index of its optimal performance
- The regular calibration of: (1) the camera (digital or film) that is used to record images captured for measurement purposes and; (2) the analytical equipment used in the measurement process.
- The documented, practiced use of a simple analytical procedure for quantitative measurements.

A computerised spread sheet may be programmed to store the data derived from the verifications and calibrations mentioned here, to insert newly made measurements on a structure and to calculate automatically individual standard uncertainties and the combined uncertainty to report.

REFERENCES

AACB (2005) Uncertainty of Measurement in Quantitative Medical Testing – a Laboratory Implementation Guide: NATA, AACB, AIMS, RCPA Joint Document, http://www.sanas.co.za/manuals/pdfs/tg_2402.pdf (accessed July 2012).

Bell, S. (2011) A Beginner's Guide to Uncertainty of Measurement, *Measurement Good Practice Guide* No. 11 (issue 2), 2011, National Physical Laboratory, http://www.wmo.int/pages/prog/gcos/documents/gruanmanuals/UK_NPL/mgpg11.pdf (accessed July 2012).

CALA (2005) Quality Assurance, http://www.cala.ca (accessed July 2012).

Cook, R.R. (2002) *Assessment of Uncertainties of Measurement for Calibration and Testing Laboratories*, 2nd edn, http://www.nata.asn.au/phocadownload/publications/Technical_publications/Uncertainity/MUbook_2002.pdf (accessed July 2012).

Crang, R.F.E. and Klomparens, K.L. (1988) *Artifacts in Biological Electron Microscopy*, Plenum Press, New York.

Gravel, J.E.J. (2011) Implementing Uncertainty, http://www.cala.ca/pr_MU-clinical_lab.pdf (accessed July 2012).

International Organisation of Legal Metrology (OIML) (2010) Evaluation of Measurement Data – Guide to the Expression of Uncertainty in Measurement, corrected version, http://www.oiml.org/publications/G/G001-100-e08.pdf (accessed July 2012).

International Organisation of Standardisation (ISO) (2009) ISO GUM Guide 98, Uncertainty of Measurement – Part 3: Guide to the Expression of Uncertainty in Measurement, http://webstore.iec.ch/preview/info_isoiecguide98-3%7Bed1.0%7Den.pdf (accessed July 2012).

International Organisation of Standardisation (ISO) (2010) ISO 29 301. *Micro Beam Analysis – Analytical Transmission Electron Microscopy – Methods for Calibrating Image Magnification by Using Reference Materials Having Periodic Structures*, 1 June.

NATA (2012) Guidelines for the Validation and Verification of Quantitative and Qualitative Test Methods, NATA Technical Note, issued August 2004, last amended and reissued March 2012, http://www.nata.asn.au/phocadownload/publications/Technical_publications/Technotes_Infopapers/technical_note_17.pdf (accessed July 2012).

NIST (2004) Guidelines for Evaluating and Expressing the Uncertainty of NIST Measurement Results, NIST Technical Note 1297, 2004 edn, http://www.nist.gov/pml/pubs/tn1297/index.cfm (accessed July 2012).

RCPA (2009) Uncertainty of Measurement, NATA Policy circular 11 – November, http://www.rcpa.edu.au/static/File/Asset%20library/public%20documents/Policy%20Manual/Guidelines/Uncertainty%20of%20measurement.pdf (accessed July 2012).

UKAS (2007) UKAS Publication M3003: The Expression of Uncertainty and Confidence in Measurement. 1st edn, December, http://www.ukas.com/library/Technical-Information/Pubs-Technical-Articles/Pubs-List/M3003.pdf (accessed July 2012).

Index

Page numbers in *italics* refer to figure captions and bold refers to table.

A *(anisotropic)* band, *94*, 94–95
Acanthamoeba, 206, 213
Accreditation, 385, 403, 432, 437
Accuracy of measurement, 431, 432
Acid haemolytic test, 294
acid maltase deficiency *see* Pompe
acidified serum lysis test, 299
ACTA skeletal actin gene defect in
  actinopathies, actin accumulation in, 100
  cap myopathy/'cap disease', 100
  core myopathies, 104
  nemaline myopathy, 100
  Zebra body myopathy, 101
Actin, *94*, *95*, 100, 120
  accumulation (actin masses), 100–101
Actinolite, **325**, 326, *327*, 333
actinopathies, 100
acute quadriplegic myopathy
  myosin loss in, 100
ADAMTS13, 24
Adenovirus, 183, 184, 187, 188
  transplant kidney, 75, 77
age related phenomena
  mitochondrial DNA deletions, 107
  capillary endothelial cell basal lamina thickness of, 97
aggressive angiomyxoma, 159
AIDS, 12, 196
Albendazole, 197
Alcian blue, 412
Aldose reductase, 25
Alport Anti-GBM Nephritis, 73
Alport's syndrome, 2, 6, 12, 19–21
Alteration of dimensions
  due to shrinkage, 434
  due to expansion, 435
Aluminosis, **325**, 336
Aluminum, **325**, 336
Aluminum hydroxide, 239
Alveolar proteinosis, 338
alveolar soft-part sarcoma, 159

Diagnostic Electron Microscopy: A Practical Guide to Interpretation and Technique, First Edition. Edited by John W. Stirling, Alan Curry and Brian Eyden.

Ammonium
  molybdate, 411
Ammonium sulphate
  precipitation, 410
Amoebae, 206
Amosite, 324, **325**,
  326, 327
Amphibole
  Commercial, 324,
    327, 330
Amyloid, 8, 10–11,
  40–42, 119, 125,
  141, 144
Amyloidosis (renal),
  432
Amyloidosis, 40–41
Amyotrophic lateral
  sclerosis (ALS), 247
anaemia
  aplastic, 294
  congenital
    dyserythropoietic
    anaemia, 293–308
    congenital dysery-
      thropoietic
      anaemia,
      diagnosis, 294,
      299, 301
    congenital dysery-
      thropoietic
      anaemia type
      I, 294–99, 301
    congenital
      dyserythropoi-
      etic
      anaemia type
      II, 299–306
  haemolytic anaemia,
    299
Analytical electron
  microscopy, 324,
  330, 331, 332
Analytical
  methodology, **325**,
  326, 331, 333, 335,
  337
Analytical procedure,
  446
ANCA-associated
  glomerulonephritis,
  18–19
anchoring fibril, 155,
  169
angiomyofibro-
  blastoma, 159
angiosarcoma, 159,
  165
anisochromasia, 306
anisocytosis, 306
Anncaliia, 203
  A. vesicularum, 203
  A. connori, 203
  A. algerae, 204
Anthophyllite, **325**,
  326, 333
Anti-GBM
  glomerulonephritis
  (Goodpasture
  syndrome), 19
Antinuclear antibodies
  (ANCA), 36–37
Apicomplexa, 206,
  207
apoptosis, 105
Arthrogryposis, renal
  dysfunction and
  cholestasis
  syndrome (ARC),
  285, 287
  cytoplasmic
    vacuoles in, 285
  VPS, 33B gene,
    285
Asbestos, 324, **325**,
  326, *327*, 330, 333,
  336
Asbestos bodies, 327,
  330
Asbestosis, 324, **325**,
  326, 327, *327*
Asbestos-related
  diseases, **325**, 326
Aspartylglucosa-
  minuria, 249
Asteroid body, *327*,
  331
astrocytoma, 159, 165
Astrovirus, 184, 188
atrophy, 130
  axolemma, 126
  axonal flow, 126
  axonal growth cone,
    125
  cytoskeleton, 120
    microtubules, 120
    neurofilaments,
      120
      neurofilament
        light
        (*NEFL*),126
  degeneration, 125
  diseases
    Charcot-Marie
      Tooth disease
      (CMT2E), 126
  regeneration, 126,
    128, 133
  sprouts, 126, 128
  unmyelinated, 124,
    133, *137*, 147
    degeneration,
      125, 128, 129
attachment plaque,
  159, 161, 173, *175*
Autoantibody, 30
autophagic vacuolar
  myopathy
  childhood, 110
  Danon Disease, 110
  infantile, 110
  with multi-organ
    involvement, 110

x-linked myopathy with excessive autophagia (XMEA), 110
autophagic vacuoles with sarcolemmal features (AVSF), 109–110
autophagic vacuoles, 109
autosomal recessive disorder, 299
Axon, 117, 120, 125

Bacillus, 192
Bacitracin, 412
Back-scattered electron imaging, **325**, *327*, 331, 332, 333, 334
Bacteria, 182, 191–193, 195
Balantidium coli, 206
Bands of Büngner, 126, 128, *129*, 142
basal lamina (endothelial), 97, 111
basal lamina, 155, 157, 159–62, 169–71
  basement membrane, 171
  external lamina, 158–62, *170*, 173, 175
basophilic erythroblast, 294, 295
Berger's disease, 30–32
Berylliosis, **325**, 337
Beryllium, 323, **325**, 337
bilayer membrane, 301
Biological safety precautions, 186, 409
Biological variation, 431, 445, 446
Bioterrorism, 185, 409
Birbeck granule, see *Langerhans cell granule*
BK Virus
  transplant kidney, 74–8, (Figures on 76, 77, 78)
Blood, 341, 368, 369
Blood vessels,
  endoneurial, *140*, 145
  epineurial, 117
  pericytes, 145
Bone Marrow
  EM preparation of, 281
bone-marrow aspirate, 294
bone-marrow particle, 294
bone-marrow smear, 301
Bowman's capsule, 2–5, 34, 40–41
Brachiola (now *Anncaliia*), 204
brancher enzyme deficiency (GSD type IV), 102, *103*
Bronchoalveolar lavage, 332–3, 337
BTB/Kelch (KBTBD13)
  in nemaline myopathy, 100
Budding (of viruses), 183
CADASIL, 269–74, 343–4, 354
Cajal bands, 124
Calcineurin Inhibitor Toxicity
  hyaline arteriolar beading, 83–4 (Figure on 84)
  thrombotic microangiopathy, 63, 67
Calcofluor stain, 197
Calibrating grid, 436
Calibration, 432, 433, 436–439, 442, 444–446
Calibration specimen, 414
Calicivirus, 183, 188
Call-Exner body, 160
Canaliculi, 124
cap myopathy/'cap disease', 100
capillaries, 110
  empty basal lamina loops, 111
  endothelial cell
    hypertrophy, 110
    shrinkage, 110
    swelling, 110
    thickened basal lamina, 97
  myofilament streaming adjacent to, 99
  tubuloreticular inclusions, 111
capillary endothelium, 4–5, 7, 14–15
Capillary lumen, 4, 15–16, 34, 39, 41, 43–4, 47
Capillary tuft, 3, 16

Carbon-coated grids, 412
carcinoid, *174*
carcinoma, 155, 158, 167, 169, 176
  adenocarcinoma, 162, 167, 169, 173
  adrenocortical, 159
  basal cell, *166*
  chromophobe renal-cell, 158–9
  signet-ring cell, 161
  spindle-cell, 158, 161, 173
  squamous-cell, 155, 167, *168*, 169
Carcinoma of the lung, 324
Cauliflower-shaped collagen fibril, 313, 320
caveola, 159, 161, 173
caveoli, 98, *98*
Caveolin, 3, 98
  autoimmunity to, 98
  in autophagic vacuoles with sarcolemmal features (AVSF), 109–110
  reduction in protein expression, 98
  secondary reductions in protein expression, 98
Cavin, 1, 98, *98*
CDA, see *anaemia, congenital dyserythropoietic anaemia*
cell, 153, 155, 156, 158, 162
  glial, 170
  haematolymphoid, 158, 162, 170
  multinucleated giant
  muscle, 169, 170, *170*, 173
  nerve-sheath, 169
  normal, 155, 158, *170*, *172*, 173, 176, *177*
  perineurial, 170
  plasma, 157
  reactive, 162, 171, *174*
  Schwann, 170
cell surface, 294, 295, 299
central core disease, 104–105
Central nervous system (CNS), 117
centriole, *172*, 173
centronuclear myopathy, 96
Cerebral autosomal dominant arteriopathy, 247
Cerium, **325**, 338
Ceroid Lipofuscinosis, 260
Ceroid lipofuscinosis, 260
  Juvenile, 260
Cervical cancer, 187
CFL2 cofilin
  in core myopathies, 104
  in nemaline myopathy, 100
Charcot-Marie-Tooth disease (CMT), 120, 126, 130, 134, 135, 140, 144, 148
Chédiak Higashi syndrome, mast cells, 145
Chediak-Higashi syndrome, 287
China, 294
chromatin bridge, 295, *295*, 301, *306*
Chronic inflammatory demyelinating polyneuropathy (CIDP), *129*, 131, *131*, 133, *138*
Chrysotile, 324, **325**, 326, 327, 440
Cilia, 157, 159, 221–6, 228–9, 231–6, 368
Ciliary beat frequency, 223
Ciliary disorders, 221–3, 225, 227, 229, 231, 233–5
Ciliated specimens, 230–1
Ciliopathies, 222, 224, 228–9
cisterna, 301, 303–5
  degenerated cisterna, 301
Clinical features, 229, 235
clover leaf, 301, *302*
Coal dust, 338
Coal worker's pneumoconiosis, 338
Coated Grids
  Carbon, 412
  Formvar, 412
  Pioloform, 412
Coated grids, 412
Cobalt, **325**, 336, 337
Coccidia, 207

Collagen, 117, 119, 160
glycosaminoglycans, 120
laminin, *120*
Collagen fibril, 309, 310, 311, 312, 313, 316, 317, 318, 319, 320, 321
Collagen IV, 3, 5, 19–21, 120
Collagen pockets, *121*, 128, *129*
Collection of specimens, 410
Complement pathway, 33, 35
Computer monitors, 427
concentric laminated bodies, 112, *112*
Confidence level (or interval), 431
congenital myopathy, 96
Congo red, 8–10, 41
Connexin, 32, *124*
Control reagents, 413
Controlled reference value, 433
cores, 104, *104*
structured and unstructured, 104
Cornea, 344–5
Coronavirus, 186, 189
costamere, 95
Cowpox, 190, 191
Coxiella, 193
crinophagy, 165
crista, 158, 160, 162
tubular 158, 160, 162
critical illness myopathy
see acute quadriplegic myopathy
Crocidolite, **325**, 326, 327
Cryoglobulinaemia, 11, 43, 45
Cryoglobulinaemic GN, 45
Cryptosporidium, 181, 206, 207, 210, 211
C. parvum, 210
C. hominis, 210
C. muris, 210
crystal, 160
Reinke, 160
Crystalline inclusions, 128, *129*, 130, *138*, 140
Cyclospora, 209, 211
C. cayetanensis, 209
cylindrical spirals, 111, *111*
Cystinosis, 246, 259
Cytomegalovirus
transplant kidney, 75, (Figure on 79)
cytoplasmic bodies, 112, *112*

Danon disease (lysosomal glycogen storage disease with maltase), 238, 245, 257
(LAMP-2 deficiency), 110
De Novo Glomerular Disease
Alport anti-GBM nephritis, 73
focal segmental glomerulosclerosis, 73
membranous nephropathy, 72
thrombotic microangiopathy, 74
Dehydrant (dehydrating agent), 346, 348–349, 364
Dehydration, 348–351, 362, 365, 367
Demyelination, 130, 132–3, *137*, 142, 147
allergic, 130, 131
primary, 130
secondary, 126, 130
Dense deposit disease, 7, 33–35
Dense tubular system (DTS), 282
deparaffinisation, 156
Deposits, 2, 5–15, 18, 19, 22, 24–30, 32–35, 37, 40, 41, 44, 45, 99, 100, 105, 106, 111
Dermatosparaxis, 309, 310, 312, 316
dermis, 169
Desmin, 173
granulomatous, *101*
in myofibrillar myopathies, 102
in normal muscle, 95
desmosome, 155, 157, 159–61, 167, *168*, 169, 171, 173
desmosomal plaque, 156, 167, 169
desmosome-like junction, 167, *168*, 169

diabetes
  capillaries in, 97
  myofibre external
    lamina in, 97
Diabetic fibrillosis, 10, 23
Diabetic
  Glomerulosclerosis
  recurrent, 66,
    72
Diabetic nephropathy, 22–24
Diabetic neuropathy see Neuropathy, diabetic
Diffraction grating, 444–5
Digital camera, 419–429
Distribution, 221
Domestic exposures, 330
Double contouring or 'tram tracking' (interpositioning), 14, 24, 33
doubling plasma membrane, 301, 303, 305, 306
Dynein arms, 222–5, 229, 234
dysferlinopathy, 97, *97*, *98*
dysplasia, 293

Ehlers-Danlos syndrome, 309, 311, 313, 315, 317, 319, 320, 321
Elastic fiber, 310, 313, 314, 315, 316, 317, 319, 320, 321
embryonal sarcoma of liver, 159
emerging infectious agents, 384
Encephalitozoon, 198, 199, 201, 202, 206
  transplant kidney,
    78–9, (Figure on 81)
  E. intestinalis, 201
  E. hellem, 198, 201
  E. cuniculi, 199,
    202
endomysium, 96, 97, 98
Endothelium, 4–5, 7, 14–15, 25, 170
Entamoeba, 206, 213
Enteric Coronavirus, 189
Enterocytozoon
  transplant kidney,
    78–9
Enterocytozoon bieneusi, 197–8, 201, 206
Environmental exposures, 330
ependymoma, 159
Epithelial cells (podocytes), 3, 16, 42
epithelioid sarcoma, 159
Epstein-Barr virus, 36
Errors
  due to experimental
    protocol, 431
erythroblast, 293–307
erythrocyte, 299, 306
erythroid lineage, 301
erythroid precursor, 293
erythropoiesis, 301
Ewing's sarcoma, 157, 159
excitation-contraction coupling, 96
Exposure categories, 330
external lamina (myofibre), 91
  lamina densa, 91, *98*
  lamina rara, 91, *98*
extraskeletal myxoid chondrosarcoma, 173

Fabry disease, 243, 248, 252
  basal lamina around
    blood vessels in,
    248, 252
Fabry disease, 130, 143
Fabry's disease, 15, 46
familial incidence, 294
Farber's disease, 136
Farber's lipogranulomatosis, 242
Fascicle, nerve, 117, 124
  endoneurium, 117,
    140, 143, 144,
    145
  epineurium, 117,
    125, *140*
  perineurium, 117,
    125, 133, 135,
    143
Feldspar, **325**, 334, 335
ferritin particle, 299
Fiber analysis, 324
Fibers, 323, 324, **325**
Fibrillary GN, 10, 41, 42, 43
Fibrin, 12, 16, 19, 24, 29, 32

thrombotic microangiopathy, 62–64, (Figure on 64)
Fibrin deposition, 277
Fibrinogen, 278
fibroblast, *163*, 165, 170
Fibroblasts, 120, *122*, 124, 125, 142, 163, 165, 170
fibronectin, 173
fibronectin fibril, 171, *175*
fibronexus, 158, 159, 161, 173
Fibrous long spacing collagen (FLSC).*140*, 143, *144*
Luse bodies, 144
Ficoll-Hypaque gradient, 279, 281, 282
filamentous bodies, 112
filaments, *95*
intermediate
intranuclear
in dermatomyositis/polymyositis, 105, *106*
in inclusion body myopathy, 105, *106*
in oculopharyngeal muscular dystrophy, 105, *106*
in oculopharyngodistal muscular dystrophy,*106*
sarcoplasmic, in rimmed vacuoles, 109
fingerprint bodies, 112, *112*
Flagella, 221–2, 224
Flower-like collagen fibril, 312, 313, 317, 319
Fluorescent viewing screen, 419
focal density, 158, 173, *175*
Focal Segmental Glomerulosclerosis
*de novo*73
recurrent, 66, 69, (Figure on 70)
Focal segmental glomerulosclerosis (FSGS), 17–18
follicular dendritic reticulum-cell sarcoma, 160, 167
Foot process, 2–6, 15, 17–18, 22, 24–6, 29, 33–4, 39, 41
formalin, 156
Fucosidosis, 248, 253
zebra bodies in Schwann cells, 253

Galactosialidosis, 238
G_{M1}, 241
G_{M2}, 242
B variant (Tay-Sachs), 247
Zebra bodies (fingerprints) in, 247, 249
GANT, see *gastrointestinal autonomic nerve tumour*
Gap junctions, 123–4, 133
gastrointestinal autonomic nerve tumour, 160, *168*, 169, *174*
gastrointestinal stromal tumour, 160
gastrointestinal tract, 171
Gaucher's disease (Type I, II & III), 242, 252, 256
Gaucher-like cells, 238
Genes, 222–3, 226, 228–9, 231, 234–5
Genetic mutations, 131, 143
EGR2, 135
MPZ,135
PMP22, 135
SH3TC2 (KIAA1985),135
GFPT1
in familial myasthenia, 109
Giant axonal neuropathy (GAN), 126, 130
Giant cell interstitial pneumonitis, 337
Giardia, 206, 209
G. intestinalis, 209
G. lamblia, 209
G. duodenalis, 209
GIST, see *gastrointestinal stromal tumour*
Glomerular basement membrane (GBM) thickness, 432, 438, 443, 445

Glomerular basement membrane (GBM), 2, 4–5, 13, 18, 21, 39, 47
  abnormal composition
  double contouring, 14
  folding, 14
  gaps, 14
  lamination, 5
  normal, 13
  subendothelial widening, 14
  texture, 13–14
  thick, 13, 23
  thin, 13
  width, 13
Glomerulus, normal, 3–5
Glow discharge, 412
glutaraldehyde, 156, 294
glycocalyx, *172*, 173
glycogen, 157, 159, 161
Glycogen granules, 128, *129*
glycogen storage disorder (GSD), 102
  Brancher enzyme deficiency (GSD type IV), 102, *103*
  McArdles (GSD type V), 102
  Pompe (GSD type II), 102, 103, 110
glycogen, 94
  and fibre type, 95
  blebs, 91
  en bloc section staining with uranyl acetate, 91
  myofibre, variation in content, 95
  particle crystalline arrays, 93
Golgi apparatus, 163
GOM, *see* Granular osmiophilic material (GOM)
Granular osmiophilic deposits (GRAD), 258, 259, 261
Granular osmiophilic material (GOM), 270, 274
granule, 299
Granulo-filamentous deposit, 310, 311, 312, 313, 315, 319
Granulomas, *327*, 330, 338
Granulomatous, 330, 334–8
Grey platelet syndrome (GPS), 285, 286
  NBEAL2 gene, 285
Grids, 354–5, 359, 361, 363–4, 378
Guamanian neurodegenerative disease, 247
Guillain-Barré syndrome (GBS), 131, 142
Gulf region, 294

haemoglobin, 29, 296, 299
Haemolytic uraemic syndrome (HUS), 24–25
haemopoietic island, 294
Hard metal lung disease, 325, 336
Heavy-chain deposition disease (HCDD), 37
hemidesmosome, 169
Henoch-Schönlein purpura nephritis, 32
Heparin, 279
Hereditary liability to pressure palsies (HNPP), 147
Hereditary metabolic storage disorders
  Fabry's disease, 46–7
  LCAT deficiency, 46–7
Hermansky-Pudlak syndrome, 287
Herpes simplex, 190
Herpesvirus-185, 189, 190
  transplant kidney, 75
heterochromatin, 156, 158, 169, 295
hibernoma, 160
Hieroglyphic-shaped collagen fibril, 310
High resolution electron microscopy, 412
Hirano bodies, 140
histology, 154–56, 173
HIV, 12, 36, 42
HIV-associated nephropathy, 18
HMB45 165
Household contacts, 330
Human immunodeficiency

virus (HIV), 181, 183
Hunter disease, 257
Hunter syndrome, 240
Hurler-Scheie syndrome, 240
hyaline bodies (myosin accumulation), 100
hyaline inclusion, 161
hyaluronic acid, 160
Hyaluronic acid globule, 310, 311, 319
hyperCKemia
idiopathic, 98
Hypersensitivity pneumonitis, **325**, 330, 331, 333

I (*isotropic*) band, 94
I cell disease, 238, 241
IgA Nephropathy
recurrent, 66, 71
IgA nephropathy (Berger's disease), 30–32
IgM kappa paraproteinaemia, 132
Immune clumping, 413
Immune deposits
complement (C3, C1q and C4), 1, 6, 8–9, 17–21, 24–5, 29, 32–3, 35–37, 42, 45
IgA, 1, 6–7, 9, 19, 22, 25, 29–33, 35–7, 45
IgG, 1, 6, 8–10, 19–20, 24–5, 29–30, 32–3, 36–7, 42, 45
IgM, 1, 9, 17–21, 24–5, 29, 32–3, 35–7, 45
Immunoelectron microscopy, 413
Antibody coating, 413
immunoelectron microscopy, 165
Immunofluorescence, 1, 10, 301
Immunofluorescent, 223, 229, 234
immunogold electron microscopy, 301
Immunogold labelling, 414
immunohistochemistry, 153, 154, 158, 173, 176
Immunolabelling, 1–2, 5, 7–10, 12, 17–20, 24–5, 29, 32–4, 36–8, 41–3, 45, 47
Immunoperoxidase, 1
Immunotactoid GN, 42–5
Inactivation of specimens, 409
Affect on virus morphology, 409
Formaldehyde, 409
Paraformaldehyde, 409
inclusion body myopathy, 105
inclusion body myositis, *106*
COX negative fibres in, 107
filamentous inclusions in, *106*
pathological overlap with MFM, 102
sarcoplasmic inclusions in, 109
inclusion, 299
filamentous, intranuclear, 105, 106
sarcoplasmic, 109
Indium, 338
Infantile Osteopetrosis, 238, 246
Infection
transplant kidney, 74
Infectious diseases, 26, 29, 31, 33
Infectious organisms, 45
Insudate, 60, (Figure on 62)
insulinoma, 171
interdigitating reticulum-cell sarcoma, 160
interferon
relationship with tubuloreticular inclusions, 111
intermediate filaments, 158–61, 167
paranuclear whorl, 160
Internode, 123–4, 128, 130
Intra-laboratory reference, 433
Iron, , 299, 300, 301, 306, **325**, 326
iron deposition, 301
iron overloading, 299
iron-overloaded mitochondria, *300*, 301, 306

Ischaemia
glomerular changes, 57, 63, (Figure on 65)
Isospora, 181, 206, 211, 212
I. belli, 211, 212
I. natalensis, 211

Jacobsen syndrome, 285–7
chromosome, 11, 286
α-granules in, 288
jaundice, 294, 301
junctional complex, 159, 161

Kaolin worker's pneumoconiosis, **325**, 334
Kaolinite, 334
Kaposi's sarcoma, 160
Kartagener syndrome, 222, 235
karyolysis, 299
karyoplasm, 295
karyorrhexis, 299, 301, *306*
KBTBD1 BTB/Kelch, 100
keratinocyte, 155, 165
Kidney Transplant (see Transplant Kidney)
Kimmelstiel, Wilson nodule, 9, 23, 33
Kinetoplast, 212, 213
Krabbe's disease, 136, *137*, 244, 249, 255

Laminations in GBM ('basket weave' appearance), 5, 13, 20, 21
Laminins
in autophagic vacuoles with sarcolemmal features (AVSF), 109
LAMP-2 (Lysosomal associated membrane protein 2), 110
LAMP-2 deficiency
Danon disease, 110
Langerhans cell granule, 156–7, 160
Langerhans cell granulomatosis, 160
Lecithin cholesterol acyltransferase (LCAT) deficiency, 46–47
leiomyoma, 158, 165, 173, *175*
leiomyosarcoma, 158, 162, 165, 173
Leishmania, 206, 212, 213
L. donovani complex, 213
Leprosy, 130, 135, 147
lepromatous leprosy, 135, *137*
tuberculoid leprosy, 135, 147
foam cells, 135
Leukoencephalopathy, 247
lifespan, 293
light chain deposits, 97
Light chains
kappa, 1, 7–9, 37, 41–2, 45
lambda, 1, 9, 37, 41–2, 45
Light-chain deposition disease (LCDD), 37, 40–41
limb deformity, 294
limb girdle muscular dystrophy (LGMD)
autophagic vacuoles in, 110
caveolin, 3 in, 98
lobulated fibres in LGMD 2A, 108
Limit of precision, 431
lipid, 91, 96, 157–60
Lipofuscin, 138
Liver, 174, 345
lobulated fibres
in LGMD 2A, 108
in Ullrich's congenital muscular dystrophy, 108
LSDs, 237–7
age of onset, 239
clinical manifestations, 239
diseases, 237–7
mechanisms, 237, 248
morphological findings, 239, 247, 248, 249, 252, 257, 258, 261
lumen 157, 159–62, 170, 171
Lung cancer, **325**, 327
Lymph nodes, 345
Lymph vessels, 117
Lymphocytes, 120, 125, 131, 133, 143, 174

Lysosomal storage diseases (LSD), 237–267
  acquired, 237, 239
  classification of, 238
  EM diagnosis of, 239, 247, 248, 249, 252, 257–8, 261
  genetically inherited LSDs, 239–246
  major organs involved, 240–246
  mimics of, 239
  with osteopetrosis, 238, 246
Lysosome, 157, 159–62, 164, 164, 165, 171, 301, 306
  function, 237
  primary, 164, *164*, 171, 174
  secondary, 164, *164*, 165, 166
Lysosomes, 110, 120, *122*, 136, 138
  in amiodarone toxicity, 138
  in Pompes, 102, *103*

M Band, *94*, *95*
macrocytotic, 299
Macrophages, 120, 130, 131, 138, 142, 154, 157, 164–6, 170
Magnification verification, 446
malignant melanoma, 155, 161, *164*, 165, *166*, 169, 173, 176
  signet-ring cell, 161
Mannosidosis, 248
Marinesco-Sjogren syndrome, 105
Maroteaux-Lamy syndrome, 241, 258
Mast cell, 120, *121*, 145
matrix, 156
McArdle disease
  GSD type V, 102
  rhabdomyolysis in, 102
Megakaryocyte, 281, 283
melanin, *164, 166*
melanocyte, 155
melanosome, 155–7, 161, 165, *166*
  compound, 165, *166*
  lattice, *166*
  lattice-deficient, 165
  type II 165, *166*
  type III 165, *166*
  type IV 165
Membranous GN, 25–29, 47
Membranous Nephropathy
  de novo, 70, 72
  recurrent, 66, 70, (Figure on 71)
meningioma, 167
Mesangial cell, 3–4, 16–17, 34
Mesangial interposition, 16, 33–4, 39
Mesangial matrix, 2–3, 5, 7, 12, 14, 17, 19, 22–3, 31, 33
Mesangiocapillary glomerulonephritis (MCGN) (membranoproliferative GN)
  Type I MCGN, 33
  Type II MCGN (dense deposit disease), 33–5
  Type III MCGN, 35–6
Mesangiocapillary Glomerulonephritis, 57, 61, 66–7
Mesaxon, 121, 124
Mesothelioma, 324, **325**, 327
mesothelioma, 160, 171, *172*, 173
Metachromatic leucodystrophy (MLD), *131*, 136, *137*
Metal-induced diseases, **325**, 335
metaplasia, 156
Methanol, 351, 363, 371
Methylamine tungstate, 411
Mica, **325**, 334, 335
Microcrystalline cellulose, 338
micronucleus, 301, *305*
Microsporidia
  transplant kidney, 74, 77–9, (Figure on 81)
Microsporidia, 181, 196–206
Microsporidial spore, 196, 198, 199, 200
Microsporidium, 205
  M. ceylonensis, 205
  M. africanum, 205
microtubule, 159–61

microvilli, 157, 159–61, 171, *172*, 173
  actin filament core, 171, 173
  in mesothelioma, 160, 171, *172*
  rootlet, 171, *172*, 173
Microvillous formation (transformation in text), 15, 17–18, 33
Mineral fibers, 326, 330
Minifascicle, *134*
Minimal change disease, 15–17
Mixed dust pneumoconiosis, **325**, 335
Molluscum contagiosum, 186, 190
Monoclonal immunoglobulin deposition disease (MIDD), 37–40
Mononuclear cell, 279
Morquito syndrome, 240
mucigen granule, 157, 161, 171
Mucolipidosis (ML), 238, 241, 251
  ML I (sialidosis), 241
  ML II (I cell disease), 241
  ML III (pseudo-Hurler polydystrophy), 241
  ML IV, 238, 241, 248, 251
Mucopolysaccharidosis (MPS), 240, 241, 250, 257, 258
multi minicore myopathy, 96, 104
multinuclearity, 299
muscular dystrophy
  Becker, 98
  Duchenne, 98
  focal breaks in, or loss of, plasmalemma, 98
Mutations, 222–3, 226, 228, 231, 233–6
Mycobacterium leprae, 135, *137*
Myelin
  development, 120, 121
  proteins, 120
  structure,
  less dense (interperiod) line, 121, 15
  major dense line, 121
  periodicity, 121
  myelin diseases see Charcot-Marie-Tooth disease
Myelin sheath, 117, 155
  crenations
  mesaxon, 121, 124
  internodal length, 121
  internode, 123
  paranode, 121, 123–4, *132*, 147
  juxtaparanode, 123, *132*, 141
myelodysplastic syndrome, 294
MYH7
  (myosin heavy chain gene defect in)
  cores myopathies, 104
  myosin
    accumulation (in hyaline bodies), 100
myoepithelioma, 160
myofibre
  affect of physical exertion on, 91
  atrophy, 99
  external lamina, 91, 96
    multilayering of, 97
  hypercontraction, 92101
  hypotrophy, 96
  lipid and glycogen content, factors affecting, 91, 95–6
  necrosis, 101, 109
  regeneration, 99, 108
  type
    lipid and glycogen content, 95–6
    proportion, 91
myofibrillar
  myopathies, 102
  autophagic vacuoles in, 110
  $\alpha\beta$-crystallin, 102
  BAG3, 102
  desmin, 102
  FHL1, 102
  filamin C, 102
  myotilin, 102

rimmed vacuoles in, 109
ZASP (Z-band alternatively spliced PDZ-motif protein), 102
myofibrils, 91, 94, 99
contraction bands, 92
disruption, 99, 104, *104*
loss of, 104
myofibroblast, 154, 162, 165, 170, 171, *175*
myofibroblastoma, 154, 161
myofibrosarcoma, 173
myofilaments, 158, 159, 161, *175*, 176
sarcomeric, 176
smooth-muscle, 158–61, 173, *175*
myonuclear membrane, 95
myonucleus, 95, 105
apoptosis, 105
chromatin, 95, 105
filamentous inclusions in, 105, *106*
internal, 90, 95
in Marinesco-Sjogren syndrome, 105
in normal muscle, 95
myophosphorylase deficiency
see McArdle disease
myosin, 95
accumulation (hyaline bodies), 100
loss, 99, *101*
in acute quadriplegic myopathy, 100
in cap myopathy/'cap disease', 100
in dermatomyositis, 100
myotendinous junction, 90, 94
rods associated with, 90

Nasal brush biopsies, 341
Nasal nitric oxide, 223, 236
NEB Nebulin, 95, 100
needle biopsy, 90
nemaline myopathy, 96, 99–100, 101
nemaline rods
and myotendinous junction, 90
in regenerating/degenerating fibres, 99
nuclear, 99, 100
sarcoplasmic, 99, 100
Neoplastic potential, 187
Nerve impulse, 117
conduction speed, 120
neuroblastoma, 161
neuroendocrine granule, 157, 159–61, 164, 171, 173, *174*
Neurofibromatosis, 133, 142, 144
Neuroma, 133, *134*
neuromuscular junction, 94
Neuron, 117, 170
motor, 117
sensory, 117
Neuronal ceroid lipofuscinosis (NCL or CLN), 245–6, 257–9, 261
Neuropathy, see also Axon, axonal diseases
perhexilene maleate, 138
paraproteinaemic, 121, 132, 143, 145
polyglucosan body disease, 128
toxic, 126
giant axonal neuropathy (GAN), 126
Niemann-Pick disease (NP), 243, 248, 251
Node of Ranvier, 117, 121, *137*
intercalated, 130, *131*
juxtaparanode, 123, *132*, 141
nodal processes, 122, *146*
paranode, 121, 123, *132*, 147
terminal loops, 126
nodular fasciitis, 173
norepinephrine, 171, *174*

Norovirus (see also Norwalk-like virus, Small Round Structured Virus, SRSV), 184, 189
Nosema, 204
  N. ocularum, 204
  N. sp., 204
Notch, 3, 269–71
nuclear envelope, 295, 297
nuclear membrane, 295, 297
nuclear pore, 295, 297, *297*
nucleolus, 158, 161, 163, 295
  rope-like, 161
nucleus (double), 299, 301, *302*
nucleus, 158
  perinuclear cisterna, 158

oculopharyngeal muscular dystrophy, 105–106
oculopharyngodistal myopathy, 106
oligocilium, 173
oligodendroglioma, 161
oncocytoma, 161, 162
Onion bulb, 133, *134*, 147
  basal laminal onion bulbs, *134*, 135, 148
open biopsy, 90
Open canalicular system (OCS), 283
Orf (contagious pustular dermatitis), 190, 192
Oropharyngeal cancer, 187
orthochromatic erythroblast, 295, *296*, 297, 299
Orthomyxovirus, 183
Orthopoxviruses, 186, 187, 190, 191, 193
osmium tetroxide, 156
Oxytalan fibres, *122*, 125, 144

Papillomavirus, 185, 187, 190
Papovavirus, 187, 190
Paramesangium, 34
Paramyxovirus, 183
Parapoxviruses, 190, 192
Particle measurement, 414
Parvovirus, 189
passive system, 400
Pathology report, 432, 444, 446
Pauci-immune glomerulonephritis, 18–19
PCD, 222–6, 228–9, 231–4
PEComa, see *tumour, perivascular epithelioid cell*
Pericytes, 270, 273–274
Perineuriosis, 133, *140*
Perineurium, 117, 133, *134*, 135, *141*, 142–3
periodic paralysis
  t-tubules in, 108
  tubular aggregates in, 109
Peripheral nervous system (PNS), 121
peripheral neuroectodermal tumour, 159, 167
Peritubular Capillary
  false reduplication of basal lamina, 60, (Figure on 61)
  transplant glomerulopathy, 58–60, (Figure on 59)
phaeochromocytoma, 171, *174*
pinocytotic vesicle, 159
pituitary adenoma, 171
plasmalemma, 91, 97, *97*
  caveoli in, 98, *98*
  focal loss or breaks in, 98, *98*
  vacuole fusion with, 109
Plasmodium, 206
Platelet, 277–289
Platelet receptors
  Glycoprotein 1b-IX-V complex, 277
  Glycoprotein Integrin αIIbβ3 complex, 277
Platelet rich plasma, 279
Platelets, 368
Plectin, 95
Pleistophora ronneafiei, 203
Pleural diseases, 324

PNET, see *peripheral neuroectodermal tumour*
Pneumoconiosis, **325**, 331, 334, 338
Pneumocystis, 181, 195, 196
  P. jiroveci, 195, 196
  P. carinii, 195
Podocytes, 3–4, 15, 17–18, 25–6, 46
  effacement (fusion), 15
  normal
poikilocyte, 299
polyacrylamide gel electrophoresis, 301
polychromatic erythroblast, 299, 300, *302, 303, 305*
polyglucosan, 102, *103*
Polygucosan bodies (PGB), 128 *see also* polygucosan body disease
polymyositis, 105
Polyomavirus, 183, 185, 187
Pompe disease
  GSD type II, 102, *103*
  lysosomes in vacuolar myopathy, 110
  rimmed vacuoles in, 109
Pompe disease (glycogen storage disease type II), 244, 257
Post mortem changes, 346
Post-infectious glomerulonephritis (PIGN), 6, 29–31
Post-Transplant Nephrotic Syndrome, 56, 82
Poxviruses, 183, 185, 187
Preparatory techniques, 324, 331, 333, 335, 337
preservation, 156, 158, 174
Primary ciliary dyskinesia, 222–3, 234–6
process, 155, 157, 159–61, 170, 171, *172, 295*
  interdigitating, 160
  microvillous, see *microvilli*
  spinous, 160
proerythroblast, 294, 295, 301
Progressive massive fibrosis, 332, 334, 336
Progressive multifocal leucoencephalopathy (PML), 185
Prokaryotic organisation, 191
proliferation, 293, 299
Promastigote, 212
protein aggregate myopathies, 100
protocol for MW, 387–9
Protozoa, 181, 206 to 213
protrusion, 295
Pseudoclumps, 413
Pseudocowpox, 190
Pseudo-Hurler polydystrophy, 238, 241
Pycnodysostosis, 246
PYGM
  myophosphorylase gene in McArdle disease, 102

Quantitative measurement, 432–4, 437, 439–441, 443–446

Rare earth pneumoconiosis, **325**, 338
Recombinant virus capsids, 185
Recurrent preterm premature rupture of fetal membrane syndrome, 319, 320
Recurrent Primary Kidney Disease
  adenosine phosphoribosyl transferase deficiency, 69
  amyloidosis, 67
  ANCA-mediated disease, 68
  anti-GBM disease, 68
  C3 nephropathy, 69
  complement factor H-related protein, 5 nephropathy 69
  cryoglobulinaemia, 69
  cystinosis, 69

Recurrent Primary Kidney Disease (*continued*)
diabetic glomerulosclerosis, 66, 72
Fabry disease, 68
fibrillary/immunotactoid glomerulonephritis, 68
fibronectin glomerulopathy, 69
focal segmental glomerulosclerosis, 66, 69, (Figure on 70)
haemolytic uraemic syndrome, 67
Henoch-Schonlein purpura, 68
IgA nephropathy, 66, 71
lecithin-cholesterol acyltransferase deficiency, 68
membranous nephropathy, 66, 70, (Figure on 71)
mesangiocapillary glomerulonephritis (types I, II & III) 66–7
monoclonal immunoglobulin deposition disease, 67
oxalosis, 69
sickle cell nephropathy, 69
systemic lupus erythematosus, 67
thrombotic microangiopathy, 67
thrombotic thrombocytopaenic purpura, 69
reducing bodies, *111*, 112
Refsum disease, 140
Reich granules, 120, *122*, 136, *129*
Remak fibre, 121, 124, *137*, 138, *138*, 140
abnormalities, 128, 135
Remyelination, 130, *132*, 133, *137*
Renal laminopathy, 432
Renal transplants, 12, 42, 43
Renaut body, *122*, 125, 142, 144
Reovirus, 189
reproductive tract, 171
rER, see *rough endoplasmic reticulum*
Resin infiltration, 349–352
Resins, 349–352, 362, 370, 376
reticulocyte, 299, 306
Retrovirus, 183
rhabdomyoblast, 162, 176
rhabdomyoma, 167, 176
rhabdomyosarcoma, 173, 176
rheumatoid arthritis
capillaries in, 97
Rheumatoid factor, 45
Ribonucleoprotein, 183
rimmed vacuoles, 106, 109
ring fibres, 101–102
rippling muscle disease
Caveolin, 3 in, 98
Rosenthal fibre, 159
ryanodine receptor, 96
RYR, 1 Ryanodine receptor gene in core myopathies, 104

S100 protein, 165
Safety cabinet, 409
Salla disease, 238, 246
Sandhoff disease, 242, 248
Sanfillipo syndrome, 240
Sapovirus, 183, 184, 188
Sarcoglycans
in autophagic vacuoles with sarcolemmal features (AVSF), 109
Sarcoidosis, **325**, 330, 331, 332, 333, 337
sarcolemma, 96, 98
costamere, 95
in atrophic fibres, 96
distinguishing from hypotrophic fibres, 96
in autophagic vacuoles with sarcolemmal features (AVSF), 109
in myotendinous junctions, 90, 94

in neuromuscular junctions, 94
in normal muscle, 91
sarcomere, 157, 176, *177*
sarcomeres, 94–5
sarcoplasm, 95
sarcoplasmic reticulum, 95, 96
swelling (artefact), 91, 93
in tubular aggregate formation, 108
satellite cells, 95
Schaumann body, 331
Schindler, 243
Schmidt-Lanterman incisures, 121, 124, 133, *146*, 147
Schwann cell (SC), 117, 120, 138, *138*, 143
basal lamina, 120, *121*, 126, 144
flattened sheets, 128, *129*
nodal processes see node of Ranvier
plasma membane (plasmalemma), 121, 128
regenerating, 142
Schwann cell/axon networks, *138*, 141
Schwannoma, 155, *164*, 165, *170*
scleroderma, 111
sclerosing epithelioid fibrosarcoma, *163*, 164
section, 158, 162
semi-thin, 158
Sectioning, 435
Sectioning technique, 352–363
seminoma, 161
semithin sections, 91
SEPN1 selenoprotein 1 core myopathies
gene defect in (multiminicore myopathy), 104
Septata (synonym for *Encephalito-zoon*)*intestinalis*, 201
Septate junctions, 123
sER, see *smooth endoplasmic reticulum*
SERCA
sarcoendoplasmic reticulum calcium ATPase, 108
serous granule, 157
Sialidosis, 241, 248
sideroblast, 301
Siderosis, **325**
siderosome, 160, 301
signet-ring cell lymphoma, 161
SIL1
and Marinesco-Sjogren syndrome, 105
Silicate pneumoconiosis, **325**, 333, 335
Silicatosis, See Silicate pneumoconiosis
Silicone, 338
Silicoproteinosis, 332, 336
Silicosis, **325**, 331, 333, 335, 337
Silicotic nodule, 332–6
Sjogren's disease, 111
skeinoid fibre, 160
Smallpox (Variola), 186, 190
smooth endoplasmic reticulum, 156, 157, 159, 160, 301, 304
Sodium channels, see channels, sodium
Sperm, 221, 231–3, 235–6, 368
splenomegaly, 294
Spontaneous cervical artery dissection syndrome, 317
Staining, 348, 349, 356–366, 370–379
Standard error of mean (SEM), 440, 441–2, 445
STED-microscopy, 402
Storage disorders, 15, 46
Subperineurial oedema, 143
surfactant, 171
Swiss cheese-like, 295, *296*
synovial sarcoma, 159
Systemic Lupus Erythematosus (SLE), 11, 36–7, 111
systems pathology, 384

Talcosis, **325**, 334, 339
Tangier disease, 136
target fibres, 105
targeted therapy, 384
Tay-Sachs, 247

Teased fibres, 130, 147
  processing, 118, 119
thalassaemia, 299
Thin basement membrane disease (TBMD), 21–2
Thin sectioning of virus-infected cells or tissues, 414
Thioflavin, 8
Thrombocytopaenia, 279
Thrombotic microangiopathy (TMA), 24–25
  transplant kidney, 57, 59, 62, 67, 74, 83, (Figure on 63)
Thrombotic thrombocytopaenic purpura (TTP), 24
Tight junction, 121, 133, 170
Titin
TNNT 1 slow troponin T1 in nemaline myopathy, 100
Tomacula, 147
tonofibril, 155–61, 167, *168*, 169, 171, 173
Toxoplasma, 206
TPM2 β-tropomyosin
    in nemaline myopathy, 100
    in 'cap disease', 100
TPM3 α-tropomyosin
    in nemaline myopathy, 100
    in 'cap disease', 100
Trachipleistophora, 199, 202
  T. hominis, 199, 202
  T. anthropophthera, 202
transdifferentiation, 156
Transplant Glomerulopathy
  differential, 61
  glomerular basement membrane, 57, (Figure on 58)
  insudate, 60, (Figure on 62)
  peritubular capillary, 57, (Figure on 59)
  subendothelial lucency, 57, (Figure on 58)
  ultrastructural features, 57–61
Transplant Kidney
  acute rejection, 56
  acute tubular necrosis, 56
  arteriolar hyalinosis, 72, 83, (Figure 84)
  calcineurin inhibitor toxicity, 63, 67, 73–4, 83, (Figure on 84)
  de novo glomerular disease, 56–7, 65, 70, 72–4
  donor-related disease, 56, 74
  hyperacute rejection, 56
  hypertension, 56
  inconclusive diagnosis by LM and/or IM, 56, 80
  indications for transmission electron microscopy, 56
  infection, 56, 74
  interstitial fibrosis, 56
  mechanical obstruction, 56
  neoplasia, 56, 83
  post-transplant nephrotic syndrome, 56, 81
  pyelonephritis, 56
  recurrent primary disease, 56–7, 62, 64–72
  reflux nephropathy, 56
  renal biopsy, 56
  reperfusion injury, 56
  thrombosis, 56
  transplant glomerulopathy, 56–64
  tubulointerstitial nephritis, 56
  vascular rejection, 81, (Figure on 82)
  vascular sclerosis, 56
Tremolite, 324, **325**, 326, *327*, 333
Triad, 222
triads, *94, 96, 97, 112*
Trichomonas 206
Trichrome stain (modified), 197
Trophoderma whippleii, 193

Trypanosoma, 206, 212
t-tubules, 95, *96*, 108
  confusion with lipid droplets, 109
  swollen by fixation delay, 93
  swollen, in necrotic fibres, 109
  t-system networks, 108
tubular aggregates, 108
  and sarcoendoplasmic reticulum calcium ATPase (SERCA) expression, 108
  in familial myasthenia, 109
  in periodic paralysis, 109
Tubules, 5, 7–11, 37, 40, 43–5
  proximal, 4, 16
Tubulinosema, 204
Tubuloreticular bodies (TRBs) or Tubuloreticular inclusions (TRIs), 12, 14, 39
tubuloreticular inclusions, 105, 111, 160
tumour, 153–158, 162, 169, 167, 171, 176
  chondroblastic, 158
  fibroblastic, 154
  granular-cell, 160, 164, *164*, 165
  granulosa-cell, 160, 167
  juxtaglomerular-cell, 160
  Leydig-cell, 160
  malignant rhabdoid, 160
  Merkel cell, 160
  myomelanocytic, 154, *165*
  osteoblastic, 158
Type 1 dermal dendrocyte, 309

Ullrich congenital muscular dystrophy lobulated fibres in, 108
Ultracentrifugation, 410
Uncertainty of measurement (UM), 431, 433, 437, 440, 442–7
uranaffin reaction, 171, *174*
Urine samples ('early morning'), 183, 184, 410

vacuolation, 299
vacuole, 161–2, 299
vacuoles, 109
  autophagic, 109
  in autophagic vacuolar myopathy (childhood), 110
  in autophagic vacuolar myopathy (infantile), 110
  in autophagic vacuoles with sarcolemmal features (AVSF), 109–110
  in necrotic fibres, 109
  in Pompe disease, 103, *110*
  rimmed, 109
Variation, 434–436, 438, 440, 444
Varicella zoster, 190
Vascular dissection, 317, 320
Vascular pole, 2–4
Vascular smooth muscle cells, 270
Vasculitis, *127, 131*, 133, *138*, 143
Vesicle fluid, 409, 410
vesicle, 295
vessel, 162, 170
vimentin, 158, *168*, 169
Viral gastroenteritis, 181, 183, 188
Virus
  BK virus in transplant kidney, 74–8, (Figures on 76, 77 and 78)
  cytomegalovirus in transplant kidney, 75, (Figure on 79)
  hepatitis C virus in transplant kidney, 57, 61, 66
  varicella zoster virus in transplant kidney (Figure on 80)
Virus artefacts, 191
Virus factory, 187, 193
Virus morphology, 183

Vittaforma corneae, 200, 204
Vittaforma-like species, 205
VMA21
gene defect in X-linked myopathy with excessive autophagia abnormal accumulation of lysosomes, 110
von Willebrand disease (VWD)
  GP1b-IX-V gene
  PT-VWD gene
  platelet VWD, 288
  Type 2B VWD, 288, 289
von Willebrand factor (VWF), 278, 282, 283
  immunogold labelling of, 282, 289

Waldenström's macroglobulinaemia, *132*
Wallerian degeneration, 126
  see also axon degeneration
wax, 156, 176
Weibel-Palade body 156, 162
Welder's pneumoconiosis, 336
Whipple's disease, 194
Winter vomiting disease, 183, 189
Wolman disease, 243, 252, 256

X-linked disease, 20
X-linked myopathy with excessive autophagia (XMEA), 110
xylene, 156

Z line streaming, 99, *104*
Z line, 94
  thickness, 95, 99, 100
ZASP Z-band alternatively spliced PDZ motif protein, 102
Z-disk, 176, *177*
Zebra bodies, 46
Zebra body myopathy, 101
  actin masses in, 100–101
zymogen granule, 171
α, 1, 4glucosidase deficiency
  see Pompe
α-Actinin, *94*, 99
α-mannosidosis, 248
α-quartz, *327*, 331, 333
β-mannosidosis, 247
  in normal muscle, 95